Lecture Notes in Physics

The Editorial Policy for Proceedings

The series Lecture Notes in Physics reports new developments in physical research and teaching – quickly, informally, and at a high level. The proceedings to be considered for publication in this series should be limited to only a few areas of research, and these should be closely related to each other. The contributions should be of a high standard and should avoid lengthy redraftings of papers already published or about to be published elsewhere. As a whole, the proceedings should aim for a balanced presentation of the theme of the conference including a description of the techniques used and enough motivation for a broad readership. It should not be assumed that the published proceedings must reflect the conference in its entirety. (A listing or abstracts of papers presented at the meeting but not included in the proceedings could be added as an appendix.)

When applying for publication in the series Lecture Notes in Physics the volume's editor(s) should submit sufficient material to enable the series editors and their referees to make a fairly accurate evaluation (e.g. a complete list of speakers and titles of papers to be presented and abstracts). If, based on this information, the proceedings are (tentatively) accepted, the volume's editor(s), whose name(s) will appear on the title pages, should select the papers suitable for publication and have them refereed (as for a journal) when appropriate. As a rule discussions will not be accepted. The series editors and Springer-Verlag will normally not interfere with the detailed editing except in fairly obvious cases or on technical matters.

Final acceptance is expressed by the series editor in charge, in consultation with Springer-Verlag only after receiving the complete manuscript. It might help to send a copy of the authors' manuscripts in advance to the editor in charge to discuss possible revisions with him. As a general rule, the series editor will confirm his tentative acceptance if the final manuscript corresponds to the original concept discussed, if the quality of the contribution meets the requirements of the series, and if the final size of the manuscript does not greatly exceed the number of pages originally agreed upon. The manuscript should be forwarded to Springer-Verlag shortly after the meeting. In cases of extreme delay (more than six months after the conference) the series editors will check once more the timeliness of the papers. Therefore, the volume's editor(s) should establish strict deadlines, or collect the articles during the conference and have them revised on the spot. If a delay is unavoidable, one should encourage the authors to update their contributions if appropriate. The editors of proceedings are strongly advised to inform contributors about these points at an early stage.

The final manuscript should contain a table of contents and an informative introduction accessible also to readers not particularly familiar with the topic of the conference. The contributions should be in English. The volume's editor(s) should check the contributions for the correct use of language. At Springer-Verlag only the prefaces will be checked by a copy-editor for language and style. Grave linguistic or technical shortcomings may lead to the rejection of contributions by the series editors. A conference report should not exceed a total of 500 pages. Keeping the size within this bound should be achieved by a stricter selection of articles and not by imposing an upper limit to the length of the individual papers. Editors receive jointly 30 complimentary copies of their book. They are entitled to purchase further copies of their book at a reduced rate. As a rule no reprints of individual contributions can be supplied. No royalty is paid on Lecture Notes in Physics volumes. Commitment to publish is made by letter of interest rather than by signing a formal contract. Springer-Verlag secures the copyright for each volume.

The Production Process

The books are hardbound, and the publisher will select quality paper appropriate to the needs of the author(s). Publication time is about ten weeks. More than twenty years of experience guarantee authors the best possible service. To reach the goal of rapid publication at a low price the technique of photographic reproduction from a camera-ready manuscript was chosen. This process shifts the main responsibility for the technical quality considerably from the publisher to the authors. We therefore urge all authors and editors of proceedings to observe very carefully the essentials for the preparation of camera-ready manuscripts, which we will supply on request. This applies especially to the quality of figures and halftones submitted for publication. In addition, it might be useful to look at some of the volumes already published. As a special service, we offer free of charge LaTeX and TeX macro packages to format the text according to Springer-Verlag's quality requirements. We strongly recommend that you make use of this offer, since the result will be a book of considerably improved technical quality. To avoid mistakes and time-consuming correspondence during the production period the conference editors should request special instructions from the publisher well before the beginning of the conference. Manuscripts not meeting the technical standard of the series will have to be returned for improvement.

For further information please contact Springer-Verlag, Physics Editorial Department II, Tiergartenstrasse 17, D-69121 Heidelberg, Germany

Vincent Rivasseau (Ed.)

Constructive Physics

Results in Field Theory, Statistical Mechanics and Condensed Matter Physics

Proceedings of the Conference
Held at Ecole Polytechnique, Palaiseau, France
25-27 July 1994

Springer

Editor

Vincent Rivasseau
Ecole Polytechnique
Centre de Physique Théorique
F-91128 Palaiseau Cedex, France

ISBN 978-3-662-14061-1 ISBN 978-3-540-49222-1 (eBook)
DOI 10.1007/978-3-540-49222-1

CIP data applied for

Originally published by Springer-Verlag Berlin Heidelberg New York in 1995
Softcover reprint of the hardcover 1st edition 1995

Typesetting: Camera-ready by the editor
SPIN: 10481088 55/3142-543210 - Printed on acid-free paper

Preface

Constructive field theory is now roughly 30 years old. It was born with the program of A. Wightman to systematically study "butchered" models of field theories (with cutoffs) in increasing order of difficulty. Over the years, from that main trunk which never stopped growing, many new branches emerged, developped, and made connections with unexpected areas not included in the initial program. Two large branches that developed roughly 20 and 10 years ago are constructive statistical mechanics and constructive solid-state physics. We expect now to see the emergence of constructive classical physics.

It is understandable that constructive field theory developped first. The ultraviolet divergences of perturbation theory are so obvious that physicists cannot ignore them. For some time some of them even believed that field theory per se was inconsistent. These controversies spurred the axiomatic and constructive programs to settle these issues rigorously. In the first decade of its existence, constructive theory obtained the first non-trivial ultraviolet limits and the first non-trivial thermodynamic limit, leading to complete construction of superrenormalizable theories which satisfy the axioms of axiomatic field theory.

Constructive statistical mechanics was born when the Euclidean approach was definitively adopted in field theory. Field theory in Euclidean space-time of dimension d was recognized to be identical to classical statistical mechanics in space dimension d. As a consequence, during their second decade constructive methods were applied with success to many areas of statistical mechanics. As typical significant results to which this period led, one could cite the proof of the Kosterlitz-Thouless transition in two dimensions and of localization in strongly disordered systems. During the time of these progresses, the main trunk of constructive field theory kept growing. Renormalization was analyzed more carefully, leading to its correct constructive formulation, which is strongly inspired by, but not identical to, the renormalization group concept: this correct formulation computes an effective action à la Wilson for relevant and marginal terms only, the irrelevant terms being kept in the form of a hardcore polymer gas (equivalently, one could say that it computes the effective action only in so-called "small-field regions"). This effort lead to the construction of renormalizable field theories.

During its third decade the constructive field activity concentrated on the ultraviolet limit of non-Abelian Yang-Mills theory, a surprisingly hard problem, and on digesting the previous results, in particular to formalize and transmit better the methods. Some solutions to the Yang-Mills problem have by now been obtained, but here also we expect that the digestion work will be long and difficult.

Constructive solid-state physics also developed in the last decade, because the renormalization group had matured enough to be formulated correctly around a Fermi surface. As mathematical physicists we can take some pride in this achievement which certainly has significance for theoretical physics. Within the next decade we expect that branch of constructive theory to lead to a complete and rigorous construction of the BCS theory, and to the solution of long-standing puzzles such as the existence of extended states in the Anderson model for electrons in a weakly disordered medium. Ultimately we expect the powerful field theoretic methods to replace completely the traditional approach to solid-state physics. This does not happen without some friction, in a domain for so long under the influence of the famous citation by Landau: *"We are talking here about theoretical physics, and therefore of course mathematical rigor is irrelevant and impossible"* (L.D. Landau, old preface of "Statistical Physics").

One can also predict that a branch now emerging from the constructive tree is promised to a bright future, namely the study of classical PDE and classical mechanics with constructive methods. To this direction belong the work of Bricmont and Kupiainen, the constructive study of stochastic quantization, and the work on the Navier-Stokes equations with renormalization group methods. Note also that the construction of global solutions for the non-linear Schrödinger equation with ϕ^4 potential by Bourgain makes use of the corresponding functional measures of constructive theory.

The full constructive tree is now much too large to be adequately described as "constructive field theory" alone. This description is clearly too narrow and nowadays perhaps also somewhat old fashioned. Therefore we suggest that it could be time to boldly call our subject "constructive physics", and to characterize it as the rigorous study of particular physical models by hard mathematical methods such as expansions, which give detailed information on the properties of the observables. The very existence of such a subject which crosses the usual boundaries inside physics is ultimately due to the incredibly universal character of Lagrangians and field theoretic methods (functional integrals) in physics. Of course some domains still seem to be out of reach: for example, constructive string theory remains a dream. But we hope and believe that sometime in the next century (or millenia?) when theoretical physicists finally agree on some "theory of everything" there will be constructivists to start their rigorous version of it!

The fast pace of advance of constructive theory in recent years made it necessary to organize some meetings and to spend some time on the diffusion of results. Since the constructive "tribe" has always kept a human size, it is still possible to meet and share our new results in the context of small workshops. The occasion which led to these proceedings came from the decision of the International Association of Mathematical Physics (IAMP) to hold its 1994 congress in Paris. A satellite conference took place at Ecole Polytechnique, where constructive theory has been studied for twenty years. We thank the Ecole Polytechnique and the Ministry of Foreign Affairs for their financial support. We thank D. Iagolnitzer, main organizer of the Paris Congress, for sharing also some of the burden of organizing the satellite conferences. Finally we thank all participants of the workshop for the pleasant atmosphere they helped to create, and we hope such meetings will continue to occur in the coming years.

This book contains the main lectures at that meeting, together with a few additional selected contributions. The goal has been to offer a rather wide spectrum of subjects covering the diversity of present day constructive theory, but with emphasis on a pedagogical presentation. In particular, several contributions are directly devoted to the basic methods of constructive theory, namely cluster expansions and the rigorous renormalization group, (contributions by A. Abdesselam and myself, David Brydges and Andreas Pordt), and we suggest that a beginner should start with these.

Among other results of a general nature, the first lecture by Yu. Zinoviev establishes a long searched for equivalence theorem between the Wightman and the Osterwalder-Schrader axioms of quantum field theory. The axiomatization of supersymmetric theories is discussed in the contribution of Konrad Osterwalder. More specific lectures are devoted to the Gross-Neveu model with a review on the construction of the ultraviolet limit in the non-renormalizable regime (Claude de Calan) and on mass generation in one (Paolo Faria da Veiga et al) and two dimensions (Christoph Kopper). In all these cases the key expansion parameter is not the ordinary coupling constant but $1/N$, where N is the number of components. The second lecture by Yu. Zinoviev is devoted to the convergence of the U(1) gauge theory on a toroidal lattice as the lattice and torus cutoffs are removed.

Constructive solid-state physics is illustrated by the contributions of Giovanni Benfatto, Joel Feldman, Detlef Lehmann, Horst Knörrer and Eugene Trubowitz. The lecture of Benfatto initiates a program for the study of Bose condensation. FKLT construct the first example of a non-trivial interacting Fermi liquid in two-dimensions, using an anisotropic Fermi surface which closes the Cooper channel. This is a very typical product of the recent developments in constructive solid-state physics.

The two dimensional Coulomb gas has been one of the most studied model in mathematical physics, from both the field theory and the statistical mechanics points of view. The lecture by Tom Hurd presents in a unified way the ultraviolet ($\beta < 8\pi$) or infrared ($\beta > 8\pi$) regime, including the proof of new results, using Brydges formalism.

The lectures of Steven Golowich, Daniel Iagolnitzer and Jacques Magnen are devoted to the construction of weakly self-avoiding polymers in four dimensions, using the asymptotic freedom of that model in the infrared regime. Constructive theory and cluster expansions are also among the useful tools for the less advanced theory of interfaces or surfaces, as shown in the review of François

Dunlop. A program for the constructive study of continuous symmetry breaking in statistical mechanics is reviewed in Tadeusz Balaban's lecture.

The contribution of Jean Bricmont and Antti Kupiainen is a beautiful review on the emerging subject alluded to above, namely the study of classical partial differential equations via renormalization group methods.

Non-Abelian gauge theory in high-energy physics,is not directly discussed in this book. But this omission is somewhat compensated for by the lecture of Edward Witten. He makes the fascinating and slightly provocative suggestion of reversing the traditional course in constructive theory: instead of attacking models closer and closer to actual physics, we may invite more pure mathematicians to join us by convincing them that field theory is the natural framework for modern geometry. This may require us to apply the constructive methods to prove results in that direction which are of interest to them; in the long run it may speed up our advance in physical problems as well. This direction is illustrated in a spectacular way by the recent results of Seiberg and Witten. Their work on Olive-Montonen duality establishes in particular a new formulation for differential invariants in four dimensions (the Donaldson invariants) by computing the exact infrared behavior of twisted supersymmetric $N = 2$ Yang-Mills theory. To relate rigorously the ultraviolet properties of that theory and the former formulation of these invariants in terms of instantons to the infrared behavior, and their formulation in terms of monopoles, is a fascinating future task for constructive theory.

In conclusion, we hope that these Lecture Notes in Physics may be in the 1990s' what the Erice Lecture Notes on constructive field theory were in the 1970s': a powerful stimulus and a guide for the next generation of students.

I would like to dedicate my contribution to this book to my friends Yves Delon and Jean-Paul Lahaye.

Palaiseau, January 1995 — Vincent Rivasseau

Contents

Condensed Matter, Statistical Mechanics

Some Questions for Constructive Field Theorists

*Edward Witten**

School of Natural Sciences
Institute for Advanced Study
Olden Lane
Princeton, NJ 08540

I was invited to share a few reflections on questions of possible interest to constructive field theorists. The single most important observation in this regard is of course what David Gross said the other day: showing that QCD exists is even more important than verifying the conjectured status of QED, because a positive result would open more doors. To this comment, I'd like mainly to add a few words of a general nature.

Broadly speaking, a major theme in twentieth century mathematics has been absorbing (or in some cases, independently formulating and generalizing some notions of) one-particle quantum mechanics.

One may well suspect that many-body quantum physics – which I hope will mainly mean relativistic quantum field theory – will play a similar role in the twenty-first century. To me the question is what can be done to accelerate the process.

The analogy suggests that, since in the twentieth century people have had to understand quantum mechanics, for instance Green's functions, elliptic operators, and index theory on manifolds, in the twenty-first century it will be necessary to understand quantum field theory, for instance YM_4, on manifolds.

Thus constructive field theorists might accelerate this process if you can understand YM_4 on manifolds – with arbitrary gauge bundles. It would seem that some of the machinery this would require may anyway be needed to construct YM_4 on flat $\mathbf{R}^4$ in the infinite volume limit. Handling instantons on $\mathbf{R}^4$, which are needed for cluster decomposition, may well raise some of the issues that would enter in working on a general gauge bundle on a general four-manifold.

In any event, I think it will take time before mathematicians whose interests are not derived from physics study YM_4 (or quantum field theory generally) on manifolds. First, it is necessary to learn that QFT is the correct subject. The present state of affairs is rather that most mathematical investigation of various things (elliptic genera, Donaldson theory, three-manifold invariants, symplectic geometry via instantons, mirror symmetry...) that could well be viewed as manifestations of quantum field theory are conducted quite independently of that interpretation. In looking at the mathematical literature on, say, mirror symmetry, the uninitiated reader would not get any clue that it is already known that

* Research supported by NSF grant #PHY92-45317

supersymmetric quantum field theory is the right setting – a fact which if it were not already known would probably be one of the crowning insights of generations of intense inquiry. Nor would one guess in looking at mathematical papers on, say, mirror symmetry and elliptic genera, that these are already known to be aspects of the *same* theories.

Of course, the reason for this is that mathematicians not motivated by physics cannot construe quantum field theory as the natural object of study since it is not well enough established in enough generality mathematically.

For mathematicians not motivated by physics to study YM_4 on four-manifolds would require learning new types of applications of QFT or becoming interested in QFT per se. There is a big gap here. The gap is much less for some other theories that have mathematical applications that are already of non-physics interest. Yet even in these examples, the QFT viewpoint cannot be properly developed and appreciated mathematically because the necessary constructive field theory has not been done.

For instance, consider in four dimensions supersymmetric Yang-Mills with $N = 2$ supersymmetry; I will call this theory $YM_{4;2}$. (In general, I'll write $YM_{k;N}$ for the k dimensional theory with N supersymmetries; note that N determines the Fermionic dimension and is a kind of analog of the bosonic dimension k.) Unlike $YM_{4;0}, YM_{4;2}$ has a mathematical manifestation that is already of intense interest (independent of physics). This is Donaldson theory, which can be considered to arise from a certain "twisted" version of $YM_{4;2}$ formulated on a four-manifold. This leads to a path integral formulation of the Donaldson invariants which has led to concrete predictions about them. But the arguments cannot be made rigorous as $YM_{4;2}$ is not rigorous.

Note that although there are infrared puzzles in $YM_{4;2}$ (I reported at the Mathematical Physics Congress last week on work with N. Seiberg on those problems) they are not immediately relevant to Donaldson theory, which is usually formulated on a compact four-manifold.

$YM_{4;2}$ is a rather hard case, even in finite volume, and as such is not really representative of what I want to say. There are other, somewhat analogous cases, where the relevant QFT's are much easier. The process of giving QFT its due role might be greatly accelerated by studying the geometrical applications of some of these.

My real purpose today is to point out a few examples of such "easy" cases in which there are neither ultraviolet nor infrared problems, but there are new types of questions not traditionally considered by constructive field theorists.

I will start with the simplest example of all, which is ordinary two-dimensional Yang-Mills theory without matter, $YM_{2;0}$. This theory is "trivial" because of having no propagating degrees of freedom, and is exactly soluble, as was pointed out by Migdal twenty years ago. The partition function on a Riemann surface of genus g and area A can be computed explicitly and is essentially

$$Z = \sum_{\mathbf{R}} \frac{1}{(\dim \mathbf{R})^{2g-2}} \exp(-e^2 A c_2(R))$$

where **R** ranges over irreducible representations of the gauge group, e is the gauge coupling, and $c_2(\mathbf{R})$ is the quadratic Casimir operator. This theory (and the above formula) can be made rigorous [F].

However, I showed a few years ago[W1] another formula for Z, essentially

$$Z = \int_{\mathcal{M}} e^{\omega + e^2 p_1(\mathcal{M})} \;\; + O(e^{-1/e^2})$$

where $\mathcal{M}$ is the moduli space of flat bundles on the Riemann surface, and ω and $p_1(\mathcal{M})$ are the usual symplectic form and the first Pontryagin class of $\mathcal{M}$. (The exponentially small terms, written above simply as $O(e^{-1/e^2})$, can be described rather explicitly.) The comparison of the two formulas gives a precise determination of some topological invariants of $\mathcal{M}$. By considering in a similar way suitable correlation functions, and not just the partition function, one can get a complete description of the intersection ring of $\mathcal{M}$. The resulting formula has not been obtained from any other point of view.

This then gives putatively the complete solution of the two dimensional analog of Donaldson theory (which, under the guise of determining "the intersection ring of the moduli space of bundles on a curve," was much studied long before the four-dimensional version was introduced by Donaldson). What is the difficulty in making rigorous this proposal for the solution of what we might somewhat ahistorically call (Donaldson)$_2$? While the first formula above for the $\mathrm{YM}_{2;0}$ partition function is rigorous, the second is not. The second formula comes from a technique of integration on symplectic manifolds that is perfectly rigorous in finite dimensions but whose application to the $\mathrm{YM}_{2;0}$ path integral is not rigorous. I believe that making this formula and its topological application rigorous would be a very nice step that would have a lot of impact.

I should not leave this example without mentioning that while above I formulated the discussion in terms of $\mathrm{YM}_{2;0}$, $\mathrm{YM}_{2;2}$ (which is not soluble) is lurking behind the scenes. A certain relation between $\mathrm{YM}_{2;0}$ and $\mathrm{YM}_{2;2}$ provides a natural setting for the analysis; as in four-dimensions, $\mathrm{YM}_{2;2}$ has a twisted version that is related to the two-dimensional analog of Donaldson theory.

$\mathrm{YM}_{2;0}$ is no doubt the simplest of nonlinear QFT's. A harder example to illustrate my general theme is $\mathrm{YM}_{3;0}$ with action (in Euclidean space)

$$\mathcal{L} = \frac{1}{4e^2}\int d^3x\ \mathrm{Tr} F_{\mu\nu}^2 + \frac{ik}{8\pi}\int d^3x\ \varepsilon^{\mu\nu\lambda}\,\mathrm{Tr}(A_\mu \partial_\nu A_\lambda + \frac{2}{3} A_\mu A_\nu A_\lambda).$$

The second term is the Chern-Simons term; k must be an integer to ensure invariance of $e^{-\mathcal{L}}$ under gauge transformations that are not continuously converted to the identity. Usual $\mathrm{YM}_{3;0}$ is the case of $k = 0$. For $k = 0$, there is an infrared problem in perturbation theory. It is believed that the $k = 0$ theory in fact has a mass gap, but this is not under good control even heuristically. Thus, the $k = 0$ theory is "too hard" for our present purposes. (Also, the issues I am about to mention are believed to be trivial for $k = 0$.)

The $k \neq 0$ theory is much easier since there is a mass gap in perturbation theory. Moreover, perturbation theory (which is really an expansion in $1/k$, as

e can be set to 1 by a choice of units) should be reliable for large enough k. Therefore one would think that, by justifying the perturbation expansion, one might be able to prove that there is a mass gap for large enough k.

If so, the following general principle should apply: any theory with a mass gap becomes a topological field theory when the metric is scaled up. That is, one considers a theory with mass gap on a closed manifold M with metric tensor g; if g is scaled by $g \to tg$, then for $t \to \infty$ the propagating modes should drop out, leaving a topological field theory. The last assertion really should be a general theorem of axiomatic field theory; I do not know whether a version of this theorem has already been proved.

The topological field theories obtained from the $t \to \infty$ limit of a massive theory are usually more or less trivial. For $YM_{3;0}$ this is not so (as one can see in perturbation theory for sufficiently big k). For instance, three is the dimension in which a circle can be knotted, and the expectation value of a Wilson line on flat $\mathbb{R}^3$ will go over for large t (that is, as the dimensions of the Wilson line are scaled up) to a knot invariant. I believe that a rigorous, manifestly three-dimensional construction of knot and three-manifold invariants along the lines I just sketched would be influential.

My other examples would be in two dimensions. There are too many relevant examples for me to give a reasonable explanation here, but I will briefly comment on

(a) WZW models;

(b) Calabi-Yau models;

(c) new problems about supersymmetric theories with polynomial interactions.

Only in (c) am I strictly keeping to my promise that the theories I will mention are easier than the ones most studied on the frontier of constructive field theory. I have included other examples because it seemed misleading to discuss the two-dimensional world in terms of (c) only.

My question about (a) is whether – since there has been progress with asymptotically free theories – the WZW model (some of whose manifestations are extensively studied out of the QFT context) can be made rigorous as a conformal field theory.

The idea would be to consider a two-dimensional field g that takes values in a group manifold G, with

$$\mathcal{L} = \frac{1}{\lambda} \int d^2x \ \mathrm{Tr}(g^{-1}\partial_\mu g)^2 + ik\Gamma(g).$$

The first term is the usual sigma model Lagrangian; Γ is the Wess-Zumino term. k must be an integer for topological reasons, so only λ is renormalized. However, the renormalization of λ depends on k.

The one-loop renormalization of λ is independent of k, and the theory is asymptotically free for all k. Thus for all k there is, conversely, an infrared problem. For $k = 0$, the theory is believed to generate a mass gap, but that is not under good control even heuristically and is presumably out of each rigorously. For $k \neq 0$, it is believed instead[W2] that the theory flows to an infrared stable fixed point at $\lambda = \lambda_0(k)$. For large k, $\lambda_0(k)$ is calculable and small, and the whole picture is under good control heuristically. Can this story be made rigorous for sufficiently large k? That would give a construction of the WZW model, which is the conformal field theory at the fixed point. Of course, there would still be many properties of the fixed point theory to understand, but existence would be a fundamental step.

My other points – (b) and to a large extent also (c) – involve conformal field theories constructed as sigma models with Calabi-Yau manifolds as target spaces. Many fascinating properties of these models have been discovered at least heuristically. The outstanding examples are, perhaps, the occurrence of continuous transformations in the topology of the target, and also mirror symmetry. One would like to study these models directly, but that may be too difficult. I would therefore like to point out [W3] that on the same renormalization group orbits as these difficult theories are some super-renormalizable polynomial field theories with N = 2 supersymmetry. Moreover (see [W3], section 3.3), there are many interesting observables in these theories that are invariant under the renormalization group flow and so can be studied directly in the polynomial, super-renormalizable theory. These include the elliptic genus, and Yukawa couplings that are sensitive to topology change. (There is no infrared problem, because one is interested in finite volume.) So one would think that, if the supersymmetric polynomial theories were under good control, there would be many interesting things to do that would involve neither infrared nor ultraviolet problems.

I have tried to indicate what I think should be one of the directions of constructive field theory approaching the next century: applying the methods to "geometric" questions that may arise in "easier" theories than those at the current frontiers. I do believe that exploring such questions may in the long run, by attracting new interest and new perspectives, accelerate progress even on the traditional questions.

References

[F] For instance, see Fine, D. "Quantum Yang-Mills theory on the two sphere," Commun. Math. Phys. **134** (1990) 273; "Quantum Yang-Mills on a Riemann surface," Commun. Math. Phys. (1991) 321-338.

[W1] Witten, E. "Two-Dimensional Gauge Theories Revisited," Jnl. Geom. Phys. **9** (1992) 303-368.

[W2] Witten, E. "Nonabelian Bosonization in Two Dimensions," Comm. Math. Phys. **92** (1984) 455.

[W3] E. Witten, E. "Phases of N=2 Models in Two Dimensions," Nucl. Phys. **B403** (1993) 159-222.

Trees, Forests and Jungles: A Botanical Garden for Cluster Expansions

Abdelmalek Abdesselam,
Vincent Rivasseau

Centre de Physique Théorique, CNRS, UPR 14
Ecole Polytechnique, 91128 Palaiseau Cedex
e-mail: abdesselam@orphee.polytechnique.fr
rivass@orphee.polytechnique.fr

Abstract
Combinatoric formulas for cluster expansions have been improved many times over the years. Here we develop some new combinatoric proofs and extensions of the tree formulas of Brydges and Kennedy, and test them on a series of pedagogical examples.

1 Introduction

Cluster expansions have a reputation of being hard to use; this is largely due to the difficulty to capture them in a single short formula. These expansions were introduced in constructive theory by Glimm, Jaffe and Spencer [GJS1-2] and they were improved or generalized over the years [BF][BaF][Bat][B1]. For many years the Ecole Polytechnique was happy using a cluster "tree formula" due to Brydges, Battle and Federbush (see [R1-2] and references therein). This formula expresses connected amplitudes more naturally as sums over "ordered trees" rather than regular trees; but a combinatoric lemma due to Battle and Federbush, shows that the sum over all ordered trees corresponding to a given ordinary Cayley tree has total weight 1. A slightly disturbing dissymmetry in this formula reveals itself in the need to order in an arbitrary way the elements on which the cluster expansion is performed, and in the particular rôle played by the first element or root of the tree.

Brydges kept digging for a truly satisfying, still more beautiful formula, and with Kennedy they obtained it in [BK] (see also [BY][B2]). This formula shows a clear conceptual progress; it does not require the use of arbitrary choices such as an arbitrary ordering of the objects to decouple, and the outcome can be written in a somewhat shorter form, involving directly standard Cayley trees. Both formulas share a positivity preserving property which is crucial in constructive theory: the corresponding interpolations of positive matrices remain positive. Both can be built by iterating inductively some perturbation step. But the first formula insists in completing the cluster containing the "first" cube, before building the next one. The second and better formula blindly derives connections, hence all clusters grow symmetrically at the same time. Therefore we prefer to call it a forest formula (see below).

However the original proof of this formula in [BK] relies on a differential equation (Hamilton-Jacobi) which is perhaps not totally transparent. The purpose of this paper is to derive a fully combinatoric or algebraic proof of this type of formula, and to show how to apply it to various examples chosen for their pedagogic value. Many situations in constructive theory in fact require several cluster expansions on top of each other, and for this case we derive a generalization of the tree or forest formula, which we call the jungle formula.

We also derive a generalization that applies not only to exponentials of two-body interaction potentials, or to perturbations of Gaussian measures. It is a general interpolation formula we call the Taylor forest formula.

Finally we propose a formula that performs, in a single move, the succession of a cluster and a Mayer expansion.

2 Two Forest Formulas and Their Combinatoric Proofs

First let us recall the Brydges-Kennedy formula [BK] under the most convenient setting for the following discussion. Let $n \geq 1$, be an integer , $I_n = \{1, \ldots, n\}$, $\mathcal{P}_n = \big\{\{i,j\}/i,j \in I_n, i \neq j\big\}$ (the set of unordered pairs in I_n). Consider $n(n-1)/2$ elements $u_{\{ij\}}$ of a commutative Banach algebra $\mathcal{B}$, indexed by the elements $\{ij\}$ of $\mathcal{P}_n$. An element of $\mathcal{P}_n$ will be called a *link*, a subset of $\mathcal{P}_n$, a *graph*. A graph $\mathfrak{F} = \{l_1, \ldots, l_\tau\}$ containing no loops, i.e. no subset $\big\{\{i_1,i_2\},\{i_2,i_3\},\ldots,\{i_k,i_1\}\big\}$ with $k \geq 3$ elements, will be called a *u-forest* (unordered forest). A sequence $F = (l_1, \ldots, l_\tau)$ of links the range of which $\{l_1, \ldots, l_\tau\}$ is a u-forest, will be called an *o-forest* (ordered forest).

A u-forest is a union of disconnected trees, the supports of which are disjoint subsets of I_n called the *connected components* or *clusters* of $\mathfrak{F}$. Isolated points also form clusters reduced to singletons, hence the total number of clusters is $n - \tau$, where $\tau = |\mathfrak{F}|$. Two points in the same cluster are said *connected by* $\mathfrak{F}$. Now the Brydges-Kennedy forest formula states:

Theorem II.1 [BK]

$$\exp\Big(\sum_{l \in \mathcal{P}_n} u_l\Big) = \sum_{\substack{\mathfrak{F}=\{l_1,\ldots,l_\tau\} \\ \text{u-forest}}} \Big(\prod_{\nu=1}^{\tau} \int_0^1 dh_{l_\nu}\Big)\Big(\prod_{\nu=1}^{\tau} u_{l_\nu}\Big) \exp\big(\sum_{l \in \mathcal{P}_n} h_l^{\mathfrak{F}}(\mathbf{h}).u_l\big) , \tag{II.1}$$

where the summation extends over all possible lengths τ of $\mathfrak{F}$, including $\tau = 0$ hence the empty forest. To each link of $\mathfrak{F}$ is attached a variable of integration h_l; and $h^{\mathfrak{F}}_{\{ij\}}(\mathbf{h}) = \inf\{h_l, l \in L_{\mathfrak{F}}\{ij\}\}$ where $L_{\mathfrak{F}}\{ij\}$ is the unique path in the forest $\mathfrak{F}$ connecting i to j. If no such path exists, by convention $h^{\mathfrak{F}}_{\{ij\}}(\mathbf{h}) = 0$.

Our second forest formula writes exactly the same, except that the definition of the $h^{\mathfrak{F}}_{\{ij\}}(h)$ is different, and it involves for each cluster a particular choice of a root. For simplicity let us slightly restrict this arbitrary choice by imposing the following rule : for each non empty subset or cluster C of I_n, choose r_C, the smallest element in the natural ordering of $I_n = \{1, \ldots, n\}$, to be the root of all the trees with support C that appear in the following expansion. Now if i is in some tree $\mathfrak{T}$ with support C we call the *height* of i the number of links in the unique path of the tree $\mathfrak{T}$ that goes from i to the root r_C. We denote it by $l^{\mathfrak{T}}(i)$. The set of points i with a fixed height k is called the k-th *layer* of the tree.

We now have

Theorem II.2

$$\exp\Big(\sum_{l\in\mathcal{P}_n} u_l\Big) = \sum_{\substack{\mathfrak{F}=\{l_1,\ldots,l_\tau\} \\ \text{u-forest}}} \Big(\prod_{\nu=1}^{\tau}\int_0^1 dw_{l_\nu}\Big)\Big(\prod_{\nu=1}^{\tau} u_{l_\nu}\Big)\exp\Big(\sum_{l\in\mathcal{P}_n} w_l^{\mathfrak{F}}(\mathbf{w}).u_l\Big)\ , \tag{II.2}$$

where the summation extends over all possible lengths τ of $\mathfrak{F}$, including $\tau = 0$ hence the empty forest. To each link of $\mathfrak{F}$ is attached a variable of integration w_l. We define the $w_l^{\mathfrak{F}}$ as follows. $w_{\{ij\}}^{\mathfrak{F}}(\mathbf{w}) = 0$ if i and j are not connected by the $\mathfrak{F}$. If i and j fall in the support C of the same tree $\mathfrak{T}$ of $\mathfrak{F}$ then

$w_{\{ij\}}^{\mathfrak{F}}(\mathbf{w}) = 0$ if $|l^{\mathfrak{T}}(i) - l^{\mathfrak{T}}(j)| \geq 2$ (i and j in distant layers)

$w_{\{ij\}}^{\mathfrak{F}}(\mathbf{w}) = 1$ if $l^{\mathfrak{T}}(i) = l^{\mathfrak{T}}(j)$ (i and j in the same layer)

$w_{\{ij\}}^{\mathfrak{F}}(\mathbf{w}) = w_{\{ii'\}}$ if $l^{\mathfrak{T}}(i) - 1 = l^{\mathfrak{T}}(j) = l^{\mathfrak{T}}(i')$, and $\{ii'\} \in \mathfrak{T}$. (i and j in neighboring layers, i' is then unique). In particular, if $\{ij\} \in \mathfrak{F}$, then $w_{\{ij\}}^{\mathfrak{F}}(\mathbf{w}) = w_{\{ij\}}$.

The proof we give of theorem II.1 relies on two lemmas.

Lemma II.1 *Let $a_0, \ldots, a_p, (p \geq 1)$ be distinct complex numbers, then:*

$$\int_{\substack{t_0\geq 0,\ldots,t_p\geq 0 \\ t_0+\ldots+t_p=1}} dt_0 \ldots dt_p\ \exp(t_0a_0 + \ldots + t_pa_p) = \sum_{i=0}^{p} \frac{e^{a_i}}{\prod_{\substack{j=0 \\ j\neq i}}^{p}(a_i - a_j)} \tag{II.3}$$

Proof: By induction. $p = 1$ is an easy computation. We assume the result is true for $p \geq 1$ thus

$$\int_{\substack{t_0\geq 0,\ldots,t_{p+1}\geq 0 \\ t_0+\ldots+t_{p+1}=1}} dt_0 \ldots dt_{p+1}\ \exp(t_0a_0 + \ldots + t_{p+1}a_{p+1}) =$$

$$\int_0^1 dt_{p+1}e^{t_{p+1}a_{p+1}}(1-t_{p+1})^p \int_{\substack{w_0\geq 0,\ldots,w_p\geq 0 \\ w_0+\ldots+w_p=1}} dw_0 \ldots dw_p\ \exp(w_0b_0 + \ldots + w_pb_p) \tag{II.4}$$

where we performed first the integration on t_{p+1} and made the following change of variables: $t_i = (1 - t_{p+1})w_i, b_i = a_i(1 - t_{p+1})$ for $0 \leq i \leq p$. By the induction hypothesis this becomes

$$\int_0^1 dt_{p+1} e^{t_{p+1}a_{p+1}} (1 - t_{p+1})^p \sum_{i=0}^{p} \frac{e^{b_i}}{\prod_{\substack{j=0 \\ j \neq i}}^{p} (b_i - b_j)} =$$

$$\sum_{i=0}^{p} \frac{e^{a_i}}{\prod_{\substack{j=0 \\ j \neq i}}^{p+1} (a_i - a_j)} + e^{a_{p+1}} \sum_{i=0}^{p} \frac{1}{(\prod_{\substack{j=0 \\ j \neq i}}^{p} (a_i - a_j))(a_{p+1} - a_i)} . \tag{II.5}$$

This boils down to the wanted expression for $p+1$ after remarking that we have the following rational fraction decomposition with simple poles:

$$\frac{1}{(X - a_0)\dots(X - a_p)} = \sum_{i=0}^{p} \frac{1}{(X - a_i)} \times \frac{1}{\prod_{\substack{j=0 \\ j \neq i}}^{p} (a_i - a_j)} ,$$

and putting $X = a_{p+1}$. ■

Lemma II.2 *With the same notation as in the beginning of this section, assume that we are given two sets of $n(n-1)/2$ indeterminates $u_{\{ij\}}$ and $v_{\{ij\}}$; then the following algebraic identity is true in the field of rational fractions $\mathbb{C}(u_{\{ij\}}, v_{\{ij\}})$.*

$$\prod_{l \in \mathcal{P}_n} v_l = \sum_{\substack{F=(l_1,\dots,l_\tau) \\ \text{o-forest}}} u_{l_1} \dots u_{l_\tau} \cdot \left(\sum_{\nu=0}^{\tau} \frac{b_\nu^F}{\prod_{\substack{\mu=0 \\ \mu \neq \nu}}^{\tau} (a_\nu^F - a_\mu^F)} \right) \tag{II.6}$$

where, for $0 \leq \nu \leq \tau$, $a_\nu^F = \sum_{\{ij\}/\nu} u_{\{ij\}}$, $b_\nu^F = \prod_{\{ij\}/\nu} v_{\{ij\}}$, $\{ij\}/\nu$ meaning that the points i and j are connected by the sub-o-forest $(l_1, \dots, l_\nu)$ of F. In particular since no points are connected by the empty forest $a_0^F = 0, b_0^F = 1$.

Proof: Both sides of this identity are polynomials in the v_l's, and we prove equality coefficient by coefficient. First consider the case of the constant monomial. We must show

$$\sum_{\substack{F=(l_1,\dots,l_\tau) \\ \text{o-forest}}} (-1)^\tau \frac{u_{l_1} \dots u_{l_\tau}}{a_1^F \dots a_\tau^F} = \begin{cases} 0 \; if \; n \geq 2 \\ 1 \; if \; n = 1 \end{cases} \tag{II.7}$$

For n=1 it is trivial, so assume $n \geq 2$ and denote $\mathcal{A}_F$ the contribution of F to the left hand side. Given an o-forest $F = (l_1, \dots, l_\tau)$, an o-forest F' of the form $(l_1, \dots, l_\tau, l_{\tau+1}, \dots, l_{\tau+\kappa})$ will be called a κ-*extension* of F. When we need not know κ we simply say an *extension* of F.

Let $\mathcal{R}_F = \mathcal{A}_F \cdot \dfrac{a_\tau^F}{a_{\text{total}}}$ where $a_{\text{total}} = \sum_{l \in \mathcal{P}_n} u_l \quad (\neq 0 \text{ since } n \geq 2)$.
Notice that

$$a_{\text{total}} = a_\tau^F + \sum_{F' \text{ 1-extension of } F} u_{l_{\tau+1}} , \tag{II.8}$$

because summing over F' is the same as summing over all links $\{ij\}$ where i and j lie in different clusters of F, while a^F_τ handles summing over links where i and j are in the same cluster.

Hence

$$a^F_\tau = a_{\text{total}} + \sum_{F' \text{ 1-extension of } F} \left(-\frac{u_{l_{\tau+1}}}{a^{F'}_{\tau+1}}\right) a^{F'}_{\tau+1} \tag{II.9}$$

i.e multiplying by $\mathcal{A}_F / a_{\text{total}}$

$$\mathcal{R}_F = \mathcal{A}_F + \sum_{F' \text{ 1-extension of } F} \mathcal{R}_{F'} \;. \tag{II.10}$$

Multiple (but finite because $\tau \leq n-1$) iteration of (II.10) yields

$$\mathcal{R}_F = \sum_{F' \text{ extension of } F} \mathcal{A}_{F'} \tag{II.11}$$

where the sum is over extensions of all possible lengths. In particular

$$\sum_F \mathcal{A}_F = \sum_{F \text{ extension of } \emptyset} \mathcal{A}_F = \mathcal{R}_\emptyset = \mathcal{A}_\emptyset . \frac{a^\emptyset_0}{a_{\text{total}}} = 0 \tag{II.12}$$

since $\mathcal{A}_\emptyset = 1$, $a^\emptyset_0 = 0$, $a_{\text{total}} \neq 0$. (II.7) is now proven.

Let us now check the other monomials in the v's. If $F = (l_1, \ldots, l_\tau)$ is a forest we say that it *creates* the partition $\mathcal{D}$ of I_n if $\mathcal{D}$ is the set of clusters of F. We then write, with a slight abuse of notation $a_\mathcal{D} = \sum_{\{ij\}/\mathcal{D}} u_{\{ij\}}$, $b_\mathcal{D} = \prod_{\{ij\}/\mathcal{D}} v_{\{ij\}}$, where $\{ij\}/\mathcal{D}$ means that i and j are in the same element or *component* of $\mathcal{D}$.

Monomials generated by (II.5) are of the form $b_\mathcal{D}$, $\mathcal{D}$ being created by a sub-o-forest of F. Remark that if $\mathcal{D}$ is created by a forest of length τ then $|\mathcal{D}| = n - \tau$. There are two cases to be treated:

Case 1: $\mathcal{D} \neq \{I_n\}$

Here the coefficient of $b_\mathcal{D}$ is zero in the left hand side, we must show that also for the right hand side. Let $\nu = n - \tau$, then (st abbreviates "such that")

$$\sum_{\tau \geq \nu} \sum_{\substack{F=(l_1,\ldots,l_\tau) \\ \text{o-forest st} \\ (l_1,\ldots,l_\nu) \text{ creates } \mathcal{D}}} \frac{u_{l_1} \ldots u_{l_\tau}}{\prod^\tau_{\substack{\mu=0 \\ \mu\neq\nu}} (a^F_\nu - a^F_\mu)} =$$

$$\sum_{\substack{F_1=(l_1,\ldots,l_\nu) \\ \text{o-forest} \\ \text{creating } \mathcal{D}}} \frac{u_{l_1} \ldots u_{l_\nu}}{\prod^{\nu-1}_{\mu=0} (a_\mathcal{D} - a^{F_1}_\mu)} \sum_{\substack{F_2=(l_{\nu+1},\ldots,l_\tau) \\ \text{st } F=(F_1,F_2) \text{ is} \\ \text{an o-forest}}} (-1)^{\tau-\nu} \frac{u_{l_{\nu+1}} \ldots u_{l_\tau}}{\prod^\tau_{\mu=\nu+1} (a^F_\mu - a_\mathcal{D})} \;. \tag{II.13}$$

Again, summation is over all possible τ's. We now arrive at the heart of the inductive argument. We show using (II.7) that F_1 being fixed the sum on F_2 vanishes. In fact, this sum is the analog of (II.7) when instead of I_n we use $\mathcal{D}$ as a point set. If a and b are two elements of the partition $\mathcal{D}$, we let $\overline{u}_{\{ab\}} = \sum_{i\in a, j\in b} u_{\{ij\}}$. Given an o-forest $F = (F_1, F_2)$ on I_n such that F_1 creates $\mathcal{D}$, $F_2 = (l_{\nu+1}, \ldots, l_\tau)$ *induces* an o-forest $\overline{F} = (\bar{l}_1, \ldots, \bar{l}_{\tau-\nu})$ on $\mathcal{D}$ in the following way: if $l_\kappa = \{ij\}$, $\nu + 1 \leq \kappa \leq \mu$, with $i \in a$ and $j \in b$, a and b elements of $\mathcal{D}$, then set $\bar{l}_{\kappa-\nu} = \{ab\}$. $\overline{F}$ is simply obtained by forgetting the details of structure inside the components of $\mathcal{D}$, which are due to the forest F_1. Demanding that the whole of (F_1, F_2) be an o-forest insures that, in the process, $\overline{F}$ remains a forest. As a result, the analog of the a_ν^F for the forest $\overline{F}$ are $\bar{a}_\rho^{\overline{F}} = a_{\rho+\nu}^F - a_{\mathcal{D}}$, for $1 \le \rho \le \mu - \nu$; and with $\bar{\tau} = \tau - \nu$ we have

$$\sum_{\substack{F_2=(l_{\nu+1},\dots,l_\tau) \\ \text{st } F=(F_1,F_2) \text{ is} \\ \text{an o-forest}}} (-1)^{\tau-\nu} \frac{u_{l_{\nu+1}} \dots u_{l_\tau}}{\prod_{\mu=\nu+1}^{\tau}(a_\mu^F - a_{\mathcal{D}})} = \sum_{\overline{F}=(\bar{l}_1,\dots,\bar{l}_{\bar{\tau}})} (-1)^{\bar{\tau}} \frac{\bar{u}_{\bar{l}_1} \dots \bar{u}_{\bar{l}_{\bar{\tau}}}}{\bar{a}_1^{\overline{F}} \dots \bar{a}_{\bar{\tau}}^{\overline{F}}} \ . \tag{II.14}$$

The last sum is zero by (II.7) because $|\mathcal{D}| \ge 2$, since we are in the case $\mathcal{D} \ne \{I_n\}$.

Case 2: $\mathcal{D} = \{I_n\}$

Here $a_{\mathcal{D}} = a_{\text{total}}$. Equating the coefficients of $b_{\mathcal{D}} = \prod_{l \in \mathcal{P}_n} v_l$ on both sides needs showing

$$1 = \sum_{\substack{F=(l_1,\dots,l_{n-1}) \\ \text{o-forest}}} \frac{u_{l_1} \dots u_{l_{n-1}}}{(a_{\text{total}} - a_0^F) \dots (a_{\text{total}} - a_{n-2}^F)} \tag{II.15}$$

This last identity is shown by induction on n. Remark that here F is a complete tree covering I_n. If $n = 1$, the only forest is the empty one, its contribution is the empty product i.e. 1.

The induction step from n to $n+1$ needs an argument similar to the former case:

$$\sum_{\substack{F=(l_1,\dots,l_n) \\ \text{o-forest on } I_n}} \frac{u_{l_1} \dots u_{l_n}}{(a_{\text{total}} - a_0^F) \dots (a_{\text{total}} - a_{n-1}^F)} =$$

$$\sum_{F_1=(l_1)} \frac{u_{l_1}}{a_{\text{total}}} \sum_{\substack{F_2=(l_2,\dots,l_n) \\ \text{st } F=(F_1,F_2) \text{ is} \\ \text{an o-forest on } I_{n+1}}} \frac{u_{l_2} \dots u_{l_n}}{(a_{\text{total}} - a_1^F) \dots (a_{\text{total}} - a_{n-1}^F)} \ . \tag{II.16}$$

The link $l_1 = \{\alpha, \beta\}$ being fixed, this determines a partition $J_n = \{\{\alpha,\beta\}\} \cup \{\{i\} | i \in I_{n+1},\ i \ne \alpha,\ i \ne \beta\}$ for which we can repeat the treatment of the partition $\mathcal{D}$ in the preceding case: introduce variables $\bar{u}_{\{ab\}} = \sum_{i \in a,\ j \in b} u_{\{ij\}}$ for a and b in J_n, and induced o-forests $\overline{F} = (\bar{l}_1, \dots, \bar{l}_{n-1})$ on J_n for any o-forest $F_2 = (l_2, \dots, l_n)$ on I_n such that $F = (F_1, F_2)$ be an o-forest on I_{n+1}. Remark that

$$\bar{a}_{\text{total}} \stackrel{\text{def}}{=} \sum_{\substack{\{ab\} \subset J_n \\ a \ne b}} \bar{u}_{\{ab\}} = \Big(\sum_{l \in \mathcal{P}_{n+1}} u_l \Big) - u_{\{\alpha\beta\}} = a_{\text{total}} - a_1^F \ , \tag{II.17}$$

and for $0 \le \rho \le n-2$, $\bar{a}_\rho^{\overline{F}} = a_{\rho+1}^F - a_1^F$, therefore $\bar{a}_{\text{total}} - \bar{a}_\rho^{\overline{F}} = (a_{\text{total}} - a_1^F) - (a_{\rho+1}^F - a_1) = a_{\text{total}} - a_{\rho+1}^F$.

Hence

$$\sum_{\substack{F_2=(l_2,\dots,l_n) \\ \text{st } F=(F_1,F_2) \text{ is} \\ \text{an o-forest on } I_{n+1}}} \frac{u_{l_2} \dots u_{l_n}}{(a_{\text{total}} - a_1^F) \dots (a_{\text{total}} - a_{n-1}^F)} =$$

$$\sum_{\substack{\bar{F}=(\bar{l}_1,\ldots,\bar{l}_{n-1}) \\ \text{o-forest on } J_n}} \frac{\bar{u}_{\bar{l}_1}\ldots\bar{u}_{\bar{l}_{n-1}}}{(\bar{a}_{\text{total}}-\bar{a}_0^{\bar{F}})\ldots(\bar{a}_{\text{total}}-\bar{a}_{n-2}^{\bar{F}})} \tag{II.18}$$

but the last sum is 1 by the induction hypothesis since $|J_n| = n$. Finally in (II.15) there remains

$$\sum_{l_1} \frac{u_{l_1}}{a_{\text{total}}} = \frac{a_{\text{total}}}{a_{\text{total}}} = 1 \ . \tag{II.19}$$

This completes the proof of Lemma II.2 . ■

Proof of Theorem II.1:

We can return now to the Brydges-Kennedy forest formula and first prove it for any complex numbers u_l, $l \in \mathcal{P}_n$ such that for all subsets X of $\mathcal{P}_n$, $\sum_{l\in X} u_l \neq 0$. In fact we can rewrite the right hand side of (II.1) using o-forests instead of u-forests. Let $\mathfrak{F}$ be a u-forest and choose an ordering $F = (l_1, \ldots, l_\tau)$ of its elements. We can slice the integral according to the ordering of the h_l's thereby giving a contribution for $\mathfrak{F}$

$$\sum_{\sigma\in\mathfrak{S}_\tau} \int_{1>h_{l_{\sigma(1)}}>\ldots>h_{l_{\sigma(\tau)}}>0} \left(\prod_{\nu=1}^{\tau} dh_{l_\nu}\right)\left(\prod_{\nu=1}^{\tau} u_{l_\nu}\right) \exp(\sum_{l\in\mathcal{P}_n} h_l^{\mathfrak{F}}(\mathbf{h}).u_l) \ , \tag{II.20}$$

$\mathfrak{S}_\tau$ is the permutation group on τ elements. For any of its elements σ let us denote the o-forest $(l_{\sigma(1)}, \ldots, l_{\sigma(\tau)})$ by F_σ. Let us define, if F is an o-forest, the function $h^F_{\{ij\}} = h_{l_\nu}$ where ν is the largest index in the ordering provided by F of the links appearing in the unique path of F connecting i and j. If no such path exists we again put by convention $h^F_{\{ij\}} = 0$.

Thus the contribution of $\mathfrak{F}$ becomes

$$\sum_{\sigma\in\mathfrak{S}_\tau} \int_{1>h_{l_{\sigma(1)}}>\ldots>h_{l_{\sigma(\tau)}}>0} \left(\prod_{\nu=1}^{\tau} dh_{l_\nu}\right)\left(\prod_{\nu=1}^{\tau} u_{l_\nu}\right) \exp(\sum_{l\in\mathcal{P}_n} h_l^{F_\sigma}(\mathbf{h}).u_l) \tag{II.21}$$

and the right hand side of (II.1)

$$I = \sum_{\substack{F=(l_1,\ldots,l_\tau) \\ \text{o-forest}}} \int_{1>h_{l_1}>\ldots>h_{l_\tau}>0} \left(\prod_{\nu=1}^{\tau} dh_{l_\nu}\right)\left(\prod_{\nu=1}^{\tau} u_{l_\nu}\right) \exp(\sum_{l\in\mathcal{P}_n} h_l^{F}(\mathbf{h}).u_l) \ . \tag{II.22}$$

Now we perform the change of variables $t_1 = h_{l_1} - h_{l_2}, \ldots, t_{\tau-1} = h_{l_{\tau-1}} - h_{l_\tau}$, $t_\tau = h_{l_\tau}$ then

$$I = \sum_{\substack{F=(l_1,\ldots,l_\tau) \\ \text{o-forest}}} u_{l_1}\ldots u_{l_\tau} \int_{\substack{t_1\geq 0,\ldots,t_\tau\geq 0 \\ t_1+\ldots+t_\tau\leq 1}} dt_1\ldots dt_\tau \ \exp(\sum_{\nu=1}^{\tau} t_\nu a_\nu^F) \ . \tag{II.23}$$

There shows up the a_ν^F when collecting in the exponential the $u_{\{ij\}}$'s multiplied by a given t_ν. Since $h_\nu = t_\nu + t_{\nu+1} + \ldots + t_\tau$, the corresponding pairs $\{ij\}$ are those for which all the links appearing in the unique path of F connecting i and

j have an index at most equal to ν, i.e. $\{ij\}/\nu$. Since $a_0^F = 0$, we can rewrite (II.23) as

$$I = \sum_{\substack{F=(l_1,\ldots,l_\tau) \\ \text{o-forest}}} u_{l_1}\ldots u_{l_\tau} \int_{\substack{t_0\geq 0,\ldots,t_\tau\geq 0 \\ t_0+\ldots+t_\tau=1}} dt_0\ldots dt_\tau \, \exp\Big(\sum_{\nu=0}^{\tau} t_\nu a_\nu^F\Big) \quad , \tag{II.24}$$

then from lemma II.1 we obtain

$$I = \sum_{\substack{F=(l_1,\ldots,l_\tau) \\ \text{o-forest}}} u_{l_1}\ldots u_{l_\tau} \cdot \left(\sum_{\nu=0}^{\tau} \frac{e^{a_\nu^F}}{\prod_{\substack{\mu=0 \\ \mu\neq\nu}}^{\tau} (a_\nu^F - a_\mu^F)} \right) \quad . \tag{II.25}$$

Then if we substitute for the indeterminates u_l the complex numbers we have now (this is allowed since the factors in the denominators $a_\nu^F - a_\mu^F$ are of the form $\pm \sum_{l\in X} u_l \neq 0$) and for the v_l the complex numbers $\exp(u_l)$, the lemma II.2 just proves (II.1). We can remove the condition $\sum_{l\in X} u_l \neq 0$ of being in a finite number of hyperplanes of $\mathbb{C}^{\mathcal{P}_n}$, by density and continuity of both sides of (II.1). To prove it for an arbitrary commutative Banach algebra $\mathcal{B}$ we need only note that the identity we are about to show, is of the form $f(u_1,\ldots,u_k) = 0$, where the function f is analytic, its value for u's in a commutative Banach algebra being defined by its power series expansion. Thus proving it for all complex numbers is enough to entail the vanishing of the coefficients of the power series and to prove the identity for any u's in $\mathcal{B}$. ∎

Proof of Theorem II.2:

We recall the notion of connected set function. If X is a finite set and ψ is a map from $\mathcal{P}(X)$, the power set of X, to a commutative algebra $\mathcal{B}$, the connected function of ψ is $\psi_c : \mathcal{P}(X) \to \mathcal{B}$ defined inductively by

$$\forall C \subset X, \;\; \psi(C) = \sum_{\substack{\Pi=\{C_1,\ldots,C_k\} \\ \text{partition of } C}} \prod_{i=1}^{k} \psi_c(C_i) \quad , \quad \psi_c(\emptyset) = 1 \quad . \tag{II.26}$$

Let $X = I_n$, $\mathcal{B}$ be the field of complex numbers, $\forall l \in \mathcal{P}_n, \epsilon_l \stackrel{\text{def}}{=} e^{u_l} - 1$, and let $\psi(C) = \prod_{\{ij\}\subset C}(1 + \epsilon_{\{ij\}})$. We make also the assumption that for any non empty subset Q of $\mathcal{P}_n$, $\sum_{l\in Q} u_l \neq 0$. It is well known [B1] that

$$\psi_c(C) = \sum_{\mathfrak{G} \text{ graph connecting } C} \Big(\prod_{l\in\mathfrak{G}} \epsilon_l\Big) \tag{II.27}$$

("connecting C" means that for every $i \neq j$ in C there exists a path in $\mathfrak{G}$ going from i to j, the empty graph connects C if $|C| \leq 1$).

We claim that

$$\psi_c(C) = \sum_{\mathfrak{T} \text{ tree connecting } C} \left(\prod_{l\in\mathfrak{T}} \int_0^1 dw_l \right) \left(\prod_{l\in\mathfrak{T}} u_l \right) \exp\left(\sum_{\{ij\}\subset C} w_{\{ij\}}^{\mathfrak{T}}(\mathbf{w}).u_{\{ij\}} \right) \quad . \tag{II.28}$$

This is trivial if $|C| \le 1$. If $|C| \ge 2$, let $\mathfrak{T}$ be a non trivial tree connecting C. We associate the sequence $\mathcal{C}^{\mathfrak{T}} = (C_\nu^{\mathfrak{T}})_{\nu\in\mathbb{N}}$ of disjoint subsets of C such that $C_\nu^{\mathfrak{T}}$ is the ν-th layer of $\mathfrak{T}$. Note that $C_0^{\mathfrak{T}} = \{r_C\}$, $\cup_{\nu\in\mathbb{N}} C_\nu^{\mathfrak{T}} = C$, and $\exists \nu_{\mathfrak{T}} \in \mathbb{N}$ such that $\nu \le \nu_{\mathfrak{T}} \Rightarrow C_\nu^{\mathfrak{T}} \ne \emptyset$ and $\nu > \nu_{\mathfrak{T}} \Rightarrow C_\nu^{\mathfrak{T}} = \emptyset$; a sequence $\mathcal{C} = (C_\nu)_{\nu\in\mathbb{N}}$ sharing those three properties will be called *admissible*.

We will compute in the right hand side of (II.28) the contribution of the trees corresponding to a given sequence $\mathcal{C}^{\mathfrak{T}} = \mathcal{C}$. Let $\{ij\}$ be a pair in C with $i \in C_{\nu_i}$, $j \in C_{\nu_j}$ and $\nu_i \ge \nu_j$. The only way $w_{\{ij\}}^{\mathfrak{T}}(\mathbf{w})$ can be non zero is that $\nu_i = \nu_j$ or $\nu_i = \nu_j + 1$. For $\nu_i = \nu_j$ we just get a factor $e^{u_{\{ij\}}} = 1 + \epsilon_{\{ij\}}$, since $w_{\{ij\}}^{\mathfrak{T}}(\mathbf{w}) = 1$.

But if $\nu_i = \nu_j + 1$, let i' be the element of C_{ν_j} such that $\{ii'\} \in \mathfrak{T}$; i' is the *ancestor* of i. Let us perform the integration explicitly in $w_{\{ii'\}}$. The pairs l such that $w_l^{\mathfrak{T}}(\mathbf{w}) \overset{\text{def}}{=} w_{\{ii'\}}$ are the pairs $\{ij\}$ with $j \in C_{\nu_i - 1}$, the corresponding contribution is

$$\int_0^1 dw_{\{ii'\}}\, u_{\{ii'\}} . \exp\Bigl(w_{\{ii'\}} . \sum_{j\in C_{\nu_i-1}} u_{\{ij\}} \Bigr) = u_{\{ii'\}} \cdot \frac{\exp\bigl(\sum_{j\in C_{\nu_i-1}} u_{\{ij\}}\bigr) - 1}{\sum_{j\in C_{\nu_i-1}} u_{\{ij\}}} . \tag{II.29}$$

Now, building a tree with prescribed $\mathcal{C}^{\mathfrak{T}}$ means linking the points of $C_1^{\mathfrak{T}}$ to the root r_C, then for each i in $C_2^{\mathfrak{T}}$ choosing an ancestor i' in $C_1^{\mathfrak{T}}$, then for each i in $C_3^{\mathfrak{T}}$ choosing an ancestor i' in $C_2^{\mathfrak{T}}$, and so on, until we exhaust C. The choice of ancestor for points in the same layer $C_\nu^{\mathfrak{T}}$ are completely independent operations. Summing the contributions (II.29) for all choices of ancestors for i is simply

$$\begin{aligned} \sum_{i'\in C_{\nu_i-1}} u_{\{ii'\}} \cdot \frac{\exp\bigl(\sum_{j\in C_{\nu_i-1}} u_{\{ij\}}\bigr) - 1}{\sum_{j\in C_{\nu_i-1}} u_{\{ij\}}} &= \exp\Bigl(\sum_{j\in C_{\nu_i-1}} u_{\{ij\}}\Bigr) - 1 \\ &= \prod_{j\in C_{\nu_i-1}} (1+\epsilon_{\{ij\}}) - 1 = \sum_{\substack{J\subset C_{\nu_i-1}\\ J\ne\emptyset}} \Bigl(\prod_{j\in J} \epsilon_{\{ij\}}\Bigr) . \end{aligned} \tag{II.30}$$

Therefore the right hand side of (II.28) is

$$\sum_{\substack{\text{admissible}\\ \text{sequence } \mathcal{C}}} \prod_{\nu\ge 1} \Biggl(\Bigl(\prod_{i\in C_\nu} \Bigl(\sum_{\substack{J\subset C_{\nu-1}\\ J\ne\emptyset}} \prod_{j\in J} \epsilon_{\{ij\}} \Bigr)\Bigr) \times \Bigl(\prod_{\{ij\}\subset C_\nu} (1+\epsilon_{\{ij\}}) \Bigr) \Biggr) , \tag{II.31}$$

and is to be compared with (II.27).

Note that, in a connected graph $\mathfrak{G}$, we can define the distance $d_{\mathfrak{G}}(i,j)$ between two points i and j by the minimal number of links in a path in $\mathfrak{G}$ going from i to j. Given a preferred point in C, namely the root r_C, and a graph $\mathfrak{G}$ connecting C, we can define an admissible sequence $\mathcal{C}^{\mathfrak{G}} = (C_\nu^{\mathfrak{G}})_{\nu\in\mathbb{N}}$, by letting $C_\nu^{\mathfrak{G}} \overset{\text{def}}{=} \{i \in C / d_{\mathfrak{G}}(i, r_C) = \nu\}$. This is consistent with our previous definition since a tree is obviously a special case of graph.

We just have now to convince the reader that for a given admissible C

$$\prod_{\nu\geq 1}\left(\left(\prod_{i\in C_\nu}\Big(\sum_{\substack{J\subset C_{\nu-1}\\ J\neq\emptyset}}\prod_{j\in J}\epsilon_{\{ij\}}\Big)\right)\times\Big(\prod_{\{ij\}\subset C_\nu}(1+\epsilon_{\{ij\}})\Big)\right)=\sum_{\substack{\mathfrak{G}\text{ graph}\\ \text{connecting } C\\ \text{such that } C^{\mathfrak{G}}=C}}\prod_{l\in\mathfrak{G}}\epsilon_l\ . \tag{II.32}$$

Note that building a graph $\mathfrak{G}$ with $C^{\mathfrak{G}}=C$ is first linking the points i of C_1 to the root r_C, then for each i in C_2 choosing a *non empty* subset $J_i\subset C_1$ of points j to which i will be linked, then for i in C_3 we again choose a non empty subset $J_i\subset C_2$ of points j to be linked with, and so on. Having built such a "skeleton" we, as a final step, have complete freedom to add all the links $\{ij\}$, for i and j in the same C_ν, that we want. Remark that in such a graph $\mathfrak{G}$, there is no link $\{ij\}$ with $i\in C_{\nu_i}$, $j\in C_{\nu_j}$ and $\nu_j+1<\nu_i$, for we would have $d_{\mathfrak{G}}(r_C,i)\leq d_{\mathfrak{G}}(r_C,j)+1=\nu_j+1<\nu_i=d_{\mathfrak{G}}(r_c,i)$, a contradiction.

Now (II.28) is proven, so the right hand side of (II.2) is

$$\sum_{\substack{\Pi=\{C_1,\ldots,C_k\}\\ \text{partition of } I_n}}\prod_{i=1}^{k}\psi_c(C_i)=\psi(I_n)=\prod_{l\in\mathcal{P}_n}(1+\epsilon_l)=\exp\Big(\sum_{l\in\mathcal{P}_n}u_l\Big)\ . \tag{II.33}$$

Finally we can remove the conditions $\sum_{l\in Q}u_l\neq 0$ and take the u_l in any commutative Banach algebra $\mathcal{B}$, as we did for the proof of theorem II.1. ■

3 A First Generalization: The Taylor Forest Formulas

In practice we not only need "algebraic" cluster expansions, but also "Taylor" cluster expansion, in which one typically makes a Taylor expansion with integral remainder to interpolate between a coupled situation (parameters set to 1) and an uncoupled one (parameters set to 0).

For instance a "pair of cubes" cluster expansion [R1] boils down to applying the following interpolation formula:

$$f(1,\ldots,1)=\sum_{I\subset I_n}\Big(\prod_{i\in I}\int_0^1 dh_i\Big)\left(\Big(\prod_{i\in I}\frac{\partial}{\partial x_i}\Big)f\right)(\psi_I(\mathbf{h})) \tag{III.1}$$

where $f(x_1,\ldots,x_n)$ is a smooth function on $\mathbb{R}^n$, and $\psi_I(\mathbf{h})$ is the vector of coordinates $(x_1,\ldots,x_n)$ defined by $x_i=0$ if $i\notin I$ and $x_i=h_i$ if $i\in I$. Such a formula is easy to prove.

The "Taylor" tree cluster expansion describes a similar interpolation formula. Let $\mathcal{S}$ be the space of smooth functions from $\mathbb{R}^{\mathcal{P}_n}$ to an arbitrary Banach space $\mathcal{V}$. An element of $\mathbb{R}^{\mathcal{P}_n}$ will be generally denoted by $\mathbf{x}=(x_l)_{l\in\mathcal{P}_n}$. The vector with all entries equal to 1 will be denoted by $\mathbb{1}$. Applied to an element H of $\mathcal{S}$, we can state two different Taylor forest formulas depending on which of theorem II.1 or II.2 we use for its derivation.

Theorem III.1 (The Brydges-Kennedy Taylor Forest Formula)

$$H(\mathbb{1}) = \sum_{\mathfrak{F}\ \text{u-forest}} \left(\prod_{l\in\mathfrak{F}} \int_0^1 dh_l\right) \left(\left(\prod_{l\in\mathfrak{F}} \frac{\partial}{\partial x_l}\right) H\right) (X_{\mathfrak{F}}^{BK}(\mathbf{h})) \quad . \tag{III.2}$$

Here $X_{\mathfrak{F}}^{BK}(\mathbf{h})$ is the vector $(x_l)_{l\in\mathcal{P}_n}$ of $\mathbb{R}^{\mathcal{P}_n}$ defined by $x_l = h_l^{\mathfrak{F}}(\mathbf{h})$, which is the value at which we evaluate the complicated derivative of H.

and

Theorem III.2 (The Rooted Taylor Forest Formula)

$$H(\mathbb{1}) = \sum_{\mathfrak{F}\ \text{u-forest}} \left(\prod_{l\in\mathfrak{F}} \int_0^1 dw_l\right) \left(\left(\prod_{l\in\mathfrak{F}} \frac{\partial}{\partial x_l}\right) H\right) (X_{\mathfrak{F}}^{r}(\mathbf{w})) \quad . \tag{III.3}$$

Here $X_{\mathfrak{F}}^{r}(\mathbf{w})$ is the vector $(x_l)_{l\in\mathcal{P}_n}$ of $\mathbb{R}^{\mathcal{P}_n}$ defined by $x_l = w_l^{\mathfrak{F}}(\mathbf{w})$, which is the value at which we evaluate the complicated derivative of H.

Proof: These formulas look more general than the algebraic forest formulas (II.1) and (II.2) (obtained by letting $H(\mathbf{x}) = \exp(\sum_{l\in\mathcal{P}_n} x_l u_l)$) but are in fact a consequence of them. For any integer $p \geq 0$, let $\mathcal{S}_p$ be the space of polynomial functions in the x_l's of total degree at most p. $\mathcal{S}_p$ can be dressed in a finite dimensional Banach space structure. Let $\mathcal{B}_p$ be the algebra of linear operators on $\mathcal{S}_p$ generated by the derivations $\partial/\partial x_l$ for $l \in \mathcal{P}_n$, equipped with the operator norm. $\mathcal{B}_p$ is a finite dimensional *commutative* Banach algebra. If we consider the $u_l \stackrel{\text{def}}{=} \partial/\partial x_l$ in $\mathcal{B}_p$ we have the right to use the multiple forest formula of the preceding section.

We then remark that, in $\mathcal{S}_p$, $\exp(\lambda\partial/\partial x_l)$ is the translation operator by the vector λe_l, where $(e_l)_{l\in\mathcal{P}_n}$ is the canonical base of $\mathbb{R}^{\mathcal{P}_n}$ and the exponential is defined by its power series, which converges because $\partial/\partial x_l$ is a *bounded* operator on $\mathcal{S}_p$. It reduces to a translation simply because Taylor's formula is finite and exact for polynomials.

Since $\exp(\sum_{l\in\mathcal{P}_n} \lambda_l\partial/\partial x_l)$ is the translation by $\lambda = \sum_{l\in\mathcal{P}_n} \lambda_l e_l$ i.e. the operator on $\mathcal{S}_p$ which maps the polynomial function H to $H' : \mathbf{x} \mapsto H(\mathbf{x} + \lambda)$, we have proven (III.2) for any H in $\mathcal{S}_p$. But p is arbitrary so we have proven the formula for any polynomial function H. By density of the polynomials in the Frechet space $\mathcal{S}$ for the C^∞ topology and the continuity relatively to that topology of both sides of (III.2), we extend the validity of the equation to all H in $\mathcal{S}$. We have proved formula (III.2) but the argument is the same for formula (III.3), just replace "h" by "w". ∎

4 A Second Generalization: The Jungle Formulas

Let $m \geq 1$ be an integer, and $(u_l^k)_{l\in\mathcal{P}_n}$, $1 \leq k \leq m$, be m families of $n(n-1)/2$ elements of a commutative Banach algebra $\mathcal{B}$. An *m-jungle* is a sequence $\mathcal{F} = (\mathfrak{F}_1, \ldots, \mathfrak{F}_m)$ of u-forests on I_n such that $\mathfrak{F}_1 \subset \ldots \subset \mathfrak{F}_m$.

The multiple forest formula states:

Theorem IV.1 (The Algebraic Brydges-Kennedy Jungle Formula)

$$\exp\Big(\sum_{\substack{l\in\mathcal{P}_n\\1\le k\le m}} u_l^k\Big) = \sum_{\substack{\mathcal{F}=(\mathfrak{F}_1,\ldots,\mathfrak{F}_m)\\ m-\text{jungle}}} \Big(\prod_{l\in\mathfrak{F}_m}\int_0^1 dh_l\Big)\Big(\prod_{k=1}^m\Big(\prod_{l\in\mathfrak{F}_k\backslash\mathfrak{F}_{k-1}} u_l^k\Big)\Big)$$

$$.\ \exp\Big(\sum_{k=1}^m\sum_{l\in\mathcal{P}_n} h_l^{\mathcal{F},k}(\mathbf{h}).u_l^k\Big) \qquad \text{(IV.1)}$$

where $\mathfrak{F}_0 = 0$ *by convention,* $\mathbf{h}$ *is the vector* $(h_l)_{l\in\mathfrak{F}_m}$ *and the functions* $h_{\{ij\}}^{\mathcal{F},k}(\mathbf{h})$ *are defined in the following manner:*

- *If* i *and* j *are not connected by* $\mathfrak{F}_k$ *let* $h_{\{ij\}}^{\mathcal{F},k}(\mathbf{h}) = 0$.
- *If* i *and* j *are connected by* $\mathfrak{F}_k$ *but not by* $\mathfrak{F}_{k-1}$ *let*

$$h_{\{ij\}}^{\mathcal{F},k}(\mathbf{h}) = \inf\Big\{h_l, l\in L_{\mathfrak{F}_k}\{ij\}\cap(\mathfrak{F}_k\backslash\mathfrak{F}_{k-1})\Big\}$$

(recall that $L_{\mathfrak{F}}\{ij\}$ *is the unique path in the forest* $\mathfrak{F}$ *connecting* i *to* j*).*

- *If* i *and* j *are connected by* $\mathfrak{F}_{k-1}$ *let* $h_{\{ij\}}^{\mathcal{F},k}(\mathbf{h}) = 1$.

Proof: By induction. The case $m = 1$ was treated in Section II. For the induction step from m to $m+1$, we sum over the last forest $\mathfrak{F}_{m+1}$:

$$\sum_{\substack{\mathcal{F}=(\mathfrak{F}_1,\ldots,\mathfrak{F}_{m+1})\\ (m+1)-\text{jungle}}} \Big(\prod_{l\in\mathfrak{F}_{m+1}}\int_0^1 dh_l\Big)\Big(\prod_{k=1}^{m+1}\Big(\prod_{l\in\mathfrak{F}_k\backslash\mathfrak{F}_{k-1}} u_l^k\Big)\Big)\exp\Big(\sum_{k=1}^{m+1}\sum_{l\in\mathcal{P}_n} h_l^{\mathcal{F},k}(\mathbf{h}).u_l^k\Big)$$

$$= \sum_{\substack{\mathcal{F}'=(\mathfrak{F}_1,\ldots,\mathfrak{F}_m)\\ m-\text{jungle}}} \Big(\prod_{l\in\mathfrak{F}_m}\int_0^1 dh_l\Big)\Big(\prod_{k=1}^{m}\Big(\prod_{l\in\mathfrak{F}_k\backslash\mathfrak{F}_{k-1}} u_l^k\Big)\Big)\exp\Big(\sum_{k=1}^{m}\sum_{l\in\mathcal{P}_n} h_l^{\mathcal{F}',k}(\mathbf{h}').u_l^k\Big)$$

$$.\ \sum_{\substack{\mathfrak{F}_{m+1}\\ \mathfrak{F}_m\subset\mathfrak{F}_{m+1}}} \Big(\prod_{l\in\mathfrak{F}_{m+1}\backslash\mathfrak{F}_m}\int_0^1 dh_l\Big)\Big(\prod_{l\in\mathfrak{F}_{m+1}\backslash\mathfrak{F}_m} u_l^{m+1}\Big)\exp\Big(\sum_{l\in\mathcal{P}_n} h_l^{\mathcal{F},m+1}(\mathbf{h}).u_l^{m+1}\Big)$$

(IV.2)

where $\mathbf{h} = (h_l)_{l\in\mathfrak{F}_{m+1}}$, $\mathbf{h}' = (h_l)_{l\in\mathfrak{F}_m}$ and we have noted that if $1\le k\le m$ then $h_l^{\mathcal{F},k}(\mathbf{h}) = h_l^{\mathcal{F}',k}(\mathbf{h}')$. To perform the summation over $\mathfrak{F}_{m+1}$, we will use the forest formula of section II and our favorite argument of forgetting the details of the tree structure up to $\mathfrak{F}_m$, to concentrate on what $\mathfrak{F}_{m+1}$ brings as new connections between the existing clusters. We introduce the partition $\mathcal{D}$ of I_n created by $\mathfrak{F}_m = \{l_1,\ldots,l_\nu\}$ and the u-forest on $\mathcal{D}$, $\overline{\mathfrak{F}}_{m+1} = \{\bar{l}_1,\ldots,\bar{l}_{\tau-\nu}\}$ induced by $\mathfrak{F}_{m+1}\backslash\mathfrak{F}_m = \{l_{\nu+1},\ldots,l_\tau\}$ with $\nu\le\tau$. The definitions are the same as in the proof of Lemma II.2 except that we have u-forests instead of o-forests.

For a link $\{ab\}$ between two elements a and b of $\mathcal{D}$, let $\overline{u}_{\{ab\}} = \sum_{i\in a, j\in b} u_{\{ij\}}^{m+1}$. Summing over $\mathfrak{F}_{m+1}$, u-forest on I_n containing $\mathfrak{F}_m$, with the "propagators" $u_{\{ij\}}^{m+1}$

is the same as summing over the u-forests $\overline{\mathfrak{F}}_{m+1}$ on $\mathcal{D}$ with the "propagators" $\overline{u}_{\{ab\}}$ i.e.

$$\sum_{\overline{\mathfrak{F}}_{m+1} \text{ u-forest on } \mathcal{D}} \left(\prod_{\overline{l}\in\overline{\mathfrak{F}}_{m+1}} \int_0^1 d\overline{h}_{\overline{l}} \right) \left(\prod_{\overline{l}\in\overline{\mathfrak{F}}_{m+1}} \overline{u}_{\overline{l}} \right) \exp\left(\sum_{\substack{\{ab\}\subset\mathcal{D}\\ a\neq b}} \overline{h}_{\{ab\}}^{\overline{\mathfrak{F}}_{m+1}}(\overline{\mathbf{h}}).\overline{u}_{\{ab\}} \right)$$

$$= \sum_{\substack{\mathfrak{F}_{m+1} \text{ u-forest on } I_n\\ \text{such that } \mathfrak{F}_m\subset\mathfrak{F}_{m+1}}} \left(\prod_{l\in\mathfrak{F}_{m+1}\setminus\mathfrak{F}_m} \int_0^1 dh_l \right) \left(\prod_{l\in\mathfrak{F}_{m+1}\setminus\mathfrak{F}_m} u_l^{m+1} \right)$$

$$\exp\left(\sum_{\substack{\{ij\}\in\mathcal{P}_n\\ \{ij\}\nparallel\mathcal{D}}} h_{\{ij\}}^{\mathcal{F},m+1}(\mathbf{h}).u_{\{ij\}}^{m+1} \right) \tag{IV.3}$$

where $\{ij\}\nparallel\mathcal{D}$ means "i and j are in different components of $\mathcal{D}$". By the forest formula of section II, the left hand side of the last equality is

$$\exp\left(\sum_{\substack{\{ab\}\subset\mathcal{D}\\ a\neq b}} \overline{u}_{\{ab\}} \right) \tag{IV.4}$$

that is

$$\exp\left(\sum_{\substack{\{ij\}\in\mathcal{P}_n\\ \{ij\}\nparallel\mathcal{D}}} u_{\{ij\}}^{m+1} \right) \ . \tag{IV.5}$$

The right hand side is almost the partial sum we want to perform on $\mathfrak{F}_{m+1}$ in (IV.2), there is missing

$$\exp\left(\sum_{\substack{\{ij\}\in\mathcal{P}_n\\ \{ij\}/\mathcal{D}}} h_{\{ij\}}^{\mathcal{F},m+1}(\mathbf{h}).u_{\{ij\}}^{m+1} \right) \ . \tag{IV.6}$$

The sum is over all pairs $\{ij\}$ in I_n such that $\{ij\}/\mathcal{D}$ i.e. i and j are connected by $\mathfrak{F}_m$. But then the definition of the functions $h_l^{\mathcal{F},k}(\mathbf{h})$ tells us $h_{\{ij\}}^{\mathcal{F},m+1}(\mathbf{h}) = 1$;. so the missing factor becomes

$$\exp\left(\sum_{\substack{\{ij\}\in\mathcal{P}_n\\ \{ij\}/\mathcal{D}}} u_{\{ij\}}^{m+1} \right) \ . \tag{IV.7}$$

In conclusion

$$\sum_{\substack{\mathfrak{F}_{m+1} \text{ u-forest on } I_n\\ \text{st } \mathfrak{F}_m\subset\mathfrak{F}_{m+1}}} \left(\prod_{l\in\mathfrak{F}_{m+1}\setminus\mathfrak{F}_m} \int_0^1 dh_l \right) \left(\prod_{l\in\mathfrak{F}_{m+1}\setminus\mathfrak{F}_m} u_l^{m+1} \right) \cdot$$

$$\exp\left(\sum_{l\in\mathcal{P}_n} h_l^{\mathcal{F},m+1}(\mathbf{h}).u_l^{m+1} \right) = \exp\left(\sum_{l\in\mathcal{P}_n} u_l^{m+1} \right) \ . \tag{IV.8}$$

The sum over the m-jungle $\mathcal{F}'$ is now $\exp(\sum_{k=1}^{m} \sum_{l\in\mathcal{P}_n} u_l^k)$ by the induction hypothesis, and the right hand side of (IV.2) is finally summed to the factor $\exp(\sum_{k=1}^{m+1} \sum_{l\in\mathcal{P}_n} u_l^k)$ as wanted. ■

We can generalize in a similar way the rooted forest formula. Given an m-jungle $\mathcal{F} = (\mathfrak{F}_1, \ldots, \mathfrak{F}_m)$, we introduce the notation $\mathbf{w}$ for the vector $(w_l)_{l\in\mathfrak{F}_m}$, and $w^{\mathcal{F},k}_{\{ij\}}(\mathbf{w})$ for the functions defined by:

- If i and j are not connected by $\mathfrak{F}_k$ let $w^{\mathcal{F},k}_{\{ij\}}(\mathbf{w}) = 0$.

- If i and j are connected by $\mathfrak{F}_{k-1}$ let $w^{\mathcal{F},k}_{\{ij\}}(\mathbf{w}) = 1$.

- If i and j are connected by $\mathfrak{F}_k$ but not by $\mathfrak{F}_{k-1}$, let r_C be the root of the cluster of $\mathfrak{F}_k$ containing i and j, define $l_k(i)$ to be the number of links of $\mathfrak{F}_k\backslash\mathfrak{F}_{k-1}$ in the unique path that goes from i to r_C, and similarly for j. Now if $|l_k(i) - l_k(j)| \geq 2$ put $w^{\mathcal{F},k}_{\{ij\}}(\mathbf{w}) = 0$, if $l_k(i) = l_k(j)$ put $w^{\mathcal{F},k}_{\{ij\}}(\mathbf{w}) = 1$, and if $l_k(i) = l_k(j) + 1$ for instance, take $w^{\mathcal{F},k}_{\{ij\}}(\mathbf{w}) = w_l$ where l is the first link of $\mathfrak{F}_k\backslash\mathfrak{F}_{k-1}$ on the path that goes from i to r_C. It is about the same definition as for theorem II.2 but we take only into account the links in $\mathfrak{F}_k\backslash\mathfrak{F}_{k-1}$.

We can now state

Theorem IV.2 (The Algebraic Rooted Jungle Formula)

$$\exp\Big(\sum_{\substack{l\in\mathcal{P}_n\\ 1\leq k\leq m}} u_l^k\Big) = \sum_{\substack{\mathcal{F}=(\mathfrak{F}_1,\ldots,\mathfrak{F}_m)\\ m-\text{jungle}}} \Big(\prod_{l\in\mathfrak{F}_m}\int_0^1 dw_l\Big)\Big(\prod_{k=1}^{m}\Big(\prod_{l\in\mathfrak{F}_k\backslash\mathfrak{F}_{k-1}} u_l^k\Big)\Big)$$

$$.\ \exp\Big(\sum_{k=1}^{m}\sum_{l\in\mathcal{P}_n} w_l^{\mathcal{F},k}(\mathbf{w}).u_l^k\Big)\ . \tag{IV.9}$$

Proof: The deduction of the jungle formula from the forest formula is the same as in the Brydges-Kennedy case. Remark that the property that was used was the following. Once $\mathfrak{F}_m$ is fixed as well as the partition $\mathcal{D}$ of I_n it creates, the numbers $w^{\mathcal{F},m+1}_{\{ij\}}(\mathbf{w})$ are unchanged if we allow i to travel freely in the component a_i of $\mathcal{D}$ it belongs to, and the same for j in a_j. Furthermore $w^{\mathcal{F},m+1}_{\{ij\}}(\mathbf{w}) = \overline{w}^{\overline{\mathfrak{F}}_{m+1}}_{\{a_i a_j\}}(\overline{\mathbf{w}})$ where $\overline{\mathfrak{F}}_{m+1}$ is the forest on $\mathcal{D}$ induced by $\mathfrak{F}_{m+1}$, and $\overline{\mathbf{w}} = (\overline{w}_{\overline{l}})_{\overline{l}\in\overline{\mathfrak{F}}_{m+1}}$ is defined by $\overline{w}_{\overline{l}} = w_l$ where l is the unique link in $\mathfrak{F}_{m+1}\backslash\mathfrak{F}_m$ inducing the link $\overline{l}$ in $\mathcal{D}$. The functions $\overline{w}^{\overline{\mathfrak{F}}}_{\overline{l}}(\overline{\mathbf{w}})$ are defined by the same algorithm as in theorem II.2 but with $\mathcal{D}$ instead of I_n as a point set, and with the following *induced choice of root*. If $\overline{C}$ is a non empty subset or cluster of $\mathcal{D}$, $\overline{C} = \{a_1, \ldots, a_\mu\}$, let $C = \cup_{\nu=1}^{\mu} a_\nu$. It is a non empty subset of I_n and already has a chosen root r_C. Then $\exists!\ \nu, 1 \leq \nu \leq \mu$ such that $r_C \in a_\nu$. So take $\overline{r}_{\overline{C}} = a_\nu \in \overline{C}$ to be the *induced root* of $\overline{C}$, that will get involved in the definition of the $\overline{w}^{\overline{\mathfrak{F}}}_{\overline{l}}(\overline{\mathbf{w}})$.

Finally, and this is the last ingredient that allows us to copy the precedent proof, we have forced $w^{\mathcal{F},m+1}_{\{ij\}}(\mathbf{w}) = 1$ if i and j fall in the same component of $\mathcal{D}$ so that we once again recover the missing factor

$$\exp\left(\sum_{\substack{\{ij\}\in\mathcal{P}_n \\ \{ij\}/\mathcal{D}}} u^{m+1}_{\{ij\}}\right) . \qquad \text{(IV.10)}$$

■

The reader must have noticed along this proof that we used a fairly general recipe to obtain a jungle formula from a forest formula. We also have a recipe to get Taylor interpolation formulas from algebraic ones: simply apply them to differentiation operators for the u_l's.

Let $\mathcal{P}_{n,m} = \mathcal{P}_n \times I_m$ and $\mathcal{S}$ be the space of smooth functions from $\mathbb{R}^{\mathcal{P}_{n,m}}$ to an arbitrary Banach space $\mathcal{V}$. An element of $\mathbb{R}^{\mathcal{P}_{n,m}}$ will be generally denoted by $\mathbf{x} = (x_l^k)_{(l,k)\subset\mathcal{P}_{n,m}}$. The vector with all entries equal to 1 will be noted $\mathbb{1}$. H is an arbitrary element of $\mathcal{S}$. We can state the following theorems, whose proofs are rewritings of that of theorem III.1.

Theorem IV.3 (The Brydges-Kennedy Taylor Jungle Formula)

$$H(\mathbb{1}) = \sum_{\substack{\mathcal{F}=(\mathfrak{F}_1,\dots,\mathfrak{F}_m) \\ m-\text{jungle}}} \left(\prod_{l\in\mathfrak{F}_m}\int_0^1 dh_l\right)\left(\left(\prod_{k=1}^m\left(\prod_{l\in\mathfrak{F}_k\setminus\mathfrak{F}_{k-1}}\frac{\partial}{\partial x_l^k}\right)\right)H\right)(X_{\mathcal{F}}^{BK}(\mathbf{h})) . \qquad \text{(IV.11)}$$

Here $X_{\mathcal{F}}^{BK}(\mathbf{h})$ is the vector $(x_l^k)_{(l,k)\in\mathcal{P}_{n,m}}$ of $\mathbb{R}^{\mathcal{P}_{n,m}}$ defined by $x_l^k = h_l^{\mathcal{F},k}(\mathbf{h})$, which is the value at which we evaluate the complicated derivative of H.

and

Theorem IV.4 (The Rooted Taylor Jungle Formula)

$$H(\mathbb{1}) = \sum_{\substack{\mathcal{F}=(\mathfrak{F}_1,\dots,\mathfrak{F}_m) \\ m-\text{jungle}}} \left(\prod_{l\in\mathfrak{F}_m}\int_0^1 dw_l\right)\left(\left(\prod_{k=1}^m\left(\prod_{l\in\mathfrak{F}_k\setminus\mathfrak{F}_{k-1}}\frac{\partial}{\partial x_l^k}\right)\right)H\right)(X_{\mathcal{F}}^{\mathcal{T}}(\mathbf{w})) . \qquad \text{(IV.12)}$$

Here $X_{\mathcal{F}}^{\mathcal{T}}(\mathbf{w})$ is the vector $(x_l^k)_{(l,k)\in\mathcal{P}_{n,m}}$ of $\mathbb{R}^{\mathcal{P}_{n,m}}$ defined by $x_l^k = w_l^{\mathcal{F},k}(\mathbf{w})$, which is the value at which we evaluate the complicated derivative of H.

Let us conclude this section by recalling the important positivity property of the Brydges-Kennedy forest and jungle formulas.

Theorem IV.5 (Positivity) *Let $\mathcal{F}$ be a m-jungle, and M^k, $1 \le k \le m$ be any sequence of m positive symmetric n by n matrices with entries m_{ij}^k. Then the interpolated matrices $M^k(\mathbf{h})$ with entries $m_{ij}^k(\mathbf{h}) = h_{ij}^{\mathcal{F},k}(\mathbf{h})m_{ij}^k$ if $i<j$ and $m_{ii}^k(\mathbf{h}) = m_{ii}^k$ are all positive.*

Proof: The definition of $M^k(\mathbf{h})$ only involves the forests $\mathfrak{F}_{k-1}$ and $\mathfrak{F}_k$ and the h_l's for l in $\mathfrak{F}_k$. Note that, for every link l, $h_l^{\mathcal{F},k}(\mathbf{h}) = h_l^{\mathfrak{F}_k}(\overline{\mathbf{h}})$, the right hand side is the $h_l^{\mathfrak{F}_k}(\overline{\mathbf{h}})$ function of the simple forest formalism of Section II, where $\overline{\mathbf{h}} = (\overline{h}_l)_{l\in\mathfrak{F}_k}$ with $\overline{h}_l = h_l$ if $l \in \mathfrak{F}_k \backslash \mathfrak{F}_{k-1}$ and $\overline{h}_l = 1$ if $l \in \mathfrak{F}_{k-1}$. In fact, for all points i and j connected by $\mathfrak{F}_{k-1}$, i.e. such that $L_{\mathfrak{F}_k}\{ij\} \subset \mathfrak{F}_{k-1}$, we have

$$h_l^{\mathfrak{F}_k}(\overline{\mathbf{h}}) = \inf\{\overline{h}_l,\ l \in L_{\mathfrak{F}_k}\{ij\}\} = \inf\{1\} = 1 \stackrel{\text{def}}{=} h_l^{\mathcal{F},k}(\mathbf{h}) \ . \tag{IV.13}$$

So we just need to prove the positivity in the simple forest formalism. This has already been done by Brydges in [B2], but we give here a proof along the lines of our derivation of the formula (II.1).

Suppose we have a forest $\mathfrak{F}$ with parameters $\mathbf{h} = (h_l)_{l\in\mathfrak{F}}$ between 0 and 1, and a positive symmetric matrix M with entries m_{ij} that is interpolated by $M(\mathbf{h})$ with entries $m_{ij}(\mathbf{h}) = m_{ij}h_l^{\mathfrak{F}}(\mathbf{h})$. Let us number the links of $\mathfrak{F}$ into an o-forest $F = (l_1,\ldots.l_\tau)$ in order that $1 \geq h_{l_1} \geq \ldots \geq h_{l_\tau} \geq 0$. We perform the same change of variables as in the formula (II.23). If $i \neq j$ let us denote by $U_{\{ij\}}$ the matrix with zero entries except at the locations (i,j) and (j,i) where we put 1. Let M_{diag} be the matrix with entries $m_{ii}\delta_{ij}$ the diagonal part of M. Then

$$M(\mathbf{h}) = M_{\text{diag}} + \sum_{l\in\mathcal{P}_n} h_l^F(\mathbf{h}).U_l m_l \tag{IV.14}$$

and for the same reason as (II.23)

$$M(\mathbf{h}) = M_{\text{diag}} + \sum_{\nu=1}^{\tau} t_\nu M_\nu^F \tag{IV.15}$$

where $M_\nu^F = \sum_{\{ij\}/\nu} U_{\{ij\}} m_{ij}$, and finally

$$M(\mathbf{h}) = (1 - \sum_{\nu=1}^{\tau} t_\nu) M_{\text{diag}} + \sum_{\nu=1}^{\tau} t_\nu (M_\nu^F + M_{\text{diag}}) \ . \tag{IV.16}$$

Now note that for $1 \leq \nu \leq \tau$, $t_\nu \geq 0$ and $\sum_{\nu=1}^{\tau} t_\nu = h_{l_1} \leq 1$ so $1 - \sum_{\nu=1}^{\tau} t_\nu \geq 0$, and $M_\nu^F + M_{\text{diag}}$ is a positive symmetric matrix. Indeed let us fix ν and consider the o-forest $F_\nu = (l_1,\ldots,l_\nu)$ completed at stage ν from the given o-forest F. We can reorder the base vectors so that M looks like a matrix with blocks corresponding to indexes in the same component of F_ν. Taking only the diagonal blocks and putting the others to 0 we get exactly the matrix $M_\nu^F + M_{\text{diag}}$. It is clear that such a process preserves the positivity of the matrix. $M(\mathbf{h})$, a sum of positive matrices with non negative multipliers, is positive. ■

Note that the rooted formulas **do not preserve positivity** as can be seen on simple examples, with 4 points for instance.

5 Some Concrete Examples

V.A) Gaussian Measures Perturbed by a Small Interaction

In constructive field theory cluster expansions are typically used to perform the thermodynamic limit of a field theory with cutoffs and a small local interaction $I_\Lambda(\phi)$ in a finite volume Λ. The cluster expansion expresses the partition function as a polymer gas with hard core conditions (see e.g. [R2]). The set I_n is then made of a partition of Λ into (unit size) cubes, and clusters are subsets of such cubes.

For instance consider the free bosonic Gaussian measure $d\mu_C$ in $\mathbb{R}^d$ defined by a covariance C with ultraviolet cutoff and good decrease at infinity. A standard example is

$$C(x,y) = \frac{1}{(4\pi)^{d/2}} \int_1^{+\infty} \frac{d\alpha}{\alpha^{d/2}} e^{-\alpha m^2 - |x-y|^2/4\alpha} \tag{V.A.1}$$

so that

$$\hat{C}(p) = \frac{e^{-(p^2+m^2)}}{p^2+m^2} \tag{V.A.2}$$

We perturb this free theory by adding an interaction such as $e^{-g\int_\Lambda \phi^4(x)}dx$, and we want to perform the thermodynamic limit $\Lambda \to \infty$, that is to define and study an intensive quantity such as the pressure

$$p = \lim_{\Lambda \to \mathbb{R}^d} \frac{1}{|\Lambda|} \log Z(\Lambda) \ , \tag{V.A.3}$$

where the partition function $Z(\Lambda)$ in a finite volume Λ is

$$Z(\Lambda) = \int d\mu_C(\phi) e^{-g\int_\Lambda \phi^4(x)dx} \ , \tag{V.A.4}$$

Let us explain how the Taylor formula (III.2) performs the task of rewriting the partition function as a dilute gas of clusters with hard core interaction. We write $\Lambda = \cup_{i\in I_n} b_i$, where each b_i is a unit cube, and define χ_b as the characteristic function of b, and $\chi_\Lambda = \sum_{i\in I_n} \chi_{b_i}$. Since the interaction lies entirely within Λ, the covariance C in (V.A.4) can be replaced by $C_\Lambda = \chi_\Lambda(x)C(x,y)\chi_\Lambda(y)$ without changing the value of $Z(\Lambda)$. Moreover C_Λ can be interpolated, defining for $l = \{i,j\} \in \mathcal{P}_n$

$$C_\Lambda((x_l)_{l\in\mathcal{P}_n})(x,y) = \sum_{i=1}^{n} \chi_{b_i}(x)C(x,y)\chi_{b_i}(y)$$

$$+ \sum_{\{i,j\}\in\mathcal{P}_n} x_{\{ij\}} \left(\chi_{b_i}(x)C(x,y)\chi_{b_j}(y) + \chi_{b_j}(x)C(x,y)\chi_{b_i}(y)\right) \tag{V.A.5}$$

Remark that $C_\Lambda(1,...,1) = C_\Lambda$. Now we apply the Taylor formula (III.2) with the function H being the partition function obtained by replacing in (V.A.3-4) the covariance C by $C_\Lambda((x_l)_{l\in\mathcal{P}_n})$. Here it is crucial to use the positivity theorem IV.5, in order for the interpolated covariance to remain positive, hence for the corresponding normalized Gaussian measure to remain well defined. From the rules of Gaussian integration of polynomials, we can compute the effect of deriving with respect to a given x_l parameter, and we obtain that (III.2) in this case takes the form

$$Z(\Lambda) = H(\mathbb{1}) = \sum_{\mathfrak{F}} \int d\mu_{C_\Lambda(X^{BK}_{\mathfrak{F}}(\mathbf{h}))}(\phi) \Big(\prod_{l \in \mathfrak{F}} \int_0^1 dh_l \Big)$$

$$\Big\{ \prod_{l=\{ij\}\in\mathfrak{F}} \int dx dy \chi_{b_i}(x) \chi_{b_j}(y) C(x,y) \frac{\delta}{\delta\phi(x)} \frac{\delta}{\delta\phi(y)} \Big\} e^{-g\int_\Lambda \phi^4(x)dx} \tag{V.A.6}$$

Since both the local interaction and the covariance as a matrix factorize over the clusters of the forest $\mathfrak{F}$, the corresponding contributions in (V.A.6) themselves factorize, which means that (V.A.6) can also be rewritten as a gas of non-overlapping clusters, each of which has an amplitude given by a *tree formula*:

$$Z(\Lambda) = \int d\mu_{C_\Lambda}(\phi) e^{I_\Lambda(\phi)} = \sum_{\substack{\text{sets } \{Y_1,\dots,Y_n\} \\ Y_i \cap Y_j = \emptyset, \cup Y_i = \Lambda}} \prod_{i=1}^{n} A(Y_i) \tag{V.A.7}$$

$$A(Y) = \sum_{\mathfrak{T} \text{ on } Y} \Big(\prod_{l \in \mathfrak{T}} \int_0^1 dh_l \Big) \int d\mu_{C_Y(X^{BK}_{\mathfrak{T}}(\mathbf{h}))}(\phi)$$

$$\Big\{ \prod_{l=\{ij\}\in\mathfrak{T}} \int dx dy \chi_{b_i}(x) \chi_{b_j}(y) C(x,y) \frac{\delta}{\delta\phi(x)} \frac{\delta}{\delta\phi(y)} \Big\} e^{-g\int_Y \phi^4(x)dx} \tag{V.A.8}$$

where b_i and b_j are the two ends of the line l, and the sum is over trees $\mathfrak{T}$ which connect together the set Y, hence have exactly $|Y|-1$ elements (if $|Y|=1$, $\mathfrak{T}=\emptyset$ connects Y). The measure $d\mu_{C_Y(\{X^{BK}_{\mathfrak{T}}(\mathbf{h})\})}(\phi)$ is the normalized Gaussian measure with (positive) covariance

$$C_Y(X^{BK}_{\mathfrak{T}}(\mathbf{h}))(x,y) = \chi_Y(x)\big(h_{\mathfrak{T}}(\mathbf{h})(x,y)\big) C(x,y) \chi_Y(y) \tag{V.A.9}$$

where $h_{\mathfrak{T}}(\mathbf{h})(x,y)$ is 1 if x and y belong to the same cube, and otherwise it is the infimum of the parameters h_l for l in the unique path $L_{\mathfrak{T}}(b(x),b(y))$ which in the tree $\mathfrak{T}$ joins the cube $b(x)$ containing x to the cube $b(y)$ containing y.

Formula (V.A.8) is somewhat shorter than the different formulas of [R1-2], and can be used in the same way to check that given any constant K, for small enough g with Re $g > 0$

$$\sum_{Y \text{ st } 0\in Y} |A(Y)| K^{|Y|} \leq 1 \tag{V.A.10}$$

Proof of (V.A.10) requires the slightly cumbersome computation of the action of the functional derivatives in (V.A.8) and a bound on the resulting functional integral. The method is identical to [R1-2]. Remark that although the full amplitudes $A(Y)$ defined in (V.A.8) must be identical to those in [R1-2], the subcontributions associated to particular trees are different.

The Mayer expansion defined below allows to deduce from (V.A.10) the existence and e.g. the Borel summability in g of thermodynamic functions such as the pressure p defined by (V.A.3).

B) The Mayer Expansion

In the cluster expansion (V.A.7), the condition that the disjoint union of all clusters is Λ is a global annoying constraint. Remark that the polymer amplitudes are translation invariant. In particular the trivial amplitude of a singleton cluster $Y = \{b\}$ is a number A_0 independent of b. Redefining $A_r(Y) = A(Y)/A_0^{|Y|}$ and $Z_r(\Lambda) = Z(\Lambda)/A_0^{|\Lambda|}$ we quotient out all the trivial clusters so that

$$Z_r(\Lambda) = 1 + \sum_{n\geq 1} \sum_{\substack{\text{sets } \{Y_1,\dots,Y_n\} \\ |Y_i|\geq 2\ ,\ Y_i\cap Y_j=\emptyset}} \prod_{i=1}^{n} A_r(Y_i) \qquad \text{(V.B.1)}$$

This is the partition function of a polymer gas: the sums over individual polymers would be independent were it not for the hard core constraints $Y_i \cap Y_j = \emptyset$. Adding in an infinite number of vanishing terms, we can replace the sum in (V.B.1) by a sum over ordered sequences $(Y_1, ..., Y_n)$ of polymers with hard core interaction and a symmetrizing factor $1/n!$ coming from the replacement of sets by sequences.

$$Z_r(\Lambda) = 1 + \sum_{n\geq 1} \frac{1}{n!} \sum_{\substack{\text{sequences } (Y_1,\dots,Y_n) \\ |Y_i|\geq 2}} \prod_{i=1}^{n} A_r(Y_i) \prod_{1\leq i<j\leq n} e^{-V(Y_i,Y_j)} \qquad \text{(V.B.2)}$$

where the hard core interaction is $V(X,Y) = 0$ if $X \cap Y = \emptyset$, and $V(X,Y) = +\infty$ if $X \cap Y \neq \emptyset$ To factorize again this formula we cannot apply directly the algebraic Brydges-Kennedy formula (II.1), because the interaction V can be infinite, but the Taylor forest formula (III.2) easily does the job. More precisely we define now I_n as our set of indices, and define $\epsilon^Y_{\{ij\}} = (e^{-V(X_i,X_j)} - 1)$, for $i \neq j$. For a fixed sequence $(Y_1, \dots, Y_n)$ of polymers, consider the function

$$H((x_l)_{l\in\mathcal{P}_n}) = \prod_{l\in\mathcal{P}_n} (1 + x_l \epsilon^Y_l) \qquad \text{(V.B.3)}$$

Rewrite (V.B.2) as

$$Z_r(\Lambda) = 1 + \sum_{n\geq 1} \frac{1}{n!} \sum_{\substack{\text{sequences } (Y_1,\dots,Y_n) \\ |Y_i|\geq 2}} H(\mathbb{1}) \prod_{i=1}^{n} A_r(Y_i)$$

$$= 1 + \sum_{n\geq 1} \frac{1}{n!} \sum_{\substack{\text{sequences} \\ (Y_1,\dots,Y_n) \\ |Y_i|\geq 2}} \prod_{i=1}^{n} A_r(Y_i) \sum_{\mathfrak{F}} \Big(\prod_{l\in\mathfrak{F}} \int_0^1 dh_l \Big) \Big(\prod_{l\in\mathfrak{F}} \epsilon^Y_l \Big) \Big(\prod_{l\notin\mathfrak{F}} (1 + h^{\mathfrak{F}}_l(\mathbf{h}) \epsilon^Y_l) \Big)$$

$$= \sum_{n\geq 0} \frac{1}{n!} \left(\sum_{k\geq 1} \frac{1}{k!} \sum_{\substack{\text{sequences } (Y_1,\ldots,Y_k) \\ |Y_i|\geq 2}} \left(\prod_{i=1}^{k} A_r(Y_i) \right) C^T(Y_1, ..., Y_k) \right)^n \quad \text{(V.B.4)}$$

where

$$C^T(Y_1, ..., Y_k) = \sum_{\substack{G \text{ connected graph} \\ \text{on}\{1,\ldots,k\}}} \prod_{l\in G} \epsilon_l^Y$$

$$= \sum_{\mathfrak{T} \text{ tree on } \{1,\ldots,k\}} \left(\prod_{l\in\mathfrak{T}} \int_0^1 dh_l \right) \left(\prod_{l\in\mathcal{P}_k,\ l\in\mathfrak{T}} \epsilon_l^Y \right) \prod_{l\in\mathcal{P}_k,\ l\notin\mathfrak{T}} (1 + h_l^{\mathfrak{T}}(\mathbf{h})\epsilon_l^Y) \quad \text{(V.B.5)}$$

where $h_l^{\mathfrak{T}}(\mathbf{h})$ is, if $l = \{ij\}$, the infimum of the parameters $h_{l'}$ for l' in the unique path $L_{\mathfrak{T}}\{ij\}$ which in the tree $\mathfrak{T}$ joins i to j.

We obtain immediately that

$$\log Z_r(\Lambda) = \sum_{k\geq 1} \frac{1}{k!} \sum_{\substack{\text{sequences } (Y_1,\ldots,Y_k) \\ |Y_i|\geq 2}} \left(\prod_{i=1}^{k} A_r(Y_i) \right) C^T(Y_1, ..., Y_k) \quad \text{(V.B.6)}$$

Formulas (V.B.5-6) are more explicit than those used in [R1-2] and have all desired advantages (every tree coefficient forces the necessary links and is bounded by 1, since $|(1 + h_l^{\mathfrak{T}}(\mathbf{h})\epsilon_l^Y)| \leq 1$). They can therefore be used together with (V.A.10) to control in a similar way the thermodynamic function $p = A_0 + \lim_{\Lambda\to\mathbb{R}^d} \frac{1}{|\Lambda|} \log Z_r(\Lambda)$.

C) A Single Formula for the Succession of a Cluster and a Mayer Expansion

Formula (V.B.6) involves trees and forests on two different kinds of objects. First are the links between boxes in our lattice that make the clusters Y_i. Then we have Mayer links between indexes i in I_k. This leads to complications, if we iterate the process, when doing a renormalization group analysis. So we propose a formula with two types of links but on the same objects, it is a 2-jungle formula. For simplicity we write it in the algebraic setting of section II, i.e. with $Z = \exp(\sum u_l)$ instead of $Z_r(\Lambda)$, and perform the apparently stupid operation $\log \exp(\sum u_l)$!

The links l are between elements of I_n, but now we introduce a new set of objects $\mathcal{D}_n = I_n \times \mathbb{N}$. If $\bar{l}$ is a link $\{(b,k),(b',k')\}$ between different elements of $\mathcal{D}_n$ we let $u_{\bar{l}} = u_{\{bb'\}}$ and $\delta_{\bar{l}} = \delta_{bb'}$ where the second delta is Kroneker's. Let $\mathcal{F} = (\mathfrak{F}_1, \mathfrak{F}_2)$ be a 2-jungle on $\mathcal{D}_n$, we say that it is *nice* if it fulfills the following requirements:

- $\mathfrak{F}_1$ is only made of horizontal links i.e. of the form $\{(b,k),(b',k)\}$;
- $\mathfrak{F}_2 \backslash \mathfrak{F}_1$ is only made of vertical links of length 1 i.e. of the form $\{(b,k),(b,k+1)\}$;
- $\mathfrak{F}_2$ is a non empty tree;
- there are no singletons among the clusters of $\mathfrak{F}_1$ in the support of $\mathfrak{F}_2$;

- there is a unique cluster of $\mathfrak{F}_1$ in the support of $\mathfrak{F}_2$ that lies in the 0-th level $I_n \times \{0\}$;

- if we fix a root r of $\mathfrak{F}_2$ in that unique cluster, then for every element (b,k) in the support of $\mathfrak{F}_2$, k is the number of vertical links on the path going from (b,k) to r (this property is independent of the choice of r).

We denote by $k(\mathcal{F})$ the number of clusters of $\mathfrak{F}_1$ in the support of $\mathfrak{F}_2$. We introduce the following lexicographical order on $\mathcal{D}_n$, $(b,k) \preceq (b',k')$ if and only if $k < k'$ or ($k = k'$ and $b \leq b'$). This is the order we use to choose a root for each non empty finite subset of $\mathcal{D}_n$ according to the rooted jungle formalism. We let $\mathbf{h} = (h_{\bar{l}})_{\bar{l}\in\mathfrak{F}_1}$ and $\mathbf{w} = (w_{\bar{l}})_{\bar{l}\in\mathfrak{F}_2\setminus\mathfrak{F}_1}$ and we take the compound $(\mathbf{h},\mathbf{w})$ instead of the $\mathbf{h}$ or the $\mathbf{w}$ vectors that were used in (IV.1) and (IV.9). The functions $h_{\bar{l}}^{\mathcal{F},1}$ and $w_{\bar{l}}^{\mathcal{F},2}$ are however defined exactly in the same manner as in section IV. We now claim that

Theorem V.C.1

$$\log Z = \sum_{\substack{\mathcal{F}=(\mathfrak{F}_1,\mathfrak{F}_2)\\ \text{nice 2-jungle}}} \frac{1}{k(\mathcal{F})} \left(\prod_{\bar{l}\in\mathfrak{F}_1} u_{\bar{l}} \int_0^1 dh_{\bar{l}}\right) \left(\prod_{\bar{l}\in\mathfrak{F}_2\setminus\mathfrak{F}_1} (-\delta_{\bar{l}}) \int_0^1 dw_{\bar{l}}\right)$$

$$\times \exp\left(\sum_{\bar{l}} h_{\bar{l}}^{\mathcal{F},1}(\mathbf{h},\mathbf{w}) u_{\bar{l}}\right) \cdot \prod_{\bar{l}\notin\mathfrak{F}_2\setminus\mathfrak{F}_1} \left(1 - w_{\bar{l}}^{\mathcal{F},2}(\mathbf{h},\mathbf{w})\delta_{\bar{l}}\right) \quad . \tag{V.C.1}$$

Proof:

For $m \in \mathbb{N}$, let $\mathcal{D}_{n,m} = I_n \times \{0,1,\ldots,m\} \subset \mathcal{D}_n$. As in (V.B.6) we have

$$\log Z = \lim_{m\to+\infty} L_m \tag{V.C.2}$$

where

$$L_m = \sum_{s=1}^{m} \frac{1}{s!} \sum_{\substack{\text{sequences } (Y_1,\ldots,Y_s)\\ Y_i\subset I_n,\ |Y_i|\geq 2}} \left(\prod_{i=1}^{s} A(Y_i)\right) C^T(Y_1,\ldots,Y_s) \tag{V.C.3}$$

and

$$A(Y) = \sum_{\mathfrak{T} \text{ tree on } Y} \left(\prod_{l\in\mathfrak{T}} u_l \int_0^1 dh_l\right) \exp\left(\sum_{l \text{ in } Y} h_l^{\mathfrak{T}}(\mathbf{h}) u_l\right) \quad . \tag{V.C.4}$$

This time we compute $C^T(Y_1,\ldots,Y_s)$ thanks to the rooted forest formalism. Note that

$$\prod_{\{ij\}\subset I_n} (1 + \epsilon_{\{ij\}}^Y) = \prod_{\{ij\}\subset I_n} \prod_{(b,b')\in Y_i\times Y_j} (1 - x_{\{ij\}}\delta_{bb'}) \tag{V.C.5}$$

with all the $x_{\{ij\}}$ set to 1. Then we apply the rooted Taylor forest formula III.3 and collect the connected parts. We get

$$C^T(Y_1,\ldots,Y_S) = \sum_{\mathfrak{T} \text{ tree on } I_s} C(s;Y_1,\ldots,Y_s;\mathfrak{T}) \tag{V.C.6}$$

where

$$C(s;Y_1,\ldots,Y_s;\mathfrak{T}) = \left(\prod_{l\in\mathfrak{T}} \int_0^1 dw_l\right) .$$

$$\left(\prod_{\{ij\}\in\mathfrak{T}} \left(\sum_{(b,b')\in Y_i\times Y_j} (-\delta_{bb'}) \prod_{\substack{(c,c')\in Y_i\times Y_j \\ (c,c')\neq(b,b')}} (1 - w^{\mathfrak{T}}_{\{ij\}}(\mathbf{w})\delta_{cc'})\right)\right)$$

$$\cdot \prod_{\{ij\}\notin\mathfrak{T}} \left(\prod_{(b,b')\in Y_i\times Y_j} (1 - w^{\mathfrak{T}}_{\{ij\}}(\mathbf{w})\delta_{bb'})\right) . \tag{V.C.7}$$

The only trees $\mathfrak{T}$ giving a non zero contribution are those where for any $i \neq j$ in the same layer of $\mathfrak{T}$, $Y_i \cap Y_j = \emptyset$. In fact for such i and j, $\{ij\} \notin \mathfrak{T}$ and $w^{\mathfrak{T}}_{\{ij\}}(\mathbf{w}) = 1$ so in (V.C.7) we get a factor

$$\prod_{(b,b')\in Y_i\times Y_j} (1 - \delta_{bb'}) = e^{-V(Y_i,Y_j)} \tag{V.C.8}$$

forcing the non overlapping condition.

Given a sequence $(Y_1,\ldots,Y_s)$ and a tree $\mathfrak{T}$ satisfying that property, we can define for $i \in I_k$, $\overline{Y}_i = Y_i \times \{l^{\mathfrak{T}}(i)\} \subset \mathcal{D}_{n,m}$. Here $l^{\mathfrak{T}}(i)$ is the height of vertex i in the tree $\mathfrak{T}$. The $\overline{Y}_i$ are all disjoint.

For a finite subset $\overline{Y}$ of $\mathcal{D}_n$ let

$$\overline{A}(\overline{Y}) = \sum_{\mathfrak{T} \text{ tree on } \overline{Y}} \left(\prod_{\overline{l}\in\mathfrak{T}} u_{\overline{l}} \int_0^1 dh_{\overline{l}}\right) \exp\Big(\sum_{\overline{l}\in\overline{Y}} h^{\mathfrak{T}}_{\overline{l}}(\mathbf{h}) u_{\overline{l}}\Big) \tag{V.C.9}$$

so that $A(Y_i) = \overline{A}(\overline{Y}_i)$ for every $i \in I_s$. Now define the (unordered) set $\mathcal{O} = \mathcal{O}(Y_1,\ldots,Y_s;\mathfrak{T}) = \{\overline{Y}_1,\ldots,\overline{Y}_s\}$. It is easy to see that $\mathcal{O}$ is a non empty set of disjoint polymers ($|\overline{Y}| \geq 2$ for each $\overline{Y} \in \mathcal{O}$) lying in $\mathcal{D}_{n,m}$, with exactly one element in the ground level $I_n \times \{0\}$, furthermore, the labels k of the occupied levels $I_n \times \{k\}$ form an interval $\{0,1,\ldots,q\}$, $q \leq |\mathcal{O}| - 1$. An $\mathcal{O}$ verifying these properties is called *admissible*.

We will sum over admissible $\mathcal{O}$'s in (V.C.3)

$$L_m = \sum_{\text{admissible } \mathcal{O}} \frac{1}{|\mathcal{O}|!} \left(\prod_{\overline{Y}\in\mathcal{O}} \overline{A}(\overline{Y})\right) \sum_{\substack{(Y_1,\ldots,Y_s);\mathfrak{T} \\ \text{st } \mathcal{O}=\mathcal{O}(Y_1,\ldots,Y_s;\mathfrak{T})}} C(s;Y_1,\ldots,Y_s;\mathfrak{T}) . \tag{V.C.10}$$

Knowing the sequence $Y_1,\ldots,Y_s)$, $\mathfrak{T}$ naturally induces a tree on $\mathcal{O}$, $\overline{\mathfrak{T}} = \overline{\mathfrak{T}}(Y_1,\ldots,Y_s;\mathfrak{T})$ by the rule $\{ij\} \in \mathfrak{T}$ if and only if $\{\overline{Y}_i,\overline{Y}_j\} \in \overline{\mathfrak{T}}$. The tree $\overline{\mathfrak{T}}$ has a natural root i.e. the unique $\overline{Y}$ of $\mathcal{O}$ lying in the 0-th level, moreover, for each $\overline{Y}$ the label of the level it belongs to is just its height in the tree $\overline{\mathfrak{T}}$ with the mentioned choice of root. Such a tree will also be called *admissible*.

Let for any tree $\overline{\mathfrak{T}}$ on $\mathcal{O}$

$$U(\mathcal{O},\overline{\mathfrak{T}}) = \prod_{L\in\overline{\mathfrak{T}}} \int_0^1 dw_L \prod_{\substack{\{\overline{Y}\ \overline{Y}'\}\subset\mathcal{O}\\ \{\overline{Y}\ \overline{Y}'\}\notin\overline{\mathfrak{T}}}} \prod_{(\overline{b},\overline{b}')\in\overline{Y}\times\overline{Y}'} (1 - w^{\overline{\mathfrak{T}}}_{\{\overline{Y}\ \overline{Y}'\}}(\mathbf{w})\delta_{\{\overline{b}\ \overline{b}'\}})$$

$$\cdot\left(\prod_{\{\overline{Y}\ \overline{Y}'\}\in\overline{\mathfrak{T}}}\left(\sum_{(\overline{b},\overline{b}')\in\overline{Y}\times\overline{Y}'}(-\delta_{\{\overline{b}\ \overline{b}'\}})\prod_{\substack{(\overline{c},\overline{c}')\in\overline{Y}\times\overline{Y}'\\ (\overline{c},\overline{c}')\neq(\overline{b},\overline{b}')}}(1 - w^{\overline{\mathfrak{T}}}_{\{\overline{Y}\ \overline{Y}'\}}(\mathbf{w})\delta_{\{\overline{c}\ \overline{c}'\}})\right)\right) . \tag{V.C.11}$$

and

$$N(\mathcal{O},\overline{\mathfrak{T}}) = \sum_{\substack{(Y_1,\dots,Y_s);\mathfrak{T}\\ \text{st } \mathcal{O}=\mathcal{O}(Y_1,\dots,Y_s;\mathfrak{T})\\ \text{and } \overline{\mathfrak{T}}=\overline{\mathfrak{T}}(Y_1,\dots,Y_s;\mathfrak{T})}} 1 . \tag{V.C.12}$$

It is clear now that

$$L_m = \sum_{\text{admissible } \mathcal{O}} \frac{1}{|\mathcal{O}|!}\left(\prod_{\overline{Y}\in\mathcal{O}} \overline{A}(\overline{Y})\right) \sum_{\overline{\mathfrak{T}}\text{tree on } \mathcal{O}} U(\mathcal{O},\overline{\mathfrak{T}}).N(\mathcal{O},\overline{\mathfrak{T}}) . \tag{V.C.13}$$

Let us admit for the moment the following

Lemma V.C.1 *The combinatoric factor $N(\mathcal{O},\overline{\mathfrak{T}})$ is $(|\mathcal{O}|-1)!$ for any admissible $\mathcal{O}$ and $\overline{\mathfrak{T}}$.*

Then

$$L_m = \sum_{\substack{\text{admissible } \mathcal{O}\\ \overline{\mathfrak{T}} \text{ tree on } \mathcal{O}}} \frac{1}{|\mathcal{O}|}\left(\prod_{\overline{Y}\in\mathcal{O}} \overline{A}(\overline{Y})\right) \sum_{\overline{\mathfrak{T}} \text{ tree on } \mathcal{O}} U(\mathcal{O},\overline{\mathfrak{T}}) . \tag{V.C.14}$$

Besides,

$$\prod_{\overline{Y}\in\mathcal{O}} \overline{A}(\overline{Y}) = \sum_{\mathfrak{F}_1}\left(\prod_{\overline{l}\in\mathfrak{F}_1} u_{\overline{l}} \int_0^1 dh_{\overline{l}}\right) \exp\left(\sum_{\overline{l}} h^{\mathfrak{F}_1}_{\overline{l}}(\mathbf{h}) u_{\overline{l}}\right) , \tag{V.C.15}$$

where we sum over all forests $\mathfrak{F}_1$ made of horizontal links $u_{\overline{l}}$ such that the connected components that are not singletons are exactly the elements of $\mathcal{O}$.

Finally remark that summing over trees $\overline{\mathfrak{T}}$ on $\mathcal{O}$ and, for every $\{\overline{Y},\overline{Y}'\} \in \overline{\mathfrak{T}}$, over $(\overline{b},\overline{b}') \in \overline{Y} \times \overline{Y}'$ is the same as summing over forests $\mathfrak{F}_2\backslash\mathfrak{F}_1$ of vertical links $-\delta_{\overline{l}}$ connecting in a tree the clusters $\overline{Y} \in \mathcal{O}$ formed by $\mathfrak{F}_1$. It is now a matter of (rather unpleasant) checking the definitions to convince oneself that (V.C.14) is just the result of formula (V.C.1), except that the clusters are confined to $\mathcal{D}_{n,m}$. However taking the limit $m \to +\infty$ removes this restriction. ■

Proof of Lemma V.C.1 : Since $\mathcal{O}$ and $\overline{\mathfrak{T}}$ are admissible, there exists a sequence $Y_1,\ldots,Y_s$ and a tree $\mathfrak{T}$ such that $\mathcal{O}=\mathcal{O}(Y_1,\ldots,Y_s;\mathfrak{T})$ and $\overline{\mathfrak{T}}=\overline{\mathfrak{T}}(Y_1,\ldots,Y_s;\mathfrak{T})$. In fact suppose we are given a one-to-one map $\sigma:\mathcal{O}\mapsto I_s$ such that $\sigma(\overline{Y}_1)=1$, where $\overline{Y}_1$ is the unique element of $\mathcal{O}$ lying in the ground level of $\mathcal{D}_n$. We can construct a sequence $(Y_1^\sigma,\ldots,Y_s^\sigma)$ and a tree $\mathfrak{T}^\sigma$ fulfilling the previous requirements. If π denotes the projection of $\mathcal{D}_n$ on I_n, we let $Y^\sigma_{\sigma(\overline{Y})}=\pi(\overline{Y})$ for all $\overline{Y}\in\mathcal{O}$, and $\mathfrak{T}^\sigma$ be the tree on I_s induced by $\overline{\mathfrak{T}}$ through the correspondence σ.

It is easy to see that every pair $((Y_1,\ldots,Y_s);\mathfrak{T})$ we are counting is obtained that way. The interesting thing is that there is a unique σ giving this pair. If σ and τ are two maps such that $((Y_1{}^\sigma,\ldots,Y_s{}^\sigma);\mathfrak{T}^\sigma)=((Y_1{}^\tau,\ldots,Y_s{}^\tau);\mathfrak{T}^\tau)$, $\sigma\circ\tau^{-1}$ must preserve the tree $\mathfrak{T}^\sigma=\mathfrak{T}^\tau$, but $\sigma\circ\tau^{-1}(1)=1$ so it preserves also the root, as a consequence it leaves the layers invariant. But, the corresponding $\overline{Y}_i$ are disjoint contained in the same level of $\mathcal{D}_n$, thus the projections Y_i are distinct and $Y_{\sigma\circ\tau^{-1}(i)}=Y_i$ forces $\sigma\circ\tau^{-1}(i)=i$ inside each layer. In conclusion $\sigma\circ\tau^{-1}=Id$. $N(\mathcal{O},\overline{\mathfrak{T}})$ is the number of σ's that is $(s-1)!$. ∎

If after all that the reader is somewhat sceptical about formula (V.C.1), he might find it amusing to check it for $n=2$.

Now to use formula (V.C.1) for more clever applications than computing the trivial expression $\log\exp(\sum u_l)$, we state the analogous result for $\log Z(\Lambda)$ in ϕ^4 theory. We introduce a copy of the field ϕ_k for each $k\in\mathbb{N}$, and write Φ for the system $(\phi_k)_{k\in\mathbb{N}}$. Then we define the interaction in $S(\mathfrak{F}_2)$, the support of $\mathfrak{F}_2$, as

$$e^{I(S(\mathfrak{F}_2))}(\Phi)=\prod_{(b,k)\in S(\mathfrak{F}_2)} e^{-\lambda\int_b\phi_k^4(x)dx} \tag{V.C.16}$$

where we have identified the boxes b in Λ with their labels in I_n. To a link $\bar{l}=\{(b,k),(b',k')\}$ we associate the operator

$$C_{\bar{l}}=\delta_{kk'}\int_b dx\int_{b'}dy\ C(x,y)\frac{\delta}{\delta\phi_k(x)}\frac{\delta}{\delta\phi_{k'}(x)} \tag{V.C.17}$$

that will play the role of $u_{\bar{l}}$. $\delta_{\bar{l}}$ is defined in the same fashion as before. Given a 2-jungle $\mathcal{F}=(\mathfrak{F}_1,\mathfrak{F}_2)$ denote by $S_k(\mathfrak{F}_2)$ the k-th slice $S(\mathfrak{F}_2)\cap(I_n\times\{k\})$ of $S(\mathfrak{F}_2)$. We define the following Gaussian measure on the fields $\phi_{k|S_k(\mathfrak{F}_2)}$ such that k is an occupied level of $\mathcal{F}_2$ (i.e. $S_k(\mathfrak{F}_2)\neq\emptyset$)

$$d\mu_C^{\mathcal{F},\mathbf{h},\mathbf{w}}(\Phi_{|S(\mathfrak{F}_2)})=\prod_{\text{occupied }k} d\mu_C^{\mathcal{F},\mathbf{h},\mathbf{w},k}(\phi_{k|S_k(\mathfrak{F}_2)}) \tag{V.C.18}$$

where $d\mu_C^{\mathcal{F},\mathbf{h},\mathbf{w},k}(\phi_{k|S_k(\mathfrak{F}_2)})$ has covariance

$$C^{\mathcal{F},\mathbf{h},\mathbf{w},k}(x,y)=C(x,y)$$

if x and y fall in the same b with $(b,k)\in S_k(\mathfrak{F}_2)$, and where it has covariance

$$C^{\mathcal{F},\mathbf{h},\mathbf{w},k}(x,y)=C(x,y).h^{\mathcal{F},1}_{\{(b,k),(b',k')\}}(\mathbf{h},\mathbf{w})$$

if $x \in b$, $y \in b'$, $b \neq b'$ $(b,k) \in S_k(\mathfrak{F}_2)$ and $(b',k) \in S_k(\mathfrak{F}_2)$.

We can now state a formula which performs a cluster and a Mayer expansion in a single move (recall that $|\Lambda| = n$)

Theorem V.C.2

$$\log Z(\Lambda) = n \log A_0 + \sum_{\substack{\mathcal{F}=(\mathfrak{F}_1,\mathfrak{F}_2) \\ \text{nice 2-jungle}}} \frac{A_0^{-|S(\mathfrak{F}_2)|}}{k(\mathcal{F})} \Big(\prod_{\bar{l} \in \mathfrak{F}_1} \int_0^1 dh_{\bar{l}} \Big) \Big(\prod_{\bar{l} \in \mathfrak{F}_2 \setminus \mathfrak{F}_1} (-\delta_{\bar{l}}) \int_0^1 dw_{\bar{l}} \Big)$$

$$\cdot \prod_{\bar{l} \notin \mathfrak{F}_2 \setminus \mathfrak{F}_1} (1 - w_{\bar{l}}^{\mathcal{F},2}(\mathbf{h},\mathbf{w})\delta_{\bar{l}}) \int d\mu_C^{\mathcal{F},\mathbf{h},\mathbf{w}}(\Phi_{|S(\mathfrak{F}_2)}) \Big(\prod_{\bar{l} \in \mathfrak{F}_1} C_{\bar{l}} \Big) e^{I(S(\mathfrak{F}_2))} \quad . \quad \text{(V.C.19)}$$

Proof: Use first $\log Z(\Lambda) = n \log A_0 + \log Z_r(\Lambda)$, then repeat exactly the same line of arguments than for Theorem V.C.1. ∎

From Theorem V.C.2 one can construct a series for thermodynamic quantities such as the pressure $\lim_{\Lambda \to \infty} |\Lambda|^{-1} \log Z(\Lambda)$ which is the sum of the trivial term $\log A_0$ plus a sum over nice 2 jungles that extend horizontally over the infinite lattice of all cubes covering $\mathbb{R}^d$, and such that the well defined unique root of $\mathfrak{F}_2$ is the particular cube containing the origin. (This requires to neglect boundary terms of order $|\Lambda|^{(d-1)/d}$ in $\log Z(\Lambda)$). For small λ this series is absolutely convergent.

D) The Resolvent Expansion

In many situations physical situations (polymers, disordered systems) we want to compute not a full theory but a single Green's function which is expressed as the inverse of some operator. This mathematical situation is formally equivalent to 0-component field theory (or to supersymmetric theories) in which the usual normalizing fermionic or bosonic determinants have been cancelled out.

Consider a finite dimensional operator M on $\mathbb{R}^n$ with matrix elements m_{ij}, and a norm strictly smaller than 1. The operator $A = \frac{1}{1+M}$ is well defined through its power series expansion. Define $M(\mathbf{x})$ as the matrix with elements $x_{\{ij\}} m_{ij}$ if $i \neq j$ and diagonal elements m_{ii} equal to those of M, and $M_{\{ij\}}$ as the matrix with 0 elements except $(M_{\{ij\}})_{ij} = m_{ij}$ and $(M_{\{ij\}})_{ji} = m_{ji}$. The Taylor forest formula (III.2) applied to the operator $H(\mathbf{x}) = \frac{1}{1+M(\mathbf{x})}$ gives

$$A = \frac{1}{1+M} = H(\mathbf{1})$$

$$= \sum_{\mathfrak{F}} \Big(\prod_{l \in \mathfrak{F}} \int_0^1 dh_l \Big) \Big(\prod_{l \in \mathfrak{F}} \oint \frac{1}{2\pi i} \frac{dx_l}{x_l^2} \Big) \frac{1}{1 + M(X_{\mathfrak{F}}(\mathbf{h})) + \sum_{l \in \mathfrak{F}} x_l M_l} \qquad \text{(V.D.1)}$$

where the action of multiple derivatives is for compactness rewritten as a multiple Cauchy integral, and the analyticity radii in (V.D.1) are small enough. This does not look like a very clever rewriting, but it allows to reblock the power series for $\frac{1}{1+M}$ according to the set of sites truly visited in this power series. In particular if we impose a given entry to the matrix A the forest formula (V.D.1) reduces to a tree formula:

$$A_{ij} = \left(\frac{1}{1+M}\right)_{ij} = \sum_{\substack{Y\subset\{1,\dots,n\}\\ i\in Y,\ j\in Y}} \sum_{\mathfrak{T} \text{ tree on } Y} \left(\prod_{l\in\mathfrak{T}} \int_0^1 dh_l\right)$$

$$\left(\prod_{l\in\mathfrak{T}} \oint \frac{1}{2\pi i}\frac{dx_l}{x_l^2}\right) \frac{1}{1_Y + M_Y(X_{\mathfrak{T}}(\mathbf{h})) + \sum_{l\in\mathfrak{T}} x_l M_{Y,l}} \qquad \text{(V.D.2)}$$

where 1_Y is the identity on R^Y, $M_Y(X_{\mathfrak{T}}(\mathbf{h}))$ is the matrix on R^Y with diagonal entries m_{ii} (for $i \in Y$) and off diagonal entries $h^{\mathfrak{T}}_{ij}(\mathbf{h})m_{ij}$ (for $i \in Y$, $j \in Y$), and $M_{Y,l}$ is the restriction of M_l to indexes in Y.

This kind of formulas, eventually in combination with a large field/small field analysis should be useful for interacting random walks [IM] and presumably for the study of disordered systems.

E) Cluster Expansions with Large/Small Field Conditions: The $m = 2$ Jungle.

Jungle formulas with several levels are interesting when there are various type of links with priority rules between them. The simplest example is a single scale cluster expansion but with so-called small/large field conditions [R2], in which one does *not* want to derive links between cubes of a large field connected component. Indeed large field regions are small for probabilistic reasons, but the vertices of perturbation theory (created by the action of functional derivatives such as those in (V.A.8)) may not be small there, hence it would be dangerous to blindly expand the clusters in the usual way.

We want to study a theory with partition function such as

$$Z(\Lambda) = \sum_{\Gamma\subset\Lambda} Z(\Lambda,\Gamma)\ , \quad Z(\Lambda,\Gamma) = \int d\mu_C(\phi)\chi_\Gamma(\phi)e^{-g\int_\Lambda I(\phi)dx}\ , \qquad \text{(V.E.1)}$$

where C is a propagator with decay at the unit scale, and $I(\phi)$ is some local interaction. Without describing a precise model we shall assume that the large field condition $\chi_\Gamma(\phi)$ is such that the functional integral $\int d\mu_C(\phi)\chi_\Gamma(\phi)e^{-g\int_\Lambda I(\phi)dx}$ can be bounded by $K^{|\Lambda-\Gamma|}c^{|\Gamma|}$ where K is fixed and c is a small constant that can tend to zero with some coupling constant in $I(\phi)$; one also assumes that the outcome of a functional derivative $\int_b dx\frac{\delta}{\delta\phi(x)}$ acting on $\chi_\Gamma(\phi)e^{-g\int_\Lambda I(\phi)dx}$ gives a small factor but only in the "small field region", hence if $b \notin \Gamma$.

The set I_n is again the set of the cubes which pave our finite volume Λ. We fix a given subset $\Gamma \subset I_n$ and define, for $l = \{ij\} \in \mathcal{P}_n$, $\epsilon^\Gamma_l = 1$ if $b_i \in \Gamma$, $b_j \in \Gamma$ and $\mathrm{dist}(b_i, b_j \leq M$. M is some constant that will be fixed to a large value.

Otherwise we put $\epsilon_l^\Gamma = 0$. We define also $\eta_l^\Gamma = 1 - \epsilon_{ij}^\Gamma$. Hence $\epsilon_l^\Gamma = 1$ means that b_i and b_j are "large field cubes" which are closer than distance M, and $\eta_l^\Gamma = 1$ means the contrary.

The level two jungle formula will at the first level create connections whose clusters are automatically the "connected components" of Γ in the generalized sense that up to distance M two cubes are considered connected. Then the second level will create ordinary connections with the propagator C. Therefore we introduce a first set of interpolation parameters $\{x_l^1\}$ and define

$$H^\Gamma(\{x_l^1\}) = \prod_{l \in \mathcal{P}_n} (x_l^1 \epsilon_l^\Gamma + \eta_l^\Gamma) \, . \tag{V.E.2}$$

Remark that $H^\Gamma(1,...,1) = 1$, so that we can freely multiply $Z(\Lambda, \Gamma)$ by the function $H^\Gamma(1,...,1)$ without changing its value. We introduce the second set of parameters $\{x_l^2\}$ on the covariance C exactly as in (V.A.5).

Then the formula (IV.11) with $m = 2$ simply gives:

$$Z(\Lambda,\Gamma) = \sum_{\substack{\mathcal{F}=(\mathfrak{F}_1,\mathfrak{F}_2)\\ 2-\text{jungle}}} \Big(\prod_{l\in\mathcal{F}} \int_0^1 dh_l\Big)\Big(\prod_{l\in\mathfrak{F}_1}\epsilon_l^\Gamma\Big)\Big(\prod_{l\notin\mathfrak{F}_1}(\eta_l^\Gamma + \epsilon_l^\Gamma h_l^{\mathcal{F},1}(\mathbf{h}))\Big)$$

$$\int d\mu_{C_{\mathcal{F}}(\mathbf{h})}(\phi) \prod_{l=\{ij\}\in\mathfrak{F}_2-\mathfrak{F}_1} \Big\{\int dx dy \chi_{b_i}(x)\chi_{b_j}(y) C(x,y) \frac{\delta}{\delta\phi(x)}\frac{\delta}{\delta\phi(y)}\Big\}$$

$$\chi_\Gamma(\phi) e^{-g\int_\Lambda I(\phi)dx} \tag{V.E.3}$$

where $d\mu_{C_{\mathcal{F}}(\mathbf{h})}$ is the normalized Gaussian measure with positive covariance

$$C_{\mathcal{F}}(\mathbf{h})(x,y) = h_{l(x,y)}^{\mathcal{F},2}(\mathbf{h}) C(x,y) \tag{V.E.4}$$

where $h_{l(x,y)}^{\mathcal{F},2}(\mathbf{h})$ is by definition 1 if x and y belong to the same cube, and is $h_l^{\mathcal{F},2}(\mathbf{h})$ if $l = \{ij\}$, $x \in b_i$ and $y \in b_j$ or $x \in b_j$ and $y \in b_i$.

The only non-zero terms in this formula are those for which the clusters associated to the forest $\mathfrak{F}_1$ are exactly the set of "connected components" Γ_a of the large field region in the generalized sense (for $M = 0$ it gives the ordinary connected components). Indeed they cannot be larger because of the factor $\prod_{l\in\mathfrak{F}_1} \epsilon_l^\Gamma$, nor can they be smaller because of the factor $\prod_{l\notin\mathfrak{F}_1} (\eta_l^\Gamma + \epsilon_l^\Gamma h_l^{\mathcal{F},1}(\mathbf{h}))$ which is zero if there are some generalized neighbors (for which $\eta_{ij}^\Gamma = 0$) belonging to different clusters (for which $h_{ij}^{\mathcal{F}}(h) = 0$). Therefore the first forest in this formula simply automatically draws connecting trees of "neighbor links" connecting each such generalized connected component, but in a symmetric way without arbitrary choices. Remark that the factor $\prod_{l\notin\mathfrak{F}_1} (\eta_l^\Gamma + \epsilon_l^\Gamma h_l^{\mathcal{F},1}(\mathbf{h}))$ is bounded by one as expected for further estimates. Then the second forest of the jungle automatically takes into account the first links, i.e. the existence of large field regions, to draw propagators connections, In this way the units connected by the full forest remain unit cubes instead of being either small field cubes or blocks of large field cubes[1]. For convergence of the thermodynamic limit one has then simply to check that all connections are summable (this is obvious for the finite range nature of the ϵ_l^Γ links), and that there is a small factor per link of the forest. For ϵ links it comes from the probabilistic factor associated to the presence of the function χ_Γ and for ordinary links, it comes either from the functional derivative localized out of Γ, or from the large distance (at least M) crossed in the case of a link between two large field cubes. This distance induces a small factor through the decay of C. In conclusion there is no need to gain a small factor from functional derivatives localized in the large field region (which is usually impossible anyway), and the whole convergence becomes much more transparent, many combinatoric difficulties being hidden in the jungle formula itself.

A concrete example in which this formula would somewhat simplify the argument is e.g. the single scale expansion of [L]; in [KMR] a three level jungle formula is used, in which the third forest hooks some cubes along the straight paths of the second forest propagators, in order to complete factorization in the context of a Peierls contour argument.

F) Cluster or Resolvent Expansions with Smooth Localizations

In some situations (for instance if momentum conservation is important), it may be inconvenient to perform cluster expansions with sharp localization functions such as χ_b in (V.A.8). But with smooth functions there is no naïve factorization in (V.A.8). This is not a serious difficulty and it can be overcome e.g. by an auxiliary expansion (sometimes called a "painting expansion") on the interaction that creates protection belts around the clusters. But in this last section we remark that the Taylor forest formula (III.2) also gives elegant solutions to this type of problems and treat the ϕ^4 interaction again as an example.

Let

$$1 = \sum_b \chi_b^2(x) \tag{V.F.1}$$

be a smooth partition of unity of $\mathbb{R}^d$ by cubes of unit size. We assume that the support of χ_b is contained in $\{x|\mathrm{dist}(x,b) \le \frac{1}{3}\}$. The set of all b's is then further restricted to the finite set $\mathcal{N}$ of all cubes which are at distance zero of our finite volume Λ (hence include a unit corridor around it). Therefore

$$\chi_\Lambda(x) = \chi_\Lambda(x) \sum_{b\in\mathcal{N}} \chi_b^2(x) \tag{V.F.2}$$

so that the corresponding sums are finite from now on.

Let us rewrite the ϕ^4 theory of section V.A in terms of an intermediate ultralocal field:

[1] These blocks lead to unpleasant additional sums for where in the blocks functional derivatives really act, etc...

$$Z(\Lambda) = \int d\mu_C(\phi) e^{-g\int_\Lambda \phi^4(x)dx} = \int d\mu_{C_\Lambda}(\phi) d\nu(V) e^{i\sqrt{g}\int_\Lambda \phi^2(x)V(x)dx} \quad \text{(V.F.3)}$$

where $d\nu$ is the normalized Gaussian measure with ultralocal covariance $\delta(x-y)$.

Inserting the identity V.F.2 we get

$$Z(\Lambda) = \int d\mu_C(\phi) d\nu(V) e^{i\sqrt{g}\int_\Lambda dx \sum_{b\in\mathcal{N}} (\phi\chi_b)^2(x)V(x)} \quad \text{(V.F.4)}$$

We define now the collection of Gaussian random variables $\{\phi_b(x)\}$, $b \in \mathcal{N}$ distributed according to the measure $d\mu(\{\phi_b\})$ with covariance $C(b,x\,;b',y) = \chi_b(x)C(x,y)\chi_{b'}(y)$, and the collection of Gaussian random variables $\{V_b(x)\}$ distributed with degenerate covariance $\Gamma(b,x\,;b',y) = \delta(x-y)$. We have

$$Z(\Lambda) = \int d\mu(\{\phi_b\}) d\nu(\{V_b\}) e^{i\sqrt{g}\int_\Lambda dx \sum_{b\in\mathcal{N}} (\phi_b)^2(x)V_b(x)} \quad \text{(V.F.5)}$$

(to prove (V.F.5), remark that the right hand side of (V.F.4) and (V.F.5) are both Borel summable functions of g with identical perturbative series). We may now change slightly the boundary condition, dropping the restriction on the interaction range of integration. In other words $Z(\Lambda)$ leads to the same thermodynamic variables than

$$\bar{Z}(\Lambda) = \int d\mu(\{\phi_b\}) d\nu(\{V_b\}) e^{i\sqrt{g}\sum_{b\in\mathcal{N}} \int dx (\phi_b)^2(x)V_b(x)} \quad \text{(V.F.6)}$$

Let us apply the Taylor forest formula (III.2) to $\bar{Z}(\Lambda)$, interpolating both the covariances $C(b,x\,;b',y)$ and $\Gamma(b,x\,;b',y)$, viewed as finite matrices with entries in $\mathcal{N}$ whose elements are operators on $L^2(\mathbb{R}^d)$.

It gives a result very similar to (V.A.7-8):

$$\bar{Z}(\Lambda) = \sum_{\substack{Y_1,\dots,Y_n \\ Y_i\cap Y_j=\emptyset,\, \cup Y_i=\mathcal{N}}} \prod_{i=1}^n A(Y_i)$$

$$A(Y) = \sum_{T \text{ on } Y} \Big(\prod_{l\in T}\int_0^1 dh_l\Big) \int d\mu_{C_Y^T(\mathbf{h})}(\phi) d\nu_{\Gamma_Y^T(\mathbf{h})}(V)$$

$$\prod_{l=\{ij\}\in T} \int dxdy \left(\chi_{b_i}(x)\chi_{b_j}(y)C(x,y)\frac{\delta}{\delta\phi_{b_i}(x)}\frac{\delta}{\delta\phi_{b_j}(y)} + \delta(x-y)\frac{\delta}{\delta V_{b_i}(x)}\frac{\delta}{\delta V_{b_j}(y)}\right)$$

$$\cdot\, e^{i\sqrt{g}\sum_{b\in Y}\int dx(\phi_b)^2(x)V_b(x)} \quad \text{(V.F.7)}$$

where b_i and b_j are the two ends of the line l, and the sum is over trees T which connect together the set Y, hence have exactly $|Y|-1$ elements. The measures $d\mu_{C_Y^T(\mathbf{h})}(\phi)$ and $d\nu_{\Gamma_Y^T(\mathbf{h})}(V)$ are normalized Gaussian measure on the restricted collections of Gaussian random variables $\{\phi_b(x)\}$ and $\{V_b(x)\}$ for $b \in Y$. $d\mu_{C_Y^T(\mathbf{h})}$ has covariance $C_Y^T(\mathbf{h})(b,x\,;b',y) = C(x,y)h_T(b,b')$, where $h^T(b,b) = 1$ and for

$b \neq b'$, $h^T(b,b')$ is the infimum of all parameters h_l for l in the unique path in T joining the cube b to the cube b'. Similarly $d\nu_{\Gamma_Y^T(\mathbf{h})}(V)$ has covariance $\Gamma_Y^T(b,x\,;b',y) = \delta(x-y)h_T(b,b')$.

Remark that the difference between (V.A.8) and (V.F.7) appears in the term $\delta(x-y)\frac{\delta}{\delta V_{b_i}(x)}\frac{\delta}{\delta V_{b_j}(y)}$ which would be zero if the functions χ_b were sharp characteristic functions. This term corresponds to "emptying" the interaction on the borders of the clusters created so that full factorization occurs, even without sharp localization functions.

Similar formulas can be written for other kind of interactions or for resolvent expansions with smooth localization functions. We leave them to the reader.

Acknowledgements We thank D. Brydges for introducing us to his formula during the Vancouver 1993 summer school; it is a pleasure to update our Ecole Polytechnique software accordingly. We thank also J. Magnen and H. Knörrer for their help and for discussions on the proof of all these formulas.

References

[BaF] G. A. Battle and P. Federbush, A phase cell cluster expansion for Euclidean field theories, Ann. Phys. 142, 95 (1982) ; A note on cluster expansions, tree graph identities, extra 1/N! factors!!! Lett. Math. Phys. 8, 55 (1984).

[Bat] G. Battle, A new combinatoric estimate for cluster expansions, Comm. Math. Phys. 94, 133 (1984).

[B1] D. Brydges, A short course on cluster expansions, in Critical phenomena, random systems, gauge theories, Les Houches session XLIII, 1984, Elsevier Science Publishers, 1986.

[B2] D. Brydges, Functional integrals and their applications, "Cours de Troisième Cycle de la Physique en Suisse Romande" taken by R. Fernandez, Université de Lausanne or University of Virginia preprint, (1992).

[BF] D. Brydges and P. Federbush, A new form of the Mayer expansion in classical statistical mechanics, Journ. Math. Phys. 19, 2064 (1978).

[BK] D. Brydges, T. Kennedy, Mayer expansions and the Hamilton-Jacobi Equation, Journ. Stat. Phys. 48, 19 (1987).

[BY] D. Brydges and H.T. Yau, Grad Φ Perturbations of Massless Gaussian Fields, Comm. Math. Phys. 129, 351 (1990).

[GJS1] J. Glimm, A. Jaffe and T. Spencer, The Wightman axioms and particle structure in the $P(\phi)_2$ quantum field model, Ann. Math. 100, 585 (1974).

[GJS2] J. Glimm, A. Jaffe and T. Spencer, The particle structure of the weakly coupled $P(\phi)_2$ model and other applications of high temperature expansions, Part II: The cluster expansion, in Constructive Quantum field theory, ed. by G. Velo and A. Wightman, Lecture Notes in Physics, Vol. 25, Springer (1973).

[KMR] C. Kopper, J. Magnen and V.Rivasseau, Mass Generation in the two dimensional Gross-Neveu Model, in preparation.

[R1] V.Rivasseau, From perturbative to constructive renormalization, Princeton University Press (1991).

[R2] V.Rivasseau, Cluster Expansions with small/large field conditions, preprint Ecole polytechnique, to appear in Proceedings of the Vancouver Summer School(1993).

Weak Perturbations of the Massless Gaussian Measure

David Brydges[*]

Department. of Mathematics, University of Virginia
Charlottesville, VA 22903

Abstract
This is a summary of some ideas and estimates that simplify the analysis of scaling limits of infinite volume limits of measures of the form

$$\frac{1}{Z} d\mu(\phi) e^{\int dx\, f(\phi,\partial\phi,\partial\partial\phi,\ldots)},$$

where $d\mu(\phi)$ is a Gaussian measure with long range correlations.

1 Introduction

This is a summary of work which has appeared in papers by J. Dimock and T. Hurd and also in collaborations by myself and: H. T. Yau, J. Imbrie, P. Mitter, G. Keller, J. Dimock, T. Hurd. Many of our ideas have been independently discovered by Pordt (ibid.) and Mack and Pordt, [MaPo89]; their point of view requires explicit formulas, because they look towards the possibility of computer algorithms.

Since a detailed version is available in [BrFe92, BDH93, BDH94, BrYa90], I am attempting in this lecture to survey the main ideas, skipping many technicalities.

The objective, which we are far from completing, is to classify, using the renormalization group (RG), scaling limits of infinite volume limits of measures on function spaces of the form

$$\frac{1}{Z} d\mu(\phi) e^{-V},$$

where $V(X) = \lambda \int_X dx\, p(\phi(x), \partial\phi(x), \partial\partial\phi(x), \ldots)$. $p \geq 0$ is a polynomial with *small* coefficients. $d\mu(\phi)$ is a Gaussian measure, supported on smooth multiscale fields.

Each model in this class of problems either has been, or could be, investigated using one of the three original organizations of the RG given by Gawedzki - Kupiainen [GaKu86], Feldman et al [FMRS87] or Balaban [Bal83]. See also

* Research supported by NSF Grant DMS 9401028.

[BCG+78]. In this sense these models are not at the frontiers of constructive field theory, but it is my hope that we will recruit some younger researchers and, perhaps some of our friends in probability, by spending time to make the RG simpler, more accessible and less model dependent. The goal should be a complete list of everything that can result from small smooth homogeneous perturbations. Moreover we should weaken the smoothness hypotheses.

In terms of problems in mathematical physics, applications of our methods so far include: bounds on correlations for the two dimensional Coulomb and Yukawa gases [DiHu91, DiHu93] and Tom Hurd's contribution (ibid.); bounds on the Green's function of polymers with self interaction on a hierarchical lattice [BEI92]; accurate upper (and lower) bounds on correlations of general composite operators for dipole gases [DiHu92b] and [BrKe94b], intended to be part of a proof that quantum Coulomb plasmas do not screen [BrKe94]; construction of superrenormalizable Euclidean quantum field theories [DiHu92a, BDH94].

2 Multiscale Fields

The typical example of a Gaussian multiscale field is a random process whose covariance is a Green's function for the Laplacian. In fact this has infinitely many scales and I want instead to consider a Gaussian measure μ with a finite number of scales, but we shall try to work uniformly in this number. Thus $\int d\mu(\phi)$ is a Gaussian integration over a Banach space $\mathcal{C}^k(\mathcal{R}^d)$ of sufficiently differentiable functions, called fields, on $\mathcal{R}^d$, with the property that it can be split into a convolution:

$$\begin{aligned} \int d\mu(\phi) F_1 &= \mu_N * \mu_{N-1} * \ldots * \mu_1 * F_1 \Big|_{\phi=0} \\ &:= \mu_N * \mu_{N-1} * \ldots * \mu_j * F_j \Big|_{\phi=0} \\ F_j &= \mu_j * F_{j-1} \text{ and } F_1 = e^{-V} \end{aligned}$$

The measures μ_j should be related to each other by scale transformations. The measure μ_1 is associated with the smallest length scale. This formula is the same as saying that $\phi = \sum \phi_j$ where the random fields ϕ_j are independent scaled copies of one field.

F, e^{-V} and K are required to be smooth functionals on the Banach space $\mathcal{C}^k(\mathcal{R}^d)$ of fields ϕ. They are also functions of subsets of $X \subset \mathcal{R}^d$, as well as ϕ. This argument is there to display the region of dependence on ϕ; more precisely the support of the measure $\partial^n/\partial\phi(x_1)\cdots\partial\phi(x_n)F(X,\phi)$ must be $X^{\times n}$. For an example, consider $V(X,\phi) := \lambda \int_X \phi^4(x)dx$.

F_1 is an exponential. This is a desirable feature because it is related to the important locality property $F_1(X \cup Y, \phi) = F_1(X,\phi)F_1(Y,\phi)$ when X and Y are disjoint. This locality is only approximately true for F_j when $j > 1$ and in fact it is very difficult to get good bounds on the logarithm of F_j when ϕ is large. This is called the "large field problem". We avoid this issue by using a more general representation

$$F_j = \left(e^{-V_j} \circ \mathcal{E}\mathrm{xp}[K_j]\right)(X,\phi) \equiv \sum_{\substack{\{X_i \subset X\} \\ \text{disjoint}}} \prod_j K(X_i,\phi) e^{-V}(X \setminus \cup X_k). \tag{1}$$

You should think of this as a functional which is almost a local exponential e^{-V_j}, but there are disjoint "holes" X_i where locality has partly broken down. We will take V_j to be the integral of a local polynomial (which contains all the monomials in the field ϕ which are "relevant" in the sense of Wilson).

The product $\circ$ is defined on functions of sets in a suitable domain $\mathcal{D}$ by

$$(A \circ B)(X,\phi) = \sum_{Y \subset X} A(X \setminus Y) B(Y).$$

The $\mathcal{E}\mathrm{xp}[K_j]$ really is an exponential, defined, provided $K_j(\emptyset) = 0$, by [2]

$$\mathcal{E}\mathrm{xp}[K] = \mathcal{I} + K + \frac{1}{2!} K \circ K + \frac{1}{3!} K \circ K \circ K + \ldots.$$

The domain $\mathcal{D}$ of sets in these definitions is required to have the property that each set X is a unique finite union of "atoms", which I shall call *cells*. The emptyset is also in $\mathcal{D}$. One of the simplest possibilities is to take finite unions of disjoint open cubes whose closures fill $\mathcal{R}^d$. Since the domain $\mathcal{D}$ has this property, provided $K(\emptyset) = 0$, $\mathcal{E}\mathrm{xp}[K_j](X)$ is always a *finite* series whose number of terms depends on how many cells there are in X.

3 RG Step

Now I will list a sequence of four operations,$\mathcal{F}, \mathcal{E}, \mathcal{B}, \mathcal{S}$ that integrate out the lowest scale in the field ϕ. Together, these operations constitute "one RG step".

$$\begin{aligned} \mu_C * e^{-V} \circ \mathcal{E}\mathrm{xp}[K] &= e^{-V'} \circ \mathcal{E}\mathrm{xp}[\mathcal{F}K] && \text{Cluster Expansion } \mathcal{F} \\ &= e^{-V''} \circ \mathcal{E}\mathrm{xp}[\mathcal{E}\mathcal{F}[K]] && \text{Extraction } \mathcal{E} \\ &= e^{-V''} \circ \mathcal{E}\mathrm{xp}_L[\mathcal{B}\mathcal{E}\mathcal{F}[K]] && \text{Reblocking } \mathcal{B} \\ &= e^{-V'''} \circ \mathcal{E}\mathrm{xp}[\mathcal{S}\mathcal{B}\mathcal{E}\mathcal{F}[K]] && \text{Rescaling } \mathcal{S} \end{aligned}$$

These operations will be explained in the next sections. From the first line you see that $\mathcal{F}$ is the actual integration out of the lowest scale. The last operation $\mathcal{S}$ is a trivial rescaling of lengths by an integral factor L which is chosen so that the next scale in ϕ becomes the same as the one just integrated out. This makes the complete RG step

$$(V_j, K_j) \to (V_{j+1}, K_{j+1}) := (V''', \mathcal{S}\mathcal{B}\mathcal{E}\mathcal{F}[K]) \tag{2}$$

[2] The identity $\mathcal{I}$ for the algebra defined by $\circ$ is given by $\mathcal{I}(X) = 1$ if $X = \emptyset, = 0$ otherwise.

an (almost) autonomous dynamical system. Without loss of generality we can assume L is large by compounding several scales.

The theorems I will be describing combine to show that when L is chosen sufficiently large depending on dimension d, then

1. there are norms $\|K_j\|_{G,\Gamma,h_j}$ with $j = 0, 1, \ldots$ defined on the space of K's.
2. Initially $\|K_0\|_{G,\Gamma,h_0} = 0$.
3. Under successive RG steps (2), $\|K_j\|_{G,\Gamma,h_j}$ remains small for as long as coefficients of V_j remain small and V_j has a well behaved lower bound, so that a condition of the form
$$\|e^{-V_j}\|_{\tilde{G}_j,\Gamma,h_j} \sim 1 \tag{3}$$
holds.

The map $V \to V'''$ is merely a map of the finite number of coefficients that determine the polynomial V. In terms of dynamical systems, these are the coordinates of the center and unstable manifolds. The map can be computed to any order in V by perturbation theory, so it is mechanical to determine the evolution of V with sufficient accuracy to check the conditions in (3). The construction of continuum limits and the determination of bounds on correlations and ultimately the complete classification of weak perturbations of scaling limits of multiscale Gaussian measures reduce to determining which initial (or final) choices of V lie on global trajectories. This is part of Wilson's [WiKo74] magnificent contribution to the understanding of functional integrals.

The parameters G and Γ appearing in the norm are model independent. G is a weight called a *large field regulator* that controls growth of $K(X, \phi)$ in ϕ when $\partial\phi$ is large. Our standard choice is
$$G(X, \phi) := \exp\{\kappa\|\partial\phi\|_X^2\}, \tag{4}$$
where $\|\partial\phi\|_X^2$ is a Sobolev norm on the functions defined on the set X. The *large set regulator* Γ controls the decay of $K(X, \phi)$ in X when the set X is either very large or very disconnected. I will define it later. Thus
$$G: \quad \|K(X)\|_G := \sup_\phi |K(X,\phi)|G^{-1}(X,\phi)$$
$$\Gamma: \quad \|K\|_\Gamma := \sup_x \sum_{X \ni x} |K(X)|\Gamma(X)$$
$$h: \quad \|K\|_{G,h} := \sum \frac{h^n}{n!} \left\| \|\frac{\partial^n K}{\partial\phi^n}\|_{\text{var}} \right\|_G$$

$\|\partial^n K(\phi)/\partial\phi^n\|_{\text{var}}$ is a variation type of norm on the measure $\partial^n K(\phi)/\partial\phi^n$; actually the variational derivatives of functionals on $C^k(\mathcal{R}^d)$ are spatial derivatives of measures, but I will oversimplify by writing equations that are appropriate for $C(\mathcal{R}^d)$ fields. In fact, we reduce $C^k(\mathcal{R}^d)$ fields to this case in [BrYa90, BrFe92, BDH93, BDH94].

From the last equation you see that h is a parameter that can be interpreted as a width of a strip of analyticity for K as a functional of ϕ, but this is intuitive:

we only use the bounds on the derivatives enforced by finiteness of the norm with parameter h. If $V(X,\phi) := \lambda \int_X \phi^4(x)dx$, then the analyticity interpretation correctly suggests that the largest possible choice for h as a function of λ is $h = \mathcal{O}(\lambda^{-1/4})$, because otherwise $\|\exp(-V)\|_h \sim \sup\{|\exp(-V)| : |\Im\phi| \leq h\}$ will blow up as $\lambda \to 0$. Thus the choice of h is dependent on the choice of V and would be different for different perturbations of the Gaussian measure.

These norms are put together in different orders, described by the order of the subscripts to obtain, for example, $\|\cdot\|_{G,\Gamma,h}$ or $\|\cdot\|_{G,h,\Gamma}$. You can find a detailed description in [BrFe92, BDH93].

4 Cluster Expansion Theorem $\mathcal{F}$

If we define

$$\Box(X) = \begin{cases} 1 \text{ if X is a cell} \\ 0 \text{ otherwise,} \end{cases}$$

then it easily follows that $e^{-V} = \mathcal{E}\text{xp}[\Box e^{-V}]$, where $(\Box e^{-V})(X) := \Box(X)e^{-V}(X)$. Therefore $e^{-V} \circ \mathcal{E}\text{xp}[K] = \mathcal{E}\text{xp}[A]$ with

$$A = \Box e^{-V} + K.$$

Next we can define $A(t)$ by

$$\mathcal{E}\text{xp}[A(t)] := \mu_{tC} * \mathcal{E}\text{xp}[A]$$

because this equation means that both sides are equal for all sets and $\mathcal{E}\text{xp}$ possesses an inverse function $\mathcal{L}\text{og}$ given by a power series with finitely many terms: $\mathcal{L}\text{og}(\mathcal{I} + A) = A - 1/2 A \circ A + \ldots$. From $A(1)$ we recover $\mathcal{F}[K]$ by making a good choice for V' and solving for $\mathcal{F}[K]$ in $A(1) = \Box e^{-V'} + \mathcal{F}[K]$. Thus $\mathcal{F}$ is uniquely defined by the equation $\mu_C * e^{-V} \circ \mathcal{E}\text{xp}[K] = e^{-V'} \circ \mathcal{E}\text{xp}[\mathcal{F}K]$ together with a choice of V'. However this definition does not easily imply bounds on $\mathcal{F}$. This is the purpose of the next theorem.

Theorem 1. *Let $G(t;X,\phi) \geq 0$, $0 \leq t \leq 1$, be a family of regulators that satisfy*

$$G(t; X \cup Y, \phi) \geq G(t;X,\phi)G(t;Y,\phi) \text{ for } X,\ Y \text{ disjoint} \tag{5}$$

$$\mu_{(t-s)C} * G(s) \leq G(t) \text{ for all } 0 \leq s < t \leq 1. \tag{6}$$

If $h' < h$ and

$$\|A\|_{G(0),\Gamma,h} \leq \frac{(h-h')^2}{16\|C\|_\theta}$$

then

$$\|A(1)\|_{G(1),\Gamma,h'} \leq \|A\|_{G(0),\Gamma,h}$$

The family of regulators $G(t; X, \phi)$ need not have the standard form (4). Eq. (6) enforces μ integrability of A.

If $G(t; X, \phi)$ is of the form $\exp\{\kappa \|\partial\phi\|_X^2\}$ then this means that the coefficients of the various L_2 norms of derivatives of ϕ are t dependent in such a way as to satisfy the hypothesis. One of the virtues of regulators of this form is that $\int(\partial\phi)^2$ is invariant under canonical scaling. It is possible to exploit this for theories with canonical scaling to show that the t dependence of coefficients can be chosen so that they return to their original values under the rescaling operation. Consequently each $\mathcal{F}$ in each RG step can use the same family $G(t; X, \phi)$. See, for example, the appendix to [BDH93] or the less general but simpler arguments in [BrFe92, BDH94].

The loss of analyticity $h \to h'$ is not a problem provided the rescaling operation restores it. If, for example, $d \geq 3$ and the scale decomposition of μ is such that ϕ has the canonical length dimension $-(d-2)/2$ then this will be the case.

The theorem provides an initial crude bound on $A(1) \equiv \Box e^{-V'} + \mathcal{F}[K]$ from which good bounds on $\mathcal{F}[K]$ follow by expanding $A(1)$ as a Taylor series in A to a few orders and using Theorem 1 to bound the remainder. The terms in the expansion which are not the remainder are explicit and can be bounded by hand. The fact that $\mathcal{F}[A]$ is analytic in A as a functional on the Banach space of A's is used to obtain a bound on the remainder by the Cauchy formula. See [BDH94] for details. Dimock and Hurd introduced many such simplifications by the systematic use of functional analyticity.

Proof. more details are given in [BrYa90]. $\mathcal{E}\mathrm{xp}[A(t)] := \mu_{tC} * \mathcal{E}\mathrm{xp}[A] \Rightarrow$

$$\frac{\partial}{\partial t}\mathcal{E}\mathrm{xp}[A(t)] = \Delta_C \mathcal{E}\mathrm{xp}[A(t)],$$

where $\Delta_C := 1/2 \int C(x,y)\partial^2/(\partial\phi(dx)\partial\phi(dy))$. Therefore,

$$\begin{aligned}\frac{\partial}{\partial t}A(t) &= (\Delta_C \mathcal{E}\mathrm{xp}[A(t)]) \circ \mathcal{E}\mathrm{xp}[-A(t)] \\ &= \Delta_C A + \\ &\quad 1/2 \int C(x,y)\frac{\partial A}{\partial\phi}(dx) \circ \frac{\partial A}{\partial\phi}(dy).\end{aligned}$$

This is implied by the integral equation

$$\begin{aligned}A(t) = \mu_{tC} * A + \int_0^t ds\, \mu_{(t-s)C} * \\ 1/2 \int C(x,y)\frac{\partial A}{\partial\phi}(dx) \circ \frac{\partial A}{\partial\phi}(dy).\end{aligned}$$

Take seminorms $(\partial^n/\partial h^n)\|\cdot\|_{G,\Gamma,h}$ of both sides. This works out well, because the norms have been designed to have good properties such as $\|A \circ B\|_G \leq \|A\|_G \circ \|B\|_G$ and $\|\partial A/\partial\phi\|_h \leq \partial/\partial h \|A\|_h$.

We find that $a(t,\zeta) := \|A(t)\|_{g(t),\Gamma,\zeta}$ is majorized term by term as a power series in ζ by the solution $u(t,\zeta)$ of a corresponding one variable integral equation

$$u(t,\zeta) = u(0,\zeta) + \int_0^t ds\, \|C\|_\theta \frac{\partial^2 u(s,\zeta)}{\partial \zeta^2},$$

where $\|C\|_\theta$ is a weighted L_1 norm. The weight θ is part of the Γ which will be defined in the next section. Solving this integral equation is equivalent to finding the unique analytic solution to

$$\frac{\partial u}{\partial t} = \|C\|_\theta \left(\frac{\partial u}{\partial \zeta}\right)^2$$
$$u(0,\zeta) = a(0,\zeta).$$

This equation is a Hamilton-Jacobi equation which can be solved easily and the theorem follows. □

The solution of the above integral equation for $A(t)$ by iteration gives a finite expansion called a cluster expansion and also some beautiful tree graph formulas which you can learn more about in the the lecture by Abdussalam. There are many different kinds of cluster expansion; this one was first published in a much more complicated guise in [GJS74].

The norm used in this theorem is really much too strong because the convolution by μ is smoothing, so analyticity should not be needed. The right result would be a norm that measures a weighted L_∞ norm of only a finite number of derivatives. There should be large simplifications in the whole scheme if a better norm were used.

5 Reblocking

Sets X are unions of *cells.*The size of the cells should be chosen comparable with the shortest scale in ϕ. Intuitively, one should choose cells as large as possible such that essentially no fluctuations occur within cells. Since the cluster expansion removes the smallest scale from ϕ we lose this relationship. If the next scale is L times larger, then our sets must be replaced by L-sets which are unions of L-cells that are L times larger than cells. We choose L-cells so that they are unions of cells, so that an L-set Z is also a set. We need an operation $\mathcal{B}$ that satisfies

$$e^{-V} \circ \mathcal{E}\mathrm{xp}[K](Z) = e^{-V} \circ_L \mathcal{E}\mathrm{xp}_L[\mathcal{B}[K]](Z),$$

where Z is viewed as a set on the left hand side and as an L-set on the right hand side. The L subscript displays which cell decomposition is to be used in the definitions of $\circ$ and $\mathcal{E}\mathrm{xp}$.

In the definition (1) of $e^{-V} \circ \mathcal{E}\mathrm{xp}[K]$ there is a sum which we rewrite by $\sum_{\{X_j\}} = \sum_{\{Z_k\}} \sum_{\{X_j\} \to \{Z_k\}}$, where $\to$ means that $\{Z_k\}$ is the finest cover of

$\{X_j\}$ by disjoint L-sets. $\sum_{\{Z_k\}}$ is the sum in $\mathcal{E}\text{xp}_L$ and $\sum_{\{X_j\}\to\{Z_k\}}$ splits into factors labeled by k. Each factor is part of $\mathcal{B}[K]$. Thus

$$\mathcal{B}[K](Z) = \sum_{\{X_j\}\to\{Z\}} e^{-V}(Z \setminus \cup X_j) \prod_j K(X_j). \tag{7}$$

The main property of $\mathcal{B}$ is that it is almost contractive in the norms we have defined. To express this we need

Definition 2. $\overline{X}$:= the ordinary closure of X and $\overline{X}^L$:= the smallest[3] L-set that contains X.

Definition 3. X is <u>small</u> iff the closure $\overline{X}$ is connected and contains less than 2^d d- dimensional cells, otherwise X is <u>large</u>.

Definition 4. V is stable if $\|e^{-V(X)}\|_{G,h} \leq 2$ for X a subset of any L-cell.

The following lemma holds with a particular choice of large set regulator Γ which depends on the parameter L. The choice is

$$\begin{aligned} \Gamma(X) &= L^{-(d+1)|X|}\Theta_L(X) \\ \Theta_L(X) &= \inf_\tau \prod_{b\in\tau} \theta_L(|b|), \end{aligned}$$

where $|X|$ is the number of d-dimensional cells in X so that $\Gamma(X)$ is large when there are many cells in X. $\Theta_L(X)$ is large when X is highly disconnected. Thus τ is a tree graph that connects the centers of cells in X and $\theta_L(|b|)$ is a carefully chosen increasing function of the length of the bond b in τ. We define $\Gamma_L(Z)$ by an analogous formula with cells replaced by L-cells.

Lemma 5. *Suppose V is stable and $\|K\|_{G,h,\Gamma}$ is sufficiently small, then*

$$\|\mathcal{B}[K]\|_{G,h,\Gamma_L} \leq \|K\|_{G,h,\Gamma} \begin{cases} O(L^d) \\ O(L^{-1}) \text{ if } K|_{\text{smallsets}} = 0 \end{cases}$$

The norm with parameters in the order G, h, Γ is the same as the norm with parameters in the order G, Γ, h if K is translation invariant.

Thus, the lemma says that if we could somehow arrange that there are no small sets for which $K \neq 0$, then we could iterate the composite operation $\mathcal{BF}$ and K would always remain small. The $\mathcal{F}$ feeds part of e^{-V} into K, but $\mathcal{B}$ would keep contracting K. This embodies Wilson's idea that only the local parts (small set parts) of the interaction are relevant.

[3] or sometimes it is better to define $\overline{X}^L$ = the smallest closed L-set that contains X.

Proof. If K is translation invariant, then

$$\|K\|_{G,h,\Gamma} := \sum_{X:O\in X} \Gamma(X)\|K(X)\|_{G,h},$$

where O is the origin. Also, if Z is an L-set, then by (7)

$$\mathcal{B}[K](Z) = \sum_{X:\overline{X}^L=Z} K(X)\,(e^{-V})(Z\setminus X)+\ldots, \tag{8}$$

where ... means terms which are higher order in K. Therefore, using properties like $\|A\circ B\|_{G,h} \le \|A\|_{G,h}\circ\|B\|_{G,h}$ and stability, we have $\|\mathcal{B}[K](Z)\|_{G,h} \le 2^{|Z|}\sum_{X:\overline{X}^L=Z}\|K(X)\|_{G,h}$. Now compute the Γ norm by using

$$\sum_{Z\ni O}\sum_{X:\overline{X}^L=Z} = \sum_{X:\overline{X}^L\ni O} = L^d\sum_{X\ni O}$$

together with a crucial property designed into Γ, namely, if $Z=\overline{X}^L$, then

$$\Gamma_L(Z)\le L^{-(d+1)}\Gamma(X) \text{ for all large sets } X.$$

This bound is true, roughly because for most sets $\overline{X}^L$ has fewer L-cells than X has cells, where we are referring to cells of dimension d. There is a "proof" in [BrYa90], but it is not quite correct because we gave a wrong definition of small set. This has been corrected in the papers by Dimock and Hurd. The terms ... in (8) are also bounded using $\|A\circ B\|_{G,h}\le\|A\|_{G,h}\circ\|B\|_{G,h}$. They are much smaller because they contain at least two factors of K and we have assumed that K is small in norm. □

6 Extraction

The K that appears in this section is really $\mathcal{F}[K]$. In the last section we learnt that $K|_{\text{small sets}}$ can expand by L^d under $\mathcal{B}$. To reduce this damage done by small set parts of K we define the operation $K\to\mathcal{E}[K]$ so that

$$\left(e^{-V'}\circ\mathcal{E}\mathrm{xp}[K]\right)(\Lambda) = \left(e^{-V''}\circ\mathcal{E}\mathrm{xp}[\mathcal{E}[K]]\right)(\Lambda)$$

and a piece

$$F(X) := \alpha(X)\int_X dx\,\phi^p,$$

where $\alpha(X)$ is independent of ϕ, is removed from K in passing to $\mathcal{E}[K]$:

$$\mathcal{E}[K] = K(X) - F(X)\,e^{-V'(X)}+\ldots$$

where ... are terms which are higher order in K and F. The operation also satisfies

$$\|\mathcal{E}[K]\|_{G,\Gamma,h}\le\mathcal{O}(1)(a^{-1}\|\alpha\|_\Gamma+\|K\|_{G,\gamma\Gamma,h}). \tag{9}$$

This can be proven, provided $\alpha|_{\text{large sets}} = 0$ and $V + \nu F$ is stable for all complex $|\nu| \le a$. $\gamma\Gamma > \Gamma$ is a worse regulator than Γ, but this is one of the technical details that I shall not discuss here. See [BDH94, BDH93].

By using this operation $\mathcal{E}$, one or more times with different choices for F, we can cancel a small set part of K, so that the first few terms in the functional McLaurin expansion of $\mathcal{E}[K]|_{\text{small sets}}$ in powers of ϕ are cancelled up to a polynomial in ϕ that scales down like L^{-p} under the operation $\mathcal{S}$. Now we want to argue that

$$\|\mathcal{SBE}[K]\|_{G,\Gamma,h} \le \mathcal{O}(L^{d-p})\|K\|_{G,\Gamma,h} \tag{10}$$

In fact a stronger estimate is required, since we need to include the $\mathcal{F}$ as well. This is achieved in [BDH93].

The first step in the proof is to split $\mathcal{E}[K]$ into a part that lives on small sets and a part that lives on large sets.

$$\mathcal{E}K = \mathcal{E}K|_{\text{large sets}} + \mathcal{E}K|_{\text{small sets}}$$

Then we have

$$\mathcal{SBE}[K] = \mathcal{SBE}K|_{\text{large sets}} + \mathcal{SBE}K|_{\text{small sets}} + \cdots$$

where, as usual, ... refers to terms that are nonlinear in $\mathcal{E}[K]$ and therefore very small, because they have two or more factors of $\mathcal{E}[K]$ which are small in norm by (9). We have a contractive estimate on the large set part from the last section which reduces us to proving (10) in the special case that the large set part of $\mathcal{E}[K]$ vanishes.

To prove (10) in this special case we would like to claim that $\mathcal{E}[K]$ is small because of the vanishing of the first few terms of the functional Taylor series of $\mathcal{E}[K]$, but this will only work when ϕ is small. The key point is that the operation $\mathcal{F}$ has removed a scale from ϕ so that it has no fluctuations within L-cells. Therefore, if ϕ is large within a small set X, then it is large within the L-closure $\overline{X}^L$. Referring to the equations (7) and (8) we see that, in the term linear in K, there is a factor $\exp(-V(\overline{X}^L \setminus X))$, which will tend to be very small if ϕ is large. It will be small, provided $V(\phi)$ has some growth[4] when ϕ is large. Terms that are nonlinear in K can cause no problems because they have more factors of K which are small in norm.

The form of the regulator (4) is critical to making this argument work and in fact this is really the only place where the regulator plays an important role. The appearance of this weight in the norm $\|K\|_G$ encodes the fact that our Gaussian measure μ is supported on fields ϕ for which $G(X, \phi)$ is bounded and this bound quantifies the lack of fluctuation inside a cell, because it is a bound on $\partial\phi$ and higher derivatives of ϕ. Details are available in [BDH94, BDH93].

[4] This is why there is a slightly different regulator $\tilde{G}_j$ in (3).

7 More on Extraction

The extraction operation has the potential to run into large field problems, which are connected with the difference between the two types of exponential $\mathcal{E}\mathrm{xp}$ and exp. In this section I want to try to explain this remark and why large field problems can easily be avoided if extraction is only done on small sets.

Consider a functional Ω which has the form $\Omega(X) = \sum_{Y \subset X} F(Y)$ for some F. Then

$$e^{\Omega}(X) = \prod_{Y \subset X} (e^{F(Y)} - 1 + 1)) = \sum_{\{Y_j\}} \prod_j J(Y), \tag{11}$$

where $J(Y) := (\exp(F(Y)) - 1)$ and the sum is over *distinct* sets Y_j. Thus these sets can overlap, and $|\prod J(Y)| \sim exp|\tau F|$, where τ is the largest number of distinct sets Y_j, $j = 1, 2, \ldots, \tau$, such that $J(Y_j) \neq 0$ and $\cap Y_j \neq \emptyset$. The important point is that this number τ is less than a fixed geometrical constant if $F(X) = 0$ for X large. Therefore (11) can be a well behaved expansion for an ordinary exponential.

Let the notation $\{X_1, \ldots, X_N\} \to X$ mean that $X = \cup X_j$ and $\{X_1, \ldots, X_N\}$ is overlap connected. Given J, let $J^+(X) = \sum_{\{X_i\} \to X} \prod_i J(X_i)$. It is easy to prove that

Lemma 6. $\sum_{\{X_i \subset X\}} \prod_i J(X_i) = \mathcal{E}\mathrm{xp}(\Box + J^+)(X)$.

From this we immediately deduce a lemma that connects the two types of exponentials

Lemma 7. *Let* $\Omega(X) = \sum_{Y \subset X} F(Y)$. *Then*

$$e^{\Omega} = \mathcal{E}\mathrm{xp}(\Box + (e^F - 1)^+).$$

Lemma 8. *Let*

$$\tilde{K}(X) = K(X) - (e^F - 1)^+(X) e^{-V(X)}.$$

Then

$$e^{-V} \circ \mathcal{E}\mathrm{xp}(K) = e^{-V+\Omega} \circ \mathcal{E}\mathrm{xp}(\tilde{K})$$

Proof. $e^{-V} \circ \mathcal{E}\mathrm{xp}[K]$

$$\begin{aligned} &= \mathcal{E}\mathrm{xp}[\Box e^{-V} + K] \\ &= \mathcal{E}\mathrm{xp}[\Box e^{-V} + (e^F - 1)^+ e^{-V}] \circ \mathcal{E}\mathrm{xp}[\tilde{K}] \\ &= \{e^{-V} \mathcal{E}\mathrm{xp}[\Box + (e^F - 1)^+]\} \circ \mathcal{E}\mathrm{xp}[\tilde{K}] \\ &= e^{-V+\Omega} \circ \mathcal{E}\mathrm{xp}(\tilde{K}). \end{aligned}$$

□

This does not quite prove what we want, because Ω is not quite the integral of a local monomial. If we do not worry about keeping V strictly local, then the lemma shows that $\mathcal{E}[K] = K(X) - (e^F - 1)^+(X) e^{-V(X)}$. The correct formula for $\mathcal{E}$ is more complicated because we keep locality, [BDH94, BDH93]. Nevertheless the argument we have presented motivates the conclusion that if $\exp(|\tau F| - V)$ is bounded by G, then so is $K(X) - (e^F - 1)^+(X) e^{-V(X)}$. This is the basis for (9).

8 Observables

This is an attempt to extend the analysis to obtain good bounds on very general observables. It has so far only been worked out [BrKe94b] in the case where $V = 0$, but this includes, for example, dipole gases and the correlations of the two dimensional Coulomb gas. See the lecture by Tom Hurd.

We say that $O(X, \phi)$ is an observable located at x iff $O(X, \phi) = 0$ when $X \not\ni x$. Under an RG step, O_1 and O_2 generate a composite observable O_{12} such that $O_{12}(X, \phi) = 0$ when $X \not\ni x_1$ and $X \not\ni x_2$. Our next equation anticipates this by including a composite observable. Observables are measured with exactly the same norm as we have used for K.

Define $\langle O_1; O_2 | O_{12} \rangle$

$$
\begin{aligned}
&:= \frac{1}{Z} \int d\mu \, \mathcal{E}\mathrm{xp}[\Box + K] \circ \{O_1 \circ O_2 + O_{12}\} \\
&- \left[\frac{1}{Z} \int d\mu \, \mathcal{E}\mathrm{xp}[\Box + K] \circ O_1\right] \left[\frac{1}{Z} \int d\mu \, \mathcal{E}\mathrm{xp}[\Box + K] \circ O_2\right] \\
&= \tau_{12} \log \int d\mu \, \mathcal{E}\mathrm{xp}[\Box + K(\lambda)] + \sum_j \tau_{12} \Omega(j),
\end{aligned}
$$

where the sum over j vanishes because the range of j is null, until some RG steps have been made, and

$$
\begin{aligned}
K(\lambda) &:= K + \lambda_1 O_1 + \lambda_2 O_2 + \lambda_1 \lambda_2 O_{12} \\
\tau_{12} &:= \frac{\partial^2}{\partial \lambda_1 \partial \lambda_2} |_{\lambda=0}
\end{aligned}
$$

The idea is that the RG map that we have discussed applies to this partition function with interaction $K(\lambda)$. A small modification of the operation $\mathcal{E}$ permits us to factor the field independent part of $K(\lambda)$ out so that it becomes the term Ω. There will be a contribution Ω_j from each RG step, in other words, from each scale that is integrated out. Ultimately all scales are integrated out (for any given finite volume Λ) and the expectation has been evaluated uniformly in Λ in terms of the λ derivatives of $\sum_j \Omega_j$, which we can calculate to any order in perturbation theory, together with bounds. The initial object looks a little unusual, but it is easy to express ordinary correlations in this form.

References

[Bal83] T. Balaban. Ultraviolet stability in field theory. The ϕ_4^3 field theory. In J. Fröhlich, editor, *Scaling and Self-similarity in Physics*, Boston, Basel, Stuttgart, 1983. Birkhäuser.

[BCG+78] G. Benfatto, N. Cassandro, G. Gallavotti, F. Nicolo, E. Olivieri, E. Presutti and E. Scacciatelli, Some probabilistic techniques in field theory. *Commun. Math. Phys.*, 59:143–166, 1978.

[BrFe92] D. C. Brydges. Functional integrals and their applications. Lecture notes taken by Roberto Fernandez. Course in the "Troisième Cycle de la Physique en Suisse Romande", May 1992. Available from: Archives and Vente de Polycopies: M.D. Reymond, Université de Lausanne, Batiment des Sciences Physiques, 1015 Lausanne-Dorigny, or as a University of Virginia Preprint, or the mathematical physics bulletin board at the University of Texas at Austin.

[BDH94] D. C. Brydges, J. Dimock, and T.R. Hurd. The short distance behavior of ϕ_3^4. McMaster University Preprint.

[BDH93] D. C. Brydges, J. Dimock, and T.R. Hurd. Weak perturbations of Gaussian measures. Preprint, lectures given by D.C. Brydges at the Mathematical Quantum Theory Conference, Vancouver , August 1993, to be published by the AMS in the conference series of the Centre de Recherches Mathematiques.

[BEI92] D. C. Brydges, S. Evans, and J. Imbrie. Self avoiding walk on a hierarchical lattice in four dimensions. *Annals of Probability*, 20:82–124, 1992.

[BrKe94] D. C. Brydges and G. Keller. Absence of Debye screening in the quantum Coulomb system. To appear in J. Stat. Phys. Summer 1994.

[BrKe94b] D. C. Brydges and G. Keller. Correlation functions of general observables in dipole type systems I: Accurate upper bounds. *Helv. Phys. Acta*, 67:43–116, 1994.

[BrYa90] D. C. Brydges and H.T. Yau. Grad ϕ perturbations of massless Gaussian fields. *Commun. Math. Phys.*, 129:351–392, 1990.

[DiHu93] J. Dimock and T. Hurd. Construction of the two-dimensional sine-Gordon model for $\beta < 8\pi$. *Commun. Math. Phys.*, 156:547–580, 1993.

[DiHu91] J. Dimock and T. R. Hurd. A renormalization group analysis of the Kosterlitz-Thouless phase. *Commun. Math. Phys.*, 137:263–287, 1991.

[DiHu92b] J. Dimock and T. R. Hurd. A renormalization group analysis of correlation functions for the dipole gas. *J. Stat. Phys.*, 66:1277–1318, 1992.

[DiHu92a] J. Dimock and T. R. Hurd. A renormalization group analysis of QED. *J. Math. Phys.*, 33:814–821, 1992.

[FMRS87] J. Feldman, J. Magnen, V. Rivasseau, and R. Seneor. Construction and Borel summability of infrared ϕ_4^4 by a phase space expansion. *Commun. Math. Phys.*, 109:437–480, 1987.

[GaKu86] K. Gawedzki and A. Kupiainen. Asymptotic freedom beyond perturbation theory. In K. Osterwalder and R. Stora, editors, *Critical Phenomena, Random Systems, Gauge Theories.* Les Houches, North Holland, 1986.

[GJS74] J. Glimm, A. Jaffe, and T. Spencer. The Wightman axioms and particle structure in the $P(\phi)_2$ quantum field model. *Ann. Math.*, 100:585, 1974.

[MaPo89] G. Mack and A. Pordt. Convergent weak coupling expansions for lattice field theories that look like perturbation series. *Rev. in Math. Phys.*, 1:47–87, 1989.

[WiKo74] K. G. Wilson and J. Kogut. The renormalization group and the ϵ expansion. *Phys. Rep. (Sect C of Phys Lett.)*, 12:75–200, 1974.

[BrD92] D. C. Brydges, Functional integrals and their applications. Lecture notes taken by Roberto Fernández, Troisième Cycle de la Physique en Suisse Romande, May 1992. Available from [illegible] M.D. Raymond, [illegible], Bâtiment des Sciences de Physique, 1015 Lausanne-Dorigny, or as University of Virginia Preprint or [illegible] physics bulletin board at the University of Texas at Austin.

[BDH94] D. C. Brydges, J. Dimock, and T. R. Hurd, The short distance behavior of [illegible]

[BDH95] D. C. Brydges, J. Dimock, and T. R. Hurd, Weak perturbations of Gaussian measures, [illegible] given by D.C. Brydges at the [illegible] Theory Conference, Vancouver, August [illegible] [illegible] the Centre de Recherches Mathématiques [illegible]

[BrY90] [illegible]

[illegible]

[GaK85] [illegible]

On Renormalization Group Flows and Polymer Algebras

Andreas Pordt

Institut für Theoretische Physik I, Universität Münster,
Wilhelm-Klemm-Str. 9, D-48149 Münster, Germany
e-mail: pordt@yukawa.uni-muenster.de

Abstract

In this talk methods for a rigorous control of the renormalization group (RG) flow of field theories are discussed. The RG equations involve the flow of an infinite number of local partition functions. By the method of exact beta-function the RG equations are reduced to flow equations of a finite number of coupling constants. Generating functions of Greens functions are expressed by polymer activities. Polymer activities are useful for solving the large volume and large field problem in field theory. The RG flow of the polymer activities is studied by the introduction of polymer algebras. The definition of products and recursive functions replaces cluster expansion techniques. Norms of these products and recursive functions are basic tools and simplify a RG analysis for field theories. The methods will be discussed at examples of the Φ^4-model, the $O(N)$ σ-model and hierarchical scalar field theory (infrared fixed points).

1 Introduction

There are several goals in constructive field theory. The first one is a rigorous definition of Euclidean functional-integrals. There exists many examples for a construction of special functional-integrals (see e. g. [GJ73, MS77, MP85, GJ87, FMRS87, GK83, GK85, B88a, B88b, BY90, DH91, R91, DH92, DH93]). But a general definition of the Euclidean functional-integral is still lacking. The second goal in constructive field theory is to compute functional-integrals by approximation methods. The functional-integrals can be represented by convergent series expansions in terms of finite-dimensional integrals (cp. [MP89]). This is in analogy to conventional perturbation theory where functional-integrals are expressed in terms of (finite-dimensional) Feynman-integrals. The definition and construction of functional-integrals lead to a proof of the existence of ultraviolet- resp. infrared fixed points. The problem is the infinite (not denumerable) number of degrees of freedom which is connected to ultraviolet and infrared problems. The large field problem is connected to the divergence of standard power series expansions. A third aim is to represent the construction in such a way that all intermediary steps can be done in a finite number of well-defined computations.

There exists several tools to perform such a program. A field theory which is represented by a functional-integral can be studied by means of Wilson's renormalization group (RG) [W71, WK74, W83]. Thereby, the original functional-integral is represented by a RG flow of effective functional-integrals. These effective functional-integrals are simpler to define than the original functional-integral. The RG flow of the effective functional-integrals has to be controlled. The effective functional-integrals can be further analyzed by methods used in statistical mechanics, especially by the introduction of polymer systems (cp. [GrK71, MP85]). A suitable defined polymer system can control large field contributions and solves the large volume problem. Thereby, the effective systems represented by effective functional-integrals will be decomposed into finite subsystems. The effective systems depend on an infinite number of degrees of freedom. Their control is reduced to the problem of analyzing the RG flow of finite subsystems. In a RG analysis, using conventional perturbation theory, one distinguishes between relevant and irrelevant parts. Likewise, there are relevant and irrelevant parts of the effective finite subsystems. The relevant part depends only on a finite number of parameters. The flow of the irrelevant part can be controlled by an application of fixed point theorems. Thereby, the control of the RG flow of effective systems is reduced to a RG flow defined in a finite-dimensional parameter space. This method is called the *method of exact beta-functions*. For a control of effective systems one introduces norms for polymer activities. A suitable definition of norms and polymer systems is the technical core in the construction of field theoretic models.

It is the aim of this talk to review old and provide new tools and definitions for such a program. A test for simplicity of these methods is the implementability on a computer.

This paper is organized as follows. We start in Sect. 2 with the introduction to general renormalization group transformations (RGT). Then, we consider the special example of the Kadanoff-Wilson (linear) block spin transformation. This RGT were firstly applied to field theory by Gawędzki and Kupiainen (cp. [GK80, GK83, GK85]). Next, we consider the definition of the nonlinear block spin RGT at the example of the nonlinear $O(N)$ σ-model.

Section 3 introduces a general polymer systems and presents the definition of polymer activities by introducing an exponential function EXP. The exponential function EXP is defined by a product $\circ$. This $\circ$-product was also used by Brydges and Yau, Dimock and Hurd [BY90, DH91] and has its origin in a product defined for problems in statistical mechanics by Ruelle [Ru69]. Ruelle's product differs from the $\circ$-product used in constructive field theory by an important point[1]. It does not allow an overlap of indices, whereas the $\circ$-product does not allow an overlap of lattice points. This property of nonoverlapping of lattice points is essential for a control of large fields. The RG flow of the effective systems is represented by the RG flow of polymer activities. The effective subsystems are defined on lattices.

One RG step can be decomposed into four steps. The first step is called in-

[1] The author thanks D. Brydges for this comment

tegration step. In this step high momentum fields are integrated out. This can be done recursively, using a $\times_{\Gamma}$-product. After this step the correlation length becomes larger. Then, the activities are defined on a coarser lattice (coarsening step). The coarsening step can be performed in a recursive way by introducing a mapping E_A. The polymer activities are defined in such a way that the localization property holds. A polymer activity $A(P|\psi)$ for a polymer P and field ψ obeys the localization property if $A(P|\psi)$ depends only on $\psi(y)$ for $y \in P$. This localization property of the polymer activities makes a third step (localization step) necessary. The localization step can be performed recursively, like the integration step, by using a $\times_{\mathcal{A}}$-product. The fourth and last step rescales the fields such that the new effective polymer activities lives on the same lattice as the polymer activities before the RG step.

Section 4 presents the general method of exact beta-functions. As examples we consider the RG flow of the Φ^4-model and hierarchical RG fixed points. Koch and Wittwer [KW86, KW91] applied the method of exact beta-functions to construct the double-well fixed point in 3 dimensions. These method is a candidate for the construction of field theories with no small coupling constants.

The split into relevant and irrelevant parts and the method of renormalization and repolymerization for the flow of effective polymer systems are studied in Sect. 5. For repolymerization a further RG step is necessary (repolymerization step).

Section 6 presents a norm for the polymer activities and shows how this norm behaves under RG steps. It will be shown that large fields are controlled by the method of exponential pinning.

2 Renormalization Group Transformations

Our main object of interest is the generating functional of Euclidean Greens functions (partition function). It is the following infinite-dimensional integral

$$Z = \int [D\phi]\, \mathcal{Z}(\phi), \qquad [D\phi] = \prod_{z \in \mathbb{R}^d} d\phi(z) \ ,$$

where $\mathcal{Z}(\phi)$ is a real-valued function, called *Boltzmannian*. We restrict our attention here to real-valued scalar fields ϕ. In renormalization group (RG) investigations the computation of Z is performed stepwise. Let us consider the definition of one RG step. Define new fields Φ and a function $P(\Phi, \phi)$ which obeys

$$\int [D\Phi]\, P(\Phi, \phi) = 1 \ . \tag{1}$$

Then, the *renormalization group transformation* is defined as follows

$$\mathcal{Z}'(\Phi) = \int [D\phi]\, P(\Phi, \phi) \mathcal{Z}(\phi) \ . \tag{2}$$

Eqs. (1) and (2) imply

$$Z = \int [D\phi]\, \mathcal{Z}(\phi) = \int [D\Phi]\, \mathcal{Z}'(\Phi) \ .$$

Thus the new Boltzmannian $\mathcal{Z}'$ and the new field Φ can be used to compute the partition function Z. This RG procedure can be repeated and the result is the *RG flow of effective Boltzmannians* :

$$\mathcal{Z} \longrightarrow \mathcal{Z}' \longrightarrow \mathcal{Z}'' \longrightarrow \dots .$$

The definition of the RG is chosen in such a way that the effective Boltzmannians depend on fewer and fewer degrees of freedom. Thus, the task of computing Z by an infinite-dimensional integral is solved by computing an infinite number of RG steps. Instead of considering the RG flow of effective Boltzmannians, it is better to consider the RG flow of effective polymer activities. In this way the RG transformations can be represented by finite-dimensional integrals.

Before coming to the definition of polymer activities, we will study two examples of RG transformations. The first example is the *Kadanoff-Wilson (linear) block spin transformation.* This method was applied to field theory by Gawędzki and Kupiainen [GK80, GK83, GK85]. The second example is the nonlinear block spin transformation applied to the nonlinear σ-model. For a definition of a RG transformation introduce new fields Φ and the function $P(\Phi, \phi)$ which obeys eq. (1). Let us suppose that the original field ϕ lives on the lattice $\Lambda := (a\mathbb{Z})^d$ and define the integral for the partition function Z by $[D\phi] := \prod_{y\in\Lambda} d\phi(y)$. Decompose the lattice Λ into hypercubes of side length La, where L is a fixed number, $L \in \{2, 3, \dots\}$. The center points of these hypercubes (blocks) build also a lattice with side length La. This lattice is called *block lattice* $\Lambda' := (La\mathbb{Z})^d$. A site y of the lattice Λ is contained in a site x of the block lattice Λ' if y is contained in the block with center point x. In this case we write $y \underline{\in} x$. For a field $\phi : \Lambda \to \mathbb{R}$, define the *block spin field* : $C\phi : \Lambda' \to \mathbb{R}$ by

$$C\phi(x) := \beta \sum_{y:\, y \underline{\in} x} \phi(y), \qquad x \in \Lambda' .$$

β is a positive real number, called *scaling parameter.* Finally, the special RG transformation, called *block spin transformation,* is defined by

$$\mathcal{Z}'(\Phi) = \int [D\phi]\, P_{\alpha,\beta}(\Phi, \phi) \mathcal{Z}(\phi) \ ,$$

where $P_{\alpha,\beta}$ is defined by

$$P_{\alpha,\beta}(\Phi, \phi) := \mathcal{N}_{\alpha,\beta} \exp\{-\frac{\alpha}{2} \sum_{x\in\Lambda'} (\Phi(x) - (C\phi)(x))^2\} \ .$$

α is a positive real parameter and $\mathcal{N}_{\alpha,\beta}$ is a normalization constant such that eq. (1) holds.

We represent the Boltzmannian by a free propagator u and interaction V

$$\mathcal{Z}(\phi) = \exp\{-\frac{1}{2}(\phi, u^{-1}\phi) - V(\phi)\} \ .$$

$(\cdot,\cdot)$ is the canonical bilinear form. Then, the RG transformation reads

$$\mathcal{Z}'(\Phi) = \mathcal{N}_{\alpha,\beta} \exp\{-\frac{1}{2}(\Phi, u'^{-1}\Phi)\} \exp\{-V'(\Phi)\} \ ,$$

where the *effective interaction* V' is defined by

$$\exp\{-V'(\Phi)\} := \int [D\zeta] \exp\{-\frac{1}{2}(\zeta, \Gamma^{-1}\zeta)\} \exp\{-V(\zeta + \mathcal{A}\Phi)\} \ .$$

The *fluctuation propagator* Γ and the *block spin propagator* u' are

$$\Gamma := (u^{-1} + \alpha C^T C)^{-1}, \qquad u' := (\alpha - \alpha^2 C \Gamma C^T)^{-1} \ .$$

The $\mathcal{A}$-operator, which maps a field defined on Λ' to fields defined on Λ, is given by $\mathcal{A} := uC^T u'^{-1}$. The normalized Gaussian measure with mean zero is defined by

$$d\mu_\Gamma(\zeta) := \mathcal{N} [D\zeta] \exp\{-\frac{1}{2}(\zeta, \Gamma^{-1}\zeta)\} \ .$$

The RG transformation, for the effective interaction V, reads

$$e^{-V'(\Psi)} = \int d\mu_\Gamma(\phi) e^{-V(\phi + \mathcal{A}\Psi)} \ .$$

Let the free propagator u be a Gaussian fixed point. Then self-similarity holds

$$u'(Ly, Ly') = L^{2-d(+\epsilon)} u(y, y'), \qquad y, y' \in \Lambda \ .$$

Replacing the field $\Phi(\cdot)$ by the rescaled one, $L^{1-\frac{d}{2}(+\frac{\epsilon}{2})}\Phi(\frac{\cdot}{L})$, we obtain the RG transformation after rescaling

$$e^{-V'(\Phi)} = \int d\mu_\Gamma(\zeta) e^{-V(\zeta + L^{1-\frac{d}{2}(+\frac{\epsilon}{2})} \mathcal{A}(\Phi(\frac{\cdot}{L})))} \ .$$

The field Φ lives on the original lattice Λ but with reduced correlation length. This procedure can be iterated and the result is the following RG flow

$$e^{-V} \longrightarrow e^{-V'} \longrightarrow e^{-V''} \longrightarrow \cdots \ .$$

Best localization properties are obtained for $\alpha = O(1)$ (see Bell and Wilson [BW74]).

We consider a second example for a RG transformation. This is the nonlinear block spin transformation at the example of the nonlinear $O(N)$ σ-model. Define a measure on the $N-1$-dimensional unit sphere S^{N-1}

$$\int_{S^{N-1}} [d\sigma] := \int \prod_{z \in \Lambda} d\sigma_1(z) \cdots d\sigma_N(z) \delta(\sigma^2(z) - 1)$$

and a partition function

$$Z = \int_{S^{N-1}} [d\sigma] e^{-V(\sigma)} \ ,$$

where the interaction V is defined by nearest-neighbor couplings

$$V(\sigma) := \beta \sum_{\mu=1}^{d} \sum_{y \in \Lambda} (1 - \sigma(y) \cdot \sigma(y + \hat{\mu})) \ .$$

$\hat{\mu}$ is a vector in μ-direction with length of one lattice spacing a. Let $\mu : \Lambda' \to S^{N-1}$ be a unit vector field on the block lattice Λ'. Then, the nonlinear block spin RG transformation is defined by

$$\mathcal{Z}'(\mu) := \int_{S^{N-1}} [d\sigma] P(\mu, \sigma) e^{-V(\sigma)} \ ,$$

where

$$P(\mu, \sigma) := \exp\{-W_\kappa(\sigma)\} \prod_{x \in \Lambda'} \exp\{\beta\kappa\mu(x) \cdot \sum_{y \underline{\in} x} \sigma(y)\} \ .$$

W_κ is defined such that

$$\int_{S^{N-1}} [d\mu] P(\mu, \sigma) = 1$$

holds. A simple computation shows

$$\begin{aligned} W_\kappa(\sigma) &= \sum_{x \in \Lambda} \ln \left(\int_{S^{N-1}} [d\mu] \exp\{\kappa\beta\mu \cdot \sum_{y \underline{\in} x} \sigma(y)\} \right) \\ &= const + \sum_{x \in \Lambda} \ln \left(\frac{I_{\frac{N-2}{2}}(\kappa\beta |\sum_{y \underline{\in} x} \sigma(y)|)}{(\kappa\beta |\sum_{y \underline{\in} x} \sigma(y)|)^{\frac{N-2}{2}}} \right) \ . \end{aligned}$$

$I_\nu(z)$ is the Bessel function represented by the series expansion

$$I_\nu(z) = \frac{z^\nu}{2^\nu} \sum_{k=0}^{\infty} (-1)^k \frac{z^{2k}}{2^{2k} k! \Gamma(\nu + k + 1)} \ .$$

We have seen how to compute partition functions iteratively by the introduction of RG transformations. RG transformations are given by infinite-dimensional integrals. For a computation of RG steps one has to reduce these infinite-dimensional integrals to finite-dimensional ones. This problem corresponds to the infinite volume problem in statistical mechanics and can be solved by cluster expansion methods (cp. [B84]) or equivalently by the introduction of polymer systems.

3 Polymer Systems, Activities, Exponentiation

Polymer systems for statistical mechanics were introduced by Gruber and Kunz [GrK71]. This section presents a polymer system for the use of a RG analysis for field-theoretic models. The definition of polymer activities presented here uses block spin RG transformations. The difference to other definitions of RG transformations is the introduction of the $\mathcal{A}$-operator. An advantage in using block spin RG is that new polymer activities are defined by finite-dimensional integrals. A further advantage is that gradients of fields can be represented by gradients of the $\mathcal{A}$-operator, $\partial(\mathcal{A}\psi) = (\partial\mathcal{A})\psi$. For estimations of such gradients one has to bound gradients of $\mathcal{A}$ and there is no need of Sobolev-inequalities. A disadvantage is that the $\mathcal{A}$-operator is non-local and has to be taken into consideration for a RG step of polymer activities. A definition, not using block spin RG, similar to the one presented here can be found in Brydges contribution to these proceedings and in Brydges and Yau [BY90] and Dimock and Hurd [DH91, DH92, DH93].

Let $\Lambda = (a\mathbb{Z})^d$ be the d-dimensional hypercubic lattice with lattice spacing a. For a subset P of Λ denote by $|P|$ the number of elements in P. The union $P \cup Q$ is denoted by $P + Q$ if P and Q are disjoint sets. The set consisting of all elements which are in P but not in Q is denoted by $P - Q$. Let $Pol(\Lambda)$ be a subset of the set of all finite subsets of Λ, $\mathcal{P}_{fin}(\Lambda) := \{P \subseteq \Lambda | \, |P| < \infty\}$, such that $P, Q \in Pol(\Lambda)$ implies $P + Q, P - Q \in Pol(\Lambda)$. $Pol(\Lambda)$ is called a *set of polymers*. Let Λ' be a block lattice of Λ. Suppose we have also defined a set of polymers $Pol(\Lambda')$. For simplicity suppose here that $Pol(\Lambda) := \mathcal{P}_{fin}(\Lambda)$ and $Pol(\Lambda') := \mathcal{P}_{fin}(\Lambda')$, i.e. polymers are finite subsets. We want to truncate the fluctuation propagator Γ and the $\mathcal{A}$-operator on polymers. For a polymer P let χ_P be the characteristic function. For $P \in Pol(\Lambda)$ and $X \in Pol(\Lambda')$ define truncated operators

$$\Gamma_P := \chi_P \Gamma \chi_P, \qquad \mathcal{A}_X := \mathcal{A}\chi_X \ .$$

For a polymer $X \in Pol(\Lambda')$ let us define a polymer $\overline{X} \in Pol(\Lambda)$ by

$$\overline{X} := \{y \in \Lambda | \, \exists x \in X : y \underline{\in} x\}$$

and for a polymer $Y \in Pol(\Lambda)$ a polymer $[Y] \in Pol(\Lambda')$ by

$$[Y] := \{x \in \Lambda' | \, \exists y \in Y : y \underline{\in} x\} \ .$$

Consider the *set of effective polymer partition functions* :

$$\mathcal{Z}(\Lambda, \mathcal{F}) := \{Z : Pol(\Lambda) \times \mathcal{F} \to \mathbb{R} | \, Z(\emptyset|\Psi) = 1\} \ , \tag{3}$$

where $\mathcal{F}$ is the set of fields, e.g. $\mathcal{F} := Fun(\Lambda) := \{F : \Lambda \to \mathbb{R}\}$. The RG transformation (without rescaling) for effective polymer partition functions is defined by

$$Z'(X|\Psi) := \int d\mu_{\Gamma_{\overline{X}}}(\zeta) Z(\overline{X}|\zeta + \mathcal{A}_X\Psi) \ ,$$

where $Z \in \mathcal{Z}(\Lambda, Fun(\Lambda))$ and $Z' \in \mathcal{Z}(\Lambda', Fun(\Lambda'))$. The RG transformation with rescaling is a mapping $\mathcal{RG} : \mathcal{Z}(\Lambda, Fun(\Lambda)) \to \mathcal{Z}(\Lambda, Fun(\Lambda))$, $Z' = \mathcal{RG}(Z)$.

The RG transformation can be performed in 4 steps. The first step is the *integration step* defined by the mapping

$$\mu_\Gamma : \mathcal{Z}(\Lambda, Fun(\Lambda)) \to \mathcal{Z}(\Lambda, Fun(\Lambda)), \quad \mu_\Gamma(Z)(Y|\phi) := \int d\mu_{\Gamma_Y}(\zeta) Z(Y|\zeta + \phi) \ .$$

The second step is the *coarsening step* defined by the mapping

$$[\,] : \mathcal{Z}(\Lambda, Fun(\Lambda)) \to \mathcal{Z}(\Lambda', Fun(\Lambda)), \quad [Z](X|\phi) := Z(\overline{X}|\phi) \ .$$

The third step is the *localization step* defined by the mapping

$$\iota_{\mathcal{A}} : \mathcal{Z}(\Lambda', Fun(\Lambda)) \to \mathcal{Z}(\Lambda', Fun(\Lambda')), \quad \iota_{\mathcal{A}}(Z)(X|\Psi) := Z(X|\mathcal{A}_X\Psi) \ .$$

The fourth and last step is the *rescaling step* defined by the mapping

$$\mathcal{R}_L : \mathcal{Z}(\Lambda', Fun(\Lambda')) \to \mathcal{Z}(\Lambda, Fun(\Lambda)), \quad \mathcal{R}_L(Z)(Y|\phi) := Z(LY|L^{1-\frac{d}{2}}\phi(\frac{\cdot}{L})) \ .$$

The mapping $\mathcal{RG}$ is therefore a composition of the four above defined mappings, $\mathcal{RG} := \mathcal{R}_L \circ \iota_{\mathcal{A}} \circ [\,] \circ \mu_\Gamma$

$$\begin{aligned} \mathcal{Z}(\Lambda, Fun(\Lambda)) &\overset{\mu_\Gamma}{\to} \mathcal{Z}(\Lambda, Fun(\Lambda)) \overset{[\,]}{\to} \mathcal{Z}(\Lambda', Fun(\Lambda)) \\ &\overset{\iota_{\mathcal{A}}}{\to} \mathcal{Z}(\Lambda', Fun(\Lambda')) \overset{\mathcal{R}_L}{\to} \mathcal{Z}(\Lambda, Fun(\Lambda)) \ . \end{aligned}$$

Polymer partition functions obey the following conditions :

(a) Locality :

$$\frac{\partial}{\partial \Psi(y)} Z(P|\Psi) = 0, \quad \forall y \notin P$$

(b) Euclidean lattice symmetry :

$$Z(RY|R\Psi) = Z(Y|\Psi),$$

$\forall R \in$ group of lattice symmetry

(c) Invariance under external symmetry transformations $\Psi \to U\Psi$:

$$Z(Y|U\Psi) = Z(Y|\Psi)$$

(d) Approximation (thermodynamic limit):

$$\lim_{Y \nearrow \Lambda} Z(Y|\Psi) = Z(\Psi)$$

For a control of polymer partition functions it is better to change to polymer activities. Polymer activities depend, like the polymer partition functions, on polymers and fields. The set of polymer activities is equivalent to the set of polymer partition functions. This means that if we know the polymer activities, then the polymer partition functions are determined and vice versa. The polymer activities have the following important property (at least for weakly coupled models) which the polymer partition functions do not share. The value of a polymer activity for a given polymer and field is small if the polymer contains a large number of elements or is of large extension. The polymer activities and polymer partition functions are related by an exponential function EXP. To define this function we will introduce a product $\circ$. The space, where this product is defined, is the *set of all polymer functions* defined by

$$\mathcal{M}(\Lambda, \mathcal{F}) := \{F : Pol(\Lambda) \times \mathcal{F} \to \mathbb{R}\} ,$$

where $\mathcal{F}$ is a set of fields. For notational simplicity we omit below the field-dependence of the polymer function. Define the $\circ$-product, for two polymer functions $U, V \in \mathcal{M}$, by

$$(U \circ V)(Y) := \sum_{\substack{P_1, P_2 \in Pol(\Lambda): \\ P_1 + P_2 = Y}} U(P_1)V(P_2) . \tag{4}$$

The sum is over all partitions of Y into two disjoint subsets. Defining an addition $+$ on $\mathcal{M}$ in the canonical way, we see that $(\mathcal{M}, \circ, +)$ is an associative algebra with unit element $\mathbf{1}$, $\mathbf{1}(P) := \delta_{P,\emptyset}$. Call this algebra a *polymer algebra* . Define on $\mathcal{M}$ a $\cdot$-multiplication by

$$(U \cdot V)(Y) := U(Y)V(Y) . \tag{5}$$

The set of all polymer activities $\mathcal{A}$ is a subset of the set of polymer functions $\mathcal{M}$, defined by

$$\mathcal{A}(\Lambda, \mathcal{F}) := \{A \in \mathcal{M} |\, A(\emptyset|\Psi) = 0\} .$$

$\mathcal{Z}, \mathcal{A} \subset \mathcal{M}$ are subalgebras of $(\mathcal{M}, \circ, +)$.

We are now in the position to define the exponential mapping $EXP : \mathcal{A} \to \mathcal{Z}$, where $\mathcal{Z}$ is the set of all polymer partition functions defined by eq. (3). For a polymer activity $A \in \mathcal{A}$, define

$$EXP(A) := \mathbf{1} + \sum_{n:\, n \geq 1} \frac{1}{n!} \underbrace{A \circ \cdots \circ A}_{n-times} . \tag{6}$$

EXP is bijective with inverse mapping $LN := EXP^{-1}$. Furthermore, $EXP : (\mathcal{A}, +) \to (\mathcal{Z}, \circ)$ is a group-isomorphism, i. e.

$$EXP(A + B) = EXP(A) \circ EXP(B) .$$

Let $Y \in Pol(\Lambda) - \{\emptyset\}$ be a nonempty polymer. Define the *set of all partitions* of Y by

$$\Pi(Y) := \bigcup_{n:\, n\geq 1} \Pi_n(Y),$$
$$\Pi_n(Y) := \{\{P_1,\ldots,P_n\} | Y = P_1 + \cdots + P_n\} \ .$$

The definition (6) of the exponential-function EXP and the fact that $(\mathcal{M}, \circ, +)$ is an associative algebra imply the following explicit representations

$$EXP(A)(Y) = \sum_{\mathbf{P}\in\Pi(Y)} \prod_{P\in\mathbf{P}} A(P)$$
$$LN(Z)(Y) = \sum_{n:\, n\geq 1} (-1)^{n-1}(n-1)! \sum_{\mathbf{P}\in\Pi_n(Y)} \prod_{P\in\mathbf{P}} Z(P) \ .$$

For the proof that EXP is a bijective mapping one uses

$$A(Y) = EXP(A)(Y) - \sum_{\substack{\mathbf{P}\in\Pi(Y) \\ P\in\mathbf{P}:\, |P|<|Y|}} \prod_{P\in\mathbf{P}} A(P)$$

and proceeds by induction in the number of elements of a polymer. The RG transformation $\mathcal{RG}$ for polymer partition function can be transformed to a RG transformation $\widetilde{\mathcal{RG}}$ for polymer activities

$$\widetilde{\mathcal{RG}} := EXP^{-1} \circ \mathcal{RG} \circ EXP : \mathcal{A} \to \mathcal{A} \ .$$

This shows that $\mathcal{RG}$ and $\widetilde{\mathcal{RG}}$ are equivalent RG transformations. The RG flow of partition functions and activities are related by the following commutative diagram

$$\begin{array}{ccccccccc} \mathcal{Z} & \overset{\mu_\Gamma}{\to} & \mathcal{Z} & \overset{[\,]}{\to} & \mathcal{Z} & \overset{\iota_\mathcal{A}}{\to} & \mathcal{Z} & \overset{\mathcal{R}_L}{\to} & \mathcal{Z} \\ EXP \uparrow & & EXP \uparrow & & EXP \uparrow & & EXP \uparrow & & EXP \uparrow \\ \mathcal{A} & \overset{\widetilde{\mu_\Gamma}}{\to} & \mathcal{A} & \overset{\widetilde{[\,]}}{\to} & \mathcal{A} & \overset{\widetilde{\iota_\mathcal{A}}}{\to} & \mathcal{A} & \overset{\widetilde{\mathcal{R}_L}}{\to} & \mathcal{A} \end{array}$$

For a control of the RG flow of the polymer activities one has to study the four mappings $\widetilde{\mu_\Gamma}$, $\widetilde{[\,]}$, $\widetilde{\iota_\mathcal{A}}$ and $\widetilde{\mathcal{R}_L}$ in more detail.

3.1 Integration Step

New polymer activities $A' = \widetilde{\mu_\Gamma}(A)$ can be explicitly represented by tree graph formulas (cp. [B84, P86]). Instead of presenting here tree graph formulas, we introduce a recursive procedure by definition of a product (cp. also [BY90]).

For a representation of the mapping $\widetilde{\mu_\Gamma}$ one introduces the set of parametrized polymer activities

$$\mathcal{A}(\Lambda, [0,1], \mathcal{F}) := \{U : Pol(\Lambda) \times [0,1] \times \mathcal{F} \to \mathbb{C}\}$$

and special parametrized polymer activities A'_s, $s \in [0,1]$, by

$$EXP(A'_s) = \mu_{s\Gamma}(EXP(A)) \ . \tag{7}$$

This implies

$$A'_0 = A, \qquad A'_1 = \widetilde{\mu_\Gamma}(A) \ .$$

Define on the set of parametrized polymer activities a $\times_\Gamma$-product. For two parametrized polymer activities $A'_s, B'_s \in \mathcal{A}(\Lambda, [0,1], \mathcal{F})$ define the product

$$\begin{aligned}(A' \times_\Gamma B')_t(Y|\Psi) :=& \\ \frac{1}{2} \sum_{\substack{Y_1,Y_2: \\ Y_1+Y_2=Y}} \sum_{\substack{y_1\in Y_1 \\ y_2 \in Y_2}} & \int_0^t ds \int d\mu_{(t-s)\Gamma}(\phi) \\ \frac{\partial}{\partial \Psi(y_1)} A'_s(Y_1|\phi+\Psi)&\Gamma(y_1,y_2)\frac{\partial}{\partial \Psi(y_2)} B'_s(Y_2|\phi+\Psi) \ .\end{aligned} \tag{8}$$

We want to compute the parametrized activity $A'_t(Y|\Psi)$ by recursion in the number of elements of Y. Use the following definitions

$$\mathbf{1}_n(Y) := \begin{cases} 1 & : \ |Y| = n, \\ 0 & : \ |Y| \neq n \end{cases} \qquad \mathbf{1}_{>n} := \sum_{k:\, k>n} \mathbf{1}_k \ .$$

Then, the recursive equation is

$$A'_t = \mathbf{1}_{>1} \cdot [(A' \times_\Gamma A')_t + A] + \mathbf{1}_1 \cdot \mu_{t\Gamma}(A) \ . \tag{9}$$

The $\cdot$-product is defined by eq. (5). The proof of eq. (9) can be done in the following way. Firstly, distinguish the parametrized polymer activities A'_t, $t \in [0,1]$ defined by eq. (9) and eq. (7). Then show that they are equal. For $t = 0$, eq. (9) obeys $A'_0 = A$. By differentiation of eq. (7) with respect to s, we derive a first order differential equation of A'_s. Differentiation of eq. (9) with respect to t implies that A'_t, defined by eq. (9), obeys the same differential equation. The initial values at $t = 0$ are the same. Therefore, the parametrized polymer activities defined by eq. (7) and (9) are the same. Thus, eq. (9) is proven for all $t \in [0,1]$.

In the definition of the $\times_\Gamma$-product eq. (8) the right hand side depends only on polymers which contain lesser elements than the polymer on the left hand side of eq. (8). This implies that applying the activities in eq. (9) to special polymers, the recursion can be solved by a finite number of steps.

3.2 Coarsening Step

For polymer activities A on the lattice Λ, the *coarsened polymer activities* $\widetilde{[A]}$ are defined on the block lattice Λ' by

$$EXP(\widetilde{[A]}) = [EXP(A)] \ .$$

For a recursive computation of the coarsened polymer activities $\widetilde{[A]}$ define the following mapping $E_A : \mathcal{A} \to \mathcal{A}$, $A \in \mathcal{A}$

$$E_A(B)(Y) := A(Y) + \sum_{P:\, y\in P\subseteq Y} \sum_{n:\, n\geq 1} \frac{1}{n!} \sum_{\substack{P_1,\dots,P_n:\,[P]\cap[P_a]\neq\emptyset \\ P+P_1+\dots+P_n=Y \quad [P_a]\cap[P_b]=\emptyset,\, a\neq b}} A(P)\prod_{i=1}^{n} B(P_i)\ .$$

In the definition of the mapping E_A we have chosen an element $y \in Y$, for all polymers Y. This choice of y is arbitrary. Define the polymer activity $\mathcal{C}A \in \mathcal{A}(\Lambda, \mathcal{F})$ by the following recursive equation

$$\mathcal{C}A = E_A(\mathcal{C}A)\ . \tag{10}$$

That this equation defines $\mathcal{C}A$ recursively can be shown in the following way. Apply the activities in eq. (10) to a polymer Y. Then, $\mathcal{C}A(Y)$ can be expressed by a sum of products of terms $\mathcal{C}A(P)$, where P are polymers with $|P| < |Y|$. Thus, if $\mathcal{C}A(P)$ is defined for all polymers P containing less than $N = |Y|$ elements, then $(\mathcal{C}A)(Y)$ is defined by eq. (10). The recursion starts with the monomer (=polymer with only one element) $\mathcal{C}A(\{y\}) := A(\{y\})$, $y \in \Lambda$. Then, the coarsened polymer activity $\widetilde{[A]}(X) \in \mathcal{A}(\Lambda', \mathcal{F})$ is

$$\widetilde{[A]}(X) = (\mathcal{C}A)(\overline{X})\ .$$

This equation can be proven in the following way. Firstly, it can be shown that the coarsened polymer activity obeys

$$\widetilde{[A]}(X) = \sum_{\substack{\mathbf{P}\in\Pi(\overline{X}): \\ \gamma(\{[P]|P\in\mathbf{P}\})\ \text{connected}}} \prod_{P\in\mathbf{P}} A(P)\ ,$$

where, for $P_1, \dots, P_n$, the Venn-diagram $\gamma(P_1, \dots, P_n)$ is defined in the following way. Each polymer P_a is represented by a vertex and draw a line $(P_a P_b)$, $a \neq b$ if P_a and P_b are not disjoint, $P_a \cap P_b \neq \emptyset$. Take an arbitrary element y of $\overline{X}$ and take the polymer $Y \in \mathbf{P}$ which contains the element y. Then, the Venn-diagram $\gamma(\{[P]|P \in \mathbf{P}\} - \{Y\})$ decomposes into connected components. This decomposition implies the recursive equation (10).

3.3 Localization Step

The localization mapping $\widetilde{\iota_A} : \mathcal{A} \to \mathcal{A}$ obeys, for all polymers $X \in Pol(\Lambda')$,

$$\sum_{X=\sum Q} \prod_{Q} A(Q|\mathcal{A}_X\phi) = \sum_{X=\sum Q} \prod_{Q} \widetilde{\iota_A}(A)(Q|\phi)\ .$$

Define a parametrized polymer activity $A'_s(Q|\phi,\psi)$, for all $Q \in Pol(\Lambda')$, and two fields ψ and ϕ, implicitly by

$$\sum_{X=\sum Q} \prod_Q A(Q|s(\mathcal{A}_X\phi-\psi)+\psi) = \sum_{X=\sum Q} \prod_Q A'_s(Q|\phi,\psi) \ .$$

The parametrized polymer activity A'_s obeys, for the special values $s=0,1$,

$$A'_0(Q|\phi,\psi) = A(Q|\psi), \qquad A'_1(Q|\phi,\psi) = \widetilde{\iota_{\mathcal{A}}}(A)(Q|\phi) \ .$$

For two parametrized polymer activities A'_s and B'_s define the $\times_{\mathcal{A}}$-product by

$$\begin{aligned}
&(A' \times_{\mathcal{A}} B')_t(X|\phi,\psi) := \\
&\frac{1}{2} \sum_{\substack{X_1,X_2: \\ X_1+X_2=X}} \sum_{\substack{y\in\overline{X_1} \\ x\in X_2}} \int_0^t ds\, \mathcal{A}(y,x)\phi(x) \frac{\partial}{\partial\psi(y)} \\
&A'_s(X_1|\phi,(t-s)(\mathcal{A}_X\phi-\psi)+\psi) \\
&B'_s(X_2|\phi,(t-s)(\mathcal{A}_X\phi-\psi)+\psi) \ .
\end{aligned}$$

Then, the following recursive equation holds

$$A'_t = \mathbf{1}_{>1} \cdot [(A' \times_{\mathcal{A}} A')_t + A] + \mathbf{1}_1 \cdot \iota_{\mathcal{A},t}(A) \ .$$

We have seen how to control the RG steps given by the mappings $\widetilde{\mu_\Gamma}$, $\widetilde{[\,]}$ and $\widetilde{\iota_{\mathcal{A}}}$ for the RG flow of polymer activities A by introducing products $\times_\Gamma$, $\times_{\mathcal{A}}$ and a recursive mapping $E_{\mathcal{A}}$. A control of the RG flow of polymer activities A is achieved by splitting A into a relevant and an irrelevant part. This method of exact beta-function is discussed in the next section.

4 Exact Beta-Function Method

In general, a RG transformation $Z' = \mathcal{RG}(Z)$ maps a partition function Z, which is described by an infinite number of parameters to a partition function Z', which depends also on a infinite number of degrees of freedom. The exact beta-function method reduces the RG flow of an infinite number of parameters to a finite number of parameters called *coupling constants* (see [P93] for an application to hierarchical models). The remaining infinite number of parameters which determine the partition functions can be controlled by the coupling constants. The extraction of a finite number of coupling constants is done by a projection operator P such that

$$Z^{rel} = Z^{rel}(\gamma_0,\ldots,\gamma_N) = \mathrm{P}(Z)$$

depends on a finite number of parameters $\gamma_0,\ldots,\gamma_N$. The *irrelevant part* of the partition function Z is

$$R = (\mathbf{1}-\mathrm{P})(Z) \ .$$

The RG flow of this irrelevant part R can be controlled by standard fixed point theorems. The irrelevant fixed point $R^* = R^*(\gamma_0, \ldots, \gamma_N)$ depends on the coupling constants $\gamma_0, \ldots, \gamma_N$ and is defined by

$$R^* = H_{\gamma_0,\ldots,\gamma_N}(R^*) := (1-\mathrm{P})\mathcal{RG}(Z^{rel}(\gamma_0,\ldots,\gamma_N)+R^*) \ .$$

Then, the exact beta-function $B : \mathbb{R}^{N+1} \to \mathbb{R}^{N+1}$ is defined by

$$Z^{rel}(B(\gamma_0,\ldots,\gamma_N)) = \\ \mathrm{P}\mathcal{RG}\left(Z^{rel}(\gamma_0,\ldots,\gamma_N)+R^*(\gamma_0,\ldots,\gamma_N)\right) \ .$$

The problem of searching fixed points $Z^* = \mathcal{RG}(Z^*)$ of the RG transformation is solved by finding fixed points $\gamma_0^*, \ldots, \gamma_N^*$ for the exact beta-function B, $B(\gamma_0^*, \ldots, \gamma_N^*) = (\gamma_0^*, \ldots, \gamma_N^*)$. Then, the fixed point Z^* is the sum of the relevant and irrelevant part at $(\gamma_0^*, \ldots, \gamma_N^*)$

$$Z^* = Z^{rel}(\gamma_0^*,\ldots,\gamma_N^*) + R^*(\gamma_0^*,\ldots,\gamma_N^*) \ .$$

For a control of the RG flow it is not necessary to determine the irrelevant fixed point R^* exactly. It is sufficient to find a neighborhood $U = U(R^*)$ of R^* which is stable under RG transformation, $H_{\gamma_0,\ldots,\gamma_N}(U) \subseteq U$. The beta-function method reduces the infinite-dimensional RG flow to a finite-dimensional RG flow of running coupling constants $\gamma = (\gamma_0, \ldots, \gamma_N)$:

$$\gamma \xrightarrow{B} \gamma' \xrightarrow{B} \gamma'' \xrightarrow{B} \cdots \ .$$

In the following subsections we will discuss the RG flow at the examples of the Φ^4-model and hierarchical fixed points.

4.1 Example: Φ_d^4-Model

Consider the RG transformations

$$V_{j-1}(\Psi) := -\ln \int d\mu_{v^j}(\Phi) \exp\{-V_j(\Phi+\Psi)\} - (\Psi = 0) \ ,$$

for all $j \in \{0, \ldots, n\}$, $n \in \mathbb{N}$. The effective interactions V_j are normalized by subtraction of a constant term, such that $V_j(0) = 0$. V_j is the effective interaction after $n-j$ RG steps. v^j is the fluctuation propagator for the $(n-j+1)$th RG step. The starting (bare) interaction for the Φ^4-model in d dimensions is given by

$$V_n(\Phi) = \frac{1}{2} m_n^2 a_n^{-2} \int_z \Phi^2(z) + \frac{1}{2}\beta_n \int_z (\nabla\Phi)^2(z) + \frac{1}{4!}\lambda_n a_n^{d-4} \int_z \Phi^4(z) \ ,$$

where $a_n := L^{-n}a$, $a =$ unit length, $L \in \{2, 3, \ldots\}$. m_0^2, β_n, and λ_n are called bare coupling constants. Since the field Φ has dimension $a^{1-\frac{d}{2}}$, ∇ has dimension a^{-1} and $\int_z$ is of dimension a^d, we see that $\int_z \nabla^m \Phi^{2n}$ is of dimension $a^{2n-m-(n-1)d}$. Therefore, the constants m_n, β_n and λ_n are dimensionless. Running coupling constants m_j^2, β_j, λ_j and r_j, $j \in \{0,\dots,n\}$ are defined by the following representation of the effective interactions

$$\begin{aligned} V_j(\Phi) &= \frac{1}{2} m_j^2 a_j^{-2} \int_z \Phi^2(z) + \frac{1}{2}\beta_j \int_z (\nabla\Phi)^2(z) \\ &+ \frac{1}{4!}\lambda_j a_j^{d-4} \int_z \Phi^4(z) + \frac{1}{6!} r_j a_j^{2d-6} \int_z \Phi^6(z) + \dots\ . \end{aligned}$$

The RG flow of the running coupling constants m_j^2, β_j, λ_j and r_j is approximatively given by

$$\begin{aligned} m_{j-1}^2 &= L^2 m_j^2 + c_{21}\lambda_j - c_{22}\lambda_j^2 + O(\lambda_j^3) \\ \beta_{j-1} &= \beta_j - c_{2'2}\lambda_j^2 + O(\lambda_j^3) \\ \lambda_{j-1} &= L^{4-d}\lambda_j - c_{42}\lambda_j^2 + O(\lambda_j^3) \\ r_{j-1} &= L^{6-2d} r_j - c_{62}\lambda_j^2 + O(\lambda_j^3). \end{aligned} \tag{11}$$

c_{mk}, $m \in \{2, 2', 4, 6\}$, $k \in \{1, 2\}$ are constants which depend on j and n (c_{21} depends only on j). For small λ_j, we see that m_j^2 is growing after RG steps. We call such a coupling constant *relevant.* The coupling constant β_j does not change (if we neglect $O(\lambda_j)$-terms) after RG steps. Such a coupling constant is called *marginal.* For $d > 3$ dimensions the coupling constant r_j becomes smaller after a RG step. Such a coupling constant is called *irrelevant.* We see by eq. (11) that the coupling constant λ_j is relevant, marginal and irrelevant for $d < 4$, $d = 4$ and $d > 4$ dimensions respectively.

Let us discuss the RG flow eqs. (11). Consider the case $2 \le d \le 4$. Let us choose

$$\lambda_n = L^{(d-4)n}\lambda$$

for the bare coupling constant. If we iterate the RG flow eq. (11) for m_j^2, we obtain $m_j^2 = m_j^2(\lambda_n, m_n^2)$. Each RG step produces a factor L^2 for m_j^2. Thus after $n-j$ RG steps we get a factor $L^{2(n-j)}$. To perform the ultraviolet limit $\lim_{n\to\infty}$ one has to dominate the factor L^{2n}. A term of order λ_n^k delivers a factor $L^{(d-4)kn}$ and therefore dominates L^{2n} if $(4-d)k > 2$. Thus terms of order λ^k for $k > \frac{2}{4-d}$ are not "dangerous" for the ultraviolet limit. For example, in $d = 2$ dimensions only the term $c_{21}\lambda_j$ on the right hand side of eq. (11) for m_j^2 produces divergent terms. To avoid such divergent terms on has to subtract this term from the starting coupling constant m_n^2. Such a subtraction term is called *counter term.* Generally, in $d < 4$ dimensions the counter terms can be expressed by a finite number of Feynman graphs. We call a model (ultraviolet) *super-renormalizable* if such a procedure is possible. The Φ_d^4-model is super-renormalizable in $d < 4$ dimensions. In $d = 4$ there are no suppression factors $L^{-\alpha}$, $\alpha > 0$, coming from terms containing powers of λ. Therefore, the counter terms cannot be expressed by a finite number of Feynman-graphs. Nevertheless, only a finite number of coupling constants are relevant. This property is called *strict renormalizability* of the Φ_4^4-model. For an existence proof of the infrared limit $\lim_{j\to-\infty}$ for the Φ_4^4-model one starts with coupling constants $m_0^2, \beta_0, \lambda_0$. The RG flow of the running coupling constant λ_j implies (for λ_0 small)

$$\lambda_j = O(\frac{1}{|j|}).$$

Therefore, $\lim_{j\to-\infty} \lambda_j = 0$ and the Φ_4^4-model becomes trivial in the infrared limit. Since $\sum_{j=0}^{-\infty} \lambda_j^k < \infty$ for $k > 1$, we see that terms of order λ^k, for $k > 1$, are "harmless" for the RG flow of running coupling constants.

4.2 Excursion: Hierarchical RG Fixed Points

A special application of the exact beta-function technique is the determination of hierarchical RG fixed points. The hierarchical RG transformations (HRGT) in d dimensions is given by a mapping $\mathcal{RG} : \{Z : \mathbb{R} \to \mathbb{R}\} \longrightarrow \{Z : \mathbb{R} \to \mathbb{R}\}$ defined by

$$Z'(\Psi) := \mathcal{RG}(Z)(\Psi) := \left[\int d\mu_\gamma(\Phi) Z(\Phi + L^{1-\frac{d}{2}}\Psi)\right]^{L^d} ,$$

where the Gaussian measure $d\mu_\gamma(\Phi)$ is defined by, $\gamma > 0$,

$$d\mu_\gamma(\Phi) := (2\pi\gamma)^{-1/2} d\Phi\, e^{-\frac{\Phi^2}{2\gamma}} .$$

There exists fixed points Z^*, $Z^* = \mathcal{RG}(Z^*)$, such that the corresponding interactions $V^* := -\ln Z^*$ are l-wells, $l \in \{2, 3, \ldots\}$. It is well-known that the l-well fixed points exist in d dimensions, $2 \le d \le d_* = \frac{2l}{l-1}$ (cp. Collet and Eckmann (ϵ-expansion)[CE77, CE78]), Felder (RGDE) [F87]). A first rigorous construction of the 3-dimensional 2-well fixed point was accomplished by Koch and Wittwer [KW86, KW91].

For a representation of partition functions choose the following coordinates, supposing $\beta \ne 1$,

$$Z(\varphi) = \sum_{n=0}^{\infty} \frac{z_n}{\gamma'^n} : \varphi^{2n} :_{\gamma'} \leftrightarrow z = (z_0, z_1, \ldots) ,$$

where $\gamma' := \frac{\gamma}{1-\beta^2}$. The normal (Wick) ordering is defined by

$$: \varphi^{2n} :_\gamma := \exp\{-\frac{\gamma}{2}\frac{\partial^2}{\partial\varphi^2}\}\, \varphi^{2n} .$$

Define an associative $\times$-product on $\mathbb{R}^\infty$ by

$$(a \times b)_l = \sum_{m,n:\, |m-n| \le l \le m+n} C_l^{mn} a_m b_n ,$$

where the structure coefficients C_l^{mn} obey

$$:\varphi^{2m}:_\gamma \cdot :\varphi^{2n}:_\gamma = \sum_{l:\,|m-n|\leq l\leq m+n} \gamma^{m+n-l} C_l^{mn} :\varphi^{2l}:_\gamma \ .$$

Then, the HRGT can be represented as an L^d-fold product (cp. [PPW94, PW94])

$$\mathcal{RG}(Z) = \mathcal{S}_\beta(\underbrace{z \times \cdots \times z}_{L^d\ factors}) \ ,$$

where $\beta = L^{1-\frac{d}{2}}$ and $\mathcal{S}_\beta$ is defined by

$$(\mathcal{S}_\beta(z))_l := \beta^{2l} z_l \ .$$

For the special case $L^d = 2$, $\beta = L^{1-\frac{d}{2}} = 2^{-\frac{d-2}{2d}}$ the fixed points z^* are solutions of the quadratic equation

$$z = z \times_\beta z := \mathcal{S}_\beta(z \times z) \ .$$

For $\beta \neq 1$ the product $\times_\beta$ is nonassociative.

Three solutions of the fixed point equation are immediately found. They are $0, 1$ and the high-temperature fixed point $z_{HT} = \mathcal{N} e^{-c_* \varphi^2}$.

There is a simple argument by ϵ-expansion to show that new non-trivial fixed points appear below $d_* = \frac{2l}{l-1}$ dimensions. Suppose that $z_* = \mathbf{1} + h$, where $\mathbf{1} := (1, 0, 0, \ldots)$, is a fixed point, i. e.

$$h = 2\mathbf{1} \times_\beta h + h \times_\beta h \ .$$

This is equivalent to

$$Uh = h \times h \tag{12}$$

where U is a diagonal matrix, $U = diag(1 - 2\beta^{2l})_{l=0,1,\ldots}$. By eq. (12) h can only be infinitesimal small if approximatively $h \in \ker U$. But the kernel of U is only non-trivial, $\ker U \neq \{0\}$, if $1 = 2\beta^{2l}$ or equivalently $d = \frac{2l}{l-1}$. In this case one sees that in lowest order $h_m = -\alpha\delta_{m,l}$, i. e. the fixed point partition function is

$$Z(\varphi) = 1 - \alpha :\varphi^{2l}:_\gamma + \ldots \ .$$

Below the critical dimensions $d = \frac{2l}{l-1}$ all terms $:\varphi^{2m}:_\gamma$ for $m \leq l$ become relevant.

Koch and Wittwer extracted the high-temperature fixed point out of the partition function. Then γ and β change to $L^{-2}\gamma$ and $L^{-2}\beta$ respectively. For their proof of the existence of the 2-well fixed point in 3 dimensions they used the norm, for the case $2\beta^2 < 1$,

$$\|z\|_\rho^{(1)} := \sum_{n=0}^{\infty} \sqrt{(2n)!}|a_n|\rho^n \ .$$

For ϵ-expansion one has to consider the case $2\beta^2 > 1$ and uses the norm (cp. Pordt and Wieczerkowski [PW94])

$$\|z\|_\rho^{(\infty)} := \sup_n (n!|a_n|\rho^n) \ .$$

In constructive field theory (cp. [BY90]) a large field regulator is used

$$\|z\|_\rho^{(c)} := \sup_{\phi\in\mathbb{R}} |e^{\rho\phi^2} Z(\phi)| \ .$$

These norms are algebra-norms for special values of ρ. This means that, for all a, b,

$$\|a \times_\beta b\|_\rho \leq \|a\|_\rho \cdot \|b\|_\rho \ .$$

The beta-function technique works for $N \geq N_0 = O(1)(= 7)$ using the projection operator $\mathrm{P}(z) := (z_0, \ldots, z_N, 0, 0, \ldots)$, for $z \in \mathbb{R}^\infty$.

5 Relevant and Irrelevant Parts, Renormalization, Repolymerization

The choice of a suitable projection operator and the definition of running coupling constants is more complicated for the full model than for the hierarchical case. The choice of running coupling constants is model-dependent and we consider in this section only the case of the Φ^4-model. The way of choosing the running coupling constants is equivalent to an implementation of renormalization conditions. It can be shown that renormalization is only necessary for small polymers (cp. Brydges and Yau [BY90], Dimock and Hurd [DH91, DH92, DH93]).

In this section the questions of renormalization of polymer activities and repolymerization of the polymer system are discussed. Firstly, specify some subsets RP_n of the polymer set $Pol(\Lambda)$ for $n \in \{0, 2, 2', 4\}$. The sets RP_n are called *renormalization parts*. Only polymers which are contained in the renormalization parts are renormalized. The renormalization parts obey the following conditions

1. $Q \in RP_n \wedge P \subseteq Q \Rightarrow P \in RP_n$.
2. $\forall y \in \Lambda\, \exists U_y \in Pol(\Lambda) : \forall P \in RP_n, y \in P : P \subseteq U_y$.
3. $RP_0 \supseteq RP_2 \supseteq RP_{2'} \supseteq RP_4 \supseteq \cdots$.
4. RP_n preserves lattice symmetry.

For a polymer $Y \in Pol(\Lambda)$, choose a function $\delta V_Y(y|\Psi)$ such that $\delta V_Y = \mathbf{1}_1 \cdot \delta V_Y$. Then, define a polymer activity $R \in \mathcal{A}$ by

$$Z(Y|\Psi) = EXP(\delta V_Y + R)(Y|\Psi) \ . \tag{13}$$

We can choose δV_Y in such a way that R obeys the following renormalization conditions

$$R(Y_0|\Psi)|_{\Psi=0} = 0, \qquad \sum_{y':y'\in Y} \frac{\partial^2}{\partial\Psi(y)\partial\Psi(y')} R(Y_2|\Psi)|_{\Psi=0} = 0, \ldots \ , \tag{14}$$

for all $Y_n \in RP_n$, $y' \in Y_n$. R is called *irrelevant activity*. The functions δV_Y correspond to perturbative counter terms. We may compute the counter terms δV_Y such that R defined by eq. (13) obeys the renormalization conditions (14). For the field $\psi = 0$ and the renormalization part $Y \in RP_0$, we have

$$Z(Y|0) = EXP(\delta V_Y)(Y|0) = \prod_{y \in Y} \delta V_Y(y|0) \ .$$

This implies

$$\delta V_Y(y|0) = \exp\{ \sum_{P:\, y \in P \subseteq Y} \frac{\widetilde{\ln Z}(P|0)}{|P|} \} \ ,$$

where the Moebius transform $\widetilde{\ln Z}$ of $\ln Z$ is implicitly defined by

$$\ln Z(Y|\psi) = \sum_{P:\, P \subseteq Y} \widetilde{\ln Z}(P|\psi) \ , \tag{15}$$

for all $Y \in Pol(\Lambda)$. Eq. (15) defines $\widetilde{\ln Z}$ uniquely. We have $\widetilde{\ln Z}(\emptyset) = \ln Z(\emptyset)$. Suppose that $\widetilde{\ln Z}(P)$ is defined for all $P \in Pol(\Lambda)$ with $|P| < N$. We want to define $\widetilde{\ln Z}(Y)$ for $Y \in Pol(\Lambda)$ with $|Y| = N$. Eq. (15) implies

$$\widetilde{\ln Z}(Y) = \ln Z(Y) - \sum_{\substack{P:\, P \subset Y \\ |P| < N}} \widetilde{\ln Z}(P) \ . \tag{16}$$

Since the terms on the right hand side of eq. (16) are uniquely defined, $\widetilde{\ln Z}(Y)$ is uniquely defined.

Eqs. (13) and (14) imply, for all $Y \in RP_2 \subseteq RP_0$,

$$\begin{aligned}
&\frac{1}{2} \sum_{y' \in Y} \frac{\partial^2}{\partial\psi(y)\partial\psi(y')} Z(Y|\psi)|_{\psi=0} = \\
&\frac{1}{2} \sum_{P:\, y \in P \subseteq Y} \sum_{y' \in P} \frac{\partial^2}{\partial\psi(y)\partial\psi(y')} A(P|\psi)|_{\psi=0} = \\
&\frac{1}{2} \sum_{y' \in Y} \frac{\partial^2}{\partial\psi(y)\partial\psi(y')} \delta V_Y(y|\psi)|_{\psi=0} \ .
\end{aligned}$$

We have used here that $\frac{\partial}{\partial\psi(y)} A(P|\psi)|_{\psi=0} = 0$, which follows from the symmetry property $A(P|-\psi) = A(P|\psi)$. This implies, for the counter term,

$$\begin{aligned}
\delta V_Y(y|\psi) = \delta V_Y(y|0) + \frac{1}{2} &\sum_{P:\, y \in P \subseteq Y} \\
&\sum_{y' \in P} \frac{\partial^2}{\partial\psi(y)\partial\psi(y')} A(P|\psi)|_{\psi=0} \, \psi^2(y) + \cdots \ .
\end{aligned}$$

We want to compute the irrelevant polymer activity R for a given counter term δV_Y. Eq. (13) implies

$$EXP(\delta V_Y) \circ EXP(R)(Y) = EXP(A)(Y) \ . \tag{17}$$

Let $\widetilde{\delta V}$ be the Moebius transform of δV

$$\delta V_Y = \sum_{P:\, P \subseteq Y} \widetilde{\delta V}_P \ . \tag{18}$$

Since $EXP(A + B) = EXP(A) \circ EXP(B)$ and $\circ$ is an associative product, eq. (17) implies

$$EXP(R)(Y) = EXP(-\delta V_Y) \circ EXP(A)(Y) \ . \tag{19}$$

This implies the following explicit formula for the irrelevant polymer activity R

$$R(Y) = \sum_{P:\, P \subseteq Y} \sum_{\substack{y \in P \mapsto P_y \subseteq Y \\ y \in P_y}} \sum_{n \geq 0} \frac{1}{n!} \sum_{\substack{P_1, \ldots P_n \subseteq Y - P:\, (P_1 + \cdots + P_n) \cup \bigcup_{y \in P} P_y = Y \\ \gamma(P_1, \ldots, P_n, P_y, y \in P) \ \mathrm{conn.}}} \prod_{y \in P} \widetilde{\delta V}_{P_y}(y) \prod_{a=1}^{n} A(P_a) \ . \tag{20}$$

The second sum on the right hand side of eq. (20) is over all polymers P_y, for all $y \in P$, such that $y \in P$. Since renormalization concerns only small polymers, we have for a polymer Y large enough, the relation $\delta V_Y = \delta V$, where

$$\begin{aligned} \delta V(y|\Psi) &:= \sum_{P:\, y \in P} \widetilde{\delta V}_P(y|\Psi) \\ &= c_0 + c_2 \Psi^2(y) + c_{2'} (\nabla \Psi)^2(y) + c_4 \Psi^4(y) \ . \end{aligned}$$

$\delta V(y|\Psi)$ is called the *relevant part* of the interaction.

If one wants stability bounds or norms with large field regulators one has to replace the counter term $\delta V(y|\Psi)$ by $\exp\{\delta V(y|\Psi)\}$. The philosophy of renormalization and repolymerization stays the same as discussed in this section.

The advantage in using this repolymerization procedure is that the relevant part depends on a fewer number of parameters. Therefore, the beta-function is defined on a space of lower dimension than the space of partition functions. The counter terms δV_Y depend on all renormalization parts which are contained in the polymer Y. The relevant part δV depend on fewer terms. For the example of the Φ^4-model it is determined by the four running coupling constants $c_0, c_2, c_{2'}, c_4$.

To control the flow of the irrelevant activites R and the running coupling constants, we introduce a 5th RG step. This step is called *repolymerization step*. It changes the polymer system only for small polymers and therefore the thermodynamic limit is unchanged. The repolymerization step is the replacement of the counter terms δV_Y on the right hand side of eq. (13) by the relevant part δV

$$Z(Y|\Psi) \to Z^{rep}(Y|\Psi) := EXP(\delta V + R)(Y|\Psi) \ .$$

For $y \in \Lambda$ let U_y be the polymer in the 2nd condition for renormalization parts. For $P \supseteq U_y$, we have $Z(P|\Psi) = Z^{rep}(P|\Psi)$. The projection operator P for the beta-function method is defined by

$$\mathrm{P}(A) = \delta V, \qquad (\mathbf{1} - \mathrm{P})(A) = R \ .$$

6 Norm Estimations

For numerically calculations one has to truncate the infinite number of irrelevant polymer activities $R(P)$, $P \in Pol(\Lambda)$. Then, one has to estimate the truncation error. The control of the RG flows can be done by using norm estimates. The most important problem for estimating polymer activities is the large field contributions or equivalently large factorials. If we would not allow large field contributions, then ordinary perturbation would converge. Consider, for example, a partition function for the 0-dimensional φ^4 field theory

$$Z(\psi) = \int d\varphi \, e^{-\frac{m^2}{2}\varphi^2 - \lambda(\varphi+\psi)^4} \ , \tag{21}$$

where λ is a non-negative constant. If we integrate over all $\varphi \in \mathbb{R}$, we see that the integral will diverge for negative λ. Therefore, the series expansion in powers of λ has zero convergence radius. If we restrict the integration over φ to the finite interval $[-K, K]$, $K < \infty$, then the integral is also convergent for negative values of λ and the power series is convergent. The large field behavior for the partition function Z is

$$Z(\psi) \sim e^{-c\psi^2}, \qquad \psi \to \infty \ , \tag{22}$$

where $c = O(\lambda^{1/2})$. Thus

$$Z(\psi) = \sum_{n=0}^{\infty} z_n \psi^{2n}, \qquad z_n \sim n!^{-1}, \qquad n \to \infty \ .$$

For $c, \kappa, \rho \in \mathbb{R}_+$, define the following norms

$$\|Z\|_{c,\kappa} := \sup_{\substack{\varphi,\psi \in \mathbb{R}: \\ |\psi| \leq \kappa}} \{ e^{c\varphi^2} |Z(\varphi + i\psi)| \}$$

and

$$\|Z\|_{\rho} := \sum_{n=0}^{\infty} \sqrt{(2n)!} |z_n| \rho^n \ .$$

For c and ρ small enough the norms of Z defined by eq. (21) are finite, $\|Z\|_{c,\kappa} < \infty$, $\|Z\|_{\rho} < \infty$. The Taylor coefficient z_n of ψ^{2n} for $Z_0(\psi) = e^{-\lambda\psi^4}$ is of order

$1/\sqrt{n!}$. Therefore, the $\|\cdot\|_\rho$-norm of $Z_0(\psi)$ is not finite, $\|Z_0\|_\rho = \infty$. Thus, we are only allowed to consider norms of partition functions after at least one RG step !

Before a norm for polymer activities can be defined, some notations for multiindices have to be introduced. Call $m \in \mathcal{M} : \Lambda \to \mathbb{N}$ a *multiindex* if

$$|m| := \sum_{x\in\Lambda} m(x) < \infty \ .$$

The *support* of the multiindex m is defined by $supp\, m := \{x \in \Lambda | \, m(x) \neq 0\}$. The factorial is defined by $m! := \prod_{y\in\Lambda} m(y)!$. Let $\varphi : \Lambda \to \mathbb{R}$ be a field and m a multiindex. Define a power by $\varphi^m := \prod_{y\in\Lambda} \varphi(y)^{m(y)}$. For two multiindices $m, n \in \mathcal{M}$ define an order relation by $m \leq n$ iff $m(y) \leq n(y)$, for all $y \in \Lambda$. Let A be a local polymer activity and write

$$A(P|\Psi) = \sum_{\substack{m\in\mathcal{M}:\\ supp\, m\subseteq P}} a(P|m)\Psi^m \ .$$

For a polymer P let $T(P)$ be the set of all tree graphs with vertex set P. The tree bound of P is defined by

$$T_\mu(P) := \sup_{\tau\in T(P)} \exp\{-\mu a^{-1} \sum_{(yy')\in\tau} \|y - y'\|\} \ , \tag{23}$$

where $\mu > 0$. $\|\cdot\|$ is the Euclidean norm in $\mathbb{R}^d$. For $0 < k_1, k_2 < 1$, $\mu, \rho > 0$ and define the norm of the polymer activity A

$$\|A\|_{k_1,k_2,\mu,\rho} := \sup_{y\in\Lambda} \left\{ \sum_{P:\, y\in P\in Pol(\Lambda)} \ \sum_{\substack{m\in\mathcal{M}:\\ supp\, m\subseteq P}} \frac{\sqrt{m!}|a(P|m)|}{k_1 k_2^{|P|-1} T_\mu(P)} \rho^{|m|} \right\} \ .$$

The definition of the norm contains two sums. A sum over all polymers P which contains an element y and a sum over all multiindices m whose support is contained in P. The terms in the sums contain a square root of the factorial m. This represents the large field behavior. The factor $k_1 < 1$ yields that terms containing a large number of polymer activities A are suppressed. The factor $k_2 < 1$ yields that $A(P)$ is small for polymers P with a large number of elements. The factor $T_\mu(P)$ yields that $A(P)$ is small for polymers P with large extension, i.e. polymers which contain elements x, y such that $\|x - y\|$ is large.

The remainder of this section is technical. It concerns the question of how to estimate general polymer activities after integration, coarsening and localization step. By these methods one can control the RG flow of the irrelevant polymer activities R.

6.1 Integration Step

For a multiindex $m : \Lambda^2 \to \mathbb{N}$, let us introduce the notations

$$m_1(x) := \sum_y m(y,x), \qquad m_2(y) := \sum_x m(y,x) \ . \tag{24}$$

In the following we will use that

$$\int d\mu_\Gamma(\phi) P(\Phi + \Psi) = \exp\{\frac{1}{2}\sum_{x,y} \Gamma(x,y) \frac{\partial^2}{\partial\psi(x)\partial\psi(y)}\} P(\Psi) \ ,$$

where P is a polynom. Series expansion of the Gaussian measure yields

$$\exp\{\frac{1}{2}\sum_{x,y} \Gamma(x,y) \frac{\partial^2}{\partial\psi(x)\partial\psi(y)}\} = \sum_{m:\Lambda^2\to\mathbb{N}} \frac{1}{2^{|m|}} \frac{\Gamma^m}{m!} \partial_\psi^{m_1+m_2} \ .$$

Define structure coefficients $C_\Gamma(m,n)$, for $m, n : \Lambda \to \mathbb{N}$ by

$$\int d\mu_\Gamma(\phi)(\phi+\psi)^n = \sum_{m:\, m\leq n} C_\Gamma(m,n)\psi^m \ .$$

We have, for $n : \Lambda \to \mathbb{N}$,

$$\begin{aligned}
\int d\mu_\Gamma(\phi)(\phi+\psi)^n &= \sum_{k:\Lambda^2\to\mathbb{N}} \frac{1}{2^{|k|}} \frac{\Gamma^k}{k!} \partial_\psi^{k_1+k_2} \psi^n \\
&= \sum_{k:\Lambda^2\to\mathbb{N}} \frac{1}{2^{|k|}} \frac{\Gamma^k}{k!} \frac{n!}{(n-k_1-k_2)!} \psi^{n-k_1-k_2} \\
&= \sum_m \sum_{\substack{k:\Lambda^2\to\mathbb{N}: \\ n=k_1+k_2+m}} \frac{1}{2^{|k|}} \frac{\Gamma^k}{k!} \frac{n!}{m!} \psi^m \ .
\end{aligned}$$

This implies

$$C_\Gamma(m,n) = \sum_{\substack{k:\Lambda^2\to\mathbb{N}: \\ n=k_1+k_2+m}} \frac{1}{2^{|k|}} \frac{\Gamma^k}{k!} \frac{n!}{m!} \ .$$

Consider the polymer activity $A(P|\psi) = \sum_{\substack{n:\Lambda\to\mathbb{N} \\ \mathrm{supp}\, n\subseteq P}} a(P|n)\psi^n$ and the Gaussian integral $A' = \mu_\Gamma(A)$. Suppose that a' are coefficients of the Taylor expansion of A'. Then

$$a'(P|m) = \sum_{n:\, m\leq n} C_\Gamma(m,n) a(P|n), \qquad a' = C_\Gamma a \ .$$

The integration step is $EXP(A') = \mu_\Gamma(EXP(A))$. The effective polymer activity are computed by an integration step $A' = \widetilde{\mu_\Gamma}(A)$. Thus

$$a'(Y|m) = \sum_{Y=\sum P} \sum_{n:\, m\leq n} C_{\Gamma,\{P\}}(m,n) \prod_P a(P|n) \ ,$$

where

$$C_{\Gamma,\{P\}}(m,n) := \sum_{\substack{k:\Lambda^2\to\mathbb{N}:\, n=k_1+k_2+m \\ \gamma(\mathrm{supp}\, k,\{P\})\ conn.}} \frac{1}{2^{|k|}} \frac{\Gamma^k}{k!} \frac{n!}{m!} \ .$$

For estimations the trick of exponential pinning is important. Exponential pinning is given by the following bound. Consider $\alpha \in \mathbb{R}_+$, $b: \Lambda \to \mathbb{R}_+$, $m: \Lambda \to \mathbb{N}$. The multinominial theorem implies

$$b^m |m|!^\alpha \leq \|b^{1/\alpha}\|^{\alpha m} m!^\alpha, \qquad \|b\| := \sum_y b(y)$$

and, for $d: \Lambda^2 \to \mathbb{R}_+$, $m: \Lambda^2 \to \mathbb{N}$,

$$\frac{d^m}{m!} \leq \frac{\|d\|^{|m|}}{\sqrt{m_1! m_2!}} \ ,$$

where m_1 and m_2 are above defined by eq. (24). For the coefficients $a(P|n)$, $\rho \in \mathbb{R}_+$ and a polymer Y define the following norm

$$\|a(Y)\|_\rho := \sum_{m\in\mathcal{M}} \sqrt{m!}\, |a(Y|m)|\, \rho^m \ .$$

For the mapping $b: \Lambda^2 \to \mathbb{R}_+$, define the norm $\|b\| := sup_y\{\sum_x b(y,x)\}$. By exponential pinning one can prove the following bound

$$\|a'(Y)\|_\rho \leq \sum_{Y=\sum P} \sup_{k:\, \gamma(\mathrm{supp}\, k,\{P\})\ conn.} \left(\frac{|\Gamma|}{b}\right)^k \prod_P \|a(P)\|_{2\sqrt{\frac{\|b\|}{2}}+\rho} \ .$$

Using the definition eq. (23) of a tree bound, we obtain

$$T_\mu(P_1)\cdots T_\mu(P_n) \leq$$
$$\inf_{\tau\in T_n} \exp\left\{\mu a^{-1} \sum_{(ab)\in\tau} dist(P_a,P_b)\right\} T_\mu(P_1\cup\cdots\cup P_n) \ .$$

Let $q>0$, $b: \Lambda^2 \to \mathbb{R}_+$ and $c := \frac{|\Gamma|}{b} e^{\|\cdot-\cdot\|} : \Lambda^2 \to \mathbb{R}_+$ and suppose that

$$4\frac{k_1}{k_2}\|c\|\, \|A\|_{e^{-qk_1},e^{-qk_2},\mu,\rho'} < q^2 \ ,$$

where $\rho' := 2\sqrt{\frac{\|b\|}{2}} + \rho$. Then

$$\|\widetilde{\mu_\Gamma}(A)\|_{k_1,k_2,\mu,\rho} \leq \frac{\|A\|_{e^{-qk_1},e^{-qk_2},\mu,\rho'}}{1 - 4q^{-2}\frac{k_1}{k_2}\|c\|\|A\|_{e^{-qk_1},e^{-qk_2},\mu,\rho'}}$$

and

$$\|1_{>1}\cdot(\widetilde{\mu_\Gamma}(A)-A)\|_{k_1,k_2,\mu,\rho}\le \frac{4q^{-2\frac{k_1}{k_2}}\|c\|\|A\|_{e^{-q}k_1,e^{-q}k_2,\mu,\rho'}}{1-4q^{-2\frac{k_1}{k_2}}\|c\|\|A\|_{e^{-q}k_1,e^{-q}k_2,\mu,\rho'}}\|A\|_{e^{-q}k_1,e^{-q}k_2,\mu,\rho'}\ .$$

After an integration step the constants k_1 and k_2 become larger by a factor e^q. Thus, for a polymer P, we loose a factor $e^{q|P|}$. This factor can become very large if $|P|$ is large. In the next subsection, we will obtain a small factor for each element in the polymer after the coarsening step in the case where polymers are not small. The constant ρ grows to ρ'. For $d>2$ dimensions this grow of ρ can be dominated by the scaling factor $L^{1-\frac{d}{2}}$ after the rescaling step. For $d=2$ one uses that the coupling constant λ (for Φ^4-model) grows by a factor L^2 after each RG step. This gives a suppression factor $L^{-\frac{1}{2}}$ for the constant ρ which dominates the growing of ρ to ρ'.

6.2 Coarsening Step

Norm estimation of the coarsening step requires to distinguish between small and large polymers. We call a polymer X of the lattice Λ' with lattice spacing La *small* if, for all $y,x\in X$, the condition $\frac{|x^\mu-y^\mu|}{La}\in\{0,1\}$, $\mu\in\{1,\dots,d\}$, holds. Let $P_1,\dots,P_n$ be polymers of Λ such that $X:=[P_1+\dots+P_n]$ is not small and the Venn-diagram $\gamma([P_1],\dots,[P_n])$ is connected. Then, there exists a positive $\epsilon=O(1)$ such that the following tree estimation holds

$$k_1\,k_2^{|P_i|-1}\prod_{i=1}^n T_\mu(P_i)\le k_1\,k_2^{|X|-1}T_\mu(X)\,(\frac{k_1}{k_2})^{n-1}\,(k_2+e^{-\mu L})^{\epsilon|P_1+\dots+P_n|}\ . \quad (25)$$

Let A be a polymer activity in Λ. The coarsened polymer activity $\widetilde{[A]}$ is given by

$$\widetilde{[A]}(X)=\sum_{n:\,n\ge1}\frac{1}{n!}\sum_{\substack{P_1,\dots,P_n\in Pol(\Lambda):\\ P_1+\dots+P_n=\overline{X},\ \gamma([P_1],\dots,[P_n])\ conn.}}\prod_{i=1}^n A(P_i)\ . \quad (26)$$

For small k_2 and large L the factor $(k_2+e^{-\mu L})^\epsilon$ is small. The tree estimation eq. (25) implies that for not small polymers $X\in Pol(\Lambda')$ the coarsened polymer activity $\widetilde{[A]}(X)$ is suppressed by $\gamma^{|X|}$ where γ is a small number. This small number can be used to suppress the extra factor e^q for k_1 and k_2 after an integration step.

We estimate firstly the contributions to the norm of the coarsened polymer $\widetilde{[A]}$ coming from not small polymers. Then, we estimate the contributions coming from small polymers and the part of the right hand side of eq. (26) where $n\ge2$. These two estimations will be discussed in this subsection. It remains to estimate contributions coming from small polymers and terms which contain at most one factor A. For these estimations one has to use the renormalization conditions to obtain suppression factors. This method is well-known and not discussed here (see for example [R91]).

Define a projection operator P_{ns} defined on the set of all polymer activities on Λ which gives zero if applied to small polymers

$$P_{ns}(A)(X) := \begin{cases} A(X) & : \ X \text{ not small} \\ 0 & : \ X \text{ small.} \end{cases}$$

For $d_1, \ldots, d_n \in \mathbb{N}$ such that $\sum_{a=1}^{n} d_a = 2(n-1)$ denote by $T_n(d_1, \ldots d_n)$ the tree graph with vertex set $\{1, \ldots, n\}$, d_i lines emerging from vertex i. Then Cayley's Theorem counts the number of tree graphs in $T_n(d_1, \ldots d_n)$

$$|T_n(d_1, \ldots d_n)| = \frac{(n-2)!}{\prod_{i=1}^{n} (d_i - 1)!} \ . \tag{27}$$

Cayley's Theorem and the estimation

$$\frac{|P_{ns}(\widetilde{[A]})(X)|}{k_1 k_2^{|X|-1} T_\mu(X)} \leq \sum_{n:\, n \geq 1} \frac{1}{(n-1)!} \sum_{\substack{d_1, \ldots, d_n: \\ \sum d_i = 2(n-1)}} \sum_{\tau \in T_n(d_1, \ldots d_n)}$$

$$\sum_{\substack{P_1, \ldots, P_n:\, x \in [P_1] \\ \gamma([P_1], \ldots, [P_n]) \supseteq \tau,\ P_1 + \cdots P_n = \overline{X}}} \prod_{i=1}^{n} \frac{A(P_i)}{k_1 k_2^{|X|-1} T_\mu(P_i)}$$

$$(\frac{k_1}{k_2})^{n-1} (k_2 + e^{-\mu L})^{\epsilon |P_1 + \ldots + P_n|} \ ,$$

imply the following bound. Suppose that $4 \frac{k_1}{k_2} L^d \|A\|_{\kappa^{-1} k_1, \kappa^{-1} k_2, \mu, \rho} < q^2$ holds, where $\kappa := e^{-q}(k_2 + e^{-\mu L})^\epsilon$. Then

$$\|P_{ns}(\widetilde{[A]})\|_{k_1, k_2, \mu, \rho} \leq \frac{\|A\|_{\kappa^{-1} k_1, \kappa^{-1} k_2, \mu, \rho}}{1 - 4q^{-2} \frac{k_1}{k_2} L^d \|A\|_{\kappa^{-1} k_1, \kappa^{-1} k_2, \mu, \rho}} \ .$$

Denote by $\widetilde{[A]}_{>1}(X)$ the part of the right hand side of eq. (26) which consists of at least two A-factors and by $\widetilde{[A]}_1(X)$ the part which consists of only one A-term

$$\widetilde{[A]}_{>1}(X) = \sum_{n:\, n \geq 2} \frac{1}{n!} \sum_{\substack{P_1, \ldots, P_n \in Pol(\Lambda): \\ P_1 + \cdots + P_n = \overline{X},\ \gamma([P_1], \ldots, [P_n]) \text{ conn.}}} \prod_{i=1}^{n} A(P_i)$$

$$\widetilde{[A]}_1(X) := \sum_{P:\, [P] = X} A(P) \ .$$

Suppose that $4C \frac{k_1}{k_2} q^{-2} L^d \|A\|_{k_1, k_2, \mu, \rho} < 1$. Then, we have

$$\|(\mathbf{1} - P_{ns})(\widetilde{[A]}_{>1})\|_{k_1, k_2, \mu, \rho} \leq \frac{4Cq^{-2} \frac{k_1}{k_2} L^d \|A\|_{k_1, k_2, \mu, \rho}}{1 - 4Cq^{-2} \frac{k_1}{k_2} L^d \|A\|_{k_1, k_2, \mu, \rho}} \|A\|_{k_1, k_2, \mu, \rho} \ .$$

6.3 Localization Step

Let $m : \Lambda \to \mathbb{N}$ be a multiindex and $\mathcal{A}$ be the operator defined in the block spin RG which maps fields defined on Λ to fields defined on Λ'. Expanding the mth power of $\mathcal{A}\phi$ yields

$$(\mathcal{A}\phi)^m = \sum_{n:\,\Lambda'\to\mathbb{N}} D(n,m)\phi^n \ ,$$

where

$$D(n,m) := \sum_{\substack{G:\,\Lambda\times\Lambda'\to\mathbb{N}\\ G_1=m,\,G_2=n}} \frac{m!}{G!}\mathcal{A}^G \ .$$

This is proven by

$$\begin{aligned}(\mathcal{A}\phi)^m &= \prod_{y\in\Lambda'}(\mathcal{A}\phi)(y)^{m(y)} = \prod_{y\in\Lambda'}\left(\sum_{x\in\Lambda'}\mathcal{A}(y,x)\phi(x)\right)^{m(y)}\\ &= \prod_{y\in\Lambda'}\left(\sum_{\substack{G_y:\,\Lambda'\to\mathbb{N}\\ |G_y|=m(y)}} \frac{m(y)!}{G_y}(\mathcal{A}(y,x)\phi(x))^{G_y(x)}\right)\\ &= \sum_{\substack{G:\,\Lambda\times\Lambda'\to\mathbb{N}\\ G_1=m}} \frac{m!}{G!}\mathcal{A}^G\phi^{G_2} \ .\end{aligned}$$

For all multiindices $m : \Lambda \to \mathbb{N}$, let $a(m)$ be a real number. Define, for a multiindex $n : \Lambda' \to \mathbb{N}$

$$a'(n) = \sum_{m:\,\Lambda\to\mathbb{N}} D(n,m)a(m) \ .$$

Then, we have

$$\begin{aligned}\|a'\|_\rho &\leq \sum_{n:\,\Lambda'\to\mathbb{N}} \sqrt{n!} \sum_{m:\,\Lambda\to\mathbb{N}} |D(n,m)|\,|a(m)|\rho^{|n|}\\ &\leq \sum_{m:\,\Lambda\to\mathbb{N}} \sum_{\substack{G:\Lambda\times\Lambda'\to\mathbb{N}\\ G_1=m,\ G_2=n}} \frac{|\mathcal{A}|^G}{G!}\sqrt{m!n!}\,\sqrt{m!}|a(m)|\rho^{|m|} \ .\end{aligned}$$

Define, for $b :\ \Lambda \times \Lambda' \to \mathbb{R}_+$, the norm

$$\|b\| := \max(\sup_{x\in\Lambda'}\sum_{y\in\Lambda}|b(y,x)|, \sup_{y\in\Lambda}\sum_{x\in\Lambda'}|b(y,x)|) < \infty \ .$$

Exponential pinning, for a multiindex $G : \Lambda \times \Lambda' \to \mathbb{N}$, yields

$$\frac{b^G}{G!}\sqrt{m!n!} \leq \|b\|^{|G|} \ ,$$

where $G_1 = m : \Lambda' \to \mathbb{N}$ and $G_2 = n : \Lambda \to \mathbb{N}$ are defined by eq. (24). Thus

$$\|a'\|_\rho \leq \sum_{m:\,\Lambda\to\mathbb{N}} \sum_{\substack{G:\Lambda\times\Lambda'\to\mathbb{N}\\ G_1=m,\ G_2=n}} \left(\frac{|A|}{b}\right)^G \|b\|^{|G|} \sqrt{m!}|a(m)|\rho^{|m|}$$

$$\leq \sup_G \left(\frac{|A|}{b}\right)^G \|a\|_{\|b\|\rho} \; .$$

Let $A = LN(Z)$ be the polymer activity of Z. Define a polymer activity A' by

$$Z(X|\mathcal{A}_X\phi) = \sum_{X=\sum Q} \prod_Q A'(Q|\mathcal{A}_Q\phi) \; .$$

Use the representations

$$A'(X|\psi) = \sum_{m:\,\Lambda'\to\mathbb{N}} a'(X|m)\psi^m$$

and

$$A(X|\phi) = \sum_{n:\,\Lambda\to\mathbb{N}} a(X|n)\phi^n \; .$$

Then, the following relation holds

$$a'(X|m) = \sum_{X=\sum Q} \sum_{n:\,\Lambda\to\mathbb{N}} D_{A,\{Q\}}(m,n) \prod_Q a(Q|n_Q) \; ,$$

where

$$D_{A,\{Q\}}(m,n) := \sum_{\substack{G:\,\Lambda\times\Lambda'\to\mathbb{N}\\ G_1=m,\, G_2=n,\ \gamma(supp\, G,\{Q\})\ conn.}} \frac{m!}{G!} A^G \; .$$

The coefficients a and a' obey the following norm inequalities

$$\|a'(X)\|_\rho \leq \sum_{X=\sum Q} \sup_{G:\,\gamma(supp\, G,\{Q\})\ conn.} \left(\frac{|A|}{b}\right)^G \prod_Q \|a(Q)\|_{\|b\|\rho} \; .$$

Define $c := \frac{|A|}{b} e^{\mu\|\cdot-\cdot\|} : \Lambda \times \Lambda' \to \mathbb{R}_+$ and suppose, for $q > 0$,

$$4\frac{k_1}{k_2} q^{-2} \|c\| \|A\|_{e^{-q}k_1, e^{-q}k_2, \mu, \|b\|\rho} < 1 \; .$$

Then

$$\|A'\|_{k_1,k_2,\mu,\rho} \leq \frac{\|A\|_{e^{-q}k_1, e^{-q}k_2, \mu, \|b\|\rho}}{1 - 4q^{-2}\frac{k_1}{k_2}\|c\|\|A\|_{e^{-q}k_1, e^{-q}k_2, \mu, \|b\|\rho}} \; .$$

7 Conclusion

The input of a RG transformation starts with a given relevant part δV depending on a finite number of (running) coupling constants and an irrelevant polymer activity R. δV is a polymer function vanishing for polymers which contain more than 1 element. The polymer activity A is the sum of these two terms, $A = \delta V + R$. A new polymer activity $\widetilde{A}$ is defined after integration, coarsening, localization and rescaling step, $\widetilde{A} := \widetilde{\mathcal{R}_L} \circ \widetilde{\iota_A} \circ \widetilde{[\,]} \circ \widetilde{\mu_\Gamma}(A)$. For a polymer X of Λ' the new irrelevant activity R' is defined by

$$EXP(\widetilde{A})(X) = EXP(\delta V_X + R')(X) \ ,$$

where the counter terms δV_X are defined by the polymer activities $\widetilde{A}$ such that R' fulfills renormalization conditions. The new (running) coupling constants determine $\delta V'$. The sum of the new relevant part $\delta V'$ and the irrelevant polymer activity R' gives the new effective polymer activity A'.

The RG flow is splited into a flow of the relevant part δV and irrelevant polymer activity R

$$\delta V \to \delta V'(\delta V, R), \qquad R \to R'(\delta V, R) \ .$$

The split of the RG flow is model-dependent. The methods for a definition and control of the irrelevant polymer activities is model-independent.

The control of the RG flow is solved by proving recursive bounds on the running coupling constants and the norms of the irrelevant polymer activities R. The definition of the norms has to contain parameters which control the size and extension of polymers and terms which control the large field behavior.

For an explicit control of the RG flow one proceeds as follows. We start with a bare interaction δV_n and an irrelevant polymer activity $R_n = 0$. For $j < n$ the effective interactions $\delta V_j^{(n)}$ and irrelevant activities $R_j^{(n)}$ are defined by the RG flows

$$\delta V_j^{(n)} \to \delta V_{j-1}^{(n)}(\delta V_j^{(n)}, R_j^{(n)}), \qquad R_j^{(n)} \to R_{j-1}^{(n)}(\delta V_j^{(n)}, R_j^{(n)}) \ .$$

$\delta V_j^{(n)}$ depends on certain coupling constants $\gamma_{0,j}^{(n)}, \ldots, \gamma_{N,j}^{(n)} \in \mathbb{R}$. For the control of the RG flow one supposes that there exists finite intervals $I_{k,j}^{(n)} \subset \mathbb{R}$, for $k \in \{0, \ldots, N\}$, $j \leq n$, and constants $k_1, k_2, \mu, \rho, \epsilon_j^{(n)}$, for $k \in \{0, \ldots, N\}$, $j \leq n$ such that

$$\gamma_{k,j}^{(n)} \in I_{k,j}^{(n)}, \qquad \|R_j^{(n)}\|_{k_1,k_2,\mu,\rho} < \epsilon_j^{(n)}$$

implies

$$\gamma_{k,j-1}^{(n)} \in I_{k,j-1}^{(n)}, \qquad \|R_{j-1}^{(n)}\|_{k_1,k_2,\mu,\rho} < \epsilon_{j-1}^{(n)}.$$

For weakly coupled models, a guess of the intervals $I_{k,j}^{(n)}$ and constants $\epsilon_j^{(n)}$ can be achieved in the following way. Set $R_j^{(n)} = 0$ and compute the RG step

$$\delta V_j^{(n)} \to \delta V_{j-1}^{(n)}(\delta V_j^{(n)}, 0)$$

using perturbation theory. This gives an approximate relation of the flow of the coupling constants $\gamma_{0,j}^{(n)}$. For a guess of the constant $\epsilon_j^{(n)}$ use the RG step

$$R_j^{(n)} = 0 \to R_{j-1}^{(n)}(\delta V_j^{(n)}, R_j^{(n)} = 0)$$

and find a bound for $\|R_{j-1}^{(n)}(\delta V_j^{(n)}, R_j^{(n)} = 0)\|_{k_1,k_2,\mu,\rho}$.

Iterating the RG equations one sees that the relevant interaction terms $\delta V_j^{(n)}$ and the irrelevant activities $R_j^{(n)}$ depend on the starting (bare) interaction δV_n. For the infrared limit one has to show that δV_n can be defined such that the limits $\lim_{j\to-\infty} \delta V_j^{(n)}(\delta V_n)$ and $\lim_{j\to-\infty} R_j^{(n)}(\delta V_n)$ exist. For the ultraviolet limit one has to show that the limits $\lim_{n\to\infty} \delta V_j^{(n)}(\delta V_n)$ and $\lim_{n\to\infty} R_j^{(n)}(\delta V_n)$ exist.

References

[B88a] T. Bałaban. *Renormalization Group Approach to Lattice Gauge Field Theories. I. Generation of Effective Actions in a Small Field Approximation and a Coupling Constant Renormalization in Four Dimensions*, Commun. Math. Phys. 109, 249 (1987).

[B88b] T. Bałaban. *Renormalization Group Approach to Lattice Gauge Field Theories. II. Cluster Expansions*, Commun. Math. Phys. 116, 1 (1988).

[BW74] T. L. Bell, K. Wilson. *Finite-lattice approximations to renormalization groups*, Phys. Rev. B, Vol. 11, 3431 (1974).

[B84] D. Brydges. *A short course on cluster expansions*, In Critical Phenomena, Random Systems, Gauge Theories 1984.

[BY90] D. Brydges, H. Yau. *Grad ϕ Perturbations of Massless Gaussian Fields*, Commun. Math. Phys. 129, 351 (1990).

[CE77] P. Collet, J. P. Eckmann. *The ϵ-expansion for the Hierarchical Model*, Commun. Math. Phys. 55, 67 (1977).

[CE78] P. Collet, J. P. Eckmann. *A Renormalization Group Analysis of the Hierarchical Model in Statistical Mechanics*, Lecture Notes in Physics, Vol. 74, Berlin, Heidelberg, New York, Springer 1978.

[DH91] J. Dimock, T. Hurd. *A Renormalization Group Analysis of the Kosterlitz-Thouless Phase*, Commun. Math. Phys. 137, 263-287 (1991).

[DH92] J. Dimock, T. Hurd. *A Renormalization Group Analysis of correlation functions for the dipole gas*, J. Stat. Phys. 66, 1277-1318 (1992).

[DH93] J. Dimock, T. Hurd. *A Renormalization Group Analysis of the Kosterlitz-Thouless Phase*, Commun. Math. Phys. 156, 547-580 (1993).

[F87] G. Felder. *Renormalization Group in the Local Approximation*, Commun. Math. Phys. 111, 101-121 (1987).

[FMRS87] J. Feldman, J. Magnen, V. Rivasseau, R. Sénéor. *Construction and Borel Summability of Infrared ϕ_4^4 by a Phase Space Expansion*, Commun. Math. Phys. 109, 137 (1987).

[GK80] K. Gawędzki, A. Kupiainen. *A Rigorous Block Spin Approach to Massless Lattice Theories*, Commun. Math. Phys. 77, 31 (1980).

[GK83] J. Gawędzki, A. Kupiainen. *Block Spin Renormalization Group for Dipole Gas and $(\nabla\Phi)^4$*, Ann. Phys. 147, 198 (1983).

[GK85] K. Gawędzki, A. Kupiainen. *Massless Lattice ϕ_4^4 Theory: Rigorous Control of a Renormalizable Asymptotically Free Model*, Commun. Math. Phys. 99, 197-252 (1985).

[GJ73] J. Glimm, A. Jaffe. *The Positivity of the φ_3^4 Hamiltonian*, Fortschr. Phys. 21, 327 (1973).

[GJ87] J. Glimm, A. Jaffe. *Quantum Physics*, Springer Verlag, Heidelberg 1987, Second Edition.

[GrK71] C. Gruber, H. Kunz. *General Properties of Polymer Systems*, Commun. Math. Phys. 22, 133 (1971).

[KW86] H. Koch, P. Wittwer. *A Non-Gaussian Renormalization Group Fixed Point for Hierarchical Scalar Lattice Field Theories*, Commun. Math. Phys. 106, 495-532 (1986)

[KW91] H. Koch, P. Wittwer. *On the Renormalization Group Transformation for Scalar Hierarchical Models*, Commun. Math. Phys. 138, 537 (1991).

[MP85] G. Mack, A. Pordt. *Convergent Perturbation Expansions for Euclidean Quantum Field Theory*, Commun. Math. Phys. 97, 267 (1985).

[MP89] G. Mack, A. Pordt. *Convergent Weak Coupling Expansions for Lattice Field Theories that Look Like Perturbation Series*, Rev. Math. Phys. 1, 47 (1989).

[MS77] J. Magnen, R. Sénéor. *Phase Space Cell Expansion and Borel Summability of the Euclidean Φ_3^4-Theory*, Commun. Math. Phys. 56, 237 (1977).

[PPW94] K. Pinn, A. Pordt, C. Wieczerkowski. *Algebraic Computation of Hierarchical Renormalization Group Fixed Points and their ϵ-Expansions*, Preprint Münster MS-TPI-94-02.

[PW94] A. Pordt, C. Wieczerkowski. *Nonassociative Algebras and Nonperturbative Field Theory for Hierarchical Models*, Preprint Münster MS-TPI-94-04.

[P86] A. Pordt. *A Tree Formula for Use in Cluster Expansions with General Multibody Interactions*, Z. Phys. C 31 (1986).

[P93] A. Pordt. *Renormalization Theory for Hierarchical Models*, Helv. Phys. Acta, Vol. 66, 105 (1993).

[R91] V. Rivasseau. *From Perturbative to Constructive Renormalization*, Princeton University Press 1991.

[Ru69] D. Ruelle. *Statistical Mechanics*, W. A. Benjamin Inc. , 1969.

[W71] K. G. Wilson. *Renormalization Group and Critical Phenomena. II. Phase Space Cell Analysis of Critical Behavior*, Phys. Rev. B4, 3184 (1971).

[W83] K. G. Wilson. *The Renormalization Group and Critical Phenomena*, Rev. Mod. Phys. 55, 583 (1983).

[WK74] K. G. Wilson, J. Kogut. *The Renormalization Group and the ϵ-expansion*, Phys. Rep. 12, 76 (1974).

Renormalizing Partial Differential Equations

Jean Bricmont[1*], *Antti Kupiainen*[2**]

1 UCL, Physique Théorique, B-1348, Louvain-la-Neuve, Belgium
e-mail: bricmont@fyma.ucl.ac.be
2 Helsinki University, Department of Mathematics,
Helsinki 00014, Finland
e-mail: ajkupiai@cc.helsinki.fi

Abstract

We explain how to apply Renormalization Group ideas to the analysis of the long-time asymptotics of solutions of partial differential equations. We illustrate the method on several examples of nonlinear parabolic equations. We discuss many applications, including the stability of profiles and fronts in the Ginzburg-Landau equation, anomalous scaling laws in reaction-diffusion equations, and the shape of a solution near a blow-up point.

1 Introduction

The development of a qualitative theory of infinite dimensional dynamical systems is a major scientific challenge. Such systems are expressed through (nonlinear) partial differential equations, and we shall concentrate on equations of the form

$$u_t = \Delta u + F(u, \nabla u, \nabla\nabla u). \tag{1}$$

where, $u(x,t) \in \mathbf{R}$ (or $\mathbf{C}$), $x \in \mathbf{R}^d$, $t \in \mathbf{R}^+$ and F is nonlinear or random. Usually, the following problems are considered:

1. Existence and regularity of the solution over a finite time interval.
2. Extension of the first problem to an infinite time interval.
3. If the solution exists for all times, what is its asymptotic behavior? How does the latter depend on the initial conditions? If the solution ceases to exist after a finite time, we may still ask about its behavior near the point where it breaks down.

The first question is often easy to deal with. In particular, perturbation methods (small initial data or small nonlinearity) tend to work well over finite (or short) time intervals. However, naive perturbation theory does not, in general, yield an answer to the last two questions. A similar situation was encountered

* Supported by EC grants SC1-CT91-0695 and CHRX-CT93-0411
** Supported by NSF grant DMS-9205296 and EC grant CHRX-CT93-0411

in the theory of critical phenomena and in quantum field theory: perturbation theory works well whenever there are enough cutoffs, but tends to diverge when the latter are removed. In order to solve these problems, the Renormalization Group (RG) method was developed and proved to be very useful. One of our goals is to develop RG ideas in the theory of PDE's. Here, the role of the cutoff is played by the finite time.

The RG method for nonlinear PDE's was developed by Goldenfeld, Oono and others [43]. Previously, Barenblatt [2] emphasized the role of self-similar solutions of these equations. The latter can be viewed as fixed points of certain RG transformations (see below). Independently, we used RG methods to deal with an equation like (1),with a linear but random F [9, 10, 11]. Here we shall explain the basic RG ideas in the framework of PDE's, which turn out to be very natural and easy, and we shall review some of the rigorous results obtained so far.

The general idea behind the RG approach is to solve the problem iteratively: first integrate the equations far away from the cutoff to be removed, i.e. here solve a finite time problem. Next, rescale the time and possibly the space and the u variables so as to produce a new problem similar to the original one. Finally, iterate and see what happens. The method will work when the new problem tends to be simpler than the original one (meaning, e.g., a weaker nonlinearity). Upon iteration, it will become simpler and simpler. Of course, this will happen only if we have chosen the scaling appropriately.

The underlying properties of the dynamical system that make the RG method work are universality and scaling. In the PDE framework, scaling amounts to the observation that many problems have solutions that behave asymptotically as

$$u(x,t) \sim t^{-\frac{\alpha}{2}} f^*(\frac{x}{t^\beta}) \tag{2}$$

Often such a limit law is *universal*: the numbers α, β and the function f^* will not depend on the initial conditions or even on the form of the equation. More precisely, pairs of initial data and equations will fall into universality classes, corresponding to given α, β and f^*. The fact that whole classes of problems may yield the same asymptotic behavior is called universality. In fact, f^* will be a fixed point of the RG transformation, while the exponents α and β are determined by the choice of the scaling. The basin of attraction of that fixed point is the universality class. The RG will lead to a dynamical system picture, with an analysis of the stable, neutral and unstable manifolds of the fixed points.

The outline of the paper is as follows: first, we give an elementary definition of the RG transformation, starting from the (linear) heat equation. Then, we show how the RG flow acts on spaces of initial data and equations for various nonlinear perturbations of the heat equation, and for data that decay at infinity. In Section 3, we consider similar questions, but for data that do not decay at infinity, which leads to the formation of patterns and fronts. Finally (Section 4), we apply the RG method to equations whose solutions blow-up in a finite time, and we give a list of possible profiles near a blow-up point. We end this review with a list of open problems.

2 The Renormalization Group

2.1 The Linear Heat Equation: Gaussian Fixed Points

To explain the RG idea, let us start with the simplest equation, the linear heat equation:

$$u_t = u_{xx} \tag{1}$$

in one dimension, with integrable initial data, $u(x,0) = f(x) \in L^1(\mathbf{R})$. The solution is

$$u(x,t) = \frac{1}{\sqrt{4\pi t}} \int dy \exp(-\frac{(x-y)^2}{4t}) f(y) dy \tag{2}$$

From this, it is easy to deduce that

$$\sqrt{t} u(\xi\sqrt{t}, t) \to \frac{e^{-\xi^2/4}}{\sqrt{4\pi}} \int f(y) dy \tag{3}$$

as $t \to \infty$ (pointwise and in $L^1(\mathbf{R})$). Let us reformulate this (trivial) result in the RG language. Notationally, it is useful to take the initial time equal to 1 instead of 0:

$$u(x,1) = f(x) \tag{4}$$

This will not change anything to the long-time asymptotics. Next, take a number $L > 1$ and define the renormalization map, acting on the initial data f as

$$(R_L f)(x) = L u(Lx, L^2) \tag{5}$$

where $u(\cdot, L^2)$ is the solution of (1) at time L^2. Thus, we solve the equation up to a finite time L^2, rescale the solution and take it as a new initial data. Here, R_L is explicit: in Fourier transform, $e^{t\Delta}$ multiplies by e^{-tk^2} and scaling x by L amounts to scaling k by $\frac{1}{L}$. Thus,

$$\widehat{R_L f}(k) = e^{-k^2(1-L^{-2})} \hat{f}(L^{-1}k) \tag{6}$$

(we solve (1) over a time interval of length $L^2 - 1$).

Observe that the set of multiples of

$$\widehat{f_0^*}(k) = e^{-k^2} \tag{7}$$

forms a line of fixed points of the RG map (6). Note also that R_L satisfies the semigroup property: Since L is arbitrary, we may replace it by L^n, and we have:

$$R_{L^n} = R_L^n. \tag{8}$$

This is simply because, if we define $u_L(x,t) = Lu(Lx, L^2t)$, then

$$(R_L f)(x) = u_L(x,1) \tag{9}$$

and u_L satisfies (1) again. Equation (1) is scale invariant.

Now, assume that our initial f belongs to the basin of attraction of the line of fixed points, i.e.

$$R_L^n f \to A f_0^* \tag{10}$$

as $n \to \infty$, for some A, in a suitable space. The basic observation is that this is equivalent to (3), by letting $t = L^{2n}$, $A = \hat{f}(0)$, and combining (5,8,10). Note that A depends on the initial data and that this dependence is the only trace left by $u(x,1)$ on the long-time asymptotics of the solution. This is why the latter is called universal.

To see what kind of functions belong to the basin of attraction of this line of fixed points, write formally $\hat{f}(k) = \hat{f}(0) + ck + \cdots$. Then, for k of order one, we have from the scaling

$$\widehat{R_L f}(k) = \hat{f}(0) f_0^*(k) + \mathcal{O}(L^{-1}) \tag{11}$$

and, for large k, we get contraction from the multiplication by $e^{-k^2(1-\frac{1}{L^2})}$ in (6).

To be more precise, we use below the following norm (inspired by [23]):

$$\|f\| = \sup_k (1 + |k|^q)(|\hat{f}(k)| + |\hat{f}'(k)|) \tag{12}$$

on a space of functions with $\hat{f} \in C^1(R)$. We denote by $\mathcal{B}$ the corresponding Banach space. The value of q will be specified when we deal with nonlinear problems. For the moment, any $q \geq 0$ will do. Any function $f \in \mathcal{B}$ can be written in Fourier transform as

$$\hat{f}(k) = \hat{f}(0) e^{-k^2} + \hat{g}(k) \tag{13}$$

with $\hat{g}(0) = 0$. Then the convergence (10) with $A = \hat{f}(0)$ will follow from

Lemma 1 *There exists a constant C independent of L so that, if $g \in \mathcal{B}$ and $\hat{g}(0) = 0$,*

$$\|R_L g\| \leq CL^{-1} \|g\| \tag{14}$$

Proof

Write $\hat{g}(\frac{k}{L}) = \int_0^{k/L} \hat{g}'(p) dp$ so, $|\hat{g}(\frac{k}{L})| \leq \frac{|k|}{L} \|g\|$. The scaling (6) brings a factor of L^{-1} in the derivative and $|k|^n e^{-k^2}$ is bounded by a constant for any $n \geq 0$. □

Remark 1. At this point, we may comment on the choice of the constant L. It is arbitrary except that we take it large enough so that $CL^{-1} < 1$ in (14). This will be assumed, wherever needed, about any other numerical constant entering the proofs. We shall denote generic constants by C, even when they change in the same equation. The only important property of such constants is that they are independent of L.
Remark 2. Using (13,14) we can prove (10), i.e. (3) for $f \in \mathcal{B}$.

2.2 Nonlinear Heat Equations

Let us now consider a less trivial example, namely

$$u_t = u_{xx} + \lambda u^p \tag{15}$$

where λ could be scaled away but is introduced for convenience. Throughout this Section, we work in one dimension (to simplify the notation), but the extension to higher dimensions is straightforward. There is no problem in proving the existence of solutions for any fixed time, provided λ is small enough, and we may define the renormalization group map as in (5). However, we do not have anymore the semigroup property (8), because the equation is not scale invariant. In fact, $u_L(x,t) = Lu(Lx, L^2t)$ satisfies

$$u_{Lt} = u_{Lxx} + \lambda L^{3-p} u_L^p \tag{16}$$

So, if we define R_L on the pair (f, λ) by

$$R_L(f,\lambda) = (u_L(x,1), \lambda L^{3-p}) \tag{17}$$

we recover the semigroup property (8), provided there is a set of pairs (f, λ) that is mapped into itself by R_L.

Now, consider $p > 3$. Then the line of fixed points $(Af_0^*, 0)$ of R_L is stable. So we may expect the long time behavior of the solutions of (15) to be again given by Af_0^* for a suitable A. This is indeed the case. However, the constant A will not depend only on f, but also on p and λ.

We have the following result which is a variant of Theorem 1 in [12]. Fix $p \in \mathbf{N}, p > 3$ in (15) and fix $q > 1$ in (12); let $\mathcal{B}_1$ be the unit ball in the corresponding Banach space $\mathcal{B}$.

Theorem 1 *There exist an $\epsilon > 0$ such that, if $|\lambda| \le \epsilon$ and $f \in \mathcal{B}_1$, equation (15) with $u(x,1) = f(x)$ has a unique solution which satisfies, for some number $A = A(f, \lambda, p)$,*

$$\lim_{t\to\infty} \|\sqrt{t}u(\sqrt{t}\cdot, t) - Af_0^*(\cdot)\| = 0.$$

Remark 1. For related results on the $p > 3$ case, see [23, 35, 42, 48].
Remark 2. The convergence in $\mathcal{B}$ implies convergence both in L^1 and in L^∞.

Let us explain the main ideas of the proof (for details, see [12]). The problem is to "construct" the constant A which is not given by an explicit formula as in (3). This construction is done inductively.

First of all, the existence of the solution for finite times is rather trivial (we take λ small and $\|f\| \leq 1$). Indeed, we can write

$$u(t) = e^{(t-1)\partial^2} f + \lambda \int_0^{t-1} ds e^{s\partial^2} u^p(t-s) \tag{18}$$

(leaving out the x dependence) and use a standard fixed point argument (see [12]; here we need only $q > 1$, because there are no derivatives in the nonlinear term). Since the first term in (18) is the linear evolution, we may write

$$Rf = R_0 f + v \tag{19}$$

where R_0 denotes the linear RG map (5) (we suppress here the L dependence of R) and $\|v\|$ is $\mathcal{O}(\lambda)$. Now, write

$$f = A_0 f_0^* + g_0 \tag{20}$$

with $\hat{g}_0(0) = 0$. This splitting of f is not preserved by R, as it was in the linear case. Thus, we shall have to change A at each step. Write

$$Rf = A_1 f_0^* + g_1 \tag{21}$$

where again $\hat{g_1}(0) = 0$. From (19) and (20), we see that

$$A_1 = A_0 + \hat{v}(0), \tag{22}$$

since $\widehat{R_0 g_0}(0) = 0$ (see (6)), and

$$g_1 = R_0 g_0 + v - \hat{v}(0) f_0^*. \tag{23}$$

Using Lemma 1 and $\|v\| = \mathcal{O}(\lambda)$, we have

$$A_1 = A_0 + \mathcal{O}(\lambda) \tag{24}$$

and

$$\begin{aligned} \|g_1\| &\leq CL^{-1}\|g_0\| + \mathcal{O}(\lambda) \\ &\leq L^{-(1-\delta)} \end{aligned} \tag{25}$$

for any $\delta > 0$, provided λ is small enough, since $\|g_0\| = \mathcal{O}(1)$ for $f \in \mathcal{B}_1$.

Now iterate: under scaling λ becomes $\lambda_1 = \lambda L^{3-p}$, and after n steps

$$\lambda_n = \lambda L^{(3-p)n} \tag{26}$$

Thus, $\|v_n\|$ will be $\mathcal{O}(\lambda_n)$ for the nth iteration, and, writing $R^n f = A_n f_0^* + g_n$, we have

$$A_{n+1} = A_n + \hat{v}_n(0) \tag{27}$$

$$g_{n+1} = R_L^0 g_n + v_n - \hat{v}_n(0) f_0^*, \tag{28}$$

with the bounds:

$$|A_{n+1} - A_n| \leq \mathcal{O}(\lambda_n) \tag{29}$$

$$\begin{aligned} \|g_{n+1}\| &\leq CL^{-1}\|g_n\| + \mathcal{O}(\lambda_n) \\ &\leq L^{-(1-\delta)n} \end{aligned} \tag{30}$$

(use (26) and $p - 3 \geq 1$ since p is an integer).

Therefore there exists a constant A such that $A_n \to A$ as $n \to \infty$ and

$$|A - A_n| \leq \mathcal{O}(\lambda_n). \tag{31}$$

Then,

$$\begin{aligned} R^n f &= A_n f_0^* + g_n \\ &= A f_0^* + (A_n - A) f_0^* + g_n \end{aligned} \tag{32}$$

and

$$\|R_L^n f - A f_0^*\| \leq L^{-(1-\delta)n} + \mathcal{O}(\lambda_n) \tag{33}$$

which goes to zero as $n \to \infty$. Using (5,8,17), this proves the claim, at least for a special sequence of times; the extension to all times is easy [12].

Remark 1. From (33,26), one gets an estimate $\mathcal{O}(t^{-(\frac{1-\delta}{2})})$ on the rate of convergence.

Remark 2. The sign of λ is not important here. It matters only for $p \leq 3$, see below.

Remark 3. R_0 is the linearization of the component of the operator R that acts on initial data, around the fixed point f_0^*. All functions of the form $k^n e^{-k^2}, n \in N$, are (Hermite) eigenvectors:

$$R_0 k^n e^{-k^2} = L^{-n} k^n e^{-k^2} \tag{34}$$

The fact that the scaling transforms the linear evolution semigroup, $e^{t\partial^2}$, which has continuous spectrum, into an operator with discrete spectrum, is one of the interesting features of this method. Using it, one can extend the above analysis and obtain higher order asymptotics of the solution in inverse powers of t, see subsection 2.5 below, and [56, 68].

2.3 The RG Flow in Spaces of Equations and Data

Of course, Theorem 1 can be extended to a much more general class of nonlinearities. Consider an equation of the form

$$u_t = u_{xx} + F(u, u_x, u_{xx}) \tag{35}$$

Then, the map R_L will be defined in general by

$$R_L = (L^\alpha u_L(\cdot, 1), F_L) \tag{36}$$

where

$$F_L(a, b, c) = L^{2+\alpha} F(L^{-\alpha} a, L^{-1-\alpha} b, L^{-2-\alpha} c) \tag{37}$$

If we consider a monomial $F(a, b, c) = a^{n_1} b^{n_2} c^{n_3}$, and the previous scaling $\alpha = 1$, we have $F_L = L^{-d_F} F$ with

$$d_F = n_1 + 2n_2 + 3n_3 - 3 \tag{38}$$

and using standard RG terminology, we call the monomial *relevant* if $d_F < 0$, *marginal* if $d_F = 0$, and *irrelevant* if $d_F > 0$.

Considering $n_i \geq 1$ or $n_i = 0$, the nonlinearities are easily classified: u^p is relevant for $p < 3$; u^3 and uu_x are marginal, and everything else is irrelevant. Then, Theorem 1 extends to the irrelevant case, provided we take $q > 3$ in (12) and $F : \mathbf{C}^3 \to \mathbf{C}$ analytic in a neighborhood of zero [12].

Remark. More general irrelevant nonlinearities (nonlocal, integro-differential terms) were considered by Taskinen [65]. Besides, as mentioned in [12], we may replace the second derivative in (15) by $-(-\Delta)^{\beta/2}$ and obtain similar results.

A nice extension to the study of waves satisfying a generalized KdV-Burgers equation was made in [5]. There, the highest order (third) derivative turns out to be irrelevant.

The two marginal terms exhibit different behaviors. For u^3, it is essential to replace the plus sign by a minus one (with $\lambda > 0$) in equation (15) (with a plus sign, and $p \leq 3$, any positive initial data leads to a solution that belows up in a finite time [34, 55]). The nonlinearity turns out to be irrelevant in the next order. The coefficient A_n satisfies a recursion of the form $A_{n+1} = A_n - \lambda\beta A_n^3$ + higher order terms, where $\beta > 0$ is an explicit constant. This implies that $A_n \sim n^{-\frac{1}{2}}$ (here we use λ positive) and since $t = L^{2n}$, this translates into the following logarithmic correction (see [12]):

$$u(\xi\sqrt{t}, t) \simeq (t \log t)^{-\frac{1}{2}} f_0^*(\xi) \tag{39}$$

On the other hand, for uu_x, we have Burgers' equation and, via the Cole-Hopf transformation, we get a new line of fixed points:

$$f^*(x) = \frac{d}{dx} \log(1 + Ae(x)) \tag{40}$$

where $e(x) = \int_{-\infty}^{x} e^{-\frac{y^2}{4}} dy$ is the error function and A is a parameter. A theorem analogous to Theorem 1 can be proven for these new fixed points [12]. Of course we can also add to these marginal terms a general irrelevant F and obtain a similar result.

2.4 Non-Gaussian Fixed Points

Let us now turn to the relevant case, $u^p, 1 < p < 3$. Actually, since we will not restrict ourselves to integer p or to everywhere positive u's, we shall consider the equation:

$$u_t = u_{xx} - u|u|^{p-1}, \tag{41}$$

but the minus sign will again be essential. Here, the parameter λ of (15) has been eliminated by a rescaling of u.

Since the nonlinear term is relevant, the Gaussian fixed point will be of no use. In order to find another fixed point, consider the transformation (36) with $\alpha = \frac{2}{p-1}$. The reason for this choice is that equation (41) is then scale invariant: $F_L = F$. It is easy to see that we shall have a fixed point of the transformation if we have a solution of the form:

$$u(x,t) = t^{-\frac{1}{p-1}} f(xt^{-\frac{1}{2}}) \tag{42}$$

where $f(\xi)$ is a function of one variable which solves:

$$f'' + \frac{1}{2}\xi f' + \frac{f}{p-1} - f^p = 0. \tag{43}$$

These solutions are called self-similar. The theory of positive solutions of (43) has been developed in [8, 35, 48]. The main result is that, for any $p > 1$, there exist smooth, everywhere positive solutions, f_γ, of (43) with $f_\gamma'(0) = 0$ and $f_\gamma(0) = \gamma$ for γ larger than a certain critical value γ_p (but not too large). The decay at infinity of these solutions is given by

$$f_\gamma(x) \sim |x|^{-\frac{2}{p-1}} \tag{44}$$

as $|x| \to \infty$ if $\gamma > \gamma_p$, while, for $\gamma = \gamma_p$, it decays at infinity as

$$f_{\gamma_p}(x) \sim |x|^{\frac{2}{p-1}-1} e^{-\frac{x^2}{4}}. \tag{45}$$

Remark. The existence of a critical γ_p can be understood intuitively by viewing (43) as Newton's equation for a particle of mass one, whose "position" as a function of "time" is $f(\xi)$. The potential is then $U(f) = \frac{f^2}{2(p-1)} - \frac{f^{p+1}}{p+1}$ and the "friction term" $\frac{\xi}{2} f'$ depends on the "time" ξ. Hence, if $f_\gamma'(0) = 0$ and $f_\gamma(0) = \gamma$ is large enough, the time it takes to approach zero is long and, by then, the friction term has become sufficiently strong to prevent "overshooting". However, as p increases, the potential becomes flatter and one therefore expects γ_p to decrease with p.

We shall now explain why these self-similar solutions are stable. Consider the initial data (taken as before at time 1 for convenience)

$$u(x,1) = f_\gamma(x) + h(x) \tag{46}$$

with $\gamma \geq \gamma_p$ and $h \in B$, where B is the Banach space of L^∞ functions equipped with the norm (with some abuse of notation!)

$$\|h\|_\infty = \operatorname*{ess\,sup}_\xi |h(\xi)(1 + |\xi|^q)|. \tag{47}$$

One can show the following [16]:

Theorem 2 *Let* $1 < p < 3$. *There exist* $\varepsilon > 0, C < \infty$ *and* $\mu > 0$ *such that, if the initial data* $u(x,1)$ *of* (41) *is given by* (46) *with* $h \in B$ *and satisfies*

$$\|h\|_\infty \leq \varepsilon, \text{ or } h \geq 0 (a.e.)$$

then, (41) *has a unique classical solution and, for all* t,

$$\|t^{\frac{1}{p-1}} u(\cdot t^{\frac{1}{2}}, t) - f_\gamma(\cdot)\|_\infty \leq C t^{-\mu} \|h\|_\infty$$

Remark. For related results on the stability of self-similar solutions, see [29, 30, 31, 35, 49].

Let us explain the main ideas of the proof. Given the initial data (46), it is convenient to rewrite (41) in terms of the variables $\xi = xt^{-\frac{1}{2}}$ and $\tau = \log t$; so, define $v(\xi, \tau)$ by:

$$u(x,t) = t^{-\frac{1}{p-1}} (f_\gamma(xt^{-\frac{1}{2}}) + v(xt^{-\frac{1}{2}}, \log t)) \tag{48}$$

where now

$$v(\xi, 0) = h(\xi). \tag{49}$$

Then, (41) is equivalent to the equation

$$v_\tau = \mathcal{L}v - \left(|f_\gamma + v|^{p-1}(f_\gamma + v) - f_\gamma^p - p f_\gamma^{p-1} v\right) \equiv \mathcal{L}v + N(v) \tag{50}$$

where we used the fact that f_γ solves (43) and we gathered the linear terms in

$$\mathcal{L} = \mathcal{L}_0 + V_\gamma,$$

with

$$\mathcal{L}_0 = \frac{d^2}{d\xi^2} + \frac{\xi}{2}\frac{d}{d\xi} + \frac{1}{p-1}, \tag{51}$$

and

$$V_\gamma(\xi) = -pf_\gamma^{p-1}(\xi). \tag{52}$$

To prove the theorem, it suffices to show that the corresponding solution of (50) goes to zero as $\tau \to \infty$. For that, the main estimate will be that the semigroup $e^{\tau\mathcal{L}}$ contracts, at least for τ large:

$$\|e^{\tau\mathcal{L}}\| \leq Ce^{-\mu\tau} \tag{53}$$

for some $\mu > 0$, $C < \infty$. Given (53), the control over the nonlinear term in (50) is standard.

There are two important ingredients in the proof of (53). The first is the fact that $e^{\tau\mathcal{L}}$ is a contraction in a suitable Hilbert space of rapidly decreasing functions. To see this, note first that $\mathcal{L}_0$ can be conjugated to the following Schrödinger operator:

$$e^{\frac{\xi^2}{8}}\mathcal{L}_0 e^{-\frac{\xi^2}{8}} = \frac{d^2}{d\xi^2} - \frac{\xi^2}{16} - \frac{1}{4} + \frac{1}{p-1}, \tag{54}$$

i.e. to minus the Hamiltonian of the harmonic oscillator. Thus, $\mathcal{L}_0$ is self-adjoint on its domain $\mathcal{D}(\mathcal{L}_0) \subset L^2(\mathbf{R}, d\mu)$, where

$$d\mu(\xi) = e^{\frac{\xi^2}{4}}d\xi.$$

$\mathcal{L}_0$ has a pure point spectrum $\{\frac{1}{p-1} - \frac{1}{2} - \frac{m}{2} \mid m = 0, 1, \ldots\}$ and the largest eigenvalue $\frac{1}{p-1} - \frac{1}{2}$ is *positive* if $1 < p < 3$. Thus $e^{\tau\mathcal{L}_0}$ is *not* contractive and, for $e^{\tau(\mathcal{L}_0+V_\gamma)}$ to contract, we need to use the potential in a non-trivial way (this is the reason why $1 < p < 3$ is harder than the $p > 3$ case).

Remarkably, it is possible to prove that $\mathcal{L} < -E < 0$ without a detailed study of the function f_γ, but only using equation (43). Indeed, one first shows that $\mathcal{L}$ is self-adjoint and that $-E_\gamma$, its largest eigenvalue, satisfies $-E_\gamma \leq -E_{\gamma_p}$. So, writing $E \equiv E_{\gamma_p}$, it is enough to show that $-E < 0$. Next, it is easy to see, using the Feynman-Kac formula [62] and the the Perron-Frobenius theorem [41] that $\mathcal{L}$ has a unique eigenvector Ω with eigenvalue $-E$ and Ω can be chosen to be strictly positive. Write (43) as

$$\mathcal{L}f = -(p-1)f^p.$$

So,

$$(\Omega, \mathcal{L}f) = -(p-1)(\Omega, f^p)$$

where $(\cdot,\cdot)$ denotes the scalar product in $L^2(\mathbf{R}, d\mu)$. By the self-adjointness of $\mathcal{L}$ and the definition of Ω, i.e. $\mathcal{L}\Omega = -E\Omega$, we have

$$-E = -(p-1)\frac{(\Omega, f^p)}{(\Omega, f)} < 0$$

since Ω and f are strictly positive. This proves our claim.

Notice that functions in $L^2(\mathbf{R}, d\mu)$ have essentially a Gaussian decay at infinity, which is much faster than what is allowed in our Banach space B, see (47). An extra work is needed, using the Mehler's formula for the kernel of $e^{\tau \mathcal{L}_0}$ (see [16] for details).

Finally, let us express this result in the RG language. The RG map has a fixed point $R_L(f_\gamma, F^*) = (f_\gamma, F^*)$ for $F^*(u) = -u|u|^{p-1}$. There are different fixed points for different values of γ, but since the decay of the functions in B is faster than the one of any of these fixed points, each of them is unique in the corresponding set $\{f_\gamma + h\}$ for h as in Theorem 2. In other words, unlike the situation for $p > 3$, we do not have to deal with a line of fixed points and here there is no constant like A in Theorem 1. This is reflected in the fact, (53), that the linear semigroup $e^{\tau \mathcal{L}}$ contracts.

Remark. The reader who is familiar with the theory of critical phenomena will recognize an analogy between the behavior of an Ising model at its critical point, when the dimension of the lattice changes and the asymptotic behavior of the solution of (41) when p varies. This is further discussed in [12]. Here, our results for $p < 3$ are non-perturbative, i.e. we do not use any "ε-expansion" (see [68] for an analysis of the bifurcation at $p = 3$).

Also, the reader may notice that our approach to the renormalization group is close to the "Wilsonian" one in the theory of critical phenomena, while the method developed in [43] is analogous to the renormalized perturbation of quantum field theory. For a rigorous treatment of Barenblatt's equation, which was one of the first example analyzed in [43], see [50].

2.5 Higher-order Asymptotics

It is interesting to reexamine the $p \geq 3$ case, using what we learned from the analysis of $p < 3$. This has been done by Wayne [68]. Namely, consider equation (15), and make the change of variable (48), but replacing f_γ by 0, i.e.

$$u(x,t) = t^{-\frac{1}{p-1}} v(xt^{-1/2}, \log t) \tag{55}$$

Then $v(\xi, \tau)$ satisfies

$$v_\tau = \mathcal{L}_0 v + \lambda v^p \tag{56}$$

with $\mathcal{L}_0$ as in (51); we have computed the spectrum of $\mathcal{L}_0$ which is $\{\frac{1}{p-1} - \frac{1}{2} - \frac{m}{2} | m = 0, 1, \cdots\}$. Now, notice that, for $p > 3$, this spectrum is entirely negative. Thus $e^{\tau \mathcal{L}_0}$ contracts exponentially and it is not hard to show that, for λ small, and $v(\xi, 0)$ bounded in a suitable norm, the same contraction holds for the solution of (56). Actually, since the largest eigenvalue of $\mathcal{L}_0$ corresponds to $m = 0$, we expect a decay like $e^{\tau(\frac{1}{p-1} - \frac{1}{2})}$, which, using $\tau = \log t$ and combining it with the $t^{-\frac{1}{p-1}}$ prefactor in (55), leads to a $t^{-1/2}$ decay of $u(x,t)$, for all $p \geq 3$. This of course has to be the case, in view of Theorem 1.

Actually, much more can be said. Take the n largest eigenvalues $\{\lambda_1, \cdots, \lambda_n\}$ of $\mathcal{L}_0$ and the corresponding subspaces. Then, using a theorem of Gallay [37],

Wayne constructs an invariant manifold (in the v variable) such that any solution (for $p > 3$ and small initial data) will converge to that invariant manifold at a rate at least of order $t^{-(\lambda_{n+1}-\delta)}$ (with, as before, δ arbitrarily small for sufficiently small initial conditions).

Since, for $n = 1$, the eigenvector of $\mathcal{L}_0$ is $e^{-\xi^2/4}$, one easily recovers Theorem 1. But all higher order asymptotics in time of $v(\xi,\tau)$, hence of $u(x,t)$, can also be obtained in terms of Hermite functions (such higher order corrections were also derived in [56]).

For $p < 3$, an invariant manifold can still be constructed but it becomes unstable ($\frac{1}{p-1} - \frac{1}{2} > 0$) and thus, it does not, in general, give information on the long-time asymptotics. Of course, we know from Theorem 2 what the situation is: instead of 0, we have to consider the solution f_γ and, presumably, a similar picture (with invariant manifolds and higher order asymptotics) holds there, since the spectrum of the operators

$\mathcal{L}$ for $p < 3$ is qualitatively similar to the one of $\mathcal{L}_0$ for $p > 3$.

2.6 An Application to Reaction–Diffusion Equations

In [4], Berlyand and Xin consider the following model which leads to a nice RG analysis and which exhibits the novel feature of anomalous exponents depending on the initial data:

$$u_t = \Delta u + vu^{p-1} \tag{57}$$

$$v_t = \Lambda^{-1}\Delta v - vu^{p-1} \tag{58}$$

where $u = u(x,t), v = v(x,t), x \in \mathbf{R}^n$ (we take $n = 1$ for simplicity) and $\Lambda > 0$ is the Lewis number. Equations (57,58) model a chemical reaction $A \to B$ where v is the mass fraction of reactant A and u the one of reactant B.

First, observe that for $p > 3$, the analysis done for (15) applies and the asymptotics of u, v is Gaussian. So let $p = 3$, so that the nonlinear terms are marginal, and let us see heuristically what happens. If u and v are positive (which is physically necessary and is preserved by (57,58) due to the maximum principle) then, again by the maximum principle, u is larger than the solution of the heat equation with the same initial data:

$$u(x,t) \geq \frac{A}{t^{1/2}} e^{-\frac{x^2}{4t}}. \tag{59}$$

where A depends on $u(x,0)$.

Then, v is less than the solution of (58) with $u^{p-1} = u^2$ replaced by its lower bound (59). Calling $\bar{v}$ this upper bound on v, $\bar{v}$ solves a linear equation

$$\bar{v}_t = \Lambda^{-1}\Delta\bar{v} - A^2 t^{-1} e^{-\frac{x^2}{2t}}\bar{v}. \tag{60}$$

It is not hard to see, by direct substitution, that (60) admits a self-similar solution of the form

$$\bar{v}(x,t) = t^{-\frac{\alpha}{2}} f_\alpha^*(\frac{x}{\sqrt{t}}). \tag{61}$$

Indeed, writing $f^*_\alpha(\xi) = e^{-\frac{A\xi^2}{8}}\Psi(\xi)$, Ψ solves

$$-\Lambda^{-1}\Psi'' + (\frac{\Lambda}{16}\xi^2 + \frac{A^2}{4\pi}e^{-\xi^2/2} + \frac{1}{4})\Psi = \frac{\alpha}{2}\Psi \tag{62}$$

We see that, requiring f^*_α to be positive means that Ψ is the ground state of the operator in the LHS of (62) i.e. of an harmonic oscillator perturbed by a positive, rapidly decaying, potential. The exponent $\frac{\alpha}{2}$ is the corresponding ground state energy. For the harmonic oscillator, $\frac{\alpha}{2} = \frac{1}{2}$ and, perturbation theory tells us that, for A small,

$$\frac{\alpha}{2} = \frac{1}{2} + \frac{A^2}{4\pi\sqrt{2\Lambda^{-1}+1}} + h.o.t. \text{ in } A > \frac{1}{2}. \tag{63}$$

Thus, α depends on A, i.e. on $u(x,0)$. Since $\alpha > 1$, $\bar{v}$, and hence v, decay strictly faster than the solution of the heat equation. Inserting this in (57) gives an upper solution for u, which solves an equation where the nonlinear term is now *irrelevant.*

These considerations lead us to expect u to have the heat equation decay:

$$u(x,t) \simeq \frac{A}{\sqrt{t}}e^{-x^2/4t} \tag{64}$$

and v to have the anomalous decay:

$$v(x,t) = \frac{B}{t^{\alpha/2}}f^*_\alpha(\frac{x}{\sqrt{t}}) \tag{65}$$

where A, B and α depend on the initial data. Upper and lower bounds of the form (64,65) are proven in [4]. From the RG point of view, one can define a map

$$R_{L,\alpha}(u,v) = (u_L(\cdot,1), v_L(\cdot,1))$$

where

$$u_L(x,t) = Lu(Lx, L^2t), \tag{66}$$
$$v_L(x,t) = L^\alpha v(Lx, L^2t). \tag{67}$$

We have thus a two-parameter family of fixed points $(Af^*_0, Bf^*_{\alpha(A)})$, and the scaling exponent varies continuously with A.

3 Patterns and Fronts

3.1 The Ginzburg–Landau Equation

In the previous section, we have seen that nonlinear equations with initial data decaying at infinity produce universal, Gaussian or non-Gaussian, diffusive profiles. Here we shall show that other types of asymptotic behavior can also exhibit such universality. To discuss a concrete example, consider the Ginzburg-Landau equation

$$u_t = u_{xx} + u - |u|^2 u \tag{1}$$

where $u : \mathbf{R} \times \mathbf{R} \to \mathbf{C}$ is complex. This equation has a two parameter family of stationary solutions

$$u_{q\theta}(x) = \sqrt{1-q^2}\, e^{i(qx+\theta)} \tag{2}$$

and a natural question is to inquire about the time development of initial data $u(x)$ which approach two such solutions at $\pm\infty$:

$$\lim_{x\to\pm\infty} |u(x) - u_{q_\pm \theta_\pm}(x)| = 0. \tag{3}$$

This problem has been extensively studied for equation (1) with u *real* and $u(-\infty) = 1$, $u(\infty) = 0$ (i.e. $q_- = 0$, $q_+ = 0$) [1, 6]. Then the solution takes the form of a propagating front. This occurs because $u(x) = 1$ is a stable stationary solution, while $u(x) = 0$ is an unstable one. For complex u, the solutions (2) are stable, under small perturbations, for $q^2 < \frac{1}{3}$ (the Eckhaus stable domain) [23].

We shall consider two questions. The first one was suggested in [24], namely we take $q_\pm$ in (3) small, belonging to the Eckhaus stable domain, but not necessarily equal. What is then the long-time asymptotics of the solution? The second question concerns the stability of the real front solutions of (1) for real or complex u. For reviews on these questions, we refer the reader to [22, 25].

3.2 Non-Gaussian Patterns

For the first question, we considered in [13] a class of initial data satisfying (3) and we showed that, for any interval I,

$$\sup_{x\in I} |u(x,t) - e^{i\sqrt{t}\phi^*} u_{q^*\theta^*}(x)| \le \frac{C_I}{\sqrt{t}} \tag{4}$$

where the constants q^*, ϕ^* and θ^* depend only on the boundary conditions (3). For a more detailed bound, see Section 3 of [13]. Graphics of the solution can be found in [25].

In order to see where (4) comes from, we write, following [24],

$$u = (1-s)e^{i\phi}. \tag{5}$$

Equation (1) becomes, in these variables,

$$\begin{aligned} s_t &= s_{xx} - 2s + 3s^2 - s^3 + \phi_x^2 - s\phi_x^2 \equiv s_{xx} - 2s + F(s,\phi_x) \\ \phi_t &= \phi_{xx} - \frac{2\phi_x s_x}{1-s} \equiv \phi_{xx} + G(s, s_x, \phi_x) \end{aligned} \tag{6}$$

with the initial data (taking again $t = 1$ as initial time)

$$\lim_{x\to\pm\infty} s(x,1) = s_\pm\,, \quad \lim_{x\to\pm\infty} |\phi(x,1) - \phi_\pm x - \theta_\pm| = 0. \tag{7}$$

where $2s_\pm = F(s_\pm, \phi_\pm)$. We proved in [13] the following asymptotics, as $t \to \infty$:

$$\phi(x,t) = \sqrt{t}\phi^*(\frac{x}{\sqrt{t}}) + \theta^*(\frac{x}{\sqrt{t}}) + \mathcal{O}(\frac{1}{\sqrt{t}}) \tag{8}$$

$$s(x,t) = s^*(\frac{x}{\sqrt{t}}) + \frac{1}{\sqrt{t}}r^*(\frac{x}{\sqrt{t}}) + \mathcal{O}(\frac{1}{t}) \tag{9}$$

for a set of initial data satisfying (7), with $\phi_\pm$, and $\theta_\pm$ small enough (see again [13] for the precise definition of the norms in which these limits take place). The functions ϕ^*, θ^*, s^*, r^* are *universal*, depending on the initial data only through the boundary conditions (7). They are smooth and therefore (expanding ϕ^* to first order, and using (2)) the phase of u will have the asymptotics (4), with

$$\phi^* = \phi^*(0), \quad q^* = \phi^{*'}(0), \quad \theta^* = \theta^*(0).$$

The behavior of s will be discussed below.

The peculiar scaling behavior exhibited by (8) can already be understood through the linear heat equation. Previously, we considered the Gaussian fixed point of the RG in a space of integrable initial data. Since the heat equation has stationary solutions of the form $a + bx$, we may also consider the solution of $\phi_t = \phi_{xx}$ with initial data $\phi(x,1) = f(x)$ satisfying (7). The solution can be computed:

$$t^{-\frac{1}{2}}\phi(\sqrt{t}x, t+1) = \frac{1}{\sqrt{4\pi t}}\int e^{-\frac{1}{4}(y-x)^2} f(\sqrt{t}y)dy,$$

$$\to_{t\to\infty} \phi_- x + (\phi_+ - \phi_-)(x + 2\frac{d}{dx})e(x) \equiv \phi_0^*(x) \tag{10}$$

where $e(x) = \int_{-\infty}^{x} e^{-\frac{y^2}{4}}dy$ is the error function. This gives the first term in (8), with ϕ^* replaced by ϕ_0^*. If we want to obtain the next correction, we write $\phi^* = \phi_0^* + \theta$, where $\theta(x,t)$ solves $\theta_t = \theta_{xx}$ with $\theta(x,1) \to \theta_\pm$ as $x \to \pm\infty$. One computes that

$$\theta(x,t) \to \theta_0^*\left(\frac{x}{\sqrt{t}}\right) = \theta_- + (\theta_+ - \theta_-)e\left(\frac{x}{\sqrt{t}}\right) \tag{11}$$

and this yields the second term in (8). It is trivial to check that ϕ_0^* (resp. θ_0^*) is a fixed point of the RG map (2.36) with $\alpha = -1$ (resp. $\alpha = 0$), and $F = 0$.

We shall explain below that ϕ^*, θ^* are small perturbations of ϕ_0^*, θ_0^*. However, as we saw, the scaling (8,4) can be understood qualitatively on the basis of the heat equation, and the boundary conditions (7). We still need to understand (9).

The latter comes from the "slaving" of s to ϕ, due to the linear $-2s$ term in (6). This will imply that s will have the right form so that (4) holds. Concretely, the slaving means that, to leading order, we may solve the algebraic equation $-2s + F(s, \phi_x) = 0$, which is equivalent to $1 - s = \sqrt{1 - (\phi_x)^2}$, and substitute the result in the equation (6) for ϕ. Since the derivative of s is proportional to the second derivative of ϕ, we obtain an equation for ϕ only, of the form:

$$\phi_t = (1 + a(\phi_x, \phi_{xx}))\phi_{xx}\ , \quad \phi(x,1) = f(x) \tag{12}$$

with the boundary conditions (7) for the initial data f. The function a is analytic around the origin.

The RG map acts again on pairs of initial data and equations and can be defined (with $\alpha = -1$) as

$$R_L(\phi, a) = (\phi_L, a_L) \tag{13}$$

where

$$\phi_L(x,t) = L^{-1}\phi(Lx, L^2t)$$

and

$$a_L(u,v) = a(u, L^{-1}v). \tag{14}$$

The semigroup property follows by observing that ϕ_L satisfies

$$\phi_{Lt} = (1 + a_L(\phi_{Lx}, \phi_{Lxx}))\phi_{Lxx}. \tag{15}$$

We may now iterate R as before, i.e. solve a finite time problem, to study the asymptotics of (12). One wants to show that there exist functions ϕ^*, a^* such that

$$R_L^n(\phi, a) \to (\phi^*, a^*) \tag{16}$$

where

$$R_L(\phi^*, a^*) = (\phi^*, a^*) \tag{17}$$

is the fixed point of the RG, corresponding to the scale-invariant equation $\phi_t = (1 + a^*)\phi_{xx}$. Then, replacing $\frac{x}{\sqrt{t}}$ by ξ, the asymptotics of the original problem is given by

$$\phi(\xi t^{1/2}, t) \sim t^{\frac{1}{2}}\phi^*(\xi). \tag{18}$$

Because of the factor L^{-1} in the second argument of a in (14), the fixed point equation is

$$\phi_t = (1 + a^*(\phi_x))\phi_{xx} \tag{19}$$

where $a^*(\cdot) = a(\cdot, 0)$. The fixed point is the scale invariant solution $\phi(x,t) = \sqrt{t}\phi^*(\frac{x}{\sqrt{t}})$. We get for ϕ^* the equation

$$(1 + a^*)\frac{d^2}{d\xi^2}\phi^* + \frac{1}{2}\xi\frac{d}{d\xi}\phi^* - \frac{1}{2}\phi^* = 0 \tag{20}$$

with $a^* = a^*(\frac{d}{d\xi}\phi^*)$ and, for small $\phi_\pm$, we look for a solution

$$\phi^* = \phi_0^* + \rho \tag{21}$$

where $\rho(\pm\infty) = 0$ and ϕ_0^* is the "Gaussian" solution (10), which solves (20) with $a^* = 0$. This is easy to solve by a fixed point argument(see Proposition 1 in [13] or the Proposition in Section 4 of [12]).

This gives us the first term in (8). Turning to θ^*, we write

$$\phi(x,t) = \phi^*(x,t) + \theta(x,t) \tag{22}$$

where, with an abuse of notation, $\phi^*(x,t) = \sqrt{t}\phi^*(\frac{x}{\sqrt{t}})$ and ϕ^* is given by (21), while ϕ solves (12). Then,

$$\theta_t = \theta_{xx} + (a\phi_{xx} - a^*\phi^*_{xx}) \tag{23}$$

with $\theta(\pm\infty) = \theta_\pm$. Now we set

$$\theta_L(x',t') = \theta(Lx', L^2t'). \tag{24}$$

Putting $x = Lx', t = L^2t'$, we have $\frac{d^l}{dx^l}\phi^*(Lx', L^2t') = L^{1-l}\frac{d^l}{dx'^l}\phi^*(x',t')$ and therefore θ_L, satisfies the equation (replacing (x',t') by (x,t)):

$$\begin{aligned}\theta_{Lt} = \theta_{Lxx} + L(a(\phi^*_x + L^{-1}\theta_{Lx}, L^{-1}\phi^*_{xx} + L^{-2}\theta_{Lxx})(\phi^*_{xx} + L^{-1}\theta_{Lxx})\\ -a^*(\phi^*_x)\phi^*_{xx}).\end{aligned} \tag{25}$$

Thus, reasoning as above, we expect

$$\theta_{L^n} \to \theta^* \tag{26}$$

where $\theta^*(x,t) = \theta^*(\frac{x}{\sqrt{t}})$ satisfies the $L \to \infty$ form of (25):

$$\theta^*_t = \theta^*_{xx} + a^*\theta^*_{xx} + (a_u(\phi^*_x, 0)\theta^*_x + a_v(\phi^*_x, 0)\phi^*_{xx})\phi^*_{xx}. \tag{27}$$

This is a linear equation, easy to solve, whose solution is, for $\theta_\pm$ small, a small perturbation of the "Gaussian" solution (11) (which solves (27) with $a = 0$). Finally, one sets (with the same abuse of notation)

$$\phi(x,t) = \phi^*(x,t) + \theta^*(x,t) + \psi(x,t) = \sqrt{t}\phi^*(\frac{x}{\sqrt{t}}) + \theta^*(\frac{x}{\sqrt{t}}) + \psi(x,t) \tag{28}$$

As for the s variable, one gets

$$s(x,t) = s^*(x,t) + r^*(x,t) + v(x,t) \tag{29}$$

where

$$s^*(x,t) = s^*(\frac{x}{\sqrt{t}}), \quad r^*(x,t) = \frac{1}{\sqrt{t}}r^*(\frac{x}{\sqrt{t}}) \tag{30}$$

are fixed points "slaved" respectively to ϕ^* and θ^*. They satisfy the boundary conditions

$$\lim_{x\to\pm\infty} |s^*(x) - s_\pm| = 0\,, \quad \lim_{x\to\pm\infty} r^*(x) = 0\,. \tag{31}$$

The equations for ψ and v are rather complicated, but are essentially of the form heat equation plus irrelevant terms, in the sense of the previous section. Since now ψ and v decay to zero at infinity, we are in the situation of that section and, taking small initial data (in a suitable norm), one shows that the corresponding solution diffuses to zero. This, then, allows us to prove equation (4).

3.3 Stability of Fronts in the Ginzburg–Landau Equation

Let us write the Ginzburg-Landau equation (1) in radial and angle variables, $u = re^{i\phi}$:

$$r_t = r_{xx} + r(1 - \phi_x^2) - r^3 \tag{32}$$

$$\phi_t = \phi_{xx} + 2r^{-1}\phi_x r_x \tag{33}$$

and let us discuss the stability of the front solutions of these equations. It is well known that these equations have real, positive, front solutions, i.e. solutions of the form

$$\phi = 0, r = r_c(x - ct) \geq 0, \tag{34}$$

such that r_c interpolates between a stable and an unstable solution of (32), i.e. $r_c \to +1$ for $x \to -\infty, r_c \to 0$, for $x \to +\infty$. Indeed, from (32,34), we see that r_c satisfies

$$r_c'' + cr_c' + r_c - r_c^3 = 0 \tag{35}$$

which, if we reinterpret the variable as "time", can be seen as Newton's equation of motion of a particle of mass one subjected to a friction term cr_c' and to a force deriving from the potential $\frac{r^2}{2} - \frac{r^4}{4}$, which is an inverted double-well. It is intuitively clear and easily proved that, for c not too small, solutions exist that satisfy the required conditions, i.e. such that r_c tends, as "time" goes to $+\infty$, to zero, the stable critical point of the potential, and to one as "time" goes to $-\infty$. For large "time" $u = x - ct, r_c(u)$ will decay exponentially, as is seen from the linearization of (35) at $r = 0$. One gets

$$r_c(u) \leq (C_1 + C_2 u)e^{-\gamma u} \tag{36}$$

where γ is given by $\gamma^2 - c\gamma + 1 = 0$ i.e.

$$\gamma_c = \frac{1}{2}(c - \sqrt{c^2 - 4}) \tag{37}$$

which is real for $c \geq 2$, in which case $\gamma_c \leq 1$ (actually, one can take $C_2 = 0$ in (36), if $\gamma < 1$). Thus, the larger the friction, the slower the decay. For $c < 2$, the solution "overshoots" the minimum at zero, i.e. r_c is no longer positive. Each of the solutions r_c with $c \geq 2$ is stable under real perturbations $(\phi = 0)$: if we start with initial data $r(x, 0)$, with $r = r_c + s$ with $0 \leq r \leq 1$, s decaying faster than $e^{-\gamma_c x}$ for $x \to +\infty$, $r(x, t)$ will converge, as $t \to +\infty$, to $r_c(x - ct)$, see [1, 6, 22].

However the solution with $c = 2, \gamma_c = 1$ is more stable than the others in the sense that any initial data $r(x, 0)$ with $0 \leq r \leq 1$ which decays faster than e^{-x} as $x \to +\infty$ (in particular, if r is of compact support) will converge, as $t \to +\infty$, to $r_2(x - 2t)$ [1, 6]. From now on, we shall concentrate on the most interesting front, namely the one with $c = 2, \gamma = 1$ and we write r for r_2 .

We consider a complex perturbation of r : $r(x,0) = r + s$ with $\phi(x,0) \neq 0$ and $\phi(x,0), s(x,0)$ are small in a suitable sense. The equations satisfied by ϕ and s are :

$$\phi_t = \phi_{xx} + 2(r_x\phi_x + s_x\phi_x)(r+s)^{-1} \tag{38}$$

$$s_t = s_{xx} + s(1-3r^2) - r\phi_x^2 - s\phi_x^2 - 3rs^2 - s^3 \tag{39}$$

Since r is a function of $x - 2t$ it is convenient to consider also the equation in the frame of reference of the front : let $u = x - 2t$ and $\phi_f(u,t) = \phi(u + 2t, t)$,$s_f(u,t) = s(u+2t,t)$; then ϕ_f and s_f satisfy equations like (38,39), with $2\phi_{fu}$ added to the RHS of (38) and $2s_{fu}$ added to the one of (39). Now $r = r(u)$ is time-independent.

To understand the expected behavior of $\phi_f(u,t)$, let us consider the linearized equation around the zero solution :

$$\phi_{ft} = \phi_{fuu} + 2r_u\phi_{fu}r^{-1} + 2\phi_{fu}. \tag{40}$$

It is convenient to rewrite this equation as a heat equation with a potential: Let

$$\phi_f(u,t) = e^{-u}r(u)^{-1}\psi(u,t). \tag{41}$$

Then, ψ satisfies

$$\psi_t = \psi_{uu} - V\psi \tag{42}$$

with

$$V = 1 + \frac{r''}{r} + 2\frac{r'}{r} = r^2. \tag{43}$$

To derive the last equality, we used eq.(35), satisfied by r. Since $r \simeq 1$ for $u \to -\infty$, $r \simeq e^{-u}$ for $u \to +\infty$, we have $V \simeq 1$ for $u \to -\infty, V \simeq 0$ for $u \to +\infty$.

So, starting with $\psi(u,0)$ localized around $u = 0$, we expect that

$$\psi(u,t) \sim \begin{cases} \frac{e^{-t-\frac{u^2}{4t}}}{\sqrt{t}} & u \to -\infty \\ \frac{e^{-\frac{u^2}{4t}}}{\sqrt{t}} & u \to +\infty. \end{cases} \tag{44}$$

Hence, using (41) and the asymptotic behavior of $r(u)$,

$$\phi_f(u,t) \sim \begin{cases} \frac{e^{-t-\frac{u^2}{4t}-u}}{\sqrt{t}} & u \to -\infty \\ \frac{e^{-\frac{u^2}{4t}}}{\sqrt{t}} & u \to +\infty \end{cases} \tag{45}$$

which can be written as

$$\phi_f(u,t) \sim \begin{cases} \frac{e^{-\frac{(u+2t)^2}{4t}}}{\sqrt{t}} & u \to -\infty \\ \frac{e^{-\frac{u^2}{4t}}}{\sqrt{t}} & u \to +\infty. \end{cases} \tag{46}$$

Since $u + 2t = x$, the first part of (46) is a diffusive wave stationary in the fixed frame. The second part represents a diffusive wave which is "carried along" by the front. This is a rough, but basically correct picture.

Let us consider the linear equation for s_f, in the front frame :

$$s_{ft} = s_{fuu} + s_{fu} + s_f(1 - 3r^2) \tag{47}$$

Writing $s_f = e^{-u}\sigma$, we get

$$\sigma_t = \sigma_{uu} - \tilde{V}\sigma \tag{48}$$

with $\tilde{V} = 3r^2$. Following the analysis leading to (44), we get

$$s(u,t) \sim \begin{cases} \frac{e^{-3t-\frac{u^2}{4t}-u}}{\sqrt{t}} = \frac{e^{-2t}}{\sqrt{t}} e^{-\frac{(u+2t)^2}{4t}} & \text{as} \quad u \to -\infty \\ \frac{e^{-\frac{u^2}{4t}}}{\sqrt{t}} e^{-u} & \text{as} \quad u \to +\infty \end{cases} \tag{49}$$

There is a "wave" which is stationary in the front frame, but exponentially decreasing in u , while the wave which stays in the fixed frame is suppressed by the factor e^{-2t}.

The rigorous results that justify this picture are of two types: Using the RG method, Gallay [35] was able to obtain very precise asymptotics on how a small perturbation of *real fronts* diffuses to zero (this improves previous results of [53, 61]). So Gallay considers equation (39) with $\phi = 0$, in the front frame, whose linearization is given by (47). He writes

$$s(u,t) = r'(u)w(u,t) \tag{50}$$

and studies the behavior of $w(u,t)$ as $t \to \infty$. Since $r(u)$ goes exponentially to 1 or 0 as u goes to $-\infty$ or $+\infty$, $r'(u)$ will be localized around $u = 0$. The main result is that w has the following universal asymptotics [36]: let $u = \xi\sqrt{t}$, then

$$w(\xi\sqrt{t}, t) \simeq At^{-3/2} f^*(\xi) \tag{51}$$

where

$$f^*(\xi) = \begin{cases} 1 & \text{if } \xi \le 0 \\ e^{-\xi^2/4} & \text{if } \xi > 0 \end{cases}$$

and A depends again on the initial conditions. The limit (51) holds in a weighted $L^1 \cap L^\infty$ norm. Going back to (47-49), the power $t^{-3/2}$ is easy to grasp intuitively: the potential $\tilde{V}$ in (48) plays the role of a barrier around $u = 0$ ($\tilde{V}$ goes exponentially to 3 for $u \to -\infty$ and exponentially to 0 far $u \to +\infty$). The simplest approximation is to replace the RHS of (48) by a Laplacian on $\mathbf{R}^+$ with Dirichlet boundary conditions at $u = 0$. And that operator leads to a $t^{-3/2}$ decay of the solution. This effect was not taken into account in (49).

The second type of results deals with the complex perturbations. In [14], we consider initial data in the Banach space of C^1-functions ϕ, s with the norm

$$\|(\phi, s)\| = \sup_x (1+ | x |)^{3+\delta}(| \phi(x) | + | \phi'(x) | +(1 + e^x)(| s(x) | + | s'(x) |)) \tag{52}$$

and prove

Theorem 3. *For any $\delta > 0$ there exists an $\epsilon > 0$ such that equations* (38, 39) *with initial data $\phi(x,1) = \phi(x), s(x,1) = s(x)$, and $\|(\phi, s)\| < \epsilon$, have a unique classical solution $\phi(x,t), s(x,t)$, for all $t \geq 1$, such that*

$$|\phi(x,t)| \leq t^{-\frac{1}{2}+\delta} \tag{53}$$

$$(1+e^u)|s(x,t)| \leq t^{-1+\delta} \tag{54}$$

Remark 1. The power laws of the decay in time are presumably optimal (except for the δ) and are different from those of Gallay, because the diffusive wave that is stationary in the fixed frame, for the ϕ variable (see (45)), goes only diffusively to zero. This in turn slows downs the decay of the s variable, due to the nonlinear term $s\phi_x^2$ in (39).

Remark 2. The nonlinear terms in (38,39) turn out to be irrelevant, in the RG sense. However, this is not a simple affair: to show it, one has to take into account the precise decay of ϕ and s both in the fixed and the front frames. This makes the proofs rather complicated.

Remark 3. Finally, let us mention that Eckmann and Wayne [28], using a completely different (and simpler) method, namely coercive functionals, have proven similar results: they can consider a larger space of perturbations (s, ϕ) than the one defined by (52), but they do not obtain explicit upper bounds on the decay in time.

4 Universality in Blow-Up for Nonlinear Heat Equations

4.1 Statement of the Problem

Let us now consider equations for which global existence results do not hold: the solutions of the initial value problem

$$u_t = u_{xx} + u^p \tag{1}$$

where $p > 1, u = u(x,t), x \in \mathbf{R}$, and $u(\cdot, 0) = u_0 \in C^0(\mathbf{R})$, will, for a large class of initial data u_0, diverge in a finite time at a single point (for reviews on this problem, see [47, 55]). Again, we limit ourselves to one space dimension, but the generalization is straightforward.

The RG ideas can be applied to the analysis of the profile of the solution at the time of blow-up. To explain what this means, let us fix the blow-up point to be 0 and the blow-up time to be T. Then, we ask whether it is possible to find a function $f^*(x)$ and a rescaling $g(t,T)$ so that

$$\lim_{t \uparrow T} (T-t)^{\frac{1}{p-1}} u(g(t,T)x, t) = f^*(x) \tag{2}$$

Moreover, we want to see how g or f^* depend on the initial data.

The prefactor $(T-t)^{\frac{1}{p-1}}$ in (2) can be understood easily: for initial data $u_0(x)$ constant in x, $u(t)$ solves the ODE $u_t = u^p$, i.e. $u(t) = ((p-1)(T-t))^{\frac{1}{1-p}}$ for $T = (p-1)^{-1}u_0^{1-p}$. In [44, 45, 46, 66] (see also [32, 33]) several possible f^*'s are discussed, and the set of initial data that will lead to a given f^* is partially characterized. In [15], we showed that there exists, in the space of initial data $C^0(\mathbf{R})$, sets $\mathcal{M}_k$ of codimension $2k$, such that, for $u_0 \in \mathcal{M}_k$, the limiting behavior (2) is obtained, in the case $k = 1$, for

$$g(t,T) = ((T-t)|\log(T-t)|)^{\frac{1}{2}} \tag{3}$$

$$f^*(x) = \left(p-1+b^*x^2\right)^{\frac{1}{1-p}} \tag{4}$$

where $b^* = \frac{(p-1)^2}{4p}$, and in the case $k > 1$ for

$$g(t,T) = (T-t)^{\frac{1}{2k}} \tag{5}$$

$$f_b^*(x) = \left(p-1+bx^{2k}\right)^{\frac{1}{1-p}}. \tag{6}$$

where now b is an arbitrary positive number. As shown in [15], one can also add suitable (irrelevant) terms to (1) without affecting the result. It was shown in [44, 46, 66, 33] that, under quite general hypotheses, (3,4) or (5,6) are the only possibilities (see also [67] for a formal analysis). Moreover, solutions that behave as in (5,6) for $k = 2$ were constructed in [44].

The codimension of $\mathcal{M}_k$ for $k = 1$ is easy to understand: since we have fixed the blow-up point (to zero) and the blow-up time (to T), we have to fix two parameters in the initial data. To reach the other profiles, $2k - 2$ additional parameters need to be fixed in the initial data. The $k = 1$ situation is therefore the most generic one.

In the RG language, f^* and f_b^* can be viewed as fixed points of a renormalization group transformation having $2k$ unstable ("relevant", in renormalization group terminology) directions. Thus, to converge towards the fixed point, one has to fine-tune $2k$ parameters (one for each unstable direction) and this explains why $\mathcal{M}_k$ is of codimension $2k$, and in what sense f^*, f_b^* are "universal". In addition, we encounter also one neutral ("marginal") mode, which, for $k = 1$, turns out to be stable when nonlinear effects are taken into account and for $k > 1$ parametrizes a curve of fixed points.

Our results are perturbative, i.e. the sets $\mathcal{M}_k$ consist of initial data that are close to the corresponding fixed point. Therefore, our results are similar to those of Bressan [6] who considers a nonlinearity e^u instead of u^p and obtains the universal profile analogous to our $k = 1$ case. The connection of blow-up and center manifold theory was used earlier in the work of Filippas, Kohn and Liu [32, 33] and of Herrero, Velazquez [44, 45, 46, 47, 66] and Galaktionov [67]. Futhermore, the scaling and the dynamical systems aspect of our work goes back to Giga and Kohn [38, 39, 40]. Rescalings as in (7) below were used as a technique for numerical computation in [3].

Let us first describe a change of variables that transforms the problem (1) into a problem of long time asymptotics: we write (1) in the "blow–up–variables":

given a $u : \mathbf{R} \times [0,T) \to \mathbf{R}$, define $\phi : \mathbf{R} \times [-\log T, \infty) \to \mathbf{R}$ by

$$u(x,t) = (T-t)^{-\frac{1}{p-1}} \phi\left(\frac{x}{(T-t)^{1/2k}}, -\log(T-t)\right). \tag{7}$$

Then u is a classical solution of (1) if and only if $\phi(\xi,\tau)$ is a classical solution of

$$\phi_\tau = L_\tau^{-2}\phi_{\xi\xi} - \frac{1}{2k}\xi\phi_\xi - \frac{1}{p-1}\phi + \phi^p \tag{8}$$

$$\phi(\xi,\tau_0) = T^{\frac{1}{p-1}} u_0(T^{\frac{1}{2k}}\xi) \tag{9}$$

where $\tau_0 = -\log T$, and

$$L_\tau = e^{\frac{1}{2}\tau(1-1/k)}. \tag{10}$$

We construct in [15] global solutions of (8), with suitable initial data, thereby establishing blow–up for (1). Note that, for $k = 1$, the scaling in (7) differs from the one used in (3) by a factor $|\log(T-t)|^{1/2} = \tau^{1/2}$. Actually, the situation where $k > 1$ is easier to understand heuristically, so let us start by discussing the latter.

4.2 Analysis of $k > 1$

For $k > 1$, as $\tau \to \infty$, the factor L_τ^{-2} in front of the second derivative in (8) leads us to consider the solutions of

$$\phi_\tau = -\frac{1}{2k}\xi\phi_\xi - \frac{1}{p-1}\phi + \phi^p \tag{11}$$

Observe that the "fixed points" $f_b^*(x) = \left(p-1+bx^{2k}\right)^{\frac{1}{1-p}}$ of (6) are stationary solutions of that equation. The latter can of course be integrated in closed form, but before doing that, let us first look at its linearization around the constant solution $\phi = (p-1)^{\frac{1}{1-p}}$. The linear problem is $\phi_\tau = \mathcal{L}_\infty \phi$, where

$$\mathcal{L}_\infty = -\frac{1}{2k}\xi\frac{d}{d\xi} + 1\,. \tag{12}$$

and so, in the space of polynomials, we have now $2k$ expanding directions corresponding to ξ^n, for $n < 2k$.

Equation (11) is solved by putting $\phi(\xi,\tau) = e^{-\frac{\tau}{p-1}} h(e^{-\tau/2k}\xi, \tau)$ whereby $\frac{d}{d\tau}h(y,\tau) = e^{-\tau}h(y,\tau)^p$ and so, for $\rho = \tau - \tau_0$

$$\phi(\xi,\tau) = \frac{e^{-\frac{\rho}{p-1}} f(e^{-\rho/2k}\xi)}{[1-(p-1)f(e^{-\rho/2k}\xi)^{p-1}(1-e^{-\rho})]^{1/p-1}} \tag{13}$$

where $\phi(\xi,\tau_0) = f(\xi)$. The stationary solutions f_b^* are stable in a suitable codimension $2k$ space: let us consider f smooth, with

$$f(0) = (p-1)^{-\frac{1}{p-1}},\ f^{(\ell)}(0) = 0 \ \ \ell < 2k,\ f^{(2k)}(0) = \beta < 0 \tag{14}$$

and

$$0 \leq f(\xi) < (p-1)^{-\frac{1}{p-1}} \quad \xi \neq 0. \tag{15}$$

Then, for all $\xi \in \mathbf{R}$,

$$|\phi(\xi,\tau) - f_b^*(\xi)| \underset{\tau\to\infty}{\longrightarrow} 0 \tag{16}$$

where

$$f_b^*(\xi) = (p-1+b\xi^{2k})^{-\frac{1}{p-1}} \tag{17}$$

for some b depending on β,k, p.

These considerations thus lead us to expect (8) to have global solutions with initial data in a suitable codimension $2k$ set in a ball around (17) in a suitable Banach space. Of course, the perturbation $L_\tau^{-2}\phi_{\xi\xi}$ in (8) is very singular, but, basically, the picture is not much modified: the unstable modes turn out to be τ-dependent Hermite functions instead of the monomials ξ^n, and one has to fine-tune the projection of the initial data on these Hermite functions instead of imposing the vanishing of the derivatives as in (14). See [15] for details.

4.3 A Non-Conventional Center Manifold Problem: $k = 1$

Consider now the case $k = 1$. There are several differences with respect to the previous one: there is no damping factor in front of the second derivative in (8), and the asymptotics is given by (4), with a "universal" b^*, and an extra logarithmic factor in the definition (3) of $g(t,T)$. To understand the dynamics of (8) in that case, let us start by considering again its linearization around the constant solution $\phi = (p-1)^{\frac{1}{1-p}}$. The linear problem is $\phi_\tau = \mathcal{L}\phi$, where now

$$\mathcal{L} = \frac{d^2}{d\xi^2} - \frac{1}{2}\xi\frac{d}{d\xi} + 1\,. \tag{18}$$

Hence, the first thing we have to do, in order to understand the stability of the constant solution, is to study the spectrum of the linear operator $\mathcal{L}$. $\mathcal{L}$ is self–adjoint on $\mathcal{D}(\mathcal{L}) \subset L^2(\mathbf{R}, d\mu)$ with

$$d\mu(\xi) = \frac{e^{-\xi^2/4}d\xi}{\sqrt{4\pi}} \tag{19}$$

The spectrum of $\mathcal{L}$ is

$$spec(\mathcal{L}) = \{1 - \frac{n}{2} \mid n \in \mathbf{N}\} \tag{20}$$

and we take as eigenfunctions multiples of Hermite polynomials

$$h_n(\xi) = \sum_{m=0}^{[\frac{n}{2}]} \frac{n!}{m!(n-2m)!}(-1)^m\xi^{n-2m} \tag{21}$$

that satisfy

$$\int h_n h_m d\mu = 2^n n! \delta_{nm} \tag{22}$$

and

$$\mathcal{L} h_n = (1 - \frac{n}{2}) h_n. \tag{23}$$

Thus the linearization of (8) at the constant solution, for $k = 1$, has two expanding ("relevant") modes, h_0 and h_1, and one neutral ("marginal") one, $h_2 = \xi^2 - 2$.

Our goal is therefore to construct a center manifold for the flow of (8), in a neighborhood of the fixed point. Formally, we would expand,

$$\phi(\xi, \tau) = (p-1)^{\frac{1}{p-1}} + \psi(\xi, \tau) \text{ as } \psi = \sum_{n=0}^{\infty} \psi_n(\tau) h_n(\xi), \tag{24}$$

and rewrite (8) for the $\psi_n(\tau)$ as an infinite set of ODE's:

$$\frac{d}{d\tau} \psi_n(\tau) = (1 - \frac{n}{2}) \psi_n(\tau) + \text{ nonlinear terms}$$

A formal solution of this flow yields (see below for the calculation): $\psi_2(\tau) \simeq C_p (\log \tau) \tau^{-1}$ with $C_p = \frac{-(p-1)^{\frac{1}{1-p}}}{4p}$ as in (4).

However, there are severe problems with this approach. Since the eigenfunction h_n of the linearization $\mathcal{L}$ *increase* at infinity, the expansion (24) is not useful for ξ large, and, in particular, we cannot use any standard infinite dimensional center manifold theorem. The key to the solution to this problem comes again from a scaling argument, which will explain the emergence of the fixed point f^*. Let

$$\phi_L(\xi, \tau) = \phi(L\xi, L^2 \tau) \tag{25}$$

Then, ϕ_L satisfies the equation

$$H(\phi_L) = L^{-2}(-\phi_{L\tau} + \phi_{L\xi\xi}),$$

where we defined

$$H(\phi) = \frac{1}{2} \xi \phi_\xi + \frac{1}{p-1} \phi - \phi^p . \tag{26}$$

Hence, as $L \to \infty$, we expect the solutions of $H(\phi) = 0$ to be relevant. These are (like the stationary solutions of (11)) given by the one-parameter family f_b^*, given by (6), with $k = 1$. Therefore, we have the following picture: instead of perturbing around the constant solution, introduce $\phi_b(\xi, \tau) = (p - 1 + b\xi^2/\tau)^{\frac{1}{p-1}}$, and write

$$\phi(\xi, \tau) = \phi_b(\xi, \tau) + \eta(\xi, \tau). \tag{27}$$

A local (i.e. for ξ small) center manifold analysis then fixes $b = b^*$, as in (4). For large ξ, namely $\frac{\xi^2}{\tau} > \mathcal{O}(1)$, the linearization $\tilde{\mathcal{L}}$ of (8) around ϕ_b differs from $\mathcal{L}$, and, actually, in that region (see below) $\tilde{\mathcal{L}} \simeq \mathcal{L} - 2$. By (20), the spectrum of $\mathcal{L} - 2$ is entirely negative. Hence, the dynamics tends to *contract* η.

We will now explain the calculation that yields b^* in (4). For simplicity of notations, we shall consider only $p = 2$, i.e.

$$\phi_b(\xi, \tau) = (1 + b\xi^2/\tau)^{-1}. \tag{28}$$

We get, using $H(\phi_b) = 0$,

$$\begin{aligned} \eta_\tau &= \eta_{\xi\xi} - H(\phi_b + \eta) + H(\phi_b) + \phi_{b\xi\xi} - \phi_{b\tau} \\ &= (\mathcal{L} + W)\eta + M(\eta) + \phi_{b\xi\xi} - \phi_{b\tau} \end{aligned} \tag{29}$$

where we introduce

$$W = 2(\phi_b - 1) \tag{30}$$

$$M(\eta) = (\phi_b + \eta)^2 - \phi_b^2 - 2\phi_b\eta. \tag{31}$$

The operator $\mathcal{L}$, given by (18), has two unstable modes. Note that, formally, (i.e., for ξ of order one) W is $\mathcal{O}(\tau^{-1})$, M is nonlinear in η and $\phi_{b\xi\xi} - \phi_{b\tau}$ is $\mathcal{O}(\tau^{-1})$. We want to construct a center manifold for (29), i.e. to see how to fix two parameters in the initial data of η, such that the flow of (29) drives η to zero. A simple calculation will show that this can only be achieved through a suitable choice of b in (28).

To see this, let us simplify further and consider η even in ξ, which will imply that we need to fix only one parameter. This example contains all the relevant features of the general case. It is convenient to write

$$\eta_0(\tau) = \frac{a}{\tau} \tag{32}$$

and define ψ by

$$\eta = \eta_0 + \psi.$$

Then ψ satisfies the equation

$$\psi_\tau = \tilde{\mathcal{L}}\psi + N(\psi) + \alpha \tag{33}$$

with

$$\tilde{\mathcal{L}} = \mathcal{L} + V$$

$$V = 2(\phi_b + \eta_0 - 1) \tag{34}$$

$$N(\psi) = (\phi_b + \eta_0 + \psi)^2 - (\phi_b + \eta_0)^2 - 2(\phi_b + \eta_0)\psi \tag{35}$$

$$\begin{aligned} \alpha &= \phi_{b\xi\xi} - \phi_{b\tau} + (\mathcal{L} + W)\eta_0 - \eta_{0\tau} + M(\eta_0) \\ &= \phi_{b\xi\xi} - \phi_{b\tau} + \eta_0 + W\eta_0 - \eta_{0\tau} + M(\eta_0). \end{aligned} \tag{36}$$

Let us decompose ψ (which is even in η) as

$$\psi = \psi_0(\tau) + \psi_2(\tau)h_2 + \psi^\perp \tag{37}$$

where $\psi^\perp$ is orthogonal to h_n, $n \leq 2$. Next we expand V and α (for $\xi = \mathcal{O}(1)$):

$$V = -\frac{2b\xi^2}{\tau} + \frac{2a}{\tau} + \mathcal{O}(\frac{1}{\tau^2}) \tag{38}$$

$$\alpha = (a-2b)\tau^{-1} + (a + a^2 + (12b^2 - b - 2ab)\xi^2))\tau^{-2} + \mathcal{O}(\tau^{-3}). \tag{39}$$

Inserting (32), (37) in (33) and retaining only the leading terms in $1/\tau$ and ψ_i, $i = 0, 2$, we get from $\psi_{i\tau} = (2^i i!)^{-1}(h_i, \psi_\tau)$ ($(\cdot,\cdot)$ is the scalar product of $L^2(\mathbf{R}, d\mu)$ and we use (22)):

$$\psi_{0\tau} = \psi_0 + (a-2b)\tau^{-1} + R_0 \tag{40}$$

$$\psi_{2\tau} = \beta\tau^{-1}\psi_2 + (12b^2 - b - 2ab)\tau^{-2} + R_2 \tag{41}$$

where $R_0 = \mathcal{O}(\tau^{-2} + \tau^{-1}|\psi| + |\psi|^2)$, $R_2 = \mathcal{O}(\tau^{-3} + \tau^{-1}|\psi_0| + \tau^{-2}|\psi_2| + |\psi|^2)$,and $\beta = 2a - \frac{1}{4}b(\xi^2 h_2, h_2) = 2a - 20b$ (coming from the $V\psi$ term in (33)). We choose now a so that the $\mathcal{O}(\tau^{-1})$ term in $\psi_{0\tau}$ vanishes i.e.

$$a = 2b \tag{42}$$

and b such that the $\mathcal{O}(\tau^{-2})$ term in $\psi_{2\tau}$ is zero:

$$b = b^* = 1/8. \tag{43}$$

Note that this choice correspond to $b = b^*$ in (4) for $p = 2$. Then $\beta = -2$ and our equations read

$$\psi_{0\tau} = \psi_0 + R_0, \quad \psi_{2\tau} = -\frac{2}{\tau}\psi_2 + R_2. \tag{44}$$

Now, keeping in mind the presence of the R_0, R_2 terms,

$$\psi_0 = \mathcal{O}(\tau^{-2}), \quad \psi_2 = \mathcal{O}((\log\ \tau)\tau^{-2}) \tag{45}$$

would be consistent solutions. Of course, we need to show that the expanding variable ψ_0 will satisfy (45) by a suitable choice of $\psi_0(\tau_0)$. This is rather easy to do, using the fact that ψ_0 is expanding; in the general case (with more than one parameter to fix), we used a topological argument.

In the rigorous proof, we set up a suitable Banach space for the function η, parametrized by the ψ_i's for $\xi^2/\tau < \mathcal{O}(1)$, and a function η_l for $\xi^2/\tau > \mathcal{O}(1)$. The function η_l will

contract under the action of $\tilde{\mathcal{L}}$: indeed, from (34,28,32), we see that the potential V tends to -2 as ξ^2/τ (and τ) $\to \infty$.

5 Open Problems

There are several patterns and fronts in dissipative equations, and their stability can probably be studied using RG methods. For example, the Cahn-Hillard equation [21]:

$$u_t = \Delta(-\Delta u - u - u^3) \tag{1}$$

is often used to study the phase separation in alloys and fluids. One would like to study the stability (and possibly the dynamics), in infinite volume, of interface solutions of that equation.

Another major open problem consists in the extension of the RG method to hyperbolic equations. The latter have their own scaling laws, see [55, 64]. A lot is known on the stability of soliton solutions of (generalized) KdV equations [59, 58], and standing waves solutions of nonlinear Schrödinger equations [63], but the stability of localized solutions of other nonparabolic equations is quite open. Also, there are open problem concerning the blow-up of solutions, most notably in the nonlinear Schrödinger equation [54].

So far, we have only considered equations whose solutions have a rather simple asymptotic behavior. Of course, it is well known that finite dimensional dynamical systems described by differential equations can have a chaotic asymptotic behavior, i.e. depend sensitively on the initial data but have also some statistical regularity in the sense that the long time average along the orbits is described by an invariant (SRB) measure [27]. For certain classes of F in (1.1), one would like to find natural invariant measures for the flow.

A class of dynamical systems, modeling PDE's, is obtained by discretizing space and time and considering a recursion

$$u(x, t+1) = F(x, u(\cdot, t)) \tag{2}$$

i.e. $u(x, t+1)$, with x being a site of a lattice, is determined by the values taken by u at time t (usually on the sites in a neighborhood of x). For a suitable class of F's such dynamical systems are called Coupled Map Lattices (CML) [51, 52]. The study of the invariant measures, and their properties, even for these "toy models" is essentially a terra incognita. In [20, 60, 17] various "high-temperatures" (weak coupling) results are obtained. Almost nothing is known about the "low temperatures" (strong coupling) side (see [18, 19, 57]). One would like to find models for which the set of invariant measures changes as the coupling is varied; this phenomenon would then be interpreted as a kind of nonequilibrium phase transition.

Acknowledgments

We would like to thank T. Gallay, N. Goldenfeld, Y. Oono, J. Taskinen, G. Wayne, and J. Xin for interesting discussions during the preparation of this review. This work was supported by NSF grant DMS-9205296, by EC grants

SC1-CT91-0695 and CHRX-CT93-0411 and was done in part during a visit of the authors to the Mittag-Leffler Institute.

References

1. Aronson, D.G., Weinberger, H.F., Multidimensional non-linear diffusion arising in population genetics, Adv.Math. **30**, 33-76 (1978).
2. Barenblatt, G.I., Similarity, Self-similarity and Intermediate Asymptotics, Consultants Bureau, New York, 1979.
3. Berger, M., Kohn, R.V., A rescaling algorithm for the numerical calculation of blowing-up solutions, Comm. Pure. Appl. Math. **41**, 841-863 (1988).
4. Berlyand, Xin, J., Large time asymptotics of solutions to a model combustion system with critical nonlinearity, preprint; Renormalization group technique for asymptotic behavior of a thermal diffusive model with critical nonlinearity, to appear in Pitman Research Notes, ed. by G.F. Roach, Longman Pub. Co., England.
5. Bona, J., Promislow, K., Wayne, G., On the asymptotic behavior of solutions to nonlinear, dispersive, dissipative wave equations, to appear in J. Math. and Computer Simulation.
6. Bramson, M., Convergence of solutions of the Kolmogorov equation to travelling waves, Memoirs of the Amer. Math. Soc., **44**, nr. 285, 1-190 (1983).
7. Bressan, A., Stable blow-up patterns, J. Diff. Eq, **98**, 57-75 (1992).
8. Brezis, H., Peletier L.A., Terman D., A very singular solution of the heat equation with absorption, Arch. Rat. Mech. Anal., **95**, 185-209 (1986).
9. Bricmont, J., Kupiainen, A., Rigorous renormalization group and disordered systems, Physica A **165**, 31-37 (1990).
10. Bricmont, J., Kupiainen, A., Renormalization group for diffusion in a random medium, Phys. Rev. Lett. **66**, 1689-1692 (1991).
11. Bricmont, J., Kupiainen, A., Random walks in asymmetric random environments, Commun. Math. Phys. **142**, 345-420 (1991).
12. Bricmont, J., Kupiainen, A., Lin, G., Renormalization group and asymptotics of solutions of nonlinear parabolic equations, Comm.Pure.Appl.Math., **47**, 893-922 (1994).
13. Bricmont, J., Kupiainen, A., Renormalization Group and the Ginzburg-Landau equation, Commun. Math. Phys., **150**, 193-208 (1992).
14. Bricmont, J., Kupiainen, A., Stability of moving fronts in the Ginzburg-Landau equation, Commun. Math. Phys., **159**, 287-318 (1994).
15. Bricmont, J., Kupiainen, A., Universality in blow-up for nonlinear heat equations, Nonlinearity, **7**, 1-37 (1994).
16. Bricmont, J., Kupiainen, A., Stable non-Gaussian diffusive profiles, to appear in Nonlinear Analysis, T.M.& A.
17. Bricmont, J., Kupiainen, A., Coupled analytic maps, to appear in Nonlinearity.
18. L.A. Bunimovich, Coupled map lattices: One step forward and two steps back, preprint (1993), to appear in the Proceedings of the "Gran Finale" on Chaos, Order and Patterns, Como (1993).

19. L.A. Bunimovich, E. Carlen, On the problem of stability in lattice dynamical systems, preprint.
20. Bunimovich, L.A., Sinai, Y.G., Space-time chaos in coupled map lattices, Nonlinearity, **1**, 491-516 (1988).
21. Cahn, J.W., Hilliard, J.I., Free energy of a nonuniform system. 1. Interfacial free energy, J. Chem. Phys. **28**, 258-267 (1958).
22. Collet, P., Eckmann, J-P., Instabilities and fronts in extended systems, Princeton Univ. Press, 1990.
23. Collet, P., Eckmann, J-P., Epstein, H., Diffusive repair for the Ginsburg - Landau equation, Helv. Phys. Acta, **65**, 56-92 (1992).
24. Collet, P., Eckmann, J-P., Solutions without phase-slip for the Ginsburg - Landau equation, Commun. Math. Phys. **145**, 345-356 (1992).
25. Collet, P., Eckmann, J-P., Space-time behavior in problems of hydrodynamic type: a case study, Nonlinearity, **5**, 1265-1302 (1992).
26. M. Cross, P. Hohenberg, Pattern formation outside of equilibrium, Rev.Mod.Phys. **65**, 851-1112 (1993).
27. J-P. Eckmann, D. Ruelle, Ergodic theory of chaos and strange attractors, Rev. Mod. Phys. **57**, 617-656 (1985).
28. Eckmann, J.P., Wayne, C.E., Non-linear stability of front solutions for parabolic partial differential equations, Commun. Math. Phys., **161**, 323-334 (1994).
29. Escobedo, M., Kavian, O., Variational problems related to self-similar solutions of the heat equation, Nonlinear Analysis T.M.& A., **10**, 1103-1133 (1987).
30. Escobedo, M., Kavian, O., Asymptotic behavior of positive solutions of a nonlinear heat equation, Houston Jl. Math., **14**, 39-50 (1988).
31. Escobedo, M., Kavian, O., Matano, H. Large time behavior of solutions of a dissipative semi-linear heat equation, preprint.
32. Filippas, S., Kohn, R.V., Refined asymptotics for the blow-up of $u_t - \Delta u = u^p$, Comm. Pure Appl. Math, **45**, 821-869 (1992).
33. Filippas, S., Liu, W., On the blow-up of multidimensional semilinear heat equations, Ann. Inst. H. Poincaré, **10**, 313-344 (1993).
34. Fujita, H., On the blowing up of solutions of the Cauchy problem for $u_t = \Delta u + u^{1+\alpha}$, J. Fac. Sci. Univ. Tokyo, **13**, 109-124 (1966).
35. Galaktionov, V.A., Kurdyumov, S.P., Samarskii, A.A., On asymptotic "eigenfunctions" of the Cauchy problem for a nonlinear parabolic equation, Math. USSR Sbornik, **54**, 421-455 (1986).
36. Gallay, T., Local stability of critical fronts in nonlinear parabolic partial differential equations, Nonlinearity, **7**, 741-764, (1994); Existence et stabilité des fronts dans l'équation de Ginzburg-Landau à une dimension, Thèse, Univ. de Genève, 1994.
37. Gallay, T., A Center-stable manifold theorem for differential equations in Banach spaces, Commun. Math. Phys. **152**, 249-268 (1993).
38. Giga,Y., Kohn, R.V., Asymptotically self-similar blow-up of semilinear heat equations, Comm. Pure Appl. Math.**38**, 297-319 (1985).
39. Giga,Y., Kohn, R.V., Characterizing blowup using similarity variables, Indiana Univ. Math. J. **36**, 1-40 (1987).
40. Giga,Y., Kohn, R.V., Nondegeneracy of blowup for semilinear heat equations, Comm. Pure. Appl. Math. **42**, 845-884 (1989).
41. Glimm, J., Jaffe, A., Quantum Physics. A functional integral point of view, Springer, New York (1981).

42. Gmira, A., Veron, L., Large time behavior of the solution of a semilinear parabolic equation in $\mathbf{R}^N$, J. Diff. Equ., **53**, 258-276 (1984).
43. Goldenfeld, N., Martin, O., Oono, Y., Intermediate asymptotics and renormalization group theory, J. Sci. Comp. **4**, 355-372 (1989);
Goldenfeld, N., Martin, O., Oono, Y., Liu, F., Anomalous dimensions and the renormalization group in a nonlinear diffusion process, Phys. Rev. Lett., **64**, 1361-1364 (1990);
Goldenfeld, N., Martin, O., Oono, Y., Asymptotics of partial differential equations and the renormalization group, in: Proc. of the NATO ARW on Asymptotics beyond all orders, H. Segur, S.Tanveer, H. Levine, eds, Plenum (1991);
Goldenfeld, N., Lectures on phase transitions and the renormalization group, Addison-Wesley, Reading (1992);
Chen, L-Y., Goldenfeld, N., Oono, Y., Renormalization group theory for global asymptotic analysis, Phys. Rev. Lett. **73**, 1311-1315 (1994).
44. Herrero, M.A., Velazquez, J.J.L., Flat blow-up in one-dimensional semilinear heat equations, Differential and Integral Equations, **5**, 973-997 (1992).
45. Herrero, M.A., Velazquez, J.J.L., Blow-up profiles in one-dimensional, semilinear parabolic problems, Comm. in P.D.E., **17**, 205-219 (1992).
46. Herrero, M.A., Velazquez, J.J.L., Blow-up behavior of one-dimensional semilinear parabolic equations, Ann. Inst. H. Poincaré, **10**, 131-189 (1993).
47. Herrero, M.A., Velazquez, J.J.L., Some results on blow-up for semilinear parabolic problems. IMA preprint 1000.
48. Kamin, S., Peletier, L.A., Large time behavior of solutions of the heat equation with absorption, Ann. Sc. Norm. Sup. Pisa, **12**, 393-408 (1985).
49. Kamin, S., Peletier, L.A., Singular solutions of the heat equation with absorption, Proc. Amer. Math. Soc., **95**, 205-210 (1985).
50. Kamin, S., Peletier, L.A., Vazquez, J.L., On the Barenblatt equation of elastoplastic filtration, Indiana Univ. Math. J., **40**, 1333-1362 (1991).
51. K. Kaneko (ed): Theory and Applications of Coupled Map Lattices, J. Wiley (1993).
52. K. Kaneko (ed): Chaos, Focus Issue on Coupled Map Lattices, Chaos **2** (1993).
53. Kirchgässner, K., On the nonlinear dynamics of traveling fronts, J. Diff. Eqns, **96**, 256-278 (1992).
54. Lemesurier, B.J., Papanicolaou, G.C., Sulem, C., Sulem, P.L., Local structure of the self-focusing singularity of the nonlinear Schrödinger equation, Physica **D32**, 210-226 (1988).
55. Levine, H.A., The role of critical exponents in blow-up theorems, SIAM Review, **32**, 262-288 (1990).
56. Lin, G., The renormalization group and large-time behavior of solutions of nonlinear parabolic partial differential equations, PhD Thesis, Rutgers University, 1993.
57. J. Miller, D.A. Huse, Macroscopic equilibrium from microscopic irreversibility in a chaotic coupled-map lattice, Phys. Rev. E, **48**, 2528-2535 (1993).
58. Miller, J., Weinstein, M.I., Asymptotic stability of solitary waves for the regularized long wave equation, in preparation.
59. Pego, R., Weinstein, M., Asymptotic stability of solitary waves, Commun. Math. Phys., **164**, 305-349 (1994).
60. Pesin, Y.G., Sinai, Y.G., Space-time chaos in chains of weakly coupled hyperbolic maps, in: Advances in Soviet Mathematics, Vol. 3, ed. Y.G. Sinai , Harwood (1991).

61. Sattinger, D.H., Weighted norms for the stability of travelling waves, J. Diff. Eq. **25**, 130-144 (1977).
62. Simon, B., Functional Integration and Quantum Physics, Academic Press, New York, 1979.
63. Soffer, A., Weinstein, M.I. Multichannel nonlinear scattering for nonintegrable equations, Commun. Math. Phys., **133**, 119-146 (1990); Multichannel nonlinear scattering for nonintegrable equations 2. The case of anisotropic potentials and data, J. Diff. Eq. **98**, 376-390 (1992).
64. Strauss, W.A., Nonlinear Wave Equations, Regional Conference Series in Mathematics, AMS, CBMS **73** (1989).
65. Taskinen, J., Diffusion equation with general polynomial perturbation, in preparation.
66. Velazquez, J.L.L., Higher dimensional blow-up for semilinear parabolic equations, IMA preprint 968.
67. Velazquez, J.L.L., Galaktionov, V.A., Herrero, M.A., The space structure near a blow-up point for semilinear heat equations: a formal approach, Comput. Maths. Math. Phys. **31**, 46-55 (1991).
68. Wayne, G., Invariant manifolds for parabolic partial differential equations on unbounded domains, preprint.

Supersymmetric Quantum Field Theory

Konrad Osterwalder

ETH, Department of Mathematics
8092 Zürich, Switzerland

Abstract
After a discussion of the notion of unitary representations of supergroups we show, that the standard Wightman axioms and the axioms for the Euclidean Greens functions of a relativistic quantum field theory can be modified and supplemented, so as to allow for supersymmetry. The reconstruction theorem then states, that starting from Euclidean Greens functions a Wightman quantum field theory with a representation of the full supersymmetry group and algebra, respectively, can be recovered. A possible ansatz for the construction of supersymmetric models using the Euclidean approach is outlined.

1 Introduction

Supersymmetry is a fascinating symmetry: it is interesting mathematically and it has lead to many surprising new structures, see e.g. Kostant [Ko], Kac [K], Witten [W]; it is a challenging feature for the physicist [F], the ultimate questions being whether it occurs in elementary particle physics or not, see e.g. Ellis [E] and whether it can be put to use in other areas of theoretical physics. It also has proven to be a beautiful tool to link mathematics and physics, e.g. relativistic quantum field theory and non–commutative geometry, see Witten [W], Jaffe, Lesniewski and Osterwalder [JLO1], and references in Jaffe, Lesniewski [JL1, 2].

In physics the motivation to study supersymmetry came from the attempts to combine the symmetries of space–time with the inner symmetries of elementary particle multiplets in a non trivial way. In 1967 Coleman and Mandula [CM] showed that under the most general assumptions motivated by physics the Lie algebra of the total symmetry has to be the direct sum of the Poincaré algebra and of the Lie algebra of the inner symmetries, the elements of the later all being Lorentz scalars.

A way to circumvent this no go theorem was proposed in 1971 by Gol'fand and Likhtman [GL]. Their idea was to relax the assumption that the symmetries of the scattering matrix should be described by a Lie algebra and to allow for Lie *superalgebras* (also called graded Lie algebras) to describe the symmetries.

A Lie superalgebra is a $\mathbf{Z}_2$–graded (real or complex) vector space

$$\mathbf{A} = \mathbf{A}_0 \oplus \mathbf{A}_1$$

with $\mathbf{A}_0$ and $\mathbf{A}_1$ being its even and odd subspaces, respectively. For any two elements R, S in $\mathbf{A}$ there exists a supercommutator $[R, S]$ in $\mathbf{A}$ which is linear in both arguments and for $X_i \in \mathbf{A}$ and $Q_j \in \mathbf{A}_1$

$$\begin{aligned} [X_1, X_2] &= -[X_2, X_1] = X_3 \\ [X_1, Q_1] &= -[Q_1, X_1] = Q_2 \\ [Q_1, Q_2] &= \ \ [Q_2, Q_1] = X \, . \end{aligned}$$

Furthermore, a generalized Jacobi identity is assumed.

A Lie superalgebra can be constructed from any associative superalgebra, by defining the supercommutator for even elements or for an even with an odd element to be the ordinary commutator and for two odd elements to be the anticommutator.

In 1975, Haag, Łopuszański and Sohnius [HLS] showed that under the assumptions of Coleman and Mandula, with the Lie algebra describing the symmetries replaced by a Lie superalgebra, the most general symmetry algebra – now called supersymmetry algebra – is generated by

1. the even elements, consisting in

- the Poincaré algebra
- the algebra of the inner symmetries
- the so called central charges

2. the odd elements, consisting in

- the supercharges Q_α, which are spinors with respect to the Lorentz group.

As a simple example we describe the unextended, $N = 1$, Poincaré superalgebra in $(3 + 1)$ space time dimensions. It has as even generators $M_{\mu\nu}, P_\mu$ and the two "supertranslation generators" Q_α as well as their adjoints $\bar{Q}_{\dot\alpha}$, in the standard notation for Weyl spinors.

The supercommutators then are as follows:

$$\begin{aligned} [M_{\mu\nu},\, Q_\alpha] &= -\tfrac{1}{2}\,(\sigma_{\mu\nu})_\alpha{}^\beta\, Q_\beta \, , \\ [\,P_\mu,\, Q_\alpha] &= 0 \, , \\ \{Q_\alpha,\, \bar{Q}_{\dot\beta}\} &= 2\,(\sigma^\mu)_{\alpha\dot\beta}\, P_\mu \, , \\ \{Q_\alpha,\, Q_\beta\} &= 0 \, . \end{aligned}$$

Here we have used the symbols [.,.] and {.,.} to denote commutator and anticommutator, respectively. In the following we will stick with this example and we will mainly adopt the conventions of [WB].

The most direct way to construct models with supersymmetry seems to be to introduce the generators P_μ, Q_α, etc. directly and to verify the super commutation relations. As is well known from earlier experience in constructive quantum field theory, this leads to difficult problems due to the fact that all operators involved are unbounded. See Jaffe, Lesniewski [JL1, 2] and references given there.

Again based on earlier experience one might suspect that an approach using the Euclidean formulation and Euclidean functional integral techniques should be more accessible. This however means that not the Lie superalgebra but rather the *Lie supergroup* will be the primary object to be studied. In other words, one would try to construct the group elements

$$e^{ia^\mu P_\mu}, \quad e^{i\epsilon^\alpha Q_\alpha}, \quad \ldots$$

and verify the supergroup relations. This appears to limit the analysis to bounded (unitary) operators only. Closer inspection however shows that this conclusion is premature. Not only is ϵ^α a Grassmann algebra valued variable but, defining the exponential as a power series,

$$e^{i\epsilon^\alpha Q_\alpha} = \sum_{k=0}^{N} \frac{1}{k!}\left(i\epsilon^\alpha Q_\alpha\right)^k$$

we see that the right hand side is a finite sum with N depending on the number of ϵ components and therefore unbounded, if Q is. The intuition for the solution of this problem comes from the observation that

$$\{Q_\alpha, \bar{Q}_{\dot\beta}\} = 2\,(\sigma^\mu)_{\alpha\dot\beta}\,P_\mu\ ,$$

hence

$$e^{i(\epsilon^\alpha Q_\alpha + \bar\epsilon^{\dot\beta}\bar{Q}_{\dot\beta})}\,e^{ia^\mu P_\mu}$$

if smeared with an appropriate test function in ϵ^α, $\bar\epsilon^{\dot\beta}$ and in a^μ should be bounded.

This will be explained in section II. In section III we show how to generalize the Wightman and the Euclidean axioms so as to include supersymmetry. In a final section we make some remarks about the construction of models using the formalism developed here.

2 Unitary Representation of Supergroups

Let $\mathcal{G}$ be an ordinary Lie group and

$$\begin{aligned} U : \mathcal{G} &\longrightarrow \mathcal{L}(\mathcal{H}) \\ g &\longmapsto U(g) \end{aligned}$$

a unitary representation of $\mathcal{G}$ on a Hilbert space $\mathcal{H}$. For a test function $\rho \in L^1(\mathcal{G})$ we define

$$U(\rho) = \int \rho(g) U(g) d\mu(g),$$

where μ is the invariant measure on $\mathcal{G}$, for simplicity assumed to be left and right invariant. This yields a representation of the group algebra of $\mathcal{G}$, more specifically we have

$$U(\rho_1)U(\rho_2) = U(\rho_1 * \rho_2), \text{ and } U(\rho)^* = U(\rho^*),$$

where

$$\rho_1 * \rho_2(g) = \int \rho_1(h)\rho_2(h^{-1}g)d\mu(h),$$
$$\rho^*(g) = \overline{\rho(g^{-1})}.$$

A basic theorem says that (continuous) unitary representations of $\mathcal{G}$ are in one to one correspondence with representations of $L^1(\mathcal{G})$. For a precise formulation and a proof of this claim see e.g. Neumark [N1].

Going back to supergroups the remark at the end of the introduction suggests that we should define the unitary representations directly as representations of the group algebra (or of some appropriately chosen set of functions on the group, which carries the structure of a Hopf algebra). More specifically, instead of studying

$$e^{i(\epsilon^\alpha Q_\alpha + \bar{\epsilon}^{\dot\beta}\bar{Q}_{\dot\beta})}\, e^{ia^\mu P_\mu}$$

the right object to look at would be

$$U(\rho) = \int e^{i(\epsilon^\alpha Q_\alpha + \bar{\epsilon}^{\dot\beta}\bar{Q}_{\dot\beta})}\, e^{ia^\mu P_\mu}\, \rho(a^\mu,\, \epsilon^\alpha,\, \bar{\epsilon}^{\dot\beta})\, da d\epsilon,$$

where the precise meaning of this last expression still has to be explained.

Let $\mathcal{P}$ denote the universal covering of the restricted Poincaré group with elements (A, a^μ), $A \in SL(2,\mathbf{C})$, $a = (a^1, \ldots a^4) \in \mathbf{R}$, and $\Lambda\mathbf{C}^4$ the 4–fold exterior product of $\mathbf{C}$. As a basis for $\mathbf{C}^4$ we choose vectors ϵ^α, $\bar{\epsilon}^{\dot\beta}$, with $\alpha, \dot\beta \in \{1,2\}$. Then

$$\epsilon^{\underline{\nu}} = (\epsilon^1)^{\nu_1}\, (\epsilon^2)^{\nu_2}\, (\bar{\epsilon}^{\dot 1})^{\nu_3}\, (\bar{\epsilon}^{\dot 2})^{\nu_4}$$

with

$$\underline{\nu} = (\nu_1, \ldots \nu_4),\ \nu_i = 0, 1,$$

span $\Lambda\mathbf{C}^4$. The ϵ^α and $\bar{\epsilon}^{\dot\beta}$ (they are not each others complex conjugate) are called odd parameters and by construction they all anticommute with each other.

Now we define the unextended $N = 1$ super Poincaré group (strictly speaking: its group algebra) by

$$\text{S-Poincaré} \ = \ C_0^\infty(\mathcal{P}) \otimes \Lambda\mathbf{C}^4$$

with elements

$$\rho(A, a, \epsilon) \ \equiv \ \rho(A, a^\mu, \epsilon^\alpha, \bar{\epsilon}^{\dot\beta}) \ = \ \sum_{\underline{\nu}} \rho_{\underline{\nu}}(A, a^\mu)\, \epsilon^{\underline{\nu}},$$

where $\rho_{\underline{\nu}}(A, a^\mu) \in C_0^\infty(\mathcal{P})$.

In order to motivate the following definitions of multiplication and comultiplication on S-Poincaré we first sketch the purely algebraic notion of a representation of the super Poincaré group (not of its group algebra). Let $A \longrightarrow V(A)$ be a unitary representation of $SL(2,\mathbf{C})$ and $P_\mu, Q_\alpha, \bar{Q}_{\dot\alpha}$ a representation of the momentum and supercharge operators with P_μ self-adjoint and $(Q_\alpha)^* = \bar{Q}_{\dot\alpha}$, such that the following relations hold:

$$\begin{aligned} V(A)\, a^\mu P_\mu\, V(A^{-1}) &= (\Lambda a)^\mu P_\mu \,, \\ V(A)\, \epsilon^\alpha Q_\alpha V(A^{-1}) &= (A\epsilon)^\alpha Q_\alpha \,, \end{aligned}$$

where $\Lambda = \Lambda(A)$ is the Lorentz transformation corresponding to $A \in SL(2,\mathbf{C})$, i.e.

$$\Lambda^\mu{}_\nu \sigma^\nu = A^T \sigma^\mu \bar{A} \,.$$

Then with

$$G(A,a,\epsilon) = e^{i(\epsilon^\alpha Q_\alpha + \bar{\epsilon}^{\dot\beta} \bar{Q}_{\dot\beta})}\, e^{ia^\mu P_\mu}\, V(A)$$

we find the following multiplication law

$$G(A_1,a_1,\epsilon_1)\, G(A_2,a_2,\epsilon_2) =$$

$$G\big(A_1 A_2,\; a_1 + \Lambda_1 a_2 + i\epsilon_1 \sigma(\bar{A}_1 \bar{\epsilon}_2) - i(A_1\epsilon_2)\sigma\bar{\epsilon}_1,\; \epsilon_1 + A_1\epsilon_2\big).$$

This multiplication is associative and it allows for an inverse

$$G(A,a,\epsilon)^{-1} = G(A^{-1}, -\Lambda^{-1}a, -A^{-1}\epsilon) \,.$$

These formulae will never be used in the following, but they motivate the definitions we are going to make.

One further (standard) definition has to be given to explain, how to combine real variables with even powers of Grassmann variables. Let $f \in C^\infty(\mathbf{R}^n)$, $x = (x^1, \ldots x^n) \in \mathbf{R}^n$ and $y = (y^1, \ldots y^n)$ an n-tupel of even elements y^μ of some Grassmann algebra. Then using the standard multiindex notation we define

$$f(x^\mu + y^\mu) = \sum_{\underline{k}} \frac{1}{\underline{k}!} \partial^{\underline{k}} f(x^\mu) \cdot y^{\underline{k}}.$$

Now we translate the above operations in the supergroup to operations on S-Poincaré , which have to be taken as *definitions*. Multiplication becomes comultiplication, which is given by

$$\begin{aligned} \Delta : \text{S-Poincaré} &\longrightarrow \text{S-Poincaré} \times \text{S-Poincaré} \\ \rho &\longmapsto \Delta\rho \end{aligned}$$

where

$$\Delta\rho(A_1,a_1,\epsilon_1;\; A_2,a_2,\epsilon_2) =$$

$$\rho\Big(A_1A_2, a_1 + \Lambda_1 a_2 + i\epsilon_1\sigma(\bar{A}_1\bar{\epsilon}_2) - i(A_1\epsilon_2)\sigma\bar{\epsilon}_1,\; \epsilon_1 + A_1\epsilon_2\Big),$$

with the counit

$$\begin{aligned} \mathbf{e} \ : \ \text{S-Poincaré} \ &\longrightarrow \text{S-Poincaré} \\ \rho &\longmapsto \mathbf{e}\rho = \rho(\mathbf{1},0,0) \ , \end{aligned}$$

and the antipode

$$\begin{aligned} \mathbf{s} \ : \ \text{S-Poincaré} \ &\longrightarrow \text{S-Poincaré} \\ \rho &\longmapsto (\mathbf{s}\rho)(A,a,\epsilon) = \rho(A^{-1}, -\Lambda^{-1}a, -A^{-1}) \ . \end{aligned}$$

Note that S-Poincaré × S-Poincaré means

$$C_0^\infty(\mathcal{P}) \times C_0^\infty(\mathcal{P}) \otimes \Lambda \mathbf{C}^4 \ \wedge \ \Lambda \mathbf{C}^4 \ .$$

A multiplication on S-Poincaré is obtained from convolution, i.e.

$$\begin{aligned} * : \text{S-Poincaré} \ \times \ \text{S-Poincaré} \ &\longrightarrow \text{S-Poincaré} \\ \rho \times \sigma &\longmapsto \rho * \sigma, \end{aligned}$$

with

$$\begin{aligned} \rho * \sigma(A,a,\epsilon) &= \int \rho(A_1,a_1,\epsilon_1)\,(\mathbf{s}\otimes\mathbf{1})\cdot\Delta\sigma(A_1,a_1,\epsilon_1;\ A,a,\epsilon)\,d\mu(A_1)da_1d\epsilon_1 \ . \\ &= \int \rho(A_1,a_1,\epsilon_1)\,\Delta\sigma(A_1^{-1}, -\Lambda_1^{-1}a_1, A_1^{-1}\epsilon_1;\ A,a,\epsilon)\,d\mu(A_1)da_1d\epsilon_1 \ . \end{aligned}$$

In this expression $d\mu(A)$ is the invariant measure on $SL(2,\mathbf{C})$, see [N2], da is Lebesgue measure on $\mathbf{R}^4$ and $d\epsilon$ means Berezin integration (i.e. expanding the integrand in powers of $\epsilon_1^\alpha, \bar{\epsilon}_1^{\dot\beta}$ and picking the coefficient of $\epsilon_1^1 \dots \bar{\epsilon}_1^2$).

Remark As the Poincaré group is not compact, we had to restrict S-Poincaré to functions with compact support to make the convolution operation meaningful. Furthermore, as we do not admit distributions, our multiplication has no unit element. However there are sequences ρ_n in S-Poincaré which approximate a unit in the sense that with respect to an appropriate topology in S-Poincaré

$$\lim_{n\to\infty} \rho_n * \sigma = \sigma.$$

Finally we introduce an involution operation on S-Poincaré by

$$\rho^*(A,a,\epsilon,\bar\epsilon) = \overline{\rho(A^{-1},\ -\Lambda^{-1}a, A^{-1}\epsilon,\ \bar{A}^{-1}\bar\epsilon)}\ ,$$

where

$$\overline{(\epsilon^1)^{\nu_1}\ (\epsilon^2)^{\nu_2}\ (\bar\epsilon^1)^{\nu_3}\ (\bar\epsilon^2)^{\nu_4}} \ = \ (\epsilon^2)^{\nu_4}\ (\epsilon^1)^{\nu_3}\ (\bar\epsilon^2)^{\nu_4}\ (\bar\epsilon^1)^{\nu_1}\ .$$

Now we are ready to define what we mean by a unitary representation of a supergroup (or its group algebra, resp.).

Definition A unitary representation of S-Poincaré is a map

$$U : \text{S-Poincaré} \longrightarrow \mathcal{L}(\mathcal{H})$$
$$\rho \longmapsto U(\rho),$$

which is continuous with respect to a properly chosen topology on S-Poincaré and which satisfies

$$U(\rho_1)U(\rho_2) = U(\rho_1 * \rho_2),$$
$$U(\rho)^* = U(\rho^*),$$
$$\lim_{n\to\infty} U(\rho_n) = \mathbf{1},$$

for a sequence ρ_n in S-Poincaré , approximating the unit as explained above. The following theorem is the main result of this section.

Theorem [LO] *A unitary representation of S-Poincaré defines in a unique way a representation of the Poincaré Lie superalgebra.*

Remark This theorem obviously holds for other supergroups, too.

To give an idea of the proof of this theorem we note that formally

$$U(\rho) = \int e^{i(\epsilon^\alpha Q_\alpha + \bar{\epsilon}^{\dot\beta}\bar{Q}_{\dot\beta})}\, e^{ia^\mu P_\mu} V(A)\, \rho(A,a,\epsilon) d\mu(A) da d\epsilon.$$

Thus with $\partial_\mu = \frac{\partial}{\partial a^\mu}$ integration by parts gives

$$P_\mu U(\rho) = iU(\partial_\mu \rho).$$

Now we take this formula as definition of the operator P_μ on the domain

$$\mathcal{D}_0 = \{\ U(\rho)\mathcal{H}\ ;\ \rho \in \text{S-Poincaré}\ \}.$$

Correspondingly we set $D_\alpha = \frac{\partial}{\partial \epsilon^\alpha} + i\sigma^\mu_{\alpha\dot\beta}\ \bar{\epsilon}^{\dot\beta}\ \partial_\mu$ to find formally

$$Q_\alpha\, U(\rho) = -iU(D_\alpha \rho)$$

and again we take this as definition of the operator Q_α on the domain $\mathcal{D}_0$. The operator $\partial/\partial\epsilon^\alpha$ is defined as usual: it maps monomials in the ϵ's that do not contain the particular ϵ^α to zero, in the other monomials first move ϵ^α to the left, taking into account all the minus signs, then eliminate it.
By analogous formulae we define $\bar{Q}_{\dot\beta}$ and the generators $M_{\mu\nu}$ of the Lorentz rotations. Then the following theorem holds:

Theorem [LO] *$\mathcal{D}_0$ is an invariant domain of essential self-adjointness for $P_\mu, Q_\alpha, \bar{Q}_{\dot\beta}, M_{\mu\nu}$ and on $\mathcal{D}_0$ the Lie superalgebra relations hold.*

3 Super Wightman Theory – Super Euclidean QFT

In this section we show how to extend the Wightman formalism so as to include supersymmetry.

With $\mathcal{S}(\mathbf{R}^n)$ denoting Schwartz space over $\mathbf{R}^n$ we define super test functions by

$$\begin{aligned} \text{S-}\mathcal{S}(\mathbf{R}^{n,m}) &= \mathcal{S}(\mathbf{R}^n) \otimes \Lambda\mathbf{C}^m \\ &= \{ f(\, x^\mu, \epsilon^\alpha\,) = \sum_{\underline{\nu}} f_{\underline{\nu}}(x^\mu)\ \epsilon^{\underline{\nu}}\ ;\ f_{\underline{\nu}} \in \mathcal{S}(\mathbf{R}^n)\ \}. \end{aligned}$$

In particular, when studying $N = 1$ supersymmetry, we will be dealing with super test functions of the form

$$f(x_1, \epsilon_1, \ldots x_n, \epsilon_n) \ \in \ \text{S-}\mathcal{S}(\mathbf{R}^{(4n,4n)}),$$

where $x_k = (x_k^0, \ldots x_k^3)$ and $\epsilon_k = (\epsilon_k^\alpha, \bar{\epsilon}_k^{\dot{\beta}})$ with $\alpha, \dot{\beta} \in \{1, 2\}$.

Super Wightman distributions will be of the form

$$\mathcal{W}_n(x_1, \epsilon_1, \ldots x_n, \epsilon_n) \ = \ \sum_{\underline{\nu}} \mathcal{W}_{n,\underline{\nu}}(x_1, \ldots x_n)\ \epsilon^{\underline{\nu}},$$

and

$$\mathcal{W}_n(f_n) \ = \ \int \mathcal{W}_n(x_1, \epsilon_1, \ldots x_n, \epsilon_n)\ f_n(x_1, \epsilon_1, \ldots x_n, \epsilon_n)\ dx d\epsilon,$$

where the ϵ integration is Berezin integration as before. Notice that in this definition we have fixed the order in which we take the distribution and the test function; clearly both orders are equally possible, but we have to choose one in order to fix the signs.

Next we introduce the action of S-Poincaré on $\text{S-}\mathcal{S}(\mathbf{R}^{(4n,4n)})$. Let $\rho(A, a, \epsilon)$ be an element in S-Poincaré and $f_n \in \text{S-}\mathcal{S}(\mathbf{R}^{(4n,4n)})$. Then we define

$$u(\rho) : \ \text{S-}\mathcal{S}(\mathbf{R}^{(4n,4n)}) \ \longrightarrow \ \text{S-}\mathcal{S}(\mathbf{R}^{(4n,4n)})$$

by

$$\Big(u(\rho)\ f_n\Big)(x_1, \epsilon_1, \ldots x_n, \epsilon_n) =$$

$$\int \rho(A, a, \delta)\, f_n\Big(\Lambda^{-1}(x_1 - a - i\delta\sigma\bar{\epsilon}_1 + i\epsilon_1\sigma\bar{\delta}),\ A^{-1}(\epsilon_1 - \delta), \ldots$$

$$\ldots \Lambda^{-1}(x_n - a - i\delta\sigma\bar{\epsilon}_n + i\epsilon_n\sigma\bar{\delta}),\ A^{-1}(\epsilon_n - \delta)\Big)\ d\mu(A) da d\epsilon.$$

Clearly we have

$$u(\rho_1)\ u(\rho_2) = u(\rho_1\ *\ \rho_2).$$

Now we are ready to formulate the super Wightman axioms. The Wightman distributions $\mathcal{W}_n$ are required to be distributions on $\text{S-}\mathcal{S}(\mathbf{R}^{(4n,4n)})$ and to be *invariant* under S-Poincaré , i.e. for all $\rho \in$ S-Poincaré and $f_n \in \text{S-}\mathcal{S}(\mathbf{R}^{(4n,4n)})$ we postulate that

$$\mathcal{W}_n(f_n) \quad = \quad \mathcal{W}_n\big(u(\rho)\, f_n\big).$$

The *locality* axiom reads as follows. Let

$$f_n(x_1, \epsilon_1, \ldots x_n, \epsilon_n) = g_1(x_1, \epsilon_1) \ldots g_k(x_k, \epsilon_k) g_{k+1}(x_{k+1}, \epsilon_{k+1}) \ldots g_n(x_n, \epsilon_n) \ ,$$

and

$$f_n^{tr}(x_1, \epsilon_1, \ldots x_n, \epsilon_n) = g_1(x_1, \epsilon_1) \ldots g_{k+1}(x_k, \epsilon_k) g_k(x_{k+1}, \epsilon_{k+1}) \ldots g_n(x_n, \epsilon_n) \ ,$$

with $g_j \in \text{S-S}(\mathbf{R}^{(4,4)})$ and suppose that *supp* g_k and *supp* g_{k+1} are compact and spacelike separated. Then

$$\mathcal{W}_n(f_n) \quad = \quad \mathcal{W}_n(f_n^{tr}) \ .$$

Notice that this automatically gives the correct commutation or anticommutation property to field components with integer and half integer spin, respectively.

The additional modifications necessary for the other axioms – in particular the positivity axiom – are (almost) obvious and will not be spelled out here. The modified Wightman reconstruction theorem then says

Theorem [LO] *A set of Wightman distributions satisfying the super Wightman axioms allows for the reconstruction of a unique Wightman Quantum Field Theory with a unitary representation $U(\rho)$ of S-Poincaré and whose field is a super field $\phi(x, \epsilon)$.*

The proof of this theorem follows the original arguments given by Wightman, except for the construction of $U(\rho)$, where some additional work is needed. This will be outlined in the following.

After having reconstructed the physical Hilbert space $\mathcal{H}$ in the standard fashion, $u(\rho)$ as defined above can be shown to leave the kernel of the map

$$\bigoplus_n \text{S-}\mathcal{S}(\mathbf{R}^{(4n,4n)}) \quad \longrightarrow \quad \mathcal{H}$$

invariant and it therefore lifts to a densely defined map $U(\rho)$ on $\mathcal{H}$. From the properties of $u(\rho)$ and super invariance of $\mathcal{W}_n$ it follows that

$$\begin{aligned} U(\rho_1)U(\rho_2) &= U(\rho_1 * \rho_2) \\ U(\rho^*) \quad &\subset U(\rho)^* \,. \end{aligned}$$

The crucial step is to establish a bound on $U(\rho)$. First we observe that for

$$\rho(A, a, \epsilon) = \rho_{(1,1,1,1)}(A, a)\ \epsilon^1 \epsilon^2 \bar{\epsilon}^{\dot{1}} \bar{\epsilon}^{\dot{2}},$$

the heuristic formula for $U(\rho)$ suggests

$$U(\rho) \ = \ \int e^{ia^\mu P_\mu} V(A)\ \rho_{(1,1,1,1)}(A, a) d\mu(A) da \ .$$

This is the classical expression from the standard Wightman theory and indeed, the old reconstruction argument immediately gives

$$\| U(\rho) \| \leq \| \rho_{(1,1,1,1)} \|_1 \ .$$

A general element $\rho \in$ S-Poincaré has to be expanded in powers of $(\epsilon^\alpha \pm \bar{\epsilon}^{\dot{\alpha}})$. Hence we write it as

$$\rho(A,a,\epsilon) = \sum_{\underline{\nu}} \hat{\rho}_{\underline{\nu}}(A,a,\epsilon)$$

where

$$\hat{\rho}_{\underline{\nu}}(A,a,\epsilon) \ = \ \tilde{\rho}_{\underline{\nu}}(A,a)(\epsilon^1 + \bar{\epsilon}^{\dot{1}})^{\nu_1} \ (\epsilon^1 - \bar{\epsilon}^{\dot{1}})^{\nu_2} \ (\epsilon^2 + \bar{\epsilon}^{\dot{2}})^{\nu_3} \ (\epsilon^2 - \bar{\epsilon}^{\dot{2}})^{\nu_4} \ .$$

Clearly

$$\| U(\rho) \| \leq \sum_{\underline{\nu}} \| U(\hat{\rho}_{\underline{\nu}}) \|$$

and it suffices to estimate each term in the sum separately. The wonderful thing is that

$$\hat{\rho}^*_{\underline{\nu}} * \hat{\rho}_{\underline{\nu}} \ (A,a,\epsilon) \ =$$

$$\int \bar{\tilde{\rho}}_{\underline{\nu}}(A_1^{-1}, -A_1^{-1}a_1) \ L_{\underline{\nu}}\tilde{\rho}_{\underline{\nu}}(A_1^{-1}A, -A_1^{-1}(a-a_1)) d\mu(A_1) da_1 \ \epsilon^1 \epsilon^2 \bar{\epsilon}^{\dot{1}} \bar{\epsilon}^{\dot{2}},$$

where $L_{\underline{\nu}}$ is a differential operator homogeneous of order $4 - \sum \nu_k$. This allows us to use the simple estimate established above to conclude that

$$\begin{aligned} \| U(\hat{\rho}_{\underline{\nu}}) \|^2 \ &= \| U(\hat{\rho}_{\underline{\nu}})^* \ \hat{\rho}_{\underline{\nu}}) \| \\ &= \| U(\hat{\rho}^*_{\underline{\nu}} * U(\hat{\rho}_{\underline{\nu}}) \| \\ &\leq \| \bar{\tilde{\rho}}_{\underline{\nu}} \ \bar{*} \ L_{\underline{\nu}}\tilde{\rho}_{\underline{\nu}} \|_1 \ , \end{aligned}$$

and the precise meaning of $\bar{*}$ follows from the preceeding formula. We have proven

Theorem [LO] *With the notations as above*

$$\| U(\rho) \| \leq \| \rho_{(1,1,1,1)} \|_1 \ + \sum_{\underline{\nu}} \ \| \bar{\tilde{\rho}}_{\underline{\nu}} \bar{*} L_{\underline{\nu}}\tilde{\rho}_{\underline{\nu}} \|_1^{1/2} \ .$$

This theorem implies that there is a bounded extension of $U(\rho)$ to the whole space $\mathcal{H}$, which has all the desired properties, i.e. it defines a unitary representation of S-Poincaré .

To conclude this section we mention that there is an analogous extension for the axioms for super Euclidean Greens functions and a corresponding reconstruction theorem, see [LO]. Let us just point out again, that the super Euclidean Greens functions are *symmetric* functions of their super arguments (x_k, ϵ_k). This automatically gives the correct commutation or anticommutation properties to the Euclidean component fields, depending on their spin, see also the next section.

4 Constructing Models

In this section we briefly discuss, how the previously explained axioms for a super Wightman theory can be put to use in the construction of models. The basic idea has a long tradition in constructive quantum field theory: to define a model in terms of its Euclidean Greens functions, to verify the Euclidean axioms and then to invoke the reconstruction theorem, see e.g. [OS1, 2] or [GJ, S]. To simplify matters we restrict our discussion to 2 space time dimensions and to a $N = 1$, chiral superfield. Variables $x, y, \ldots$ will be in $\mathbf{R}^2$, the odd parameters $\epsilon = (\epsilon^1, \epsilon^2)$ will have 2 degrees of freedom only as $\bar{\epsilon} = \epsilon\gamma^0$.

4.1 Free Theory

The free Euclidean theory for a super field $\Phi(x, \epsilon)$ of mass m is given by its two point function

$$\begin{aligned} < \Phi(x,\epsilon)\, \Phi(y,\delta) > &= S_2(x,\epsilon; y,\delta) \\ &= (2\pi)^{-1} \int e^{-ip(x-y)+\bar{\epsilon}(ip\gamma - m)\delta + m/2\ (\bar{\epsilon}\epsilon + \bar{\delta}\delta)}\ (p^2+m^2)^{-1}\, dp\,. \end{aligned}$$

The Euclidean superfield can be expanded in powers of ϵ.

$$\Phi(x,\epsilon) = \phi(x) + \bar{\epsilon}\psi(x) + \tfrac{1}{2}\bar{\epsilon}\epsilon F(x)\ .$$

Here ϕ is a free boson field of mass m, ψ is a free Weyl spinor of mass m and F is an auxiliary field, as we will see. Expanding the two point function in powers of ϵ and δ we find

$$\begin{aligned} < F(x)\ \phi(y) > &= m\ \ < \phi(x)\ \phi(y) >\ , \\ < F(x)\ F(y) > &= -\delta(x-y) + m^2 < \phi(x)\ \phi(y) >\ . \end{aligned}$$

Diagonalization of the two point function is achieved by the substitution

$$F(x) = iG(x) + m\phi(x)\ .$$

Then

$$\begin{aligned} < G(x)\ G(y) > &= \delta(x-y)\ , \\ < G(x)\ \phi(y) > &= 0\ . \end{aligned}$$

This shows that the "white noise" field G has *no physical significance* : δ - function contributions to the Euclidean Greens functions drop out in the reconstruction of the physical Hilbert space and of the relativistic theory.

4.2 Interaction

Formally interaction may be introduced by means of a superpotential

$$V(\Phi) = i \int P(\Phi(x,\epsilon))\, dx\, d\epsilon\ ,$$

where P is some polynomial. The Schwinger functions of the interacting theory are then given in terms of the free Euclidean superfield by

$$S_n(x_1,\epsilon_1,\ldots x_n,\epsilon_n) = N \left\langle \Phi(x_1,\epsilon_1)\ldots\Phi(x_n,\epsilon_n)e^{-V(\Phi)}\right\rangle ,$$

where $N^{-1} = < e^{-V(\Phi)} >$. This expression may be understood as a Fock space expression or as a super functional integral. Formally these Schwinger functions satisfy the superinvariance axiom, the (super)symmetry axiom, and the super reflection positivity axiom.

Continuing with formal arguments, we compute the superpotential as a function of the component fields. Expanding the integrand in powers of ϵ and integrating over the odd parameters gives

$$\begin{aligned} i \int P(\Phi(x,\epsilon))\, dx d\epsilon &= i \int \left[P'(\phi(x)) \cdot \tfrac{1}{2}\bar{\epsilon}\epsilon F(x) + \frac{1}{2}P''(\phi(x)) \cdot (\bar{\epsilon}\psi(x))^2\right] dx d\epsilon \\ &= \int P'(\phi(x))F(x)dx \ - \tfrac{1}{2} \int P''(\phi(x))\bar{\psi}\psi(x)dx\ . \end{aligned}$$

Substituting $F(x) = iG(x) + m\phi(x)$ gives

$$i \int P(\Phi(x,\epsilon))dx d\epsilon =$$

$$i \int P'(\phi(x))G(x)dx \ + m\int P'(\phi(x))\phi(x)dx - \tfrac{1}{2} \int P''(\phi(x))\bar{\psi}\psi(x)dx\ .$$

As G appears only linearly in the exponent it can be integrated out explicitly, i.e.

$$i^n\Big\langle G(x_1)\ldots G(x_n)e^{-i\int P'\left(\phi(x)\right)G(x)dx} \Big\rangle_G$$

$$= P'(\phi(x_1))\ldots P'(\phi(x_n))e^{-\int\left(P'(\phi(x))\right)^2 dx} + \text{terms with factors } \delta(x_i - x_j)\ .$$

Introducing

$$\tilde{\Phi}(x,\epsilon) = \phi(x) + \bar{\epsilon}\psi(x) + \tfrac{1}{2}\bar{\epsilon}\epsilon\left(P'(\phi(x)) + m\phi(x)\right)\ ,$$

we find

$$S_n(x_1,\epsilon_1\ldots x_n,\epsilon_n) = N\langle\Phi(x_1,\epsilon_1)\ldots\Phi(x_n,\epsilon_n)e^{-V(\Phi)}\rangle$$

$$= N \left\langle \tilde{\Phi}(x_1,\epsilon_1)\ldots\tilde{\Phi}(x_n,\epsilon_n)e^{-\int\left[\left(P'(\phi(x))\right)^2 - m\ P'(\phi(x))\cdot\phi(x) + \frac{1}{2}P''\left(\phi(x)\right)\bar{\psi}\psi(x)\right]dx}\right\rangle .$$

To make these things mathematically rigorous we have to regularize the theory by introducing a volume and a ultraviolet cutoff. Some of the nice formal properties of the above expressions are destroyed by the cutoffs and they have to be recovered in the limit as the cutoffs are removed. A particular difficulty comes from the ultraviolet regularization which means replacing the superfield Φ in the interaction by

$$\Phi_\chi = \Phi * \chi,$$

where e.g. $\chi(x,\epsilon) = \bar{\epsilon}\epsilon\chi_{12}(x)$, and χ_{12} is a smooth approximation to the δ-function. This choice has the advantage that it does not mix the component fields. However the regularized G-field, G_χ, has a covariance $\chi * \chi$ instead of a δ-function. This means that integrating out the G-field we obtain a term

$$\int P'\left(\phi_\chi(x)\right)\ \chi * \chi(x-y)\ P'\left(\phi_\chi(y)\right) dxdy\ .$$

This expression is non local and this makes it hard to prove *stability.*

The problem of proving stability for slightly non local interactions has been studied by Jaffe, Lesniewski and Osterwalder [JLO]. Let

$$H = H_0 + \int : P\left(\phi_\kappa(x)\right) v(x-y) P\left(\phi_\kappa(y)\right) : dxdy$$

be a Hamiltonian in 1 space dimension and ϕ_κ the regularized $t = 0$ field on S^1. We say *H is weakly non local* if there are a, b, c and a sufficiently small ϵ such that

$$(1 + a \mid p \mid)^{-\epsilon} \leq \tilde{v}(p) \leq c(1 + b \mid p \mid)^{-\epsilon}.$$

Then

Theorem [JLO] *Suppose H is weakly non local. Then*

(a) H is bounded from below,

(b) $Tr\left(e^{-tH}\right) \leq \infty$, for $t \geq 0$,

(c) $H+Q$ is bounded from below for a large class of Q's which are polynomials in the field of lower order than twice the order of P.

The program for the future is to complete the development of an intrinsically supersymmetric formalism for the constructions of supersymmetric quantum field theory models.

References

[CM] Coleman, S. and Mandula, J.: All Possible Symmetries of the S–matrix, Phys. Rev. 159 (1967), 1251 – 1256.

[E] Ellis, J.: Hopes grow for supersymmetry, Nature, 313 (1985), 626 – 627.

[F] Ferrara, S. (ed.): *Supersymmetry*, North Holland/ World Scientific, 1987

[GJ] Glimm, J. and Jaffe, A.: *Quantum Physics*, Second Edition, Springer: New York 1987.

[GL] Gol'fand, Yu.A., Likhtman, E.P.: Extension of the Algebra of Poincaré Group Generators and the Violation of P Invariance, JETP Lett.13 (1971), 323–326.

[HLS] Haag, R., Łopuszański, J. and Sohnius, M.: All Possible Generators of Supersymmetries of the S–Matrix, Nucl. Phys. B88 (1975), 257–274.

[JL1] Jaffe, A. and Lesniewski, A.: Geometry of Supersymmetry, in *Constructive Quantum Field Theory II*, G.Velo and A.S.Wightman, Editors, Plenum, New York, 1990.

[JL2] Jaffe, A. and Lesniewski, A.: Supersymmetric Field Theory and Infinite Dimensional Analysis, in *Nonperturbative Quantum Field Theory*, G. 't Hooft *et al.*, Editors, Plenum, New York, 1988.

[JLO1] Jaffe, A., Lesniewski, A. and Osterwalder, K.: Quantum K–Theory I: The Chern Character, *Commun. Math. Phys.* **178** (1987), 313-329.

[JLO2] Jaffe, A., Lesniewski, A. and Osterwalder, K.: Stability for a Class of Bilocal Hamiltonians,*Commun. Math. Phys.*, **155** (1993), 183-197.

[K] Kac, V.G.: Lie superalgebras, Adv. in Math. 26 (1977), 8 – 96.

[Ko] Kostant, B.: Graded manifolds, graded Lie theory and prequantization, Springer Lecture Notes in Mathematics, 570 (1977), 177 – 306.

[LO] Lesniewski, A. and Osterwalder, K.: Euclidean Axioms for Supersymmetric Quantum Field Theories, to appear

[N1] Neumark, M. A.: *Normierte Algebren*, (Deutscher Verlag der Wissenschaften, 1959)

[N2] Neumark, M. A.: *Lineare Darstellungen der Lorentz Gruppe*, (Deutscher Verlag de r Wissenschaften, 1958)

[OS1] Osterwalder, K. and Schrader, R.: Axioms for Euclidean Green's functions, Commun. Math.Phys. 31 (1973), 83–112.

[OS2] Osterwalder, K. and Schrader, R.: Axioms for Euclidean Green's functions II, Commun. Math.Phys. 42 (1975), 281–305.

[S] Simon, B.:*The $P(\Phi)_2$ Euclidean(Quantum) Field Theory*, (Princeton University Press, Princeton, 1974).

[WB] Wess, J. and Bagger, J.: *Supersymmetry and Supergravity*, (Princeton University Press, Princeton, 1983).

[W] Witten, E.: Supersymmetry and Morse Theory, *J. Diff. Geom.***17** (1982), 661-692.

Equivalence of the Euclidean and Wightman Field Theories

Yu. M. Zinoviev[*]

Steklov Mathematical Institute, Vavilov St. 42,
Moscow 117966, GSP-1, Russia
e–mail: zinoviev@qft.mian.su

Abstract The new inversion formula for the Laplace transformation of the tempered distributions with supports in the closed positive semiaxis is obtained. The inverse Laplace transform of the tempered distribution is defined by means of the limit of the special distribution constructed from this distribution. The weak spectral condition on the Euclidean Green's functions requires that some of the limits needed for the inversion formula exist for any Euclidean Green's function with even number of variables. We prove that the initial Osterwalder–Schrader axioms [1] and the weak spectral condition are equivalent with the Wightman axioms.

1 Introduction

In 1973 K.Osterwalder and R.Schrader [1] claimed to have found necessary and sufficient conditions under which Euclidean Green's functions have analytic continuations whose boundary values define a unique set of Wightman distributions. The principal idea of the Osterwalder–Schrader paper [1] was to consider the Euclidean Green's functions to be the distributions. Usually the Euclidean Green's functions were considered to be the analytic functions. Later R.Schrader [2] found the counter example for the crucial lemma of the paper [1]. In 1975 K.Osterwalder and R.Schrader proposed additional "linear growth condition" under which Euclidean Green's functions, satisfying the Osterwalder–Schrader axioms [1], define the Wightman theory. But these new extended axioms for the Euclidean Green's functions may be not equivalent with the Wightman axioms. It is possible to restore the equivalence theorem by adding the new condition [2] that the Euclidean Green's functions are the Laplace transforms of the tempered distributions with supports in the positive semiaxis with respect to the time variables. The equivalence theorem becomes trivial [2]. This new condition contradicts the principal Osterwalder–Schrader idea to consider the Euclidean Green's functions to be the distributions and it is not suitable for applications because it seems difficult to check it up. This paper is an attempt to understand

* Supported by the Russian Foundation of Fundamental Researches under Grant No. 93–011–147.

the mathematical foundation of the Osterwalder–Schrader results. Our aim is to find the additional reasonable condition which allows to prove that the extended Osterwalder–Schrader axioms are equivalent with the Wightman axioms.

One of the Osterwalder–Schrader axioms is the positivity condition. If we consider the simplest case and neglect the space variables we can write the positivity condition in the form

$$\int_0^\infty dt \int_0^\infty ds f(t+s)\overline{\phi(t)}\phi(s) \geq 0 \tag{1.1}$$

Due to [3, Lemma A] the positivity condition (1.1) for the distribution $f(t) \in D'(\mathbf{R}_+)$, where $\mathbf{R}_+$ is the open positive semiaxis, implies the condition in $\mathbf{R}_+$

$$\sum_{m,n} a_m \bar{a}_n \frac{d^{m+n} f}{dt^{m+n}}(t) \geq 0 \tag{1.2}$$

for all finite sequences of the complex numbers a_m. Corollary C from [3] implies that the distribution $f(t) \in D'(\mathbf{R}_+)$, satisfying the condition (1.2) for all terminating sequences of complex numbers a_m, is the restriction to the semiaxis of a function $A(x+iy)$ analytic in the tube $\mathbf{R}_+ + i\mathbf{R}$. To explain the difficulties which this way encounters in proving the Osterwalder–Schrader theorem we cite here an extract from the remarkable paper [3] : "The Euclidean Green's functions satisfying the Osterwalder–Schrader postulates can be shown to be restrictions of the functions analytic in the whole Wightman causal domain and to satisfy the positivity condition there in a sence to be presently explained. The author has, however, not been able to show the tempered growth of those analytic functions near the real Minkowski space boundary and believes at present that this is impossible to achieve without further assumptions on the growth properties of Schwinger functions s_n with respect to the index n. This is suggested by the fact that in order to reach the real Minkowski space by analytic completion for a given s_n an infinite number of steps are required, each of which involves the other functions s_m via the Schwartz inequality with higher and higher values of m". Our way of proving the equivalence theorem doesn't use the analytic functions at all. Due to the Osterwalder–Schrader idea we consider the Euclidean Green's functions to be the distributions.

S.Bernstein [4] called a function exponentially convex if it satisfies the positivity condition (1.1). We shall prove that a tempered distribution $f(t) \in S'(\mathbf{R}_+)$ is exponentially convex iff a tempered distribution $g(t) = f(-t) \in S'(\mathbf{R}_-)$ is absolutely monotonic, i. e.

$$\frac{d^m g}{dt^m}(t) \geq 0 \tag{1.3}$$

for all $m = 0, 1, \ldots$. The following counter example: $f(t) = \exp\{t\}$ shows that this theorem is wrong for the distributions from $D'(\mathbf{R}_+)$. S.Bernstein [4] studied the absolutely monotonic functions. It is natural to have a try to generalize the Bernstein result. We shall prove that if for a distribution $f(t) \in D'(\mathbf{R}_+)$ a distribution $g(t) = f(-t) \in D'(\mathbf{R}_-)$ is absolutely monotonic then

$$f(t) = \int_0^\infty e^{-ts} d\mu(s), \tag{1.4}$$

where the positive measure $\mu(s)$ has tempered growth. The measure $\mu(s)$ explicitly depends on the distribution $f(t)$. It is the sum of two limits of the special distributions constructed from the distribution $f(t)$. By using the generalized Bernstein theorem it is possible to obtain the new inversion formula for the Laplace transformation of the tempered distributions with supports in the closed positive semiaxis. Our weak spectral condition on the Euclidean Green's functions requires that some of the limits needed for the inversion formula exist for any Euclidean Green's function with even number of variables. We shall prove that the initial Osterwalder–Schrader axioms [1] and the weak spectral condition are equivalent with the Wightman axioms.

In the next section we study the absolutely monotonic distributions. The generalization of the Bernstein theorem [4] is proved. The new inversion formula for the Laplace transformation of the tempered distributions with supports in the closed positive semiaxis is obtained. The third section is devoted to study the exponentially convex tempered distributions and the tempered distributions satisfying the Osterwalder–Schrader positivity condition which includes the space variables. In the fourth section the revised Osterwalder–Schrader theorem is proved.

2 Absolutely Monotonic Distributions

$D(\mathbf{R}_+)$ denotes the subspace of $D(\mathbf{R})$ of functions with support in the positive semiaxis $\overline{\mathbf{R}}_+ = [0, \infty)$, given the induced topology. Similarly $D(\mathbf{R}_-)$ denotes the subspace of $D(\mathbf{R})$ of functions with support in the negative semiaxis $\overline{\mathbf{R}}_- = (-\infty, 0]$, given the induced topology. If the function $\phi(x) \in D(\mathbf{R}_-)$ then the function $\phi(-x) \in D(\mathbf{R}_+)$.

The distribution $f(x) \in D'(\mathbf{R}_-)$ is said to be absolutely monotonic if for all natural numbers $m = 0, 1, ...$ the distribution $\frac{d^m f}{dx^m}(x)$ is positive.

If a function $\phi(x) \in D(\mathbf{R})$ then for sufficiently large positive t the function $\phi(x+t) \in D'(\mathbf{R}_-)$.

Lemma 2.1. *Let the distribution $f(x) \in D'(\mathbf{R}_-)$ be absolutely monotonic then for any function $\phi(x) \in D(\mathbf{R})$*

$$\lim_{t \to -\infty} (f(x), \phi(x-t)) = L_0^{-1}[f] \int_{-\infty}^{\infty} \phi(x) dx \tag{2.1}$$

$$\lim_{t \to -\infty} t^k \left(\frac{d^k f}{dx^k}(x), \phi(x-t)\right) = 0, k = 1, 2, ... \tag{2.2}$$

Moreover if the distribution $f(x) \in D'(\mathbf{R}_-)$ is absolutely monotonic the distribution $(-x)^{-1} f(x) \in D'(\mathbf{R}_-)$ is also absolutely monotonic, and the constant $L_0^{-1}[(-x)^{-1} f(x)] = 0$.

Proof. Let the function $\phi(x) \in D(\mathbf{R})$ be positive and its support be in the interval $[a,b]$. The function $f(t;\phi) \equiv (f(x), \phi(x-t))$ is defined on the semiaxis $(-\infty, -a]$. It is infinitely differentiable. Since the distribution $f(x)$ is absolutely monotonic the positivity of the function $\phi(x)$ implies

$$\frac{d^n}{dt^n} f(t;\phi) = (\frac{d^n f}{dx^n}(x), \phi(x-t)) \geq 0, n = 0,1,... \tag{2.3}$$

Hence for every $n = 1,2,...$ we get

$$\frac{d^n}{dt^n} f(t;\phi) \leq 2|t|^{-1} \int_t^{t/2} \frac{d^n}{dy^n} f(y;\phi) dy = \tag{2.4}$$
$$\leq 2|t|^{-1} \left[\frac{d^{n-1}}{dy^{n-1}} f(y;\phi) \mid_{y=t/2} - \frac{d^{n-1}}{dt^{n-1}} f(t;\phi) \right],$$

where $t < 0, -a$. Due to the inequalities (2.3) the function $f(t;\phi)$ is positive and non–decreasing on the semiaxis $(-\infty, -a]$. Therefore the limit (2.1) exists. Then it follows from the inequality (2.4) for $n = 1$ that the limit (2.2) equals zero for $k = 1$. By using the induction and the inequality (2.4) it is easy to prove the equalities (2.2) for $k = 1,2,...$ and for any positive function $\phi(x) \in D(\mathbf{R})$.

Let a function $\phi(x) \in D(\mathbf{R})$ and the number $M = \sup_{x \in \mathbf{R}} |\phi(x)|$. Let a positive function $h(x) \in D(\mathbf{R})$ be equal to one on the support of the function $\phi(x)$. The function $\phi(x)$ is the difference of the positive functions $1/2(Mh(x) \pm \phi(x))$. This decomposition implies the equalities (2.2) for the function $\phi(x)$ since the equalities (2.2) are valid for the positive functions from $D(\mathbf{R})$.

Let the integral of a positive function $h(x) \in D(\mathbf{R})$ be equal to one. Then any function $\phi(x) \in D(\mathbf{R})$ may be rewritten as

$$\phi(x) = h(x) \int_{-\infty}^{\infty} \phi(y) dy + \frac{d\psi}{dx}(x), \tag{2.5}$$

where $\psi(x) \in D(\mathbf{R})$. The limit (2.1) exists for any positive function from $D(\mathbf{R})$. Hence the decomposition (2.5) and the equality (2.2) for $k = 1$ imply the equality (2.1).

If the distribution $f(x) \in D'(\mathbf{R}_-)$ is absolutely monotonic the distribution $(-x)^{-1} f(x) \in D'(\mathbf{R}_-)$ is also absolutely monotonic. It follows from the relations (2.1) for the distributions $f(x)$ and $(-x)^{-1} f(x)$ that

$$\lim_{t \to -\infty} t((-x)^{-1} f(x), \phi(x-t)) = \tag{2.6}$$
$$-L_0^{-1}[(-x)^{-1} f(x)] \int_{-\infty}^{\infty} x\phi(x) dx - L_0^{-1}[f] \int_{-\infty}^{\infty} \phi(x) dx$$

Since the limit (2.1) for the distribution $(-x)^{-1} f(x)$ exists the limit (2.6) may exist when the constant $L_0^{-1}[(-x)^{-1} f(x)] = 0$.

Let a positive function $h(x) \in D(\mathbf{R})$ have the integral equal one and its support be in the positive semiaxis. Let us construct the infinitely differentiable function with finite support for every $T < 0$

$$h_T(x) = \int_x^{x-T} h(y)dy. \tag{2.7}$$

For a distribution $f(x) \in D'(\mathbf{R}_-)$ we define a functional on the space $S(\mathbf{R})$ by the following relation

$$(L_c^{-1}[f](-x;n,T), \phi(x)) = (f(x), L_c^{-1}[\phi](-x;n,T)) =$$
$$(n!)^{-1}((-x)^n \frac{d^{n+1}f}{dx^{n+1}}(x), \theta(-x)h_T(x)\phi(-nx^{-1})), \tag{2.8}$$

where n is a positive integer and the function $h_T(x)$ is given by the equality (2.7). It is easy to show that the tempered distribution $L_c^{-1}[f](-x;n,T) \in S'(\mathbf{R})$ is positive and its support is in the positive semiaxis.

Proposition 2.2. *Let a distribution $f(x) \in D'(\mathbf{R}_-)$ be absolutely monotonic. Then in the topological space $S'(\mathbf{R})$ there exists the limit*

$$\lim_{n\to\infty} \lim_{T\to-\infty} L_c^{-1}[f](-x;n,T) = L_c^{-1}[f](-x) \tag{2.9}$$

The tempered distribution $L_c^{-1}[f](-x) \in S'(\mathbf{R})$ is positive and its support is in the positive semiaxis.

This proposition is proved in the paper [5] .

By the definitions (2.8) and (2.9) the tempered distribution $L_c^{-1}[f](-x) \in S'(\mathbf{R})$ is positive and its support is in the positive semiaxis. Now Theorem 2 from [6, Chapter 2, Section 2.2] implies that the positive tempered distribution is given by the positive measure. This positive measure has tempered growth and its support is in the positive semiaxis.

Theorem 2.3. *For any absolutely monotonic distribution $f(x) \in D'(\mathbf{R}_-)$ the following representation*

$$(f(x), \phi(x)) = L_0^{-1}[f] \int_{-\infty}^{\infty} \phi(x)dx + \int_{-\infty}^{\infty} L_c^{-1}[f](-p)dp \int_{-\infty}^{\infty} e^{px}\phi(x)dx \tag{2.10}$$

is valid for any function $\phi(x) \in D(\mathbf{R}_-)$. Here the positive number $L_0^{-1}[f]$ is given by the relation (2.1), *the positive measure $L_c^{-1}[f](-p)$ with tempered growth and support in the positive semiaxis is defined by the relations* (2.8), (2.9).

This theorem is proved in the paper [5] . Theorem 2.3 is the analogue of the Bochner–Schwartz theorem [6, Chapter 2, Section 3.3, Theorem 3] for the Laplace transformation.

For O an open set in $\mathbf{R}^n$, $S(O)$ denotes the subspace of $S(\mathbf{R}^n)$ of functions with support in the closure $\overline{O}$, given the induced topology. For example $S(\mathbf{R}_-)$ denotes the subspace of $S(\mathbf{R})$ of functions with support in the negative semiaxis, given the induced topology. Similarly $S(\mathbf{R}_+)$ denotes the subspace of $S(\mathbf{R})$ of functions with support in the positive semiaxis, given the induced topology. It follows from Theorem 2.3 that every absolutely monotonic distribution $f(x) \in D'(\mathbf{R}_-)$ may be extended to the tempered distribution from $S'(\mathbf{R}_-)$.

We remind that a tempered distribution $g(t) \in S'(\mathbf{R}_-)$ is called absolutely monotonic if it satisfies the conditions (1.3) for all $m = 0, 1, \ldots$.

Corollary 2.4. *The tempered distribution $f(x) \in S'(\mathbf{R}_+)$ is the Laplace transform of a tempered distribution with support in the positive semiaxis if and only if there exists the natural number k such that*

$$(-x)^{-k} f(-x) = g_1(x) - g_2(x), \tag{2.11}$$

where the tempered distributions $g_j(x) \in S'(\mathbf{R}_-)$, $j = 1, 2$, are absolutely monotonic.

Proof. Due to Theorem from [7, Section 3.8] a tempered distribution with support in the positive semiaxis may be written as

$$g(x) = \sum_{m=0}^{k-1} \frac{d^m}{dx^m} \mu_m(x), \tag{2.12}$$

where $\mu_m(x)$ are the measures with tempered growth and with supports in the positive semiaxis.

It is easy to verify that $(m!)^{-1}(\frac{d}{dx})^{m+1} x_+^m = \delta(x)$. Hence the relation (2.12) implies

$$g(x) = \frac{d^k}{dx^k}[\sum_{m=0}^{k-1} ((k-m-1)!)^{-1} x_+^{k-m-1} * \mu_m(x)], \tag{2.13}$$

where $*$ denotes the convolution of two tempered distributions with supports in the positive semiaxis. If we represent the measure in the right–hand side of the equality (2.13) as the difference of two positive measures with tempered growth and with supports in the positive semiaxis we get

$$g(x) = \frac{d^k}{dx^k}[\nu_1(x) - \nu_2(x)]. \tag{2.14}$$

Taking the Laplace transform of the equality (2.14) and dividing it by x^k we obtain the equality (2.11), where x is replaced by $-x$.

It is straightforward to show that the equality (2.11) and Theorem 2.3 imply that the tempered distribution $f(x)$ is the Laplace transform of a tempered distribution with support in the positive semiaxis.

Let x denote a point in $\mathbf{R}^4$ with coordinates $(x^0, x^1, x^2, x^3) = (x^0, \mathbf{x})$. A point in $\mathbf{R}^{4n}$ will be written as $\underline{x} = (x_1, \ldots, x_n)$, $x_i \in \mathbf{R}^4$. We will use the following open set $\mathbf{R}_+^{4n} = \{\underline{x} \in \mathbf{R}^{4n} | x_j^0 > 0, j = 1, \ldots, n\}$.

Theorem 2.5. *Let the tempered distribution $f(x) \in S'(\mathbf{R}_+^{4n})$ be the Laplace transform with respect to the time variables of a tempered distribution with support in the closure $\overline{\mathbf{R}}_+^{4n}$. Then there is the natural number K such that for any integers $k > K$, $1 \leq j_1 < \cdots < j_l \leq n$, $1 \leq l \leq n$ and for all test functions*

$\phi_j(x) \in S(\mathbf{R}^4)$, $j \in \{j_1, ... j_l\}$; $\phi_i(x) \in S(\mathbf{R}^4_+)$, $1 \leq i \leq n$, $i \neq j_1, ... j_l$, *there exists the limit*

$$\int d^{4n}x L_c^{-1}[(\prod_{m=1}^{l} x_{j_m}^0)^{-k} f]_{x_{\underline{j}}^0}(\underline{x}) \prod_{i=1}^{n} \phi_i(x_i) =$$

$$\lim_{n_1,...,n_l \to \infty, n_i \in \mathbf{Z}} \lim_{T_1,...,T_l \to -\infty, T_i \in \mathbf{R}} \int d^{4n}x (\prod_{m=1}^{l} x_{j_m}^0)^{-k} f(\underline{x}) \times$$

$$(\prod_{i=1, i\neq j_1,...,j_l}^{n} \phi_i(x_i)) \prod_{m=1}^{l} L_c^{-1}[\phi_{j_m}]_{x_{j_m}^0}(x_{j_m}; n_m, T_m), \tag{2.15}$$

$$L_c^{-1}[\phi]_{x^0}(x; n, T) = (n!)^{-1} (\frac{\partial}{\partial x^0})^{n+1} ((x^0)^n \theta(x^0) h_T(-x^0) \phi(n(x^0)^{-1}, \mathbf{x})),$$

where the function $h_T(x^0)$ *is given by the equality* (2.7).

The limit (2.15) *defines the inversion formula for the Laplace transformation:*

$$(f(\underline{x}), \prod_{i=1}^{n} \phi_i(x_i)) = \int d^{4n}x ((\prod_{m=1}^{l} \frac{\partial}{\partial x_{j_m}^0})^k L_c^{-1} \times$$

$$[(\prod_{m=1}^{l} x_{j_m}^0)^{-k} f]_{x_{\underline{j}}^0}(\underline{x})) \int_0^\infty dy_{j_1}^0 ... \int_0^\infty dy_{j_m}^0 \exp\{-\sum_{m=1}^{l} x_{j_m}^0 y_{j_m}^0\} \times$$

$$(\prod_{m=1}^{l} \phi_{j_m}(y_{j_m}^0, \mathbf{x}_{j_m})) \prod_{i=1, i\neq j_1,...,j_l}^{n} \phi_i(x_i) \tag{2.16}$$

for all functions $\phi_i(x) \in S(\mathbf{R}^4_+)$, $i = 1,..., n$, *and for any integer* $k > K$.
This theorem is proved in the paper [5] .

3 Exponentially Convex Distributions

Due to S.Bernstein [4] we call a tempered distribution $f(x) \in S'(\mathbf{R}_+)$ exponentially convex if it satisfies the positivity condition (1.1) for any function $\phi(x) \in S(\mathbf{R}_+)$.

Proposition 3.1. *For any exponentially convex tempered distribution* $f(x) \in S'(\mathbf{R}_+)$ *the tempered distribution* $f(-x) \in S'(\mathbf{R}_-)$ *is absolutely monotonic.*
Proof. Let us introduce the convolution function

$$\overline{\phi} * \phi(x) = \int_{-\infty}^{\infty} \overline{\phi(x-y)} \phi(y) dy. \tag{3.1}$$

For $\phi(x) \in S(\mathbf{R}_+)$ the function $\overline{\phi} * \phi(x) \in S(\mathbf{R}_+)$. The condition (1.1) may be rewritten as $(f(x), \overline{\phi} * \phi(x)) \geq 0$. Let $F[\phi](p)$ be the Fourier transform of the function $\phi(x) \in S(\mathbf{R}_+)$. The definition (3.1) implies the following equality $F[\overline{\phi} * \phi](p) = \overline{F[\phi](-p)}F[\phi](p)$.

By $FS(\mathbf{R}_+)$ we denote the space of all analytic in the open upper half plane and infinitely differentiable in the closed upper half plane functions $\psi(z)$ such that the seminorms of the form

$$\sup_{\mathrm{Im} z \geq 0} (1 + |z|)^n \left|\frac{d^m \psi}{dz^m}(z)\right| \tag{3.2}$$

are finite for all positive integers m and n. The topology of $FS(\mathbf{R}_+)$ is given by the set of the seminorms (3.2).

Let us prove that the Fourier transformation defines an isomorphism between two topological spaces: $S(\mathbf{R}_+)$ and $FS(\mathbf{R}_+)$. Fourier transform $F[\phi](x)$ of a function $\phi(x) \in S(\mathbf{R}_+)$ has an analytical continuation $F[\phi](z)$ into the open upper half plane. The function $F[\phi](z)$ is infinitely differentiable in the closed upper half plane. The inequality $y^k \exp\{-yt\} \leq C(k)t^{-k}$, valid for $t > 0$, $y \geq 0$ implies the following estimation

$$\sup_{x \in \mathbf{R}, y \geq 0} \left|x^{k_1} y^{k_2} \frac{d^m F[\phi]}{dz^m}(x + iy)\right| \leq \tag{3.3}$$

$$C \sup_{t > 0, 0 \leq l \leq k_1} (1 + t^2) t^{-k_2 - l} \left|\frac{d^{k_1 - l}}{dt^{k_1 - l}}(t^m \phi(t))\right|.$$

Therefore the Fourier transformation defines the continuous mapping of the space $S(\mathbf{R}_+)$ into the space $FS(\mathbf{R}_+)$. For a function $\psi(z) \in FS(\mathbf{R}_+)$ its restriction $\psi(x)$ on the real axis belongs to the Schwartz space $S(\mathbf{R})$. A straightforward application of Cauchy's theorem shows that the inverse Fourier transform $F^{-1}[\psi](p)$ of the function $\psi(x)$ may be rewritten for $y > 0$ as

$$F^{-1}[\psi](p) = (2\pi)^{-1} e^{py} \int_{-\infty}^{\infty} \exp\{-ipx\} \psi(x + iy) dx. \tag{3.4}$$

Since any seminorm (3.2) is finite we get $F^{-1}[\psi](p) = 0$ for $p < 0$ by tending y in (3.4) to infinity. Hence $F^{-1}[\psi](p) \in S(\mathbf{R}_+)$. The inverse Fourier transformation is the topological isomorphism of the Schwartz space $S(\mathbf{R})$. Any seminorm of the Schwartz space $S(\mathbf{R})$ on the subspace $FS(\mathbf{R}_+)$ is majorized by a corresponding seminorm of the type (3.2). Thus the inverse Fourier transformation is the mapping of the space $FS(\mathbf{R}_+)$ into the space $S(\mathbf{R}_+)$.

For any natural number k and for any number $\alpha > 0$ we define the function

$$\begin{aligned} &F[\chi_\alpha](z) = ((\alpha + z/i)^{1/2})^k \exp\{-\alpha(1 + z/i)^{1/2}\}, \\ &(a + z/i)^{1/2} = (x^2 + (a + y)^2)^{1/4} \exp\{-i/2 \arctan[x(a + y)^{-1}]\}, \end{aligned} \tag{3.5}$$

holomorphic in the open upper half plane. Due to the estimation $\mathrm{Re}(1+z/i)^{1/2} \geq (|z|/2)^{1/2}$, valid in the closed upper half plane, the function (3.5) belongs to the space $FS(\mathbf{R}_+)$. Hence its inverse Fourier transform $\chi_\alpha(x)$ belongs to the space $S(\mathbf{R}_+)$, as the function $\chi_\alpha(x-t)$ for any positive number t. Therefore the convolution function $\overline{\chi}_\alpha * \chi_\alpha(x-2t) = \chi_\alpha(\bullet - t) * \chi_\alpha(\bullet - t)(x)$ belongs to the space $S(\mathbf{R}_+)$ for any $t \geq 0$. Now the positivity condition (1.1) for the exponentially convex tempered distribution $f(x) \in S'(\mathbf{R}_+)$ implies the inequality $(f(x), \overline{\chi}_\alpha * \chi_\alpha(x-2t)) \geq 0$ for any $t \geq 0$. By integration of this inequality with positive function $1/2\phi(2t) \in S(\mathbf{R}_+)$ we obtain

$$(f(x), (\overline{\chi}_\alpha * \chi_\alpha) * \phi(x)) \geq 0. \tag{3.6}$$

In view of the definition (3.5) we get $\overline{F[\chi_\alpha](-x)} = F[\chi_\alpha](x)$. Now it is easy to show that

$$(\overline{\chi}_\alpha * \chi_\alpha) * \phi(x) = (2\pi)^{-1} \int_{-\infty}^{\infty} dp e^{-ipx} F[\phi](p)(F[\chi_\alpha](p))^2 \tag{3.7}$$

When $\alpha \to +0$ the functions $F[\phi](z)(F[\chi_\alpha](z))^2$ converge to $(z/i)^k F[\phi](z)$ in the topology of the space $FS(\mathbf{R}_+)$. Then the left-hand side of the equality (3.7) converges to the function $\frac{d^k\phi}{dx^k}(x)$ in the topology of the space $S(\mathbf{R}_+)$ as $\alpha \to +0$. This implies that the inequality (3.6) converges as $\alpha \to +0$ to the inequality

$$(-1)^k(\frac{d^k f}{dx^k}(x), \phi(x)) \geq 0. \tag{3.8}$$

It follows from the inequalities (3.8) for arbitrary natural numbers k and for arbitrary positive functions $\phi(x) \in S(\mathbf{R}_+)$ that the tempered distribution $f(-x) \in S'(\mathbf{R}_-)$ is absolutely monotonic.

The distribution $f(x) \in S'(\mathbf{R}^4_+)$ satisfies the Osterwalder–Schrader positivity condition, if for any function $\phi(x) \in S(\mathbf{R}^4_+)$

$$\int d^4x d^4y f(x^0 + y^0, \mathbf{x} - \mathbf{y}) \overline{\phi(x)} \phi(y) \geq 0. \tag{3.9}$$

Therefore the Osterwalder–Schrader positivity condition (3.9) is the condition of exponentially convexity with respect to the time variable and it is the condition of positively definiteness with respect to the space variables. By using the proof of Proposition 3.1 and the proof of Theorem 1 from [6, Chapter 2, Section 3.1] it is possible to show that the Fourier transform $F_{\mathbf{x}}[f](x^0, \mathbf{x})$ with respect to the space variables of the distribution $f(x) \in S'(\mathbf{R}^4_+)$, satisfying the condition (3.9), satisfies the following condition

$$(-\frac{\partial}{\partial x^0})^n F_{\mathbf{x}}[f](x^0, \mathbf{x}) \geq 0 \tag{3.10}$$

for any natural number $n = 0, 1, \ldots$.

Lemma 3.2. *Let tempered distribution $f(x) \in S'(\mathbf{R}^4_+)$ satisfy the Osterwalder–Schrader positivity condition* (3.9) *then for any function $\phi(x) \in S(\mathbf{R}^4_-)$*

$$\lim_{t \to -\infty} (f(x), \phi(-x^0 - t, \mathbf{x})) = \int_{\mathbf{R}^3} L_0^{-1}[f]_{x^0}(\mathbf{x}) d^3\mathbf{x} \int_{-\infty}^{\infty} \phi(x^0, \mathbf{x}) dx^0, \tag{3.11}$$

where the tempered distribution $L_0^{-1}[f]_{x^0}(\mathbf{x}) \in S'(\mathbf{R}^3)$ is positively definite, and for any natural number $k = 1, 2, \ldots$

$$\lim_{t \to -\infty} t^k \left(\left(\frac{\partial}{\partial x^0} \right)^k f(x), \phi(-x^0 - t, \mathbf{x}) \right) = 0. \tag{3.12}$$

If the tempered distribution $f(x) \in S'(\mathbf{R}^4_+)$ satisfies the Osterwalder–Schrader positivity condition (3.9) *the tempered distribution $(x^0)^{-1} f(x) \in S'(\mathbf{R}^4_+)$ satisfies the inequalities* (3.10) *and the limits* (3.11), (3.12) *for this distribution are equal to zero.*

Proof. Let the Fourier transform $F_{\mathbf{x}}[\phi](x^0, \mathbf{x})$ with respect to the space variables of a function $\phi(x) \in S(\mathbf{R}^4_-)$ be a positive function. Then by the straightforward application of the inequalities (3.10) and by the proof of Lemma 2.1 we can prove the relations (3.12) and the existence of the limit (3.11).

Due to the Theorem 2 from [6, Chapter 2, Section 2.2] the inequalities (3.10) imply the following estimation for any function $\phi(x) \in S(\mathbf{R}^4_-)$

$$\left| \left(\left(\frac{\partial}{\partial x^0} \right)^n f(x), \phi(-x^0, \mathbf{x}) \right) \right| \leq C \sup_{x^0 \leq 0, \mathbf{x} \in \mathbf{R}^3} (1 + |x|^2)^p |F_{\mathbf{x}}[\phi](x^0, \mathbf{x})|, \tag{3.13}$$

where the numbers C and p depend on the natural number $n = 0, 1, 2, \ldots$. Let $\alpha(\mathbf{x})$ be an infinitely differentiable positive function with a compact support and let it be equal to one into some neighborhood of zero. For a function $\phi(x) \in S(\mathbf{R}^4_-)$ we define $M = \sup |F_{\mathbf{x}}[\phi](x)|$. The difference of two positive functions from $S(\mathbf{R}^4_-)$:

$$\begin{aligned} F_{\mathbf{x}}[\phi_{1m}](x) &= \exp\{m^{-1}(x^0 + (x^0)^{-1})\}\theta(-x^0)\alpha(m^{-1}\mathbf{x})M \\ F_{\mathbf{x}}[\phi_{2m}](x) &= \exp\{m^{-1}(x^0 + (x^0)^{-1})\}\theta(-x^0)\alpha(m^{-1}\mathbf{x})(M - F_{\mathbf{x}}[\phi](x)) \end{aligned} \tag{3.14}$$

converges as $m \to \infty$ to the function $F_{\mathbf{x}}[\phi](x) \in S(\mathbf{R}^4_-)$ in the norm (3.13). Due to the inequality (3.13) it implies now the relations (3.12) and the existence of the limit (3.11) for any function $\phi(x) \in S(\mathbf{R}^4_-)$.

Let the integral of a positive function $h(x^0) \in S(\mathbf{R}_-)$ be equal to one. Any function $\phi(x) \in S(\mathbf{R}^4_-)$ may be represented as

$$\phi(x) = h(x^0) \int_{-\infty}^{\infty} \phi(y^0, \mathbf{x}) dy^0 + \frac{\partial \psi}{\partial x^0}(x), \tag{3.15}$$

where the function $\psi(x) \in S(\mathbf{R}^4_-)$. The existence of the limit (3.11) and the equalities (3.15) and (3.12) for $k = 1$ provide the equality (3.11) for any function $\phi(x) \in S(\mathbf{R}^4_-)$. In virtue of the inequality (3.10) for $n = 0$ the tempered distribution $L_0^{-1}[f]_{x^0}(\mathbf{x})$ is positively definite. The proof of the last part of Lemma 3.2 follows the arguments of Lemma 2.1.

Let a positive function $h_T(x^0)$ be given by the relation (2.7) for some positive function $h(x^0) \in D(\mathbf{R}_-)$, having the integral equal one. For a tempered distribution $f(x) \in S'(\mathbf{R}^4_+)$ satisfying the Osterwalder–Schrader positivity condition (3.9) we define a functional on the space $S(\mathbf{R}^4)$ by the following relation

$$(L_c^{-1}[f]_{x^0}(x;n,T),\phi(x)) = (f(x), L_c^{-1}[\phi]_{x^0}(x;n,T)) = \tag{3.16}$$
$$(n!)^{-1}((x^0)^n(-\frac{\partial}{\partial x^0})^{n+1}f(x^0,\mathbf{x}),\theta(x^0)h_T(-x^0)\phi(n(x^0)^{-1},\mathbf{x})),$$

where n is a positive integer and the function $h_T(x^0)$ is given by the equality (2.7). It is easy to prove that the tempered distribution $F_{\mathbf{x}}[L_c^{-1}[f]_{x^0}(\bullet;n,T)](x)$ is positive and its support is in the closure of $\mathbf{R}_+^4$.

Proposition 3.3. *Let a tempered distribution $f(x) \in S'(\mathbf{R}_+^4)$ satisfy the Osterwalder-Schrader positivity condition* (3.9). *Then in the topological space $S'(\mathbf{R}^4)$ there exists the limit*

$$\lim_{n\to\infty}\lim_{T\to-\infty} L_c^{-1}[f]_{x^0}(x;n,T) = L_c^{-1}[f]_{x^0}(x) \tag{3.17}$$

The tempered distribution $F_{\mathbf{x}}[L_c^{-1}[f]_{x^0}(\bullet)](x) \in S'(\mathbf{R}^4)$ is positive and its support is in the closure of $\mathbf{R}_+^4$.

This proposition is proved in the paper [5] as the following theorem.

Theorem 3.4. *Consider a tempered distribution $f(x) \in S'(\mathbf{R}_+^4)$ which satisfies the Osterwalder–Schrader positivity condition* (3.9). *Then the following representation*

$$(f(x),\phi(x)) = \int_{\mathbf{R}^3} L_0^{-1}[f]_{x^0}(\mathbf{x})d^3\mathbf{x}\int_{-\infty}^{\infty}\phi(p^0,\mathbf{x}))dp^0 + \tag{3.18}$$
$$\int_{\mathbf{R}^4} L_c^{-1}[f]_{x^0}(x)d^4x\int_{-\infty}^{\infty}\exp\{-x^0p^0\}\phi(p^0,\mathbf{x}))dp^0$$

is valid for any function $\phi(x) \in S(\mathbf{R}_+^4)$. Here the positively definite tempered distribution $L_0^{-1}[f]_{x^0}(\mathbf{x})\in S'(\mathbf{R}^3)$ is defined by the equality (3.11) *and the tempered distribution $L_c^{-1}[f]_{x^0}(x)\in S'(\mathbf{R}^4)$ defined by the relations* (3.16), (3.17). *The distribution $F_{\mathbf{x}}[L_c^{-1}[f]_{x^0}(\bullet)](x) \in S'(\mathbf{R}^4)$ is the positive measure with tempered growth and support in the closure of $\mathbf{R}_+^4$.*

4 Revised Osterwalder–Schrader Theorem

We deal with the theory of one Hermitian scalar field. By using the below results and Chapter 6 of the paper [1] it is possible to formulate the extended Osterwalder–Schrader axioms and to prove the revised Osterwalder–Schrader theorem for the theories of arbitrary spinor fields.

We introduce some notation from the papers [1] and [2]. We define the following open sets in $\mathbf{R}^{4n}$: $\mathbf{R}_<^{4n} = \{\underline{x} \in \mathbf{R}^{4n} | \; x_{j+1}^0 > x_j^0, j = 1,...,n-1\}$ and $\mathbf{R}_0^{4n} = \{\underline{x} \in \mathbf{R}^{4n} | \; x_i \neq x_j, 1 \leq i < j \leq n\}$. For O an open set in $\mathbf{R}^{4n}$, the space $S(O)$ is defined above. On $S(\mathbf{R}^{4n})$ we define two involutions

$$f^*(x_1,...,x_n) = \overline{f}(x_n,...,x_1) \tag{4.1}$$
$$\theta f(x_1,...,x_n) = f(\theta x_1,...,\theta x_n),$$

where $\theta x = (-x^0, \mathbf{x})$ and $\overline{f}$ means complex conjugation. The space $S(\mathbf{R}^{4n}_<)$ is invariant under the involution $f \to \theta f^*$. Let $f \in S(\mathbf{R}^{4n})$, $R \in SO_4$ be an element in the rotation group, $a \in \mathbf{R}^4$ and $\pi \in P_n$ be an element in the group of all permutations of n objects (the letter S_n will be used elsewhere). Then we define $f_{(a,R)}$ and f^π by $f_{(a,R)}(x_1,...,x_n) = f(Rx_1+a,...,Rx_n+a)$ and $f^\pi(x_1,...,x_n) = f(x_{\pi(1)},...,x_{\pi(n)})$.

We remind the Osterwalder–Schrader axioms [1] for the Schwinger functions (Euclidean Green's functions). The set of the Schwinger functions $\{s_n\}$ is a sequence of distributions $s_n(x_1,...,x_n)$ with the following properties

E0. *Distributions*

$s_0 \equiv 1$, $s_n \in S'(\mathbf{R}^{4n}_0)$ and

$$\overline{(s_n, f)} = (s_n, \theta f^*) \tag{4.2}$$

for all functions $f \in S(\mathbf{R}^{4n}_<)$.

E1. *Euclidean invariance*

$$(s_n, f_{(a,R)}) = (s_n, f) \tag{4.3}$$

for all $R \in SO_4$, $a \in \mathbf{R}^4$ and $f \in S(\mathbf{R}^{4n}_0)$.

E2. *Positivity*

$$\sum_{n,m} (s_{n+m}, \theta f_n^* \otimes f_m) \geq 0 \tag{4.4}$$

for all finite sequences of the functions $f_n \in S(\mathbf{R}^{4n}_< \cap \mathbf{R}^{4n}_+)$, where the function $(f \otimes g)(x_1,...,x_{n+m}) = f(x_1,...,x_n)g(x_{n+1},...,x_{n+m})$ is defined for all functions $f \in S(\mathbf{R}^{4n})$ and $g \in S(\mathbf{R}^{4m})$.

E3. *Symmetry*

$$(s_n, f^\pi) = (s_n, f) \tag{4.5}$$

for all permutations $\pi \in P_n$ and for all functions $f \in S(\mathbf{R}^{4n}_0)$.

E4. *Cluster property*

$$\lim_{t\to\infty} (s_{n+m}, \theta f_n^* \otimes (g_m)_{(ta,1)}) = (s_n, \theta f_n^*)(s_m, g_m) \tag{4.6}$$

for all $f_n \in S(\mathbf{R}^{4n}_< \cap \mathbf{R}^{4n}_+)$, $g_m \in S(\mathbf{R}^{4m}_< \cap \mathbf{R}^{4m}_+)$, $a = (0, \mathbf{a})$, $\mathbf{a} \in \mathbf{R}^3$.

Let us consider the restriction of the distribution $s_n \in S'(\mathbf{R}^{4n}_0)$ on the test functions from the space $S(\mathbf{R}^{4n}_<)$. Then the translation invariance (4.3) implies

$$s_n(x_1,...,x_n) = S_{n-1}(x_2 - x_1,...,x_n - x_{n-1}), \tag{4.7}$$

where the distribution $S_{n-1}(\underline{x}) \in S'(\mathbf{R}_+^{4(n-1)})$. We note that for the function $g(x_1,...,x_{n+1}) = f(x_2-x_1,...,x_{n+1}-x_n)$ the definitions (4.1) imply the equality $\theta g^*(x_1,...,x_{n+1}) = (\theta_p f^*)(x_2-x_1,...,x_{n+1}-x_n)$, where the involution

$$\theta_p f(x_1,...,x_n) = f(-\theta x_1,...,-\theta x_n) \tag{4.8}$$

leaves the space $S(\mathbf{R}_+^{4n})$ invariant.

We substitute into the inequality (4.4) the sequence consisting of single function

$$f_{m+1}(x_1,...,x_{m+1}) = \begin{cases} \bar{\phi}_1(x_1)\phi_m(x_2-x_1,...,x_{m+1}-x_m), & m>0 \\ \bar{\phi}_1(x_1), & m=0, \end{cases} \tag{4.9}$$

where the functions $\phi_1 \in S(\mathbf{R}_+^4)$ and $\phi_m \in S(\mathbf{R}_+^{4m})$. Then by using the definitions (4.7) and (4.8) we can rewrite the inequality (4.4) for $m=0$ in the form (3.9) and for $m>0$ in the following form

$$\int d^4x d^4y S_{2m+1}(\theta_p\phi_m^*, x-\theta y, \phi_m)\bar{\phi}_1(x)\phi_1(y) \geq 0. \tag{4.10}$$

Here we introduce the distribution

$$\int d^4x S_{n+m+1}(f_n, x, f_m) f_1(x) = \int d^{4(n+m+1)}x S_{n+m+1}(\underline{x}) f_n \otimes f_1 \otimes f_m(\underline{x}), \tag{4.11}$$

constructed from the distribution $S_{n+m+1}(\underline{x}) \in S'(\mathbf{R}_+^{4(n+m+1)})$ and the test functions $f_n \in S(\mathbf{R}_+^{4n})$, $f_m \in S(\mathbf{R}_+^{4m})$. For $n=0$ or $m=0$ the distribution (4.11) is defined in an obvious way.

The inequalities (4.10) show that the distributions $S_{2m+1}(\theta_p\phi_m^*, x, \phi_m)$ are extremely significant. We formulate the new axiom exactly for these distributions.

E5. *Weak spectral condition*

Let $S_{2m+1}(\underline{x}) \in S'(\mathbf{R}_+^{4(2m+1)})$, $m = 1,2,...$, be any distribution defined by the relation (4.7). Then there is the natural number K such that for any integers $k > K$, $1 \leq l \leq m$ and for all test functions $\psi_i(x) \in S(\mathbf{R}^4)$, $i = 1,...,l$, $\psi_j(x) \in S(\mathbf{R}_+^4)$, $j = l+1,...,m+1$, there exists the limit

$$\lim_{n_1,...,n_l\to\infty, n_i\in\mathbf{Z}} \lim_{T_1,...,T_l\to-\infty, T_i\in\mathbf{R}} \int d^{4(2m+1)}x \Big(\prod_{i=1}^{l} x_i^0 x_{2m+2-i}^0\Big)^{-k} S_{2m+1}(\underline{x}) \times$$
$$[(L_c^{-1}[\psi_1\otimes\cdots\otimes\psi_m]_{x_1^0,...,x_l^0}(\bullet;\underline{n},\underline{T}))\otimes\psi_{m+1}\otimes$$
$$\theta_p(L_c^{-1}[\psi_1\otimes\cdots\otimes\psi_m]_{x_1^0,...,x_l^0}(\bullet;\underline{n},\underline{T}))^*](\underline{x}), \tag{4.12}$$

where the function

$$L_c^{-1}[\psi_1 \otimes \cdots \otimes \psi_m]_{x_1^0,...,x_l^0}(\underline{x}; \underline{n}, \underline{T})) = (\prod_{i=1}^{l} L_c^{-1}[\psi_i]_{x_i^0}(x_i; n_i, T_i))(\prod_{i=l+1}^{m} \psi_i(x_i)) \tag{4.13}$$

and the function $L_c^{-1}[\psi]_{x^0}(x; n, T)$ is given by the equality (2.15).

Theorem 2.5 clarifies this weak spectral condition.

Let us remind the Wightman axioms [8] for the Wightman distributions. The set of the Wightman distributions $\{w_n\}$ is a sequence of distributions with the following properties

R0. *Temperedness*

$w_0 \equiv 1$, $w_n \in S'(\mathbf{R}^{4n})$ and

$$\overline{(w_n, f)} = (w_n, f^*) \tag{4.14}$$

for all $f \in S(\mathbf{R}^{4n})$.

R1. *Relativistic invariance*

$$(w_n, f_{(a,\Lambda)}) = (w_n, f) \tag{4.15}$$

for all vectors $a \in \mathbf{R}^4$, for all Lorentz transformations $\Lambda \in L_+^\uparrow$ and for all functions $f \in S(\mathbf{R}^{4n})$, where the function $f_{(a,\Lambda)}(\underline{x}) = f(\Lambda^{-1}(x_1-a), ..., \Lambda^{-1}(x_n-a))$.

R2. *Positivity*

$$\sum_{n,m} (w_{n+m}, f_n^* \otimes f_m) \geq 0 \tag{4.16}$$

for all finite sequences of the functions $f_n \in S(\mathbf{R}^{4n})$.

R3. *Local commutativity*

For all natural numbers $n > 0$ and $j = 1, ..., n-1$

$$w_n(x_1, ..., x_{j+1}, x_j, .., x_n) = w_n(x_1, ..., x_j, x_{j+1}, .., x_n) \tag{4.17}$$

if the vector $x_{j+1} - x_j \in \mathbf{R}^4$ is spacelike: $(x_{j+1} - x_j, x_{j+1} - x_j) \equiv (x_{j+1}^0 - x_j^0)^2 - \sum_{i=1}^3 (x_{j+1}^i - x_j^i)^2 < 0$.

R4. *Cluster property*

$$\lim_{\lambda \to \infty} w_{n+m}(x_1, ..., x_n, x_{n+1} + \lambda a, ..., x_{n+m} + \lambda a) = w_n(x_1, ..., x_n) w_m(x_{n+1}, ..., x_{n+m}) \tag{4.18}$$

for all natural numbers $n, m > 0$ and for all spacelike vectors $a \in \mathbf{R}^4$.

R5. *Spectral condition*

For all natural numbers $n > 1$ there exists the tempered distribution $W_{n-1}^{\sim} \in S'(\mathbf{R}^{4(n-1)})$ with support in $\overline{V}_+^{\times n}$, where $\overline{V}_+$ is the closed forward light cone, such that

$$w_n(\underline{x}) = \int d^{4(n-1)}p W^{\sim}_{n-1}(\underline{p}) \exp\{i \sum_{j=1}^{n-1} (p_j, (x_{j+1} - x_j))\}. \tag{4.19}$$

Now we are able to formulate the revised Osterwalder–Schrader theorem.

Theorem 4.1. *To a given sequence of Wightman distributions satisfying R0–R5, there corresponds a unique sequence of Schwinger functions with the properties E0–E5. To a given sequence of Schwinger functions satisfying E0–E5, there corresponds a unique sequence of Wightman distributions with the properties R0–R5.*

Proof. We start from a relativistic field theory given by a sequence of Wightman distributions, satisfying the axioms R0–R5. Due to Theorem 3.5 from [8] the Wightman distribution w_n is the boundary value of the Wightman function $w_n(z_1, ..., z_n) = W_{n-1}(z_2 - z_1, ..., z_n - z_{n-1})$, where the function $W_{n-1}(z_1, .., z_{n-1})$ is analytic in the tube $T_{n-1} = \{z_1, ..., z_{n-1} | \operatorname{Im} z_i \in V_+, i = 1, ..., n-1\}$. The Wightman function $w_n(z_1, ..., z_n)$ is Lorentz invariant (Lorentz covariant for the theories of arbitrary spinor fields). The Bargmann Hall Wightman theorem [8, Theorem 2.11] implies that the function $W_n(z_1, ..., z_n)$ allows a single valued $L_+(\mathbf{C})$ invariant ($L_+(\mathbf{C})$ covariant for the theories of arbitrary spinor fields) analytic continuation into the extended tube $T_n' = \cup_{A \in L_+(\mathbf{C})} AT_n$. By using Theorem 3.6 from [8] we conclude that the function $w_n(z_1, ..., z_n)$ has an $L_+(\mathbf{C})$ invariant, single valued, symmetric under the permutations analytic continuation into the domain $IT_n^p = \{z_1, ..., z_n | (z_{\pi(2)} - z_{\pi(1)}, ..., z_{\pi(n)} - z_{\pi(n-1)}) \in T_{n-1}'$ for some permutation $\pi(1), ..., \pi(n)$ of the numbers $1, ..., n\}$. (For the theories of arbitrary spinor fields this function has an $L_+(\mathbf{C})$ covariant, single valued analytic continuation into the domain IT_n^p with obvious symmetry properties under the permutations.) The set IT_n^p contains the set of the Euclidean points $E_n = \{z_1, ..., z_n | \operatorname{Re} z_k^0 = 0, \operatorname{Im} \mathbf{z}_k = 0, z_k \neq z_j \text{for all } 1 \leq k, j \leq n, k \neq j\}$. The restriction of the Wightman functions to Euclidean points defines the Schwinger functions

$$s_n(x_1, ..., x_n) = w_n((ix_1^0, \mathbf{x}_1), ..., (ix_n^0, \mathbf{x}_n)). \tag{4.20}$$

The derivation of the extended Osterwalder–Schrader axioms E0–E5 from the Wightman axioms follows the arguments of the paper [1] and of Theorem 2.5.

Let $\{s_n\}$ be a sequence of distributions satisfying the extended Osterwalder–Schrader axioms E0–E5. If we substitute into the inequality (4.4) the sequence consisting of single function (4.9) for $m = 0$ we get the inequality (3.9) for the distribution $S_1(x)$. Due to Theorem 3.4 this distribution is the Laplace transform with respect to the time variable of a tempered distribution with support in the closure $\overline{\mathbf{R}_+^4}$. Let us substitute into the inequality (4.4) the sequence consisting of two functions $f_{n+1}(x)$ and $f_{m+1}(x)$ of type (4.9) with the same function $\phi_1(x)$. Then we obtain the following inequality

$$\int d^4x d^4y S\{\phi_n, \phi_m\}(x - \theta y) \overline{\phi_1(x)} \phi_1(y) \geq 0, \tag{4.21}$$

where the distribution

$$S\{\phi_n,\phi_m\}(x)=S\{\phi_m,\phi_n\}(x)=S_{2m+1}(\theta_p\phi_m^*,x,\phi_m)+S_{2n+1}(\theta_p\phi_n^*,x,\phi_n)+$$
$$S_{m+n+1}(\theta_p\phi_m^*,x,\phi_n)+S_{m+n+1}(\theta_p\phi_n^*,x,\phi_m). \tag{4.22}$$

This definition may be easily modified for the case $n=0$ or $m=0$

$$S\{\lambda,\phi_m\}(x)=S\{\phi_m,\lambda\}(x)= \tag{4.23}$$
$$S_{2m+1}(\theta_p\phi_m^*,x,\phi_m)+|\lambda|^2S_1(x)+\lambda S_{m+1}(\theta_p\phi_m^*,x)+\overline{\lambda}S_{m+1}(x,\phi_m),$$

where λ is a complex number. The equality (4.22) implies that $S\{\phi_n,\phi_n\}(x)=4S_{2n+1}(\theta_p\phi_n^*,x,\phi_n)$. Hence the inequality (4.10) is the particular case of the inequality (4.21) for $m=n$. It follows from the definitions (4.22), (4.23) that the distribution (4.11) is the linear combination of the distributions (4.22) and (4.23)

$$S_{m+n+1}(\phi_m,x,\phi_n)=1/2S\{\phi_n,\theta_p\phi_m^*\}(x)+i/2S\{\phi_n,i\theta_p\phi_m^*\}(x)-$$
$$(1+i)/2S_{2m+1}(\phi_m,x,\theta_p\phi_m^*)-(1+i)/2S_{2n+1}(\theta_p\phi_n^*,x,\phi_n). \tag{4.24}$$

In particular for $m=0$ or $n=0$ and $\phi_0=1$ we get

$$S_{n+1}(x,\phi_n)=1/2S\{1,\phi_n\}(x)+i/2S\{i,\phi_n\}(x)-$$
$$(1+i)/2S_{2n+1}(\theta_p\phi_n^*,x,\phi_n)-(1+i)/2S_1(x), \tag{4.25}$$

$$S_{n+1}(\phi_n,x)=1/2S\{1,\theta_p\phi_n^*\}(x)+i/2S\{1,i\theta_p\phi_n^*\}(x)-$$
$$(1+i)/2S_{2n+1}(\phi_n,x,\theta_p\phi_n^*)-(1+i)/2S_1(x). \tag{4.26}$$

The inequalities (4.21) imply that for any function $\phi_n\in S(\mathbf{R}_+^4)$ every of four distributions, depending on the variable x, in the right–hand side of the equality (4.25) is proportional to the distribution from $S'(\mathbf{R}_+^4)$, satisfying the Osterwalder–Schrader positivity condition (3.9). Due to Lemma 3.2 the limits (3.11) and (3.12) are equal to zero for the distribution $(x^0)^{-k}S_{n+1}(x,\phi_n)\in S'(\mathbf{R}_+^4)$ if the integer $k>0$. It follows from Proposition 3.3 that for the distribution $S_{n+1}(\underline{x})\in S'(\mathbf{R}_+^{4(n+1)})$ there exists the limit (2.15) for the integers $l=1$, $j_1=1$, $K=0$. By the definition the support of the distribution $L_c^{-1}[(x_1^0)^{-k}S_{n+1}]_{x_1^0}(x_1,...,x_{n+1})$ with respect to the first variable x_1 is in the closure $\overline{\mathbf{R}}_+^4$. Theorem 3.4 and Lemma 3.2 imply that for all functions $\psi_i\in S(\mathbf{R}_+^4)$, $i=1,...,n+1$, and for any integer $k>0$ the following relation holds

$$\int d^{4(n+1)}xS_{n+1}(\underline{x})\prod_{i=1}^{n+1}\psi_i(x_i)=\int d^{4(n+1)}x(\frac{\partial}{\partial x_1^0})^kL_c^{-1}[(x_1^0)^{-k}S_{n+1}]_{x_1^0}(\underline{x})\times$$
$$\int dy_1^0\exp\{-x_1^0y_1^0\}\psi_1(y_1^0,\mathbf{x}_1)\prod_{i=2}^{n+1}\psi_i(x_i) \tag{4.27}$$

In view of the equality (4.26) all above results are valid for the distribution $S_{n+1}(\phi_n, x) \in S'(\mathbf{R}^4_+)$, where the function $\phi_n \in S(\mathbf{R}^{4n}_+)$.

The weak spectral condition E5 implies the existence of the limit

$$\lim_{m\to\infty}\lim_{T\to-\infty}\int d^{4(2n+1)}x S_{2n+1}(\underline{x})(x_1^0)^{-k}L_c^{-1}[\psi_1]_{x_1^0}(x_1;m,T)\times \tag{4.28}$$
$$(\prod_{i=2}^{n+1}\psi_i(x_i))\theta_p((\prod_{i=2}^{n}\overline{\psi}_i(x_{2n+2-i}))(x_{2n+1}^0)^{-k}L_c^{-1}[\overline{\psi}_1]_{x_{2n+1}^0}(x_{2n+1};m,T))$$

for some positive integer k and for all functions $\psi_1(x) \in S(\mathbf{R}^4)$, $\psi_i(x) \in S(\mathbf{R}^4_+)$, $i = 2, ..., n+1$. Here the function $L_c^{-1}[\psi_1]_{x^0}(x;m,T)$ is defined by the relation (2.15). The linear functional (4.28) with respect to the function $\psi_{n+1}(x) \in S(\mathbf{R}^4_+)$ is the tempered distribution from the space $S'(\mathbf{R}^4_+)$. It satisfies the Osterwalder–Schrader positivity condition (3.9). The similar arguments may be applied for the distributions $S\{1, \theta_p(\prod_{i=1}^n \psi_i(x_i))^*\}(x)$ and $S\{1, i\theta_p(\prod_{i=1}^n \psi_i(x_i))^*\}(x)$ in the right–hand side of the equality of type (4.26). Thus the limit (2.15) $\left(L_c^{-1}[(x_1^0)^{-k}S_{n+1}]_{x_1^0}(\underline{x}), (\prod_{i=1}^{n+1}\psi_i(x_i))\right)$, the existence of which is proved above, for some positive integer k and for all functions $\psi_1(x) \in S(\mathbf{R}^4)$, $\psi_i(x) \in S(\mathbf{R}^4_+)$, $i = 2, ..., n+1$, has the decomposition of type (4.26) into four distributions with respect to the function $\psi_{n+1}(x) \in S(\mathbf{R}^4_+)$. These distributions are proportional to the distributions from $S'(\mathbf{R}^4_+)$ satisfying the Osterwalder–Schrader positivity condition (3.9). Now Proposition 3.3 implies that for the distribution $S_{n+1}(\underline{x}) \in S'(\mathbf{R}^{4(n+1)}_+)$ there exists the limit (2.15) for $l = 2$, $j_1 = 1$, $j_2 = n+1$ and for some positive integer k. Due to the definition the supports of this limiting distribution $L_c^{-1}\ [(x_1^0x_{n+1}^0)^{-k}S_{n+1}]_{x_1^0,x_{n+1}^0}\ (\underline{x})$ with respect to the first and the last variables are in the closure $\overline{\mathbf{R}}^4_+$. Theorem 3.4, Lemma 3.2 and the relation (4.27) imply that for sufficiently large positive integer k and for all functions $\psi_i(x) \in S(\mathbf{R}^4_+)$, $i = 1, ..., n+1$, the following relation holds

$$\int d^{4(n+1)}x S_{n+1}(\underline{x})\prod_{i=1}^{n+1}\psi_i(x_i) \tag{4.29}$$
$$= \int d^{4(n+1)}x(\frac{\partial^2}{\partial x_1^0\partial x_{n+1}^0})^k L_c^{-1}[(x_1^0x_{n+1}^0)^{-k}S_{n+1}]_{x_1^0,x_{n+1}^0}(\underline{x})\times$$
$$\int dy_1^0 dy_{n+1}^0 \exp\{-\sum_{i=1,n+1} x_i^0y_i^0\}(\prod_{i=1,n+1}\psi_i(y_i^0,\mathbf{x}_i))\prod_{i=2}^{n}\psi_i(x_i).$$

By using the weak spectral condition E5 and the equalities (4.24) it is possible to prove step by step that there exists the limit (2.15) for the distribution $S_{n+1}(\underline{x}) \in S'(\mathbf{R}^{4(n+1)}_+)$, for $l = n+1$ and for some positive integer k. By the definition the supports of this limiting distribution $L_c^{-1}\ [(\prod_{i=1}^{n+1} x_i^0)^{-k}S_{n+1}]_{\underline{x}^0}\ (\underline{x})$ with respect to any variable is in the closure $\overline{\mathbf{R}}^4_+$. Since the weak convergence in the space S' implies the convergence in the topology of the space S' (see [7, Section 3.7]) the limit $(L_c^{-1}\,[(\prod_{i=1}^{n} x_i^0)^{-k} S_{n+1}]_{\underline{x}^0}\,(\underline{x}), (\prod_{i=1}^{n+1} \psi_i(x_i)))$ is continuous in each function $\psi_i(x) \in S(\mathbf{R}^4)$. Hence the nuclear theorem [6, Chapter 1, Section 1, Theorem 6] implies that $L_c^{-1}\,[(\prod_{i=1}^{n+1} x_i^0)^{-k} S_{n+1}]_{\underline{x}^0}\,(\underline{x}) \in S'(\mathbf{R}^{4(n+1)})$. Its support is in the closure $\overline{\mathbf{R}}_+^{4(n+1)}$. The application step by step of Theorem 3.4 and Lemma 3.2 gives for sufficiently large positive integer k and for all functions $\psi_i(x) \in S(\mathbf{R}_+^4)$, $i = 1, ..., n+1$ the following relation

$$\int d^{4(n+1)}x S_{n+1}(\underline{x}) \prod_{i=1}^{n+1} \psi_i(x_i) \tag{4.30}$$

$$= \int d^{4(n+1)}x \Big(\frac{\partial^{n+1}}{\partial x_1^0 \cdots \partial x_{n+1}^0}\Big)^k L_c^{-1}[(\prod_{i=1}^{n+1} x_i^0)^{-k} S_{n+1}]_{\underline{x}^0}(\underline{x}) \times$$

$$\int dy_1^0 \cdots dy_{n+1}^0 \exp\{-\sum_{i=1}^{n+1} x_i^0 y_i^0\}\Big(\prod_{i=1}^{n+1} \psi_i(y_i^0, \mathbf{x}_i)\Big).$$

Therefore for any $n = 1, 2, ...$ the distribution $S_n(\underline{x}) \in S'(\mathbf{R}_+^{4n})$ is the Laplace transform of the tempered distribution from $S'(\mathbf{R}^{4n})$ with support in the closure $\overline{\mathbf{R}}_+^{4n}$.

Now the derivation of the Wightman axioms R0–R5 from the Osterwalder–Schrader axioms E0–E4 follows the arguments of the paper [1] .

References

1. Osterwalder, K., Schrader, R. : Axioms for Euclidean Green's Functions. Commun. Math. Phys. **31**, 83–112 (1973)
2. Osterwalder, K., Schrader, R. : Axioms for Euclidean Green's Functions 2. Commun. Math. Phys. **42**, 281–305 (1975)
3. Glaser, V. : On the Equivalence of the Euclidean and Wightman Formulation of Field Theory. Commun. Math. Phys. **37**, 257–272 (1974)
4. Bernstein, S. : Sur les fonctions absolument monotones. Acta Math. **52**, 1–66 (1929)
5. Zinoviev, Yu.M. : Equivalence of the Euclidean and Wightman Field Theories. (submitted to Commun. Math. Phys.) hep-th/9408009
6. Gel'fand, I.M., Vilenkin, N.Ya. : Generalized Functions, Vol. 4. New York : Academic Press 1964
7. Vladimirov, V.S. : Methods of Theory of Functions of Many Complex Variables. Cambridge MA : MIT Press 1966
8. Streater, R.F., Wightman, A.S. : PCT, spin and statistics and all that. New York, Amsterdam : Benjamin 1964

Construction of the Gross–Neveu Model in Dimension 3

Claude de Calan

Centre de Physique théorique
Ecole Polytechnique
91128 Palaiseau Cedex - France

Abstract We review the construction of a non-renormalizable field theory, the Gross-Neveu model in three Euclidean dimensions with many components.

1 Introduction

We give in this paper the main steps of the construction of the Gross-Neveu model in three Euclidean dimensions, as roughly described in [1] and explained in [2]. Such a four-fermion interaction model has already been built in two dimensions [3]. But the situation for the dimension $d = 3$ is very different, and the main new features of our result can be summarized in the following way :

i) The usual perturbative expansion - power series of the renormalized coupling constant - is not renormalizable, the ultra-violet degree of divergence increases with the order of perturbation. The construction of some perturbatively non-renormalizable models has been performed [4], but these models are expected to violate the Osterwalder-Schrader reflection-positivity axiom. Conversely, we have no reason to expect such a violation in our model, though a complete proof of the axioms would be difficult, and is unfortunately not achieved at the present time.

The way to overcome perturbative non-renormalizability follows the ideas suggested in [5] : as explained below, the "chains of bubbles", which are the only graphs surviving in the limit $N \to \infty$ of an infinite number of components, are summed first. This organization leads to formal renormalizability. In other words, the phase-space expansion we perform does not start from the free fermion model, but from the solvable (non trivial but Gaussian) $N \to \infty$ model.

ii) In contrast with most of the previously constructed models, the three-dimensional Gross-Neveu model is not asymptotically free. As already remarked in [5], it presents a non-trivial ultra-violet fixed point. However, in the frame of the organization mentioned above, this fixed point goes to zero when N goes to infinity. Therefore $1/N$ is the only small parameter we are left with, and the convergence of the phase-space expansion can be proved only for N large enough (but finite).

In section 2 we give our precise starting point, after some formal manipulations, and we state the existence theorem. In section 3 we sketch the general scheme of the phase space expansion, and we comment on the main specific difficulties, with the appropriate modifications, in section 4 and 5. Finally section 6 shows the flow equations and indicates the argument for the validity of the ansatzes given in section 2.

2 Definition of the Model

The Gross-Neveu model corresponds to the Lagrangian density

$$\mathcal{L} = \bar{\psi}(i\zeta\partial\!\!\!/ + m)\psi + \frac{\lambda}{2N}(:\bar{\psi}\psi:)^2 \tag{1}$$

where ψ is a fermion field with N components ("colors").

As announced in the introduction, for $d = 3$ Euclidean dimensions, the model is not perturbatively renormalizable : the ultra-violet divergences of the usual Feynman amplitudes would be worst and worst with increasing perturbation orders. This is a hopeless situation from the traditional viewpoint. But let us assume that the "bubble" given by the second order graph

$$\pi(p) = \frac{\lambda}{N}\int \frac{d^3k}{(k\!\!\!/ + m)(p\!\!\!/ + k\!\!\!/ + m)} \tag{2}$$

is renormalized by a counterterm $a(:\bar{\psi}\psi:)^2$, and that the chains of renormalized bubbles $\pi^{\mathrm{ren}}(p)$ are summed first, by grouping together all the Feynman graphs having the same structure, modulo the chains of bubbles. This summation introduces

$$C(p) = \frac{\lambda}{N} + \frac{\lambda}{N}\pi^{\mathrm{ren}}(p) + \frac{\lambda}{N}[\pi^{\mathrm{ren}}(p)]^2 + \cdots = \frac{\lambda/N}{1+\pi^{\mathrm{ren}}(p)} \tag{3}$$

where $C(p)$ can be considered as the (non local) covariance, or propagator, of a boson field σ. Since $\pi^{\mathrm{ren}}(p)$ increases linearly with p at large momentum,

$$C(p) \underset{p\to\infty}{\sim} \frac{1}{p} \tag{4}$$

Then the Gross-Neveu model is equivalent to the model defined by

- a fermion field ψ with covariance $S(p) = \frac{1}{\zeta p\!\!\!/ + m}$ (5)
- a boson field σ with covariance $C(p) = \frac{1}{1+\pi^{\mathrm{ren}}(p)}$ (6)
- a boson-fermion-fermion vertex with coupling constant $\sqrt{\frac{\lambda}{N}}$ (7)

The power-counting in this new formulation leads to formal renormalizability, the degree of ultra-violet divergence being $3-n$, where n is the number of the new three-bodies vertices. (Of course this number has nothing to do with the usual power of the coupling constant λ, λ appearing also in the denominator of the covariance C). Furthermore the 3-bosons Schwinger function is not divergent : one power is gained from a parity argument (or due to the vanishing trace of an odd number of γ-matrices, if we work with 4×4 Dirac matrices).

Since the chains of bubbles are the dominant graphs for N large, we see that this new organization amounts to start from the solvable Gaussian model at $N \to \infty$.

How to implement the bubble summation ? The simplest way is to use the Matthews-Salam formalism, where the exponential of the interaction

$$e^{-\frac{\lambda}{2N}\int(:\bar{\psi}\psi:)^2} \tag{8}$$

is replaced by

$$\int d\mu_1(\sigma)e^{\sqrt{\frac{\lambda}{N}}\int :\bar{\psi}\psi:\sigma} \tag{9}$$

In this preliminary formula, σ is an "ultra-local" boson field with covariance $C_1(x,y) = \delta(x-y)$. It is convenient to rescale C_1 by a constant ν (defined in the following) and to define g by $\lambda = \frac{g^2}{\nu}$. Then we can explicitly integrate over the fermion field, and we obtain for the (unnormalized) n-point Schwinger function :

$$\mathcal{S}^{(n)} = \int d\mu_{\nu^{-1}}(\sigma)\det{}_2(1+K)Tr(\wedge\frac{1}{1+K}X) \tag{10}$$

where the boson covariance is

$$C_2(x,y) = \frac{1}{\nu}\delta(x-y) \tag{11}$$

and

$$K = \frac{g}{\sqrt{N}}S\sigma \tag{12}$$

$$S = \frac{1}{\zeta \not{p} + m} \tag{13}$$

$$\det{}_2(1+K) = e^{Tr\ \ell n(1+K) - Tr\ K} \tag{14}$$

and X is a projector on the external states.

Now we consider the bubble

$$\sigma\pi\sigma = Tr\ K^2 \tag{15}$$

and we introduce

$$e^{-\frac{1}{2}Tr\ K^2}e^{+\frac{1}{2}Tr\ K^2} \tag{16}$$

the first factor being used to change the C_2 covariance into the non-local covariance

$$C = \frac{1}{\nu + \pi(p)}\ . \tag{17}$$

Therefore the n-point Schwinger function reads

$$\mathcal{S}^{(n)} = \int d\mu_c(\sigma)\det{}_3(1+K)Tr(\Lambda\frac{1}{1+K}X) \tag{18}$$

where

$$\det{}_3(1+K) = e^{Tr\ \ell n(1+K) - Tr\ K + \frac{1}{2}Tr\ K^2} \tag{19}$$

In these formulas, ζ, m, g, ν are the bare parameters. To give a sense to these formal expressions, we have to introduce a volume cut-off and an ultra-violet cut-off by

$$S(x,y) \to S_{L,\rho}(x,y) = \chi_L(x) S_\rho(x,y) \chi_L(y) \tag{20}$$

In the momentum representation

$$S_\rho(p) = \frac{\eta_\rho(p)}{\zeta_\rho \not{p} + m} \tag{21}$$

where η_ρ is a C^∞ function with compact support :

$$\eta_\rho(p) = \begin{cases} 1 \text{ for } |p| < M^\rho \\ 0 \text{ for } |p| > \frac{4}{3} M^\rho \end{cases} \tag{22}$$

$M > 1$ being some fixed constant, and χ_L the characteristic function of some box L. For the boson covariance, it is convenient (see below the "painting expansion") to replace C by

$$C \to C_{L,\rho} = \chi_L C_\rho^{1/2} \chi_L C_\rho^{1/2} \chi_L \tag{23}$$

with

$$C_\rho(p) = \frac{\eta_\rho}{\nu_\rho + \pi_\rho(p)} \tag{24}$$

Moreover, we distinguish between the contributions to the fermion-mass renormalization coming from the two-fermions functions (included in m_ρ) and from the one-boson functions, explicitly isolated in a $T_\rho \sigma$ counterterm.

Then our well-defined starting point is

$$S^{(n)} = \int d\mu_{C_{L,\rho}}(\sigma) e^{\int T_\rho \sigma} \det{}_3(1 + K_{L,\rho}) Tr(\Lambda \frac{1}{1 + K_{L,\rho}} X) \tag{25}$$

where

$$K_{L,\rho} = \frac{g_\rho}{\sqrt{N}} S_{L,\rho} \;\; \sigma \tag{26}$$

Our result is the following

Theorem.— *There exist convenient choices of ansätze for the bare parameters, such that the (normalized) Schwinger functions have finite limits when the ultra-violet and volume cut-offs go to infinity.*

We will examine the flow equations at the end, and we will claim that the theorem is true with the following bare parameters (up to factors exponentially going to 1 at infinite ρ) :

$$\begin{cases} m_\rho = m_{\text{ren}} M^{\alpha\rho} \\ \nu_\rho = \nu_{\text{ren}} M^{(1+\varepsilon)\rho} \\ g_\rho = g_{\text{ren}} M^{\gamma\rho} \\ \zeta_\rho = \zeta_{\text{ren}} M^{\delta\rho} \end{cases} \tag{27}$$

α, ε, γ, δ being functions of N going to zero when N goes to infinity.

3 The Phase Space Expansion

To see that the above ansatzes are convenient, we have to perform a renormalization group analysis, i.e. to look at the various scales of momenta. A practical way of doing this is to use a discrete version of the renormalization group equations, by dividing the range of momenta from 0 to M^ρ into "slices" of momenta, essentially between M^{i-1} and M^i, $i \leq \rho$.

Let $\eta^i(p)$ be a C^∞ function with compact support (roughly in $M^{i-1} < |p| < M^i$) such that

$$\sum_{i=1}^{\rho} \eta^i(p) = \eta_\rho(p) \tag{28}$$

Then

$$C_\rho = \sum_{i=1}^{\rho} C^i \quad ; \quad C^i(p) = \frac{\eta^i(p)}{\nu_\rho + \pi_\rho(p)} \tag{29}$$

and we write similar formulas for the fermion covariance S_ρ.

We do not give here the complete description of the phase space expansion, which can be found for example in [6] (with other references therein). This expansion is essentially what is often called the "block-spin" formalism. Let us simply summarize the main steps and properties of the expansion :

3.1 The Cluster (or "Horizontal") Expansion

For each slice i, the whole box L is divided into cubes with side M^{-i}, and a cluster expansion is performed over the cubes by introducing a set of parameters h. The covariance C^i (and similarly S^i) is replaced by an interpolated covariance $C^i(\{h\})$ such that

- for all h being equal to 1, we recover the original covariance C^i.

- for any $h = 0$, there is a decoupling between some regions of space, and the Schwinger functions factorize into functions with fields localized in these regions.

Then, for each parameter h, we write

$$\mathcal{S}^{(n)}\,|_{h=1} = \mathcal{S}^{(n)}\,|_{h=0} + \int_0^1 dh \frac{d}{dh} \mathcal{S}^{(n)}(h) \tag{30}$$

The differential operations $\frac{d}{dh}$ generate explicit covariances C^i or S^i with the corresponding localizations. We call these covariances the "horizontal links".

3.2 The Momentum–Decoupling (or "Vertical") Expansion

The interaction kernel

$$K = \sum_{i,j,k} K^{ijk} , \tag{31}$$

where the upper indices $i,\ j,\ k$ label the slices of the three fields at each vertex, is interpolated by introducing parameters v_i :

$$K = \sum_{i,j,k} K^{ijk} \prod_{\ell=\inf(i,j,k)}^{\sup(i,j,k)} v_\ell \tag{32}$$

in such a way that

- for all v being equal to 1, we recover the original K.

- for any $v_i = 0$, there is a decoupling between the fields in slices higher than i and the fields in slices lower than i, and a corresponding factorization of the Schwinger functions.

Then for each parameter v we write a formula similar to (30), up to fourth order :

$$\mathcal{S}^{(n)}\,|_{v=1} = \mathcal{S}^{(n)}\,|_{v=0} + \frac{d}{dv}\mathcal{S}^{(n)}\,|_{v=0} + \frac{1}{2}(\frac{d}{dv})^2\mathcal{S}^{(n)}\,|_{v=0} + \frac{1}{6}(\frac{d}{dv})^3\mathcal{S}^{(n)}\,|_{v=0} + \mathcal{R}^{(n)} \tag{33}$$

$$\mathcal{R}^{(n)} = \int_0^1 dv \frac{(1-v)^3}{6}(\frac{d}{dv})^4\mathcal{S}^{(n)} \tag{34}$$

The Taylor formula (33) is taken up to fourth order because we want to isolate explicitly the divergent contributions with 0, 1, 2 or 3 legs. The differential operations $\frac{d}{dv_i}$ generate explicit vertices with fields in slices higher than i **and** fields in slices lower than i ; these vertices are called "vertical links".

3.3 Additional Expansions

i) The normalization of the Schwinger functions by the 0-point (or vacuum) Schwinger function can be made explicit by a third expansion, called **the "Mayer expansion"** (see [6]), which generates "Mayer links".

ii) Each time a factor $\frac{1}{1+K}$ is produced by a differential operation, this factor is expanded up to second order, to make explicit the bubble cancellations.

iii) Actually in our particular model, a new kind of expansion, called **the "painting expansion"**, must be performed **before** all the other expansions. The motivation for this, and some details, are given in the next section.

Altogether, the various expansions express the Schwinger functions as sums of products of "multimers" with space and momentum localizations, and internal ("horizontal", "vertical" or "Mayer") links. All these multimers are ultra-violet convergent (i.e. decrease exponentially when the difference of slice indices between the internal and external fields increases), except

- the 0-leg multimers, which are cancelled, in the Mayer expansion, by the normalization of the Schwinger functions.

- the 1-boson multimers, which are cancelled by the linear $T_\rho\sigma$ counterterm.

- the 2-bosons (resp. 2-fermions ; resp. 1-boson and 2-fermions) multimers, which are written as their local part plus their renormalized (ultra-violet convergent) part. The local part, if the lowest slice is i (by $v_i = 0$) defines the partial counterterm $\delta\nu_i$ (resp. δm_i and $\delta\zeta_i$; resp. δg_i), which is used to generate the evolution of ν (resp. m and ζ ; resp. g) from ν_ρ to $\nu_1 = \nu_{\text{ren}}$:

$$\nu_i = \nu_\rho + \sum_{j=i+1}^{\rho} \delta\nu_j \tag{35}$$

(and similar equations for m_i, ζ_i, g_i).

We may get a first idea of the mechanism by looking at the first order in $1/N$. To this order, there is no T counterterm, nor $\delta m_i, \delta\zeta_i$ or δg_i contributions. The only surviving evolution is the evolution of ν from the bubbles. Now

$$C^i(p) = \frac{\eta^i(p)}{\nu_\rho + \sum_{j=1}^{\rho} \pi^j(p)} \tag{36}$$

where $\pi^j(p)$ is the sum of the bubbles in which the lowest fermion slice is j. Thus the denominator in (36) may be written

$$\nu_\rho + \sum_{j=i+1}^{\rho} \pi^j(0) + \sum_{j=i+1}^{\rho} [\pi^j(p) - \pi^j(0)] + \sum_{j=1}^{i} \pi^j(p) \tag{37}$$

and the last sum disappears (except for one or two slices) by momentum conservation, thanks to the $\eta^i(p)$ factor. Neglecting the fermion mass, we get

$$\pi_j(0) = -aM^j \tag{38}$$

where a is a positive constant given by a straightforward computation, and therefore

$$\pi^j(0) = \pi_j(0) - \pi_{j-1}(0) = -a(M-1)M^{j-1} \tag{39}$$

We now demand that

$$\delta\nu_i = \pi^i(0) \tag{40}$$

$$\nu_i = \nu_\rho + \sum_{j=i+1}^{\rho} \delta\nu_j \tag{41}$$

and we obtain

$$C^i(p) \simeq \frac{\eta^i(p)}{\nu_i + \sum_{j=i+1}^{\rho} \pi^j_{\text{ren}}(p)} \tag{42}$$

where

$$\pi^j_{\text{ren}}(p) = \pi^j(p) - \pi^j(0) \tag{43}$$

is positive. If we choose as the right ansatz

$$\nu_\rho = aM^\rho + b \tag{44}$$

where a is the constant defined in (38), we find

$$\nu_i = aM^\rho + b - a(M-1)(M^{\rho-1} + M^{\rho-2} + \cdots + M^i) = aM^i + b \tag{45}$$

which shows the stability of the ansatz. We will see in our last section the general case (for finite N) and the stability of the ansätze for ν, m, ζ, g.

4 The Painting Expansion

Along the usual arguments (see the general theorems in [6]) the convergence of the phase space expansion is essentially derived from the following properties :

- "horizontal" decreasing along the horizontal links, which comes from the decreasing at large distances of C and S. Here, due to the compact cut-off functions η^i, we use only a power behavior at large distances, with the right scaling law :

$$C^i(x,y) \underset{\text{large}|x-y|}{\sim} \frac{C^t}{(1+M^i|x-y|)^r} \tag{46}$$

- "vertical" decreasing along the vertical links, which comes from the ultraviolet convergence of the remaining multimers (or renormalized multimers).

- functional integrability leading to a bound by a small constant per cube of the multimers. The small constant is provided by the $\frac{1}{N}$ factors, for N large enough, but the functional integrability is hard to prove in our model. The main problem is : how to bound the action ?

The best uniform bound on a determinant gives

$$|\det{}_2(1+K)| < e^{\frac{1}{2}Tr\ KK^*} \tag{47}$$

leading to

$$|\det{}_3(1+K)| < e^{\frac{1}{2}Tr\ KK^* + \frac{1}{3}Tr\ K^2} = e^{\frac{1}{2}\sigma\pi_{\text{ren}}\sigma} \tag{48}$$

Looking at formula (42), one may think that this positive exponent can be controlled by the π_{ren} part of the covariance C (the ν_i part of the denominator in C^i is enough for the right power-counting). But after the phase space expansion is performed

$$\pi_{\text{ren}} \quad \text{is replaced by} \quad \chi\pi_{\text{ren}}\chi \tag{49}$$

$$\frac{1}{\nu+\pi_{\text{ren}}} \quad \text{is replaced by} \quad \chi\frac{1}{\nu+\pi_{\text{ren}}}\chi \tag{50}$$

where χ is the characteristic function of some region, the support of the considered multimer. Now (50) can control (49) (using eventually a **uniformly bounded** fraction of ν) only for regions with a large enough volume.

In order to get this favorable situation, we perform, **before** the various steps of the phase space expansion, what we call the "painting" expansion. This expansion defines in each slice "painted" and "unpainted" cubes, in such a way that the (square root of the) boson covariance vanishes if it has one end in an unpainted cube, while preserving the positivity of the modified covariance. We introduce a parameter s_Δ for each cube Δ in the box L and we interpolate the covariance by

$$C(\{s_\Delta\}) = \chi_L C^{1/2}(\sum_{\Delta\in L} s_\Delta\chi_\Delta)C^{1/2}\chi_L \tag{51}$$

Evidently, for all s_Δ being equal to 1, we recover the original covariance (23). Then for each s_Δ we write

$$S^{(n)}|_{s_\Delta=1} = S^{(n)}|_{s_\Delta=0} + \int_0^1 ds_\Delta \frac{d}{ds_\Delta}S^{(n)} \tag{52}$$

When we choose the first term in the right-hand side of (52), we say that the cube Δ is unpainted, and C vanishes if one end of $C^{1/2}$ lies inside Δ. When we choose the second term, we say that the cube Δ is painted. Thus the Schwinger function is written as a sum of terms, each of which correspond to a "painting", i.e. to the set of the choices for every cube in the box. (At the end of the expansions, this further sum amounts to a factor 2 per cube, which is compensated by taking N twice larger).

Then, for a given painting, we define the "cells" as the maximal connected sets of painted cubes, surrounded by a given number of unpainted cubes. And we start the cluster expansion as before, except that :

- the h-interpolation is introduced in $C^{1/2}$ instead of C.

- the basic regions for the localizations are not the cubes, but the cells of the painting.

Therefore we produce horizontal links only between distant painted cubes, while each differential operation $\frac{d}{ds_\Delta}$ gives us one $\frac{\delta}{\delta\sigma}\frac{dC}{ds_\Delta}\frac{\delta}{\delta\sigma}$, hence one vertex and one $\frac{1}{N}$ factor, per painted cube.

Remark : This delicate game must not be broken by the vertical expansion, which in the usual way would give back sharp localizations. This leads to introduce vertical parameters v, not for each cube, but for each polymer of the cluster expansion. Such a reduced set of vertical links is large enough for the vertical decreasing and the ultra-violet convergence.

5 The Domination Problem

As explained in [6], the divergence of the perturbative series, even in the absence of any ultra-violet trouble, is due to the $q!$ behavior of the qth order term, and comes from the fact that too many fields are produced at non-distant points : precisely the disease which is cured by using the cluster expansion instead of the perturbative expansion.

Now in our multiscale analysis, there is some remnant of this disease : we produce only a bounded number of fields per cube, but since there are $M^{3(i-j)}$ cubes of side M^{-i} in a cube of side M^{-j}, we may accumulate in a given cube of slice j, by the vertical expansion, many low-momentum fields produced in a higher momentum slice. Therefore we need a "domination" technique for controlling such low-momentum fields by an effective potential. For example in the $\lambda\varphi^4$ models, the effective potential is provided by $\exp(-\lambda\varphi^4)$ [6]. Here we can extract from our interaction Lagrangian the same kind of effective potential which has been used in the Yukawa model [7]. If σ, localized in a given cube Δ, is replaced by its mean value in this cube :

$$\sigma \to \bar{\sigma} = \frac{1}{|\Delta|}\int_\Delta \sigma \tag{53}$$

and if the fermion covariance S^i is replaced by a covariance S^i_P with periodic conditions in Δ, K is changed into $\widehat{K}$. One can then make an explicit computation and find for $\det_3(1+\widehat{K})$ a bound which is essentially of the following type (at leading order in $1/N$) :

$$|\det_3(1+\widehat{K})| < \begin{cases} \exp(-M^{-2i}\bar{\sigma}^2) \text{ if } \bar{\sigma} > M^i\sqrt{N} \\ \exp(-\frac{M^{-4i}}{N}\bar{\sigma}^4) \text{ if } \sqrt{N} < \bar{\sigma} < M^i\sqrt{N} \end{cases} \tag{54}$$

In the second case ($\bar{\sigma} < M^i\sqrt{N}$) we loose a factor $N^{1/4}$ for each low-momentum dominated field $\bar{\sigma}$, which leads to use for some multimers a convenient balance between domination and Gaussian integration. On the other hand, more technical work is required for bounding $\delta S = S - S_P$ and $\delta\sigma = \sigma - \bar{\sigma}$. If the mean value of σ is taken in a cube of side M^{-i}, $\delta\sigma$ behaves like a field of momentum M^i (even if σ has a lower momentum) and can be integrated with the Gaussian measure.

6 The Flow Equations

Here we describe only the evolution of the effective parameters in a generic slice i, with $1 \ll i \ll \rho$, neglecting the deviations which appear in the lowest or the highest slices. Moreover, assuming the behavior announced in (27), we forget the fermion masses $m_i \sim M^{\alpha i}$ which are negligible compared to the momenta $p \sim M^i$ for N large (hence α small). In the frame of these approximations, it can be seen from the definition of the local parts of the divergent multimers that $\frac{\delta\nu_i}{\nu_i}$, $\frac{\delta g_i}{g_i}$ and $\frac{\delta\zeta_i}{\zeta_i}$ are only functions of the ratios $\frac{\nu_j}{\nu_k}$, $\frac{\zeta_j}{\zeta_k}$ and of the following important quantities

$$\beta_j = \frac{g_j^2 M^j}{\nu_j \zeta_j^2} \tag{55}$$

for values of j and k larger than i.

From (55), one can interpret $\sqrt{\frac{\beta_j}{N}}$ as the conveniently scaled (by $M^{j/2}$) and conveniently normalized (by $\nu_j^{1/2}\zeta_j$) effective boson-fermion-fermion coupling constant of our model.

We show the stability of the ansatz inductively : assuming (27) is true with ρ replaced by j , $j \geq i+1$, we have to show that it remains true with ρ replaced by i. Actually we assume (27) (up to scale $i+1$) with the further relation

$$2\gamma = 2\delta + \varepsilon \tag{56}$$

From (56) we see that β_j is independent of j. The invariant quantity $\frac{\beta_j}{N} = \frac{\beta}{N}$ is the non-trivial fixed point we announced in the introduction.

To check the stability of the ansatz is equivalent to show that

$$\begin{cases} \frac{\delta g_i}{g_i} = \gamma \\ \frac{\delta\zeta_i}{\zeta_i} = \delta \\ \frac{\delta\beta_i}{\beta_i} = \ell n M + 2\frac{\delta g_i}{g_i} - \frac{\delta\nu_i}{\nu_i} - 2\frac{\delta\zeta_i}{\zeta_i} = 0 \end{cases} \tag{57}$$

Now if the ansatz is true in higher slices, we have for $j, k > i$:

$$\beta_j = \beta \; ; \; \frac{\nu_j}{\nu_k} = M^{(1+\varepsilon)(j-k)} \; ; \; \frac{\zeta_j}{\zeta_k} = M^{\delta(j-k)} \tag{58}$$

Functions of the differences $j-k$, when summed over all slices larger than i, become independent of i in the limit $\rho \gg i$. Therefore we have just to find numbers β, ε, δ, satisfying

$$\begin{cases} F_1(\beta,\varepsilon,\delta) = \delta + \frac{\varepsilon}{2}(=\gamma) \\ F_2(\beta,\varepsilon,\delta) = \delta \\ \ell n M + 2F_1(\beta,\varepsilon,\delta) - F_3(\beta,\varepsilon,\delta) - 2F_2(\beta,\varepsilon,\delta) = 0 \end{cases} \tag{59}$$

where F_1, F_2, F_3 are prescribed functions, given by the definition of the local parts of the divergent multimers.

The fact that the system of equations (58) has a solution, for N large enough, follows from the fact that it has a solution at infinite N (described at the end of section 3).

We stop here the summary of the construction : the detailed description will appear elsewhere, together with the complete proofs for the bounds and the convergence. Though our construction seems to involve many technicalities, it rests on rather simple ideas, already exhibited in [5]. But a proof that the Osterwalder-Schrader axioms are satisfied is still lacking, and because of the technicalities such a proof remains a difficult challenge. Let us also mention that the same kind of techniques could be applied to a problem which has many similarities with the present one : the infra-red behavior of the massless φ^4model in 3 dimensions.

References

[1] C. de Calan, P.A. Faria da Veiga, J. Magnen, R. Sénéor, Phys. Rev. Lett. vol.66 n°25, 3233 (1991).

[2] P.A. Faria da Veiga, Doctor Thesis, Université de Paris-Sud, July 1991.

[3] K. Gawedzki, A. Kupiainen, Commun. Math. Phys. 102, 1 (1985), J. Feldman, J. Magnen, V. Rivasseau, R. Sénéor, Commun. Math. Phys. 103, 67 (1986).

[4] G. Felder, Commun. Math. Phys. 102, 139 (1985), K.Gawedzki, A. Kupiainen, Phys. Rev. Lett. 55, 363 (1985) and Nucl. Phys. B 262, 33 (1985).

[5] G. Parisi, Nucl. Phys. B 100, 368 (1975), D. Gross, in "Methods in Field Theory", Les Houches 1975, R. Balian and J. Zinn-Justin eds, North-Holland (1976), B. Rosenstein, B. Warr, S. Park, Phys. Rev. Lett. 62, 1433 (1989) and Phys. Lett. B 219, 469 (1989), B. Rosenstein, B. Warr, Phys. Lett. B 218, 465 (1989).

[6] V. Rivasseau, "From perturbative to constructive renormalization" Princeton series in Physics, Princeton University Press (1991)

[7] J. Magnen, R. Sénéor, 3rd Conference on Collective Phenomena, J. Lebowitz et al. ed., The NY Academy of Sciences (1980).

Mass Generation in a One–Dimensional Fermi Model

P.A. Faria da Veiga[1*],
T.R. Hurd[2],
D.H.U. Marchetti[3**]

1 Instituto de Ciências Matemáticas
Universidade de São Paulo, C.P. 668
13560 São Carlos, SP, Brazil
2 Department of Mathematics & Statistics
McMaster University
Hamilton, Canada L8S 4K1
3 Instituto de Física
Universidade de São Paulo, C.P. 20516
01452 São Paulo, SP, Brazil

Abstract
We consider the one-dimensional Gross-Neveu model and, for a large number N of flavor components, prove the fermion develops a mass dynamically without breakdown of the discrete chiral symmetry.

1 Introduction

Given an even integer L and, for $\Lambda \subset \mathbf{Z}$, with $|\Lambda| = L$, we consider the joint probability measure $\mu_\Lambda(\phi)$ on $\mathbf{R}^L$, whose density $g_\Lambda(\phi)$ is defined as follows

$$g_\Lambda(\phi) = \frac{1}{Z_\Lambda} e^{-\frac{1}{2}|\phi|^2_\Lambda} F_\Lambda(\phi)^{2N} \tag{1}$$

where $|\phi|^2_\Lambda = \sum_{i=1}^L \phi_i^2$, Z_Λ is the partition function and, for $L \times L$ matrices

$$\Gamma_\Lambda = \begin{pmatrix} 0 & 1 & 0 \cdots 0 \\ -1 & 0 & 1 \cdots 0 \\ 0 & -1 & 0 \cdots 0 \\ \vdots & \vdots & \vdots \ddots \vdots \\ 0 & 0 & 0 \cdots 0 \end{pmatrix} ; \Phi_\Lambda = \begin{pmatrix} \phi_1 & 0 & 0 & \cdots & 0 \\ 0 & \phi_2 & 0 & \cdots & 0 \\ 0 & 0 & \phi_3 & \cdots & 0 \\ \vdots & \vdots & \vdots & \ddots & \vdots \\ 0 & 0 & 0 & \cdots & \phi_L \end{pmatrix} \tag{2}$$

and $N \in \mathbf{N}$, $N > 1$, we have

$$F_\Lambda(\phi) = \det(\Gamma_\Lambda + \frac{\zeta}{\sqrt{N}} \Phi_\Lambda). \tag{3}$$

---

* Research supported in part by the Conselho Nacional de Desenvolvimento Científico e Tecnológico, CNPq-Brazil.
** Supported in part by the Fundação de Amparo à Pesquisa no Estado de São Paulo, FAPESP, Brazil 69.

In this note, we give a brief report on the results derived in [FHM]. Assuming N large, we analyze the limiting measure μ, obtained from μ_Λ as $|\Lambda| \to \infty$. A byproduct of this analysis is the proof that a mass gap may develop dynamically in Fermion models where "symmetry breakdown" does not occur.

One can see that μ_Λ arises by applying the usual Matthews-Salam formalism to the one-dimensional Gross-Neveu model [GN]. On the one-dimensional lattice $a\Lambda$, with spacing $a > 0$, this is given by the functional action

$$\mathcal{A}(\overline{\psi}, \psi) = a \left\{ \sum_{x \in \Lambda} \overline{\psi}_x \cdot (i\mathbf{D}_\Lambda \psi)_x + \frac{\lambda^2}{2N} \sum_{x \in \Lambda} \left(\overline{\psi}_x \cdot \psi_x \right)^2 \right\} \tag{4}$$

on the algebra generated by Grassmann fields $(\overline{\psi}, \psi)$ and $\overline{\psi} \cdot \phi = \sum_{j,\alpha} \overline{\psi}^j_\alpha \phi^j_\alpha$ where $j = 1, \ldots, N$ and $\alpha = 0, 1$ are flavor and spin indices respectively. Also, $i\mathbf{D}_\Lambda$ is the symmetric difference operator contracted with the diagonal Pauli matrix, i.e., $(i\mathbf{D}_\Lambda \psi)_x = \frac{1}{a} \begin{pmatrix} \psi_{x-a} - \psi_{x+a} \\ \psi_{x+a} - \psi_{x-a} \end{pmatrix}$ and the coupling constant $a\lambda^2 = \zeta^2$, $\lambda > 0$, is a dimensionless quantity.

Note that, as Γ_Λ is skew symmetric, in (1) we have made use of the relation

$$\det \left(\Gamma_\Lambda + \frac{\lambda}{\sqrt{N}} \Phi_\Lambda \right)^2 = \det \left(i\mathbf{D}_\Lambda + \frac{\lambda}{\sqrt{N}} \mathbf{\Phi}_\Lambda \right) \tag{5}$$

with $\mathbf{\Phi}_\Lambda = \begin{pmatrix} \Phi_\Lambda & 0 \\ 0 & \Phi_\Lambda \end{pmatrix}$.

Besides its own features that justify studying this problem, at the origin, our main interest in this model was to have a laboratory to study rigorously the phenomenon of fermion mass generation in higher dimensions. For the two-dimensional model, this phenomenon was verified to occur as a consequence of breakdown of the discrete chiral symmetry [GN]. Recently, mass generation was shown in the large N two-dimensional model with special choice of ultraviolet cutoff [KMR].

The advantage of (1) over the Fermionic representation can be seen if we restrict $\mu_\Lambda(\phi)$ to constant configurations $\{\phi_x = \varphi,\ x \in \Lambda\}$. If $h(\varphi)$ denotes the resulting conditioned density, we have:

Proposition 1. (i) $h(\varphi) = h(-\varphi)$,

(ii) *$h(\varphi)$ achieves its maximum at $\varphi = \pm \frac{m\sqrt{N}}{\lambda}$ where m is the non-trivial solution of the gap equation*

$$m = \lambda^2 \frac{1}{L} \frac{f'(m)}{f(m)} \tag{6}$$

where $f_\Lambda(m) = F_\Lambda \left(\frac{m\sqrt{N}}{\lambda} \mathbf{1} \right)$ is a monotonic increasing function.

(iii) *$\varphi = 0$ is a local minimum of $h(\varphi)$ since $h''(0) = \lambda^2 \frac{f''(0)}{f(0)} - L > 0$ (provided $\lambda = \lambda(L)$ is sufficiently large).*

The above Proposition ensures the effective potential $V(\varphi) = -\ln\, h(\varphi)$ is symmetric and double-well shaped, with minima apart by $O(\sqrt{N})$ from each other, and separated by an $O(N)$ barrier at $\varphi = 0$. Thus, as in [G, GJS], for large enough N, one could expect the lowest lying excitations of this model to be Peierls contours.

The success of the Peierls method is determined by a favorable energy-entropy balance. Thus, one of the main needs here is an estimate that puts into evidence a small factor related to (jump) discontinuities in field configurations in $\mu(\phi)$. For this, our second result goes as follows. Given an Ising-like field configuration, $\sigma_\Lambda = \{\sigma_x = \pm m\}_{x\in\Lambda}$, if σ is a diagonal matrix with entries σ_x, corresponding to a configuration that has p jumps, and $f_\Lambda(\sigma) = F_\Lambda(\sqrt{N}/\lambda\ \sigma)$, the following estimate holds:

Proposition 2. *There exists $\delta = \delta(m) < 1$, uniform in Λ, such that*

$$\left|\frac{f_\Lambda(\sigma)}{f_\Lambda(m)}\right| \le \delta^p .$$

Proposition 2 shows that this model behaves as the short range one dimensional Ising model. This result represents a preliminary step towards a rigorous control of the mass generation mechanism in higher dimensions. Of course, by the same analogy, we do *not* expect symmetry breakdown to occur in dimension one, as the entropy factor always overcome the Peierls estimate in this case.

To control the measure $\mu_\Lambda(\phi)$ and learn about its limiting behavior as $|\Lambda| \to \infty$, the fluctuations around the Ising configurations are also of concern. We can also prove stability bounds and study two of its main correlations (moments).

We now present estimates on the partition function Z_Λ.

Theorem 3. *Given $\zeta \in (0,\infty)$, there exist $N_0 = N_0(\zeta)$ and, uniformly in L, constants C_1 and C_2, with $0 < C_1 < C_2 < \infty$, such that, if $N > N_0$*

$$\left(C_1\, e^{\nu N}\right)^L \le Z_\Lambda \le \left(C_2\, e^{\nu N}\right)^L \tag{7}$$

for $\nu > 0$ given by

$$\nu = \sup_{(x,y)\in I} f(x,y) - \ln \frac{e}{2\zeta^2} > 0$$

where $I = \left\{(z,z') \in [0,1]^2 : z + z' \le 1\right\}$,

$$f(x,y) = g(x) + g(y) - (1-x-y)\ln(1-x-y) \tag{8}$$

and

$$g(z) = (1-z)\ln(1-z) - z\left(\ln z - \ln\frac{e}{2\zeta^2}\right).$$

Remark: $f(x,y)$ is a concave symmetric function, with maximum value at

$$x = y = \frac{1}{2}\left(1+\frac{1}{\zeta^2}\right) - \frac{1}{2}\left(1+\frac{1}{\zeta^4}\right)^{\frac{1}{2}} \tag{9}$$

The r.h.s. above varies from 0 to $\frac{1}{2}$ as ζ goes from ∞ to 0.

The positivity of ν leads to a "free energy " $G = \lim_{L\to\infty} \frac{1}{L} \ln Z_L = O(N)$. If symmetry breaking occurred, there would be long range order, the correlation between ϕ_i and ϕ_j would not decay as $|i-j| \to \infty$ and the measure would be attained to one of its maxima with $O(1)$ fluctuations around it. The next theorem implies the absence of long range order in the $l \to \infty$ limit.

Theorem 4. *For any $\zeta \in (0,\infty)$, there exists $N_1 = N_1(\zeta)$, such that, for $N > N_1$, the following upper bound on the covariance of μ_Λ*

$$\mathrm{Cov}\,(\phi_i, \phi_j) = \int d\mu_\Lambda(\phi)\phi_i\ \phi_j \le C\ e^{-\frac{\ell}{2N}|i-j|}$$

holds with $\ell = \ell(\zeta) > 0$ and $C = C(N,\zeta) < \infty$, uniformly in L.

Recall that the one-dimensional Ising model has exponential decay of the correlations at any temperature. In (1) ζ plays the role of the inverse temperature β.

We now turn to the Fermi correlation. The two-point function is defined by

$$\left\langle \overline{\psi}^i_{x,\beta} \psi^j_{y,\delta} \right\rangle_L = \frac{1}{Z_\Lambda} \int D\overline{\psi} D\psi\ e^{A(\overline{\psi},\psi)} \overline{\psi}^i_{x,\beta} \psi^j_{y,\delta}; \qquad x, y \in \Lambda \tag{10}$$

A fermion mass gap is implied by:

Theorem 5. *Under the conditions of Theorem 4, we have*

$$\left\langle \overline{\psi}^i_{x,\beta} \psi^j_{y,\delta} \right\rangle_L \le C\ e^{-\widetilde{m}|x-y|/a}$$

uniformly in L, with $\widetilde{m} = \widetilde{m}(\zeta) > 0$.

The detailed proofs of above Propositions and Theorems are given in [FHM] and will be omitted here. They are based on two ingredients. Firstly, the determinant $F_\Lambda(\phi)$ is expanded as a gas of (non overlapping) monomers and dimers. Secondly, performing Gaussian integration over the fields ϕ, we rewrite this system using a transfer matrix formalism.

The expansion for $F_\Lambda(\phi)$ is described below. Dimers correspond to an *occupied* lattice link plus its two endpoints whereas monomers are isolated lattice sites. A monomer at site i has activity proportional to ϕ_i. Since in (1) the determinant $F_\Lambda(\phi)$ appears raised to the power $2N$, we have to deal with $2N$ copies of this expansion. We organize these terms by fixing the occupation number of dimers in each link. Taking into account the combinatorics involved in this resummation procedure, and the one for Gaussian integration, we arrive at a transfer matrix for the system. When N is large, the main contribution to the partition function is governed by the principal eigenvalue, obtained from the Laplace asymptotic method. Correlations may be written as ratios of two such systems and are dealt in an equivalent way.

To end, we would like to mention that we are on the way to extend the procedure to the Bethe lattice. This is, in some sense, topologically equivalent to dimension one (there is only one path connecting two different points on a tree) while, from other viewpoints, such as the diffusive features of its Dirac operator, it presents higher dimensional behavior. We are aiming at showing a mean-field limit (hence, symmetry breaking) as the coordination number grows to infinity.

At this point, we also have a formula generalizing the determinant expansion for Euclidean dimension two. This includes cycles besides simply monomers and dimers. Investigation along this line will follow in a near future.

2 The Expansion for the Determinant

Let $\mathcal{B}$ be the collection of bonds in Λ. To each bond b we associate a variable $s_b \in [0,1]$ and write $s = s_\Lambda = \{s_b\}_{b\in\mathcal{B}}$. Let $\Gamma_\Lambda(s)$ denote Γ_Λ with all the entries $a_{i,i+1} = 1$ being replaced respectively by s_i, $i = 1,\ldots,L-1$. So, $\Gamma_\Lambda(1) = \Gamma_\Lambda$. We then set

$$F_\Lambda(s;\phi) = \det\left(\Gamma_\Lambda(s) + \frac{\zeta}{\sqrt{N}}\Phi_\Lambda\right) \tag{11}$$

and notice that it interpolates between the fully decoupled determinant, namely $F_\Lambda(0;\phi) = \prod_{i=1}^{L}\left(\frac{\lambda}{\sqrt{N}}\phi_i\right)$ and $F_\Lambda(1;\phi) = F_\Lambda(\phi)$, as defined in (3).

For each bond $b \in \mathcal{B}$ we introduce operators I_b and J_b on differentiable functions $G : [0,1]^{|\mathcal{B}|} \longrightarrow \mathbf{R}$. For $s^{\check{}b} = \{s_{b'}\}_{b'\in\mathcal{B}/\{b\}}$, they are given by

$$(I_bG)(s^{\check{}b}) = G(s)|_{s_b=0} \quad ; \quad (J_bG)(s^{\check{}b}) = \int_0^1 ds_b\, \frac{\partial G}{\partial s_b}(s). \tag{12}$$

Our expansion, as in [GJ], follows from the identity

$$\begin{aligned} F_L(\phi) &= \textstyle\prod_{b\in\mathcal{B}}(I_b + J_b)\, F_L(s;\phi) \\ &= \textstyle\sum_{X\subseteq\mathcal{B}} I_{\mathcal{B}/X}\; J_X F_L(s;\phi) \end{aligned} \tag{13}$$

where $I_X = \prod_{b\in X} I_b$ (analogously to J_X).

Next, for $x, w \in \Lambda$, $x < w$, let $\langle x,w\rangle = [x,w] \cap \mathbf{Z}$ and also let $F_{\langle x,w\rangle}(s;\phi)$ denote $F_{\langle x,w\rangle}(s_{\langle x,w\rangle};\phi_{\langle x,w\rangle})$ given by (11) with the lattice Λ replaced by $\langle x,w\rangle$. For $b = (zz')$ being a bond with $z, z' \in \langle x,w\rangle$ and with $F_{\langle x,x-1\rangle} = F_{\langle w+1,w\rangle} = 1$, one can check that $I_bF_{\langle x,w\rangle}(s;\phi) = F_{\langle x,z\rangle}(s;\phi)\, F_{\langle z',w\rangle}(s;\phi)$ and $J_bF_{\langle x,w\rangle}(s;\phi) = F_{\langle x,z-1\rangle}(s;\phi)\, F_{\langle z'+1,w\rangle}(s;\phi)$.

These relations can be used to evaluate (13). To this end, it is convenient to think of a bond in Λ as a dimer, i.e., an object consisting of a bond and its endpoints. A collection $X \subset \mathcal{B}$ of dimers (we use the same notation as for the set of bonds in Λ) is *compatible* if their elements are non-overlapping. Given a compatible set X, the set of monomers, U_X, is the collection of sites in Λ which are *not covered* by dimers in X. With this, equation (13) can be written as

$$F_\Lambda(\phi) = \sum_{\text{compatible } X\subseteq\mathcal{B}}\ \prod_{j\in U_X}\left(\frac{\zeta}{\sqrt{N}}\phi_j\right) \tag{14}$$

which is the generating function of a stochastic monomer-dimer problem on Λ (see e.g. ref. [L]) whose monomer activity at site i is given by the Gaussian random variable $\zeta/\sqrt{N}\,\phi_i$.

Now, for $\mathcal{L} = [1,L] \cap \mathbf{Z}$, given $\Gamma \subseteq \mathcal{L}$, we decompose $\Gamma = \cup_i \gamma_i$, using disjoint subsets (clusters) γ_i whose elements are consecutives sites such that $\mathrm{dist}(\gamma_i, \gamma_j) > 1, i \neq j$. Then, for $\alpha = (\zeta^2/2N)^N(2N!/N!)$ and using (14), it can be shown that the partition function Z_L can be written as

$$Z_L = \alpha^L \sum_{\Gamma \subseteq \mathcal{L}} \prod_i K_{\gamma_i} ; \tag{15}$$

The activity K_γ of a cluster γ is zero if $|\gamma| = 1$. When $|\gamma| = s+1 \geq 2$, and for $\mathbf{n} = (n_1, \ldots, n_s)$, we have

$$K_\gamma = \sum_{\mathbf{n} \in \{0,\ldots,N\}^s} T(0,n_1)T(n_1,n_2)\ldots T(n_s,0) = T^{s+1}(0,0) \tag{16}$$

where T is a $(N+1)\times(N+1)$ symmetric"transfer" matrix whose elements are $T(n,n') = S(n)R(n,n')S(n')$ with

$$S(n) = \left(\frac{2N^2}{e\zeta^2}\right)^n \left(\frac{2(N-n)!}{2N!\;2n!}\right)^{1/2} \tag{17}$$

and, for χ denoting the characteristic function,

$$R(n,n') = \left(\frac{e}{N}\right)^{n+n'} \frac{N!}{(N-n-n')!}\, \chi(1 \leq n+n' \leq N) \tag{18}$$

For large enough N, and for $n, n' = 0, 1, \ldots, N$, applying the Stirling formula to (16) yields

$$T(n,n') = L(n,n')\; e^{Nf(n/N,n'/N)} \tag{19}$$

where $L(n,n') = e^{O(\ln N)}$, and the function $f(x,y)$ is given in (8).

When N is large, the sum over $\mathbf{n}$ appearing in (16) can be approximated by an integral which, as mentioned in the previous section, can be controlled by means of the Laplace asymptotic method.

3 Acknowledgements

We thank Prof. A. Jaffe, Harvard University and McMaster University for support that allowed the beginning of our collaboration. P. da Veiga thanks Prof. V. Rivasseau and the organizers of the XI IAMP Congress and the Conference on Constructive Results for invitation. The Fundação de Amparo à Pesquisa no Estado de São Paulo and the Ecole Polytechnique are thanked for their financial support.

References

[FHM] Faria da Veiga, P.A.; Hurd, T.R. and Marchetti, D.H.U.; pre-print IFUSP/1994.

[G] Griffiths, R. B.; Phys. Rev. **A136**, 437 (1964).

[GJ] Glimm, J. and Jaffe, A.; *Quantum Physics: A Functional Integral Point of View*, Springer-Verlag, 2nd edition, New York (NY), 1989.

[GJS] Glimm, J.; Jaffe, A. and Spencer, T.; Ann. Phys. **101**, 610 (1976).

[GN] Gross, D. and Neveu, A.; Phys. Rev. **D10**, 3235 (1974).

[KMR] Kopper, C., Magnen, J. and Rivasseau, V.; pre-print Ecole Polytechnique, Palaiseau, and also contribution to this Conference.

[L] Lieb, E.H.; *Models in Statistical Mechanics*, Lectures in Theoretical Physics, Gordon and Breach (1969).

Mass Generation in the Large N Gross–Neveu Model

Christoph Kopper

Institut für Theoretische Physik
Universität Göttingen, D-37073 Göttingen

Abstract

Abstract: We study the infrared behaviour of the Euclidean Gross-Neveu-Model with discrete chiral symmetry. Imposing a suitable UV-cutoff we prove that for a large (but finite) number of Fermion components the model has (at least) two pure phases, realized by suitable boundary conditions and that the Fermion two-point function decays exponentially.
(Joint work with Jacques Magnen and Vincent Rivasseau, Ecole Polytechnique)

1 Introduction

We report about joint work with Jacques Magnen and Vincent Rivasseau [1] on the infrared behaviour of the large N Gross-Neveu model in two-dimensional Euclidean space. [1] may serve as a general reference for all what follows. In these notes we intend to make visible the line of the argument at the prize of being imprecise with many technical details. The Lagrangian of the model is

$$\mathcal{L} = \overline{\psi} i \partial\!\!\!/ \psi - \frac{\lambda}{2N} (\overline{\psi}\psi)^2 . \tag{1}$$

Here $\partial\!\!\!/ = \partial_\mu \gamma_\mu$ and $\{\gamma_\mu, \gamma_\nu\} = -2\delta_{\mu\nu}$. The γ_μ are the antihermitian two-dimensional Euclidean Dirac matrices. The coupling constant λ is assumed to be of order 1, to be definite we set

$$1/10 \leq \lambda \leq \pi . \tag{2}$$

N is the number of Fermion flavour components: $\overline{\psi} = (\overline{\psi}_{11}, \overline{\psi}_{12}, \cdots, \overline{\psi}_{N1}, \overline{\psi}_{N2})$ and similarly for ψ. We will also introduce the scaled coupling

$$g := \sqrt{\frac{\lambda}{N}} . \tag{3}$$

Our proof of mass generation works for

$$N \geq N_0 >> 1 \tag{4}$$

and N_0 grows exponentially with $1/\lambda_{\min}$. Our bounds have not been optimized with respect to N_0 and require in fact that $N_0^{-\frac{1}{5}}\, e^{\frac{3\pi}{\lambda_{\min}}} << 1$ (e.g. $\leq 1/100$ would do with our crude estimates). The accent is on the fact that the $1/N$-expansion for the model is not misleading and the limit $N \to \infty$ is regular.
The Lagrangian (1) is invariant under discrete chiral symmetry transformations

$$\psi \to \gamma_5 \psi\,, \quad \overline{\psi} \to -\overline{\psi}\gamma_5\,, \quad (\gamma_5 = i\gamma_0\gamma_1)\,. \tag{5}$$

This invariance would be broken by an explicit mass term $m\overline{\psi}\psi$. Thus our results imply that this symmetry is spontaneously broken since we will prove that the model has an exponentially decaying two point function. Presently we concentrate on the IR properties of the model and impose for simplicity an UV cutoff which of course has to respect the chiral symmetry. The scale of the cutoff will be fixed to equal 1.
As a motivation for studying this model we mention the following facts:

(i) Models with four Fermion interactions have a long and impressing history in field theory and many body physics.

(ii) For $N \geq 2$ the present model shares UV asymptotic freedom with four dimensional nonabelian gauge theories. This was first established at the perturbative level for $N \to \infty$ by Gross and Neveu [2] and for the model with explicit mass term by Mitter and Weisz [3]. The explicitly massive model was then rigorously constructed [4,5] for small renormalized coupling which proved asymptotic freedom beyond perturbation theory. Note also the construction of nonabelian Yang Mills theory in a small volume based on UV asymptotic freedom [6] and the related work of Balaban [7].

(iii) On the infrared side there is again some similarity to four dimensional nonabelian gauge theories on the heuristic level based on the general belief that both models have a nontrivial vacuum structure and nonperturbative mass generation. For the large N Gross-Neveu model with UV cutoff this belief is shown to be true in the following [1]. Note however that our construction does not say anything about the so-called chiral Gross-Neveu models [2] which are invariant under continuous chiral transformations

$$\psi \to e^{i\alpha\gamma_5}\psi\,, \quad \overline{\psi} \to \overline{\psi}e^{i\alpha\gamma_5}\,.$$

(iv) The three-dimensional massive large N Gross-Neveu model has been constructed as the first example of a nonrenormalizable field theory by de Calan, Faria da Veiga, Magnen and Sénéor with the aid of its nontrivial UV fix point.

2 Heuristic Analysis for $N \to \infty$

The object of interest will be in the following the two point function

$$S_2(x,y) = < \psi_i(x)\,\overline{\psi}_i(y) > \quad (\,\overline{\psi}_i = (\overline{\psi}_{i1}, \overline{\psi}_{i2})) \tag{6}$$

[1] We remark that the work of Patrascioiu and Seiler [8] tends to cast doubt on the standard belief, in particular for gauge theories. They emphasize that a study of UV behaviour using an artificial IR cutoff- and vice versa- might be misleading.

formally given by

$$
\begin{aligned}
S_2(x,y) &\sim \int D(\overline{\psi},\psi)\,\psi_i(x)\overline{\psi}_i(y)\,e^{-\int(\overline{\psi}\not{p}\psi - \frac{g^2}{2}(\overline{\psi}\psi)^2)} \\
&\sim \int D(\overline{\psi},\psi,\sigma)\,\psi_i(x)\overline{\psi}_i(y)\,e^{-\int(\overline{\psi}\not{p}\psi + \frac{1}{2}\sigma^2 + g\sigma(\overline{\psi}\psi))} \\
&\sim \int D(\sigma)\,(\frac{1}{\not{p}+g\sigma})_i(x,y)\,e^{-\int V(\sigma)}
\end{aligned}
\tag{7}
$$

where

$$
\begin{aligned}
e^{-\int V(\sigma)} &= \det(1+\frac{1}{\not{p}}\,g\sigma)\,e^{-\int \frac{1}{2}\sigma^2}\,, \\
\int V(\sigma)\,d^2x &= -Tr\ln(1+\frac{1}{\not{p}}\,g\sigma) + \frac{1}{2}\int \sigma^2\,d^2x\,.
\end{aligned}
\tag{8}
$$

In (7) we used the well-known Matthews-Salam representation in which an ultralocal interpolating scalar field σ is introduced. Formally integrating out the σ-field by completing the square leads back to the original form whereas integrating out the Fermi fields leads to a nonlocal bosonic theory, where the main part of the action is the Fredholm determinant. We did not care about normalization factors. Both presentations produce the same results in perturbation theory. Since we expect nonperturbative effects, only the bosonic presentation is viable. The interpretation of Fermionic functional integrals is essentially limited to perturbation theory.

The "potential" $V(\sigma)$ appearing in (8) is in general a complicated functional of the possible σ-field configurations. To get some insight into its structure it is helpful to restrict to σ-fields constant over all over the volume taking values $\sigma \equiv \overline{\sigma} \in \mathbf{R}$. In this case we can calculate

$$-Tr\ln(1+\frac{1}{\not{p}}g\overline{\sigma}) = -Tr\frac{1}{2}\ln(1+\frac{1}{p^2}g^2\overline{\sigma}^2) =$$

$$= -|\Lambda|\,N\int_0^1 \frac{d^2p}{(2\pi)^2}(\ln(p^2+g^2\overline{\sigma}^2) - \ln p^2) =$$

$$= -|\Lambda|\,N\,\frac{1}{4\pi}\,[(1+g^2\overline{\sigma}^2)\ln(1+g^2\overline{\sigma}^2) - g^2\overline{\sigma}^2\ln g^2\overline{\sigma}^2]\,.$$

$|\Lambda|$ is the (measure of the) volume and we used a sharp UV-cutoff at energy 1. The first equality uses the fact that an odd product of γ-matrices has vanishing trace. From this expression and (8) we obtain the following qualitative picture for $V(\overline{\sigma})$:

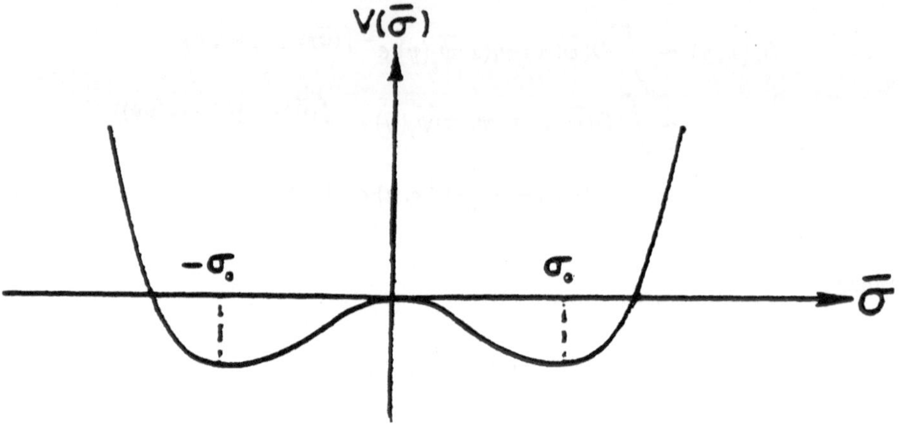

Fig.1-The Potential $V(\overline{\sigma})$. Here $\sigma_0 \sim N^{1/2}\, e^{-\frac{\pi}{\lambda}},\ V(\sigma_0) \sim -\sigma_0^2 \sim -N\, e^{-\frac{2\pi}{\lambda}}$.

In the limit $N \to \infty$ we thus have two infinitely deep and infinitely separate minima at $\pm\sigma_0$ and therefore we find that the model has two pure phases with $<\sigma> = \pm\sigma_0$. According to the Peierls argument the boundary conditions should decide which one is realized. On the level of the σ-field chiral symmetry is the symmetry under $\sigma \to -\sigma$. This symmetry is therefore broken spontaneously and the Fermion is massive, the Fermion mass being given by $g\sigma_0 \sim e^{-\frac{\pi}{\lambda}}$.
The heuristic picture could be misleading for finite N for two related reasons:
(i) For finite N the minima are no more infinitely deep and separate.
(ii) The important field configurations in the functional integral are not only those where the σ-field is constant all over the volume.
To get control of the second problem obviously requires most effort. The subsequent analysis will prove however that the heuristic picture is basically correct for sufficiently large (but finite) N.

3 Rigorous Definition of the Regularized Model

As noted before we will use the bosonic presentation of the model. For a rigorous definition we introduce the following regulators.
IR: We will restrict the σ-field to a finite volume Λ, for simplicity a square centered at the origin, and we will take the thermodynamic limit $\Lambda \to \mathbf{R}^2$ in the end.
UV: We introduce UV-cutoffs at scale 1 which respect the chiral symmetry of the model and keep them fixed from now on (we do not take the UV limit). Specifically we replace $\not{p} \to \not{p}\, e^{1/2\, p^2} =: \not{p}_r$. We also need an UV-cutoff for the

σ-field. [2] The σ-field is ultralocal, i.e. its position space propagator is $\delta^{(2)}(x-y)$. We replace $\delta^{(2)}(x-y) \rightarrow (\frac{1}{1+f})(x-y)$ where the kernel $(\frac{1}{1+f})(x-y)$ is smooth and of compact support: $(\frac{1}{1+f})(x-y) \equiv 0$, if $|x-y| > 1$. The kernel can be constructed from a function f which in momentum space has the following properties: $\tilde{f}(0) = 0$, $\tilde{f}(p) \geq 0$, $\tilde{f}$ depends on p^2 only, $\tilde{f} \in C^\infty(\mathbf{R}^2)$, $\tilde{f}(p) \sim O(|p|^L)$ with $L >> 1$ for $p \rightarrow 0$ and $\tilde{f}(p) \sim |p|^4$ for $|p| \rightarrow \infty$. This is shown in Lemma 1 in [1].

With these preparations we obtain the following rigorous expression for the normalized two point function

$$S_2(x,y) = \frac{\int d\mu_f(\sigma)\,(\frac{1}{\not{p}_r + g\sigma_\Lambda})_i(x,y)\,\det(1+\frac{1}{\not{p}_r}\,g\sigma_\Lambda)}{\int d\mu_f(\sigma)\,\det(1+\frac{1}{\not{p}_r}\,g\sigma_\Lambda)}\,. \tag{9}$$

Here $d\mu_f(\sigma)$ is the Gaussian measure with covariance $(\frac{1}{1+f})(x-y)$ supported in (a subspace of) $\mathcal{L}^2(\mathbf{R}^2)$ [10] and $\sigma_\Lambda(x) = \sigma(x)\,\chi_\Lambda(x)$, and χ_Λ is the standard characteristic function of the region Λ. In the following we will suppress the regulators r and Λ in the notation.

The boundary conditions will be chosen as follows: We cut up the volume Λ into a lattice of unit squares $\triangle$ and demand

$$\overline{\sigma}(\triangle) := \int_\Delta \sigma\, d^2x \geq 0\,, \quad \text{if } \Delta \cap \partial\Lambda \neq \emptyset\,, \tag{10}$$

i.e. we favour positive mean value for σ on the boundary of Λ. More precisely condition (10) will be implemented by a smoothed step function which suppresses σ-configurations with negative mean value on the boundary squares. There is much arbitrariness in the precise form of the boundary condition as long as it favours sufficiently one of the two minimum values $\pm\sigma_0$ of the σ-field.

The σ-field presentation allows to some degree an intuitive understanding of the model in terms of probabilistic concepts. But the propagators $1/\not{p}$ are still remnants of the Fermionic description. Their appearance in the Fredholm determinant makes the presentation unsuited for an analysis, which tries to separate the dominant from the suppressed field configurations. This is because $1/\not{p}$ is not hermitian and has no positivity properties. Therefore the following proposition is a valuable tool for the subsequent considerations:

Proposition:

$$\det(1+\frac{1}{\not{p}}\,g\sigma) = \det{}^{1/2}(1+A+B)\,\det(1+\frac{m}{\not{p}})$$

---

[2] In a purely perturbative treatment it would suffice if we regulated the Fermions only: The high momentum flow through any possibly divergent Feynman diagram is suppressed automatically if it is suppressed in the Fermion propagators since any vertex contains exactly one σ-field (see (7)). In a nonperturbative treatment also σ has to be regularized independently (see remark (i) after (21)).

for $\sigma \in \mathcal{L}^2(\Lambda)$, where

$$A = g\sigma \frac{1}{p^2+m^2} g\sigma - \frac{m^2}{p^2+m^2}, \quad B = g\sigma \frac{1}{p^2+m^2} \not{p} - \not{p} \frac{1}{p^2+m^2} g\sigma, \quad (11)$$

The parameter $m > 0$ is arbitrary at this point.

Remarks on the Proof: The determinants are well-defined due to the regulators. In the proof we use again that $Tr(\frac{1}{\not{p}})^{2n+1} = 0$. For given σ and g small enough we may expand the determinants using $\det = \exp Tr \ln$ and then introduce $g\tau = g\sigma - m$. One finds out that $\det(1 + \frac{1}{\not{p}}(g\tau + m)) = \det(1 - (g\tau + m)\frac{1}{\not{p}})$ and $\det(1 \pm \frac{1}{\not{p}}(g\tau + m)) = \det(1 \pm \frac{m}{\not{p}}) \det(1 + \frac{1}{\pm\not{p}+m} g\tau)$. Multiplying the two determinants $\det(1 + \frac{1}{\not{p}+m} g\tau)$ and $\det(1 + \frac{1}{-\not{p}+m} g\tau)$, reintroducing the variable σ and taking the square root one arrives easily at the proposition. By a standard analyticity argument [11] one may obtain the result for any g (independent of σ).

The factor of $\det(1 + \frac{m}{\not{p}}) = \det(1 - \frac{m}{\not{p}})$ in the Proposition is a global normalization factor which contributes to the minimum value of $V(\sigma)$. Since it will be isolated from the determinant independently of whether we use the presentation of the proposition or not (see also (19) below) it just drops out on performing the division in (9). The presentation of the Fredholm determinant in terms of A and B will be useful in the large field domains which we will define now. The operator B will not play an important role in the following. It will be just one of many correction terms and can be bounded similarly as Q below (see (24)) using again the fact that an odd product of γ-matrices has vanishing trace.

4 Small and Large Field Regions, Bounds on the Action

The operator A from (11) is bounded (for given σ), traceclass and > -1 in finite volume. It turns out to give the essential contribution to the Fredholm determinant and the size of its (well-defined) Hilbert-Schmidt norm in a fixed square Δ provides us with a suitable measure for the weight of the respective field configuration.

Definition *For given $\sigma \in \mathcal{L}^2(\Lambda)$ and $\Delta \in \Lambda$ we say:*
Δ is a large field (l) square, if $||A||_\Delta := ||\chi_\Delta A \chi_\Delta||_{HS} > N^{1/10}$,
Δ is a small field (s) square, if $||A||_\Delta < N^{1/10}$.
Δ is a +square if $\overline{\sigma}(\Delta) > 0$ and a −square if $\overline{\sigma}(\Delta) < 0$. More precisely all these conditions are again implemented by smoothed step functions. Remember that

$$||\chi_\Delta A \chi_\Delta||_{HS} = (Tr(\chi_\Delta A \chi_\Delta)^2)^{1/2} = (\sum_i \lambda_{i,\Delta}^2)^{1/2},$$

where $\lambda_{i,\Delta}$ are the eigenvalues of $\chi_\Delta A \chi_\Delta$.

We thus obtain, for given σ-field, regions of + and −squares separated from each other by closed contours. All squares on the boundary of Λ belong by definition to +. It turns out that two adjacent squares belonging to + and − respectively should always be counted as if they were l-squares, i.e. in the suppressed region, even if the above criterion attaches them to the s-region. This is because it costs so to say a lot of energy to change from one minimum of $V(\sigma)$ into the other [3]. The enlarged l-region to which we add those squares across the boundary of which there is a change of sign is called l'. Finally it turns out useful to associate also all squares Δ with

$$dist(\Delta, l') \leq 2\, e^{\pi/\lambda} \ln N \tag{12}$$

to the large field region. The once more enlarged version is called l'':

$$l \subset l' \subset l''\,. \tag{13}$$

l'' contains by definition in particular all "fattened" contours separating + from −.

The essential reason why l is correctly named the large field region is in the following inequality:

$$\det(1 + A_l) \leq e^{Tr\, A_l}\, e^{-O(1) N^{1/5} |l|}\,, \quad (A_l = \chi_l A \chi_l)\,. \tag{14}$$

$|l|$ is the number of l-squares. Forgetting for the moment about the first factor (on which we comment below) (14) shows shows that the l-region is strongly suppressed for $N >> 1$. For a proof of (14) one first realizes that

$$Tr\, \ln(1 + A_l) \leq Tr\, A_l \; - \; \frac{1}{2} Tr \frac{A_l^2}{2 + A_l}\,,$$

which follows from $\ln(1+x) \leq x - \frac{1}{2}\frac{x^2}{2+x}$ for $x > -1$. The second term may then be evaluated in an eigenbasis of $\mathcal{L}^2(l)$ using the l-condition. Note that the eigenvalues are $2N$- fold degenerate (flavour- and spinor indices). We choose in any square Δ in l a complete system of orthonormal eigenfunctions of $A_\Delta := \chi_\Delta A \chi_\Delta$ with eigenvalues $\lambda_{i,\Delta}(\sigma_\Delta)$. We find

$$Tr \frac{A_l^2}{2 + A_l} = \sum_{i,\Delta} \frac{\lambda_{i,\Delta}^2}{2 + \lambda_{i,\Delta}} = \sum_{\Delta} \Big(\sum_{\lambda_{i,\Delta} > 1/10} \frac{\lambda_{i,\Delta}^2}{2 + \lambda_{i,\Delta}} + \sum_{\lambda_{i,\Delta} \leq 1/10} \frac{\lambda_{i,\Delta}^2}{2 + \lambda_{i,\Delta}} \Big) \geq$$

$$\sum_{\Delta} \Big(\sum_{\lambda_{i,\Delta} > 1/10} \frac{\lambda_{i,\Delta}}{21} + 10/21 \sum_{\lambda_{i,\Delta} \leq 1/10} \lambda_{i,\Delta}^2 \Big) \geq \frac{10}{21} |l|\, N^{1/5}\,.$$

The last inequality is true -square per square- if the first sum does not vanish due to the $2N$-fold degeneracy of the eigenvalues. If it vanishes the second sum

[3] The reader should think of the s-region as being made up of two (small) neighbourhoods of the points $\pm\sigma_0$ in the figure. Field configurations in a square Δ where $\overline{\sigma}(\Delta)$ is not close to σ_0 are in the l-region. This includes in particular values of $\overline{\sigma}(\Delta)$ close to 0. But for two adjacent $\pm$-squares we can always find a subsquare Δ' in their union such that $\overline{\sigma}(\Delta') = 0$. This is the basic reason why the contribution of the two squares is at least as suppressed as that of a single l-square.

fulfills the inequality due to the large field condition. This implies the statement (14).

In the $s\pm$-regions it would be unwise to expand around $\sigma = 0$ which is not near the minima of the action. We therefore perform a local translation of the field variable through

$$\zeta(x) := \sigma(x) - \frac{m}{g}\chi_{s+}(x) + \frac{m}{g}\chi_{s-}(x)\,, \quad \frac{m}{g} =: \sigma_0\,. \tag{15}$$

The parameter m will now be fixed through the gap equation

$$Tr(\frac{1}{\not p + m} g\zeta_{s+}) = \frac{m}{g}\int_{s+} \zeta(x)\, d^2x \tag{16}$$

A similar equation holds in $s-$. The gap equation is the extremal condition which says that the action should not contain a term linear in $\zeta_{s\pm}$. The term on the l.h.s. of (16) is the linear contribution in ζ to

$$\det(1 + \frac{1}{\not p} g\sigma_{s+}) = \det(1 + \frac{1}{\not p + m} g\zeta_{s+})\,\det(1 + (\frac{m}{\not p})_{s+})\,, \tag{17}$$

the one on the r.h.s. of (16) is (-1) times the linear contribution in ζ coming from $1/2\int_{s+}\sigma^2\, d^2x$ on performing (15). For the normalization factor $\det(1 + (\frac{m}{\not p})_{s+})$ see the remark at the end of sect.3. From (16) we obtain

$$m = e^{-\pi/\lambda}(1+\delta)\,, \quad (|\delta| << 1) \tag{18}$$

as expected (see sect.2). δ depends on the details of the UV-cutoff.

We now come back to the first factor on the r.h.s. of (14). It is again related to the global normalization of the functional integral. The contribution of the ultralocal term in the l-region $e^{-\frac{1}{2}\int \sigma_l^2}$ (see (8)) together with e^{TrA_l} gives

$$e^{-\frac{1}{2}\int \sigma_l^2}\, e^{TrA_l} = e^{-\frac{1}{2}\int_l \sigma_0^2}\,, \quad \sigma_0 = m/g\,. \tag{19}$$

This normalization factor is dependent on the size of the l-region. Nevertheless it drops out on performing the division in (9) since exactly the same factor arises also in the small field region where we replace σ by ζ (15): Writing σ^2 in terms of ζ^2 gives the same field independent contribution as (19). The attentive reader might ask at this stage whether it is legitimate to absorb part of the ultralocal σ^2-contributions into normalization factors once they have been modified by the UV-cutoff function f. The point is that there are corrections due to f which however are well under control if f satisfies certain conditions which were given in sect.3. For the present purpose it is important that $\tilde f$ vanishes of high order at the origin of p-space; suitable bounds on the correction terms are then obtained if the translation of variable (15) is also smoothed at the boundary of the s-region. Here we do not give details.

Another change of presentation is still required to obtain a convergent expansion in the small field region: We also have to perform a change of covariance such that the new covariance contains all terms quadratic in ζ (up to small corrections). Looking at (17) we find that the quadratic term to be absorbed in the covariance is the following (in the region $s+$):

$$\frac{1}{2}Tr(\frac{1}{\not p+m}g\zeta_{s+})^2 = \frac{1}{2} < \zeta_{s+}, \pi\, \zeta_{s+} > . \tag{20}$$

Here π is defined to be the Fermion bubble, i.e. in momentum space

$$\pi(p) = N\, g^2\, Tr_2 \int \frac{d^2q}{(2\pi)^2} \frac{1}{\not p+m} \frac{1}{\not p+\not q+m} .$$

The factor N stems from the flavour trace and Tr_2 denotes a trace in spinor space only. Putting $\pi(p)$ together with $\mathbf{1}$ from the ultralocal term we obtain the following expression for the new covariance $C_{l\pm}$

$$C_{l\pm}^{-1} = {}_{s+}(1+\pi)_{s+} + {}_{s-}(1+\pi)_{s-} + \varepsilon_l \tag{21}$$

where ε_l is a small term to assure positivity of the covariance also in the l-region. The subscripts $s+, s-$ denote restriction to these regions by characteristic functions. More details on $1+\pi$ are given in the end of this section. Again the attentive reader might remember that the ultralocal term has already been smoothed out by the UV cutoff. As before this leads to f-dependent corrections in (21) [4]. We note:

(i) Since the covariance is different for different $s\pm$ and l regions there arises a difference of normalization depending on the size of the large field region. The relative normalization factor with respect to the situation where l' is empty is given by $Z_l = \det(C_{l\pm}/C_\emptyset)$ (see e.g. [12]). Here there is no flavour trace to be taken! It is bounded by $e^{N^{1/10}|l|}$ and therefore beaten by the suppression factor (14). We should mention that this bound on Z_l relies on the fact that the σ-field has been suitably regularized. This is the (only) reason why we had to regularize the σ-field in addition to the Fermion regularization.

(ii) From the analyticity properties of $\pi(p)$ we find that $C_{l\pm}$ decreases exponentially with mass $\mu > m$ in the $s\pm$-regions (see below).

As a result of the previous manipulations our formula for the two point function takes the following form

$$S_2(x,y) = \frac{\sum_{l\pm} Z_l \int d\mu_l(\zeta)\, (\frac{1}{\not p+g\sigma})_l(x,y)\, G_{l\pm}(\zeta)}{\sum_{l\pm} Z_l \int d\mu_l(\zeta)\, G_{l\pm}(\zeta)} \tag{22}$$

---

[4] To obtain suitable bounds on these corrections we have in fact to change the covariance once more replacing (21) by

$$\sqrt{1+\pi}(1-\chi_{l''})\sqrt{1+\pi} + \varepsilon_l + f\text{-dependent terms} .$$

This implies in turn that we have to bound the difference between (21) and the new expression. Here the security belt $l''-l$ around the l-squares comes into play. In the following we will suppress this change and its consequences but mention it to give the reader a flavour of the amount of technicalities haunting us throughout our construction.

where $d\mu_l(\zeta)$ is the Gaussian measure with covariance $C_{l\pm}$. $G_l(\zeta)$ is the interaction term. It contains the smoothed step functions implementing the boundary and the possible $l\pm$- conditions over which we have to sum. The crucial entries are

$$\det{}_2(1+A_l)\det{}_3(1+\frac{1}{\not p+m}g\zeta_{s+})\det{}_3(1+\frac{1}{\not p-m}g\zeta_{s-}) \tag{23}$$

Here $\det_2(1+A) = \det(1+A)\,e^{-Tr\,A}$, $\det_3(1+A) = \det_2(1+A)\,e^{1/2 Tr\,A^2}$. Remember that the Tr-terms subtracted in $\det_2, \det_3$ have already been taken account of.
Finally there are small correction terms, and also terms of the form $\det(1+Q)$ where e.g.

$$Q = (A - A_l - A_{s+} - A_{s-})\,\frac{1}{1+A+A_l+A_{s+}+A_{s-}}\,.$$

The origin of these terms is in the nonlocality of A or $\frac{1}{\not p\pm m}g\zeta$ which implies that the different regions $l, s+, s-$ are coupled among each other so that e.g.

$$\det(1+A) = \det(1+A_l)\,\det(1+A_{s+})\,\det(1+A_{s-})\,\det(1+Q)\,.$$

These terms are easy to bound: We have $Tr\,Q^{2n+1} = 0$ since Q has only nonvanishing matrix elements between mutually disjoint regions. From this we obtain

$$|\det(1+Q)| \le \det{}^{1/2}(1-Q^2) \le 1\,. \tag{24}$$

As regards (23) the $\det_2(1+A_l)$ is analyzed above (14), the $\det_3$-terms have a convergent expansion due to the small field condition. Any term contributing to $\det_3$ contains g to at least the power of 3, which leaves a factor of $N^{-1/2}$ after taking into account the flavour trace. Of course this is not sufficient to verify this statement since we also need a bound on the field variable ζ in $s\pm$ which appears in $\det_3$ (23). Such a bound can be deduced using the small field condition. We cannot give all details but note the following: In terms of ζ the operator A takes the form

$$A = g\zeta\frac{1}{p^2+m^2}(g\zeta+m)\,.$$

Due to the UV-cutoff the position space kernel of $\frac{1}{p^2+m^2}$ is bounded around 0 by $O(1)$, and also strictly positive if the argument $x-y$ is bounded by $O(1)$ in modulus. Using this fact and the small field condition for A in a given s-square $\triangle$ we obtain a bound of the form

$$O(1)\,(\int_\triangle g\zeta_\pm\,d^2x)^2 + \int_\triangle g\zeta_\pm\,d^2x \le O(1)N^{-2/5}\,,$$

where $2\zeta_\pm = |\zeta| \pm \zeta$. The factor $N^{-2/5}$ arises from the small field condition on remembering the factor N from the flavour degeneracy: $((N^{1/10})^2\,N^{-1})^{1/2} = N^{-2/5}$. The restriction of the integrals to the positive or negative part of ζ is a consequence of restricting to those contributions of the trace which is involved in the HS-norm whose support is that of ζ_+ or ζ_-. The previous inequality implies now

$$\int_\Delta g|\zeta|\, d^2x \le O(1)N^{-2/5}\,.$$

Taking the third power of this factor together with the flavour trace we obtain a sufficient bound for $\det_3$:

$$\det_3(1+\frac{1}{\not p+m}g\zeta_{s+}) \le e^{O(1)N^{-1/5}|s+|}\,.$$

We close this section by showing that all nonlocal kernels appearing in the presentation (22) have exponential decrease with masses $\ge m$ in the small field region. This obviously is of crucial importance for the proof of mass generation. The statement is true for the kernels of $\frac{1}{\not p\pm m}$, $\frac{1}{p^2+m^2}$ (also when taking into account our exponential UV-cutoff), and the various manipulations (11),(15),(17) leading from $\frac{1}{\not p}$ to $\frac{1}{\not p+m}$ etc. have been performed to make this exponential fall-off - guessed from the heuristic analysis - explicit. But it is also true for the covariance C in the $s\pm$-region where C is of the form $\frac{1}{1+\pi}$ (21): We write

$$1+\pi(p) \;=\; \mu^2 \,+\, \pi_{ren}(p)\,, \quad \mu^2 \;=\; 1+\pi(0)\,,\; \pi_{ren}(p) \;=\; \pi(p)-\pi(0)$$

and find

$$1+\pi(0) \;=\; \frac{\lambda}{\pi}(1+O(m^2))\,.$$

$\pi_{ren}(p)$ is (given by a convergent integral also in the absence of an UV-cutoff and) analytic in the strip $|Im\, p| \le 2m$. One can then also convince oneself that $\mu^2 + \pi_{ren}$ stays away from 0 in the strip $|Im\, p| \le \sqrt{2}m$ which assures a fall-off stronger than that of $e^{-\sqrt{2}m|x-y|}$ for large $|x-y|$. It is also important to note that $\mu^2 + \pi_{ren}(p) > 0$ since otherwise it would not provide us with a well-defined Gaussian measure.

5 Adapted Cluster and Mayer Expansions

We apply the inductive cluster expansions in the form presented in [13], [14] or also in the new version [15], [16], [17]. Earlier applications of low temperature expansions to different models are in [18], [19]. We shall first describe the situation when there are no large field constraints and then we explain how we deal with the complications induced by these constraints.

The cluster expansion is a Taylor expansion around the situation where all squares in Λ are decoupled. The simplest example is a volume made up of two squares $1,2$ and a functional integral in which there appears some kernel $K(x,y)$ with support in $1\cup 2$. Then

$$\begin{aligned}\chi_{1\cup 2}K(x,y)\chi_{1\cup 2} \;&=\; \chi_1 K(x,y)\chi_1 \,+\, \chi_2 K(x,y)\chi_2\\ &\quad + s(\chi_1 K(x,y)\chi_2 \,+\, \chi_2 K(x,y)\chi_1)|_{s=1}\,,\end{aligned}$$

and we expand the functional integral around $s = 0$ using $f(s = 1) = f(0) + \int_0^1 f'(s)\, ds$. For our purposes it will suffice to expand to first order in s. The derivative creates a link between the squares 1 and 2, in the term with $s = 0$ the squares are decoupled.

Applying such an expansion to any of the nonlocal kernels appearing in our presentation of the model, namely

$$\frac{1}{\not p \pm m}, \frac{1}{p^2 + m^2}, C_{l\pm}$$

and with respect to the whole lattice of squares Δ in Λ, we start writing for any kernel K:

$$K = s_1 K + (1 - s_1) K_{D,1} \,|_{s_1=1} . \tag{25}$$

Here $K_{D,1}$ contains only the diagonal terms of K:

$$K_{D,1} = \sum_{\Delta} \chi_\Delta K \chi_\Delta . \tag{26}$$

Then we perform a first Taylor step. The derivative w.r.t. s_1 creates a link l between to different squares, where we have to sum over the possible links. In the second step we interpolate again in any of the items of the sum, indexed by l, writing

$$s_1 K = s_2 K + (s_1 - s_2) K_{D,2}^{(l)} \,|_{s_2=s_1} . \tag{27}$$

In $K_{D,2}$ we now suppress all non-diagonal terms coupling clusters. The clusters are defined as the previous single squares plus the union of the two squares linked by l. We again Taylor expand w.r.t. s_2 and we iterate until the squares are exhausted. Without giving further details which are in the references we note that as a result of this expansion we may present the numerator and denominator in the expression for $S_2(x, y)$ as sums over products of factorized contributions. The latter are called amplitudes of polymers $A(Y)$. The polymers Y are unions of squares linked by the derivative links as indicated above. The amplitudes include a sum over the various choices as to how the squares in Y may be linked. This sum can be realized to be indexed by trees. They include also a sum over the various choices on what the links could be. Namely we said that we introduced the s-parameters in all nonlocal kernels in (22), in particular in the covariance and also in the $(1/\not p \pm m)$ and $(1/p^2 + m^2)$-kernels appearing in the $\det_2, \det_3, \det(1 + Q)$-terms (23), (24). The s-derivatives then apply to all those terms. Finally the amplitudes $A(Y)$ also include integrals over all Taylor parameters s_i associated with the links and appearing in all the interpolated kernels. The support of all these kernels in $A(Y)$ is restricted to Y.

The logarithm of the partition function (i.e. of the denominator in (9)) or the free energy should diverge linearly in the volume by translation invariance. The presentation in terms of polymers is therefore not sufficient to take the thermodynamic limit. The standard solution of the problem is to join a second expansion, the Mayer expansion. It is an expansion around the situation where there are no hardcore (nonoverlap) constraints among the polymers. Remember that these are mutually disjoint so far. This expansion may be organized similarly as the previous one with the aid of trees. We skip again the details. The important achievement is that on performing the Mayer expansion we can factorize the denominator in the numerator of (9) and perform the division to obtain a presentation of $S_2(x,y)$ of the following form:

$$S_2(x,y) = \sum_{M=(x,y)-configuration} a^T(M)\,, \quad a^T(M) = T(M)\frac{1}{(q-1)!}\prod_{i=1}^{q} a(Y_i)\,. \tag{28}$$

Here $a(Y_i)$ is the amplitude of the polymer Y_i which has been normalized by division through $\prod_{\Delta\in Y_i} A(\Delta)$. M is a sequence of overlapping(!) polymers $Y_1,\ldots,Y_q$ the first of which (by convention) contains x,y and the factor $(1/(\not{p}+g\sigma))_i(x,y)$ from (22). $T(M)$ is a combinatoric factor- again a sum over tree structures which join the polymers among each other. Since translation invariance is now broken through x,y and the condition that all polymers overlap, (28) has a chance to converge in the thermodynamic limit. A well-known sufficient condition for convergence is the following inequality

$$|\sum_{Y,0\in Y} a(Y)\,e^{|Y|}| < 1\,. \tag{29}$$

$|Y|$ is number of squares in Y. The condition $0\in Y$ says that translation invariance has been broken. (Otherwise the sum will not converge as noted above.) The proof of (29) requires in general two ingredients:
(i) The nonlocal kernels must have summable decay: $|K(x,y)| \le const|x-y|^{-(d+a)}$, $a>0$, in d dimensions.
(ii) Per derivative link one needs a small factor $<< 1$ to beat the combinatoric factors generated by all the choices involved (sum over where to apply the derivatives, sum over trees, ...) and also to beat the weight of the sums over all the squares in Y together with their distance factors (see (i)).

Coming now back to the presentation (22) of $S_2(x,y)$ we realize that the conditions (i),(ii) are fulfilled in the small field region: We have even exponential decay of the propagators, and the way of presenting $S_2(x,y)$ has been judiciously chosen such that any small field derivative generates a factor of $N^{-\frac{1}{10}}$ or less (though we did not give all the details to verify this statement). The derivatives may act in two ways on the integral (22). If we apply the s_i-derivative to the measure $d\mu_l(s_i,\zeta)$ with covariance $C_{l,\pm}(s_i)$ we use the integration by parts formula [12]

$$\frac{d}{ds_i}\int d\mu_l(s_i,\zeta)\ldots = \int d\mu_l(s_i,\zeta)\int_{x,y}\frac{\delta}{\delta\zeta(x)}\Big(\frac{d}{ds_i}C_{l,\pm}(s_i,x-y)\Big)\frac{\delta}{\delta\zeta(y)}\cdots$$

where the x,y-integral only extends over the squares Δ, Δ' linked by s_i due to the support properties of $\frac{d}{ds_i}C_{l,\pm}(s_i,x-y)$. Then we have to perform the functional ζ-derivatives. Or we apply the s-derivative directly to the kernels

$(1/\not p \pm m)(s_i, x-y)$ etc. appearing in (23). The discussion of the $\det_3$-terms in the previous section should reveal that the derivatives applied to $\det_3$ produce small factors $\leq N^{-1/10}$.

In the large field region we have neither exponential decay nor small factors per derivative. But it is intuitively clear that the cluster expansion has not a good chance to converge in a region where the squares are strongly coupled. What we have to do instead is to regard the connected components of the region l'' (13) as single entities or blocks in the cluster expansion. Then our cluster expansion is an expansion w.r.t. to the coupling of the whole of these blocks among each other (where a single small field square may also be a block). The exponential decay of the kernels in the small field region which also have to cross the security belts around the l-region and the strong exponential suppression proportional to the surface(!) of l (14) then produce even smaller contributions than in the $s\pm$-regions, and convergence is assured for the amplitudes of large polymers.

But we have to pay a price for this change of the expansion rules. Since the way of expanding depends on the $(l, s\pm)$-assignments we can no more factorize the denominator in the numerator of (22) due to the sums over the assignments. Factorization only holds term per term in the sum. Therefore we do not get a formula as (28) where the division was performed. To get rid of this nuisance is straightforward as regards the sum over the possible l-assignments. This sum can easily be interchanged with the sum over polymers: For a given polymer Y we may sum over all large field assignments $l \subset Y$ such that $l''(l) \subset Y$, i.e. basically the squares in l have to keep a distance $M = \frac{2}{m} \ln N$ (12) from the boundary of Y. But there is still another constraint, which is of topological origin and therefore of infinite range: In order to know whether some given s-square is to be assigned $s+$ or $s-$ we have to know how often a path joining $\partial\Lambda$ to this square crosses the contours separating $+$ from $-$. And this information is not given by the knowledge of the polymer Y. Due to this topological constraint the polymer amplitudes are therefore not factorized. Note also that the amplitude of the polymer will really depend on whether a single square is assigned $+$ or $-$ since all squares within Y are coupled among each other. The symmetry $\sigma \to -\sigma$ is a global one and the amplitudes are left invariant only if we flip the signs all squares in Y simultaneously.

The solution is as follows: First note that any contour belongs to a single block in the expansion, namely some connected component of l'' with "thickness" $\geq 2M$ (see (12)). Now we regard the links of a given polymer. If any of these links covers a distance $> 2M$ we join by a suitable rule of choice a minimum number of intermediate squares to the polymer in question, i.e. we cut up the link in question into several ones, such that no link joins any more directly two squares with $dist > 2M$. We include of course all of the polymers to which the added squares were attached into the previous one and thus melt several polymers into a single one while maintaining the fact that the squares in the new polymers are still connected together through tree structures. We can afford introducing these additional pseudo-links (which neither have decay nor small factors associated with them) since the "long" links covering distances $> 2M$ produce sufficiently small factors and decay also for the new links. With this new rule we are sure that our polymers contain all the contours crossed by any of their links. Therefore we know for any polymer its sign assignments, possibly up to a global sign flip (if the polymer does not touch $\partial\Lambda$). This sign flip does not affect its amplitude however, by the symmetry $\sigma \to -\sigma$.

These are the modifications of the standard scheme necessary to get convergence and factorization. Now we can verify (29) and obtain an expression for $S_2(x,y)$ of the form (28) which converges in the thermodynamic limit.

We end this section with a technical remark on how we treat the terms generated by derivatives acting on the term $\det(1+Q)$ (24) which couples the l- and s-regions and to which our previous arguments on estimating the derivative terms do not really apply. We find

$$\partial_{s_1}\ldots\partial_{s_n}\det(1+Q) \;=\; \det(1+Q)\,\times$$

$$[n!\,Tr\frac{1}{1+Q}\partial_{s_1}Q\wedge\frac{1}{1+Q}\partial_{s_2}Q\wedge\ldots\frac{1}{1+Q}\partial_{s_n}Q\;+\;rd]$$

where $\wedge$ denotes the antisymmetric tensor product [11] and rd stands for rederived terms when several ∂_{s_i} apply to the same Q. Thus the derivatives generate again determinant structures and one obtains the following bound (see also [20])

$$|n!\,Tr\frac{1}{1+Q}\partial_{s_1}Q\wedge\frac{1}{1+Q}\partial_{s_2}Q\wedge\ldots\frac{1}{1+Q}\partial_{s_n}Q\,\det(1+Q)|\;\le$$

$$(\prod_{j=1}^{n}(1+\lambda_j)^{-1}\prod_{i=1}^{n}Tr|\partial_{s_i}Q|)\det(1+Q)$$

where we denote $Tr|Q| = Tr(QQ^+)^{1/2}$, Q^+ denoting the adjoint, and λ_j, $1\le j\le n$, are the n smallest eigenvalues of Q (which are real). It is a consequence of the antisymmetric tensor product that we do not get the n-th power of the smallest eigenvalue. The r.h.s. may be bounded further using

$$\prod_{j=1}^{n}(1+\lambda_j)^{-1}\det(1+Q) \;=\; \prod_{j=n+1}^{\infty}(1+\lambda_j)\le e^n\,.$$

The last inequality follows from the following two facts:
(i) $\sum_{j=1}^{\infty}\lambda_j = TrQ = 0$ since Q is skew-symmetric.
(ii) The definition of Q through the operator A, and the fact that $A>-1$ in finite volume (which follows directly from the definition of A (11)) imply also $Q>-1$ and therefore for the $\lambda_j>-1$.

It then remains to bound the $\partial_{s_i}Q$. Without going into details we remark that the derivative applies to the nonlocal kernel $\frac{1}{p^2+m^2}$ appearing in Q which decreases exponentially and that the derivative also generates a factor small with N ($\sim N^{-2/5}$) in the small field region due to the small field condition. These conditions were addressed above as being crucial for the convergence of the cluster expansion. When the σ-fields appearing in $\partial_{s_i}Q$ become large field configurations we have to use the large field suppression to bound their contribution. It may even not be sufficient to use (14). For very large σ it is necessary to use directly the exponential suppression factor $e^{-1/2\int \sigma^2}$ to control the σ-terms generated by the expansion derivatives.

6 Results

For the verification of (29) we needed a small factor per small field derivative (it can be shown to be smaller than $m^{3/2}\,N^{-1/10}$), the large field suppression factor $e^{-O(1)N^{-1/5}|l|}$ (14) and the exponential decay of the kernels in the small field domain. Due to this decay we obtain for any tree T connecting the blocks in the polymer Y into one connected component a factor

$$e^{-mr(T,Y)} \tag{30}$$

where $r(T,Y)$ is essentially the length of the tree, namely

$$r(T,Y) = \sum_{(i,j)=l\in T} dist(\Delta_i,\Delta_j) \tag{31}$$

{(i,j)=l} being the links of the tree.
For the proof of the bound (29) (which is basically a convergence proof for the sum over polymers in the thermodynamic limit) it is allowed to use up completely the decay factor (30). Now we want to show also that $S_2(x,y)$ decays exponentially with a mass m (18) up to corrections small in N. This means that we have to isolate the decay factor from the presentation (28) of $S_2(x,y)$ without invalidating the convergence proof. To do so we have to make use of the fact that the polymer Y_1 contains (x,y) and the factor $(1/(\not{p}+m))(x,y)$. We take out of the sum over trees and polymers a minimal subtree connecting together (the squares) of x and y. Now we first assume that all blocks connected by the subtree are small field squares, e.g. in $s+$. Then for all possible choices the kernels corresponding to the links have always exponential decay with mass $\geq m$. Note in particular that the factor

$$\frac{1}{\not{p}+g\sigma} = \frac{1}{\not{p}+m+g\zeta}$$

from (22) may be expanded around $\zeta = 0$ due to the small field condition. The decay of the subtree then produces the exponential fall-off of $S_2(x,y)$ but we have to make sure of course that on taking out a minimal subtree we are still left with decay factors sufficient for the convergence proof. This can be shown to be true indeed by using the distance decay between the squares of the subtree (for fixed position of those) and the other ones. The basic point is of course the following: For the convergence proof only one square has to be fixed to break translation invariance. Here we are given two fixed squares so that their distance is left to be exploited. Finally we note that for nonempty large field assignments the exponentially small distance factors are replaced by the strong exponential suppression factors in the surface of l (14), which are even (much) smaller. Our findings can be summarized in the following

Theorem The infinite volume two-point function decays exponentially:

$$|S_2(x,y)| \leq O(1)\exp\{-m'|x-y|\}\,, \quad m' = m(1+o(N^{-1/10}))\,.$$

Furthermore we obtain for the expectation value of σ in a given square Δ

$$<\sigma_\Delta> = \frac{m}{g}(1+o(N^{-1/20}))\,.$$

This shows that the model is massive and that chiral symmetry is spontaneously broken. The model has (at least) two pure phases. The one with negative mean value for σ is realized through opposite boundary conditions.

References

[1] C. Kopper, J. Magnen, V. Rivasseau: Mass generation in the large N Gross-Neveu-Model, preprint to appear in Commun.Math.Phys.

[2] D. Gross, A. Neveu: Dynamical symmetry breaking in asymptotically free field theories, Phys. Rev.D **10**(1974) 3235.

[3] P. Mitter, P. Weisz: Asymptotic scale invariance in a massive Thirring model with $U(n)$-symmetry, Phys.Rev.**D8**(1973) 4410.

[4] K. Gawedzki, A. Kupiainen: Gross-Neveu model through convergent perturbation expansions, Commun.Math.Phys.**102**(1986) 1.

[5] J. Feldman, J. Magnen, V. Rivasseau, R. Sénéor: A Renormalizable Field Theory: The Massive Gross-Neveu Model in two Dimensions, Commun.Math.Phys.**103**(1986) 67.

[6] J. Magnen, V. Rivasseau, R. Sénéor: Construction of YM_4 with an infrared cutoff, Commun.Math.Phys.**155** (1993) 325.

[7] T. Balaban: Large Field Renormalization II: Localization, Exponentiation and bounds for the R-operation Commun.Math.Phys.**122** (1989) 355, and references cited there.

[8] A. Patrascioiu, E. Seiler: Super-Instantons and the reliability of Perturbation theory in Non-Abelian Models, MPI-preprint 93/87.

[9] P.A. Faria da Veiga: Construction de modèles non renormalisables en Théorie quantique des champs, thesis, Ecole Polytechnique 1991, and C.de Calan, these proceedings.

[10] M.C. Reed in: Constructive Field Theory, Lecture Notes in Physics 25 (1973), Proc. Erice 1973.

[11] B. Simon: Trace Ideals and their Applications, London Mathematical Society Lecture Note Series 35, Cambridge Univ.Press (1979).

[12] J. Glimm, A. Jaffe: Quantum Physics, Springer-Verlag, New York 1987.

[13] V. Rivasseau: From Perturbative to Constructive Renormalization, Princeton University Press 1991.

[14] D. Brydges in: Critical Phenomena, Random Systems, Gauge Theories, Proc. Les Houches 1984. North Holland (1986).

[15] D. Brydges and T. Kennedy: Mayer Expansions and the Hamilton-Jacobi Equation, Journ.Stat.Phys. **48** (1987) 19.

[16] D. Brydges and H.T. Yau: GradΦ Perturbations of massless Gaussian Fields, Commun.Math.Phys.**129** (1990) 351.

[17] A. Abdesselam and V. Rivasseau: Trees, Forests and Jungles, a Botanical Garden for Cluster Expansions, preprint Ecole Polytechnique June 1994.

[18] J. Glimm, A. Jaffe and T. Spencer: Phase transitions of φ_2^4 quantum fields, Commun.Math.Phys.**45** (1975) 203.

[19] T. Balaban and K. Gawedzki: A low-temperature expansion for the pseudo-scalar Yukawa model of quantum fields in two space-time dimensions, Ann.Inst.Henri Poincaré **A36** (1982) 271.

[20] E. Seiler: Schwinger functions for the Yukawa Model in two space time dimensions with space time cutoff, Commun.Math.Phys.**42** (1975) 163.

$U(1)$ Gauge Theory on a Torus

Yu. M. Zinoviev *

Steklov Mathematical Institute, Vavilov St. 42,
Moscow 117966, GSP-1, Russia
e-mail: zinoviev@qft.mian.su

Abstract $U(1)$ gauge theory with the Villain action on a cubic lattice approximation of three- and four-dimensional torus is considered. As the lattice spacing approaches zero, provided the coupling constant correspondingly approaches zero, the naturally chosen correlation functions converge to the correlation functions of the $\mathbf{R}$-gauge electrodynamics on three- and four-dimensional torus. When the torus radius tends to infinity these correlation functions converge to the correlation functions of the $\mathbf{R}$-gauge Euclidean electrodynamics.

1 Introduction

The compact lattice gauge field theory models introduced by K. Wilson [1] preserve the differential geometric structures of the continuum theory. This paper is concerned with the case where the gauge group is $U(1) = \mathbf{R}/2\pi\mathbf{Z}$. Let $h(\theta)$ be a real twice continuously differentiable even periodic function with period 2π. Any such function will be called an energy function. The main examples of interest are the Wilson [1] energy function $h(\theta) = 1 - \cos\theta$ and the Villain [2] energy function

$$\exp[-\beta h_\beta(\theta)] = c_\beta \sum_{n=-\infty}^{\infty} \exp[-\beta(\theta - 2\pi n)^2/2] \tag{1.1}$$

where $\beta > 0$ and c_β is a constant chosen so that the right-hand side is one for $\theta = 0$.

Let e_i, $i = 1, ..., d$ be the standard unit vectors in $\mathbf{R}^d$, and p be a non-negative integer less than d. The p-cells based at $\mathbf{m} \in \mathbf{Z}^d$ are the formal symbols: $(\mathbf{m}; e_{i_1}, ..., e_{i_p})$ where the unit vectors differ from each other.

Let G be one of three Abelian groups: $\mathbf{Z}$, $\mathbf{R}$ and $U(1) = \mathbf{R}/2\pi\mathbf{Z}$. A p-cochain with the coefficients in G is a G-valued function on p-cells $f(\mathbf{m}; e_{i_1}, ..., e_{i_p}) \equiv f_{i_1 \cdots i_p}(\mathbf{m})$ which is antisymmetric under the permutations of the indices $i_1, ..., i_p$.

* Supported by the Russian Foundation of Fundamental Researches under Grant 93-011-147.

Let $\Lambda = \{\mathbf{m} \in \mathbf{Z}^d : N_1 \le m_i \le N_2, i = 1, ..., d\}$ be a cube in $\mathbf{Z}^d$ for some integers N_1 and N_2. The free boundary conditions are equivalent to setting that $f_{i_1 \cdots i_p}(\mathbf{m})$ vanishes except for $\{\mathbf{m} \in \mathbf{Z}^d : N_1 \le m_i \le N_2, i \ne i_1, ..., i_p; N_1 \le m_{i_k} \le N_2 - 1, k = 1, ..., p\}$. Dirichlet boundary conditions correspond to setting that $f_{i_1 \cdots i_p}(\mathbf{m})$ vanishes except for $\{\mathbf{m} \in \mathbf{Z}^d : N_1 + 1 \le m_i \le N_2, i \ne i_1, ..., i_p; N_1 \le m_{i_k} \le N_2, k = 1, ..., p\}$. The conditions

$$f_{i_1 \cdots i_p}(m_1, ..., m_j + N, ..., m_d) = f_{i_1 \cdots i_p}(\mathbf{m}), \tag{1.2}$$

where $N = N_2 - N_1 + 1$, for every $j = 1, ..., d$ and $\mathbf{m} \in \mathbf{Z}^d$ correspond to the choice of periodic boundary conditions. For the periodic boundary conditions we define the boundary operator

$$(\partial f)_{i_1 \cdots i_{p-1}}(\mathbf{m}) = \sum_{\epsilon=0,1} \sum_{i_0=1}^{d} (-1)^{\epsilon+1} f_{i_0 i_1 \cdots i_{p-1}}(\mathbf{m} - \epsilon e_{i_0}) \tag{1.3}$$

and the coboundary operator

$$(\partial^* f)_{i_1 \cdots i_{p+1}}(\mathbf{m}) = \sum_{\epsilon=0,1} \sum_{k=1}^{p+1} (-1)^{\epsilon+k} f_{i_1 \cdots \widehat{i_k} \cdots i_{p+1}}(\mathbf{m} + \epsilon e_{i_k}) \tag{1.4}$$

For Dirichlet or free boundary conditions we need to modify the definitions (1.3) and (1.4), respectively, in an obvious way.

For the p-cochains with coefficients in $\mathbf{Z}$ and $\mathbf{R}$ the inner product is defined by

$$(f, g) = \sum_{i_1 < \cdots < i_p} \sum_{\mathbf{m} \in \Lambda} f_{i_1 \cdots i_p}(\mathbf{m}) g_{i_1 \cdots i_p}(\mathbf{m}). \tag{1.5}$$

For a smooth differential p-form $f(\mathbf{x}) = \sum_{i_1 < \cdots < i_p} f_{i_1 \cdots i_p}(\mathbf{x}) dx^{i_1} \wedge \cdots \wedge dx^{i_p}$ on $\mathbf{R}^d$ we introduce two lattice approximations: $(f_a)_{i_1 \cdots i_p}(\mathbf{m}) = f_{i_1 \cdots i_p}(a\mathbf{m})$ and

$$(f^a)_{i_1 \cdots i_p}(\mathbf{m}) = \int_{[0,a]^{\times p}} f_{i_1 \cdots i_p}(a\mathbf{m} + \sum_{k=1}^{p} s_k e_{i_k}) d^p s. \tag{1.6}$$

The energy function h and 1-cochain θ on Λ provide two 2-cochains on Λ with real coefficients. These 2-cochains are defined for any indices $i_1 < i_2$ by the following relations $(h(\partial^*\theta))_{i_1 i_2}(\mathbf{m}) \equiv h((\partial^*\theta)_{i_1 i_2}(\mathbf{m}))$, $(h'(\partial^*\theta))_{i_1 i_2}(\mathbf{m}) \equiv h'((\partial^*\theta)_{i_1 i_2}(\mathbf{m}))$. By 1 we denote 2-cochain $(1)_{i_1 i_2}(\mathbf{m}) = 1$ for any indices $i_1 < i_2$.

The finite volume Gibbs state in a cube $\Lambda \subset \mathbf{Z}^d$, at inverse temperature β and with energy function h is given by

$$\langle F \rangle_{\Lambda,\beta} = Z^{-1} \left[\prod_{\mathbf{m} \in \Lambda i=1,...,d} \int_{-\pi}^{\pi} d\theta_i(\mathbf{m}) \right] F(\theta) \exp[-\beta(h_\beta(\partial^*\theta), 1)]. \tag{1.7}$$

Here θ is a 1-cochain on Λ with coefficients in $U(1)$ and θ satisfies the periodic boundary conditions. For Dirichlet or free boundary conditions some $\theta_i(\mathbf{m})$ are equal to zero and the corresponding integrations in (1.7) are omitted. The measure $d\theta_i(\mathbf{m})$ is Lebesgue measure on $[-\pi, \pi]$. Z is the normalization constant and F is a function of the bond variables $\theta_i(\mathbf{m})$.

Let $\langle\rangle_\beta$ be any translation invariant infinite volume limit Gibbs state. L.Gross [3] proved that for Wilson and Villain energy functions h and for every smooth differential real 1-form j on $\mathbf{R}^3$ the following equality holds

$$\lim_{a\to 0}\langle \exp[i(h'(\partial^*\theta), \partial^* j_a)]\rangle_{(ag^2)^{-1}} = \exp[-g^2(dj, dj)/2], \tag{1.8}$$

where g is a strictly positive real number and d is a differential operator on the differential 1-forms on $\mathbf{R}^3$. The inner product of the differential 2-forms on $\mathbf{R}^d$ is similar to the inner product (1.5).

Let ψ be a smooth real differential 3-form on $\mathbf{R}^3$ and r be any number in $[1, \infty)$. L.Gross [3] proved also that for the Villain energy function

$$\lim_{a\to 0}\langle |(h'_{(ag^2)^{-1}}(\partial^*\theta), \partial\psi_a)|^r\rangle_{(ag^2)^{-1}} = 0. \tag{1.9}$$

In the four dimensional case B.Driver [4] proved that for the Wilson energy function and "for all but at most countable numbers of $g > 0$"

$$\lim_{a\to 0}\langle \exp[i(h'(\partial^*\theta), \partial^* j^a)]\rangle_{g^{-2}} = \exp[-\alpha g^2(dj, dj)/2], \tag{1.10}$$

where $\langle\rangle_{g^{-2}}$ is any translation and 90°- rotation invariant infinite volume Gibbs state, j is any real smooth differential 1-form on $\mathbf{R}^4$, j^a is its lattice approximation (1.6) and the number $\alpha \geq 0$ is independent of the particular choice $\langle\rangle_{g^{-2}}$.

This paper is concerned with the case of the Villain energy function and the periodic boundary conditions. We study the correlation functions: $\langle \exp[i(j, \theta)]\rangle_{\Lambda,\beta}$ where j is a 1-cochain on Λ with the integer coefficients. The inner product (j, θ) is not defined for a 1-cochain θ on Λ with coefficients in $U(1) = \mathbf{R}/2\pi\mathbf{Z}$, but $\exp[i(j, \theta)]$ is well defined. It is easy to show that $\langle \exp[i(j, \theta)]\rangle_{\Lambda,\beta} = 0$ if $j \neq \partial\phi$ for some 2-cochain ϕ on Λ with the integer coefficients (see, for example, [5]). In view of the periodic boundary conditions we can identify the opposite vertices of the cube $[N_1, N_2 + 1]^{\times d}$ and obtain a lattice approximation $\mathbf{T}^d_N$ of the torus $\mathbf{T}^d$ of radius R.

Let $f_{i_1\cdots i_p}(\mathbf{x})$ be the coefficients of a real smooth differential p-form on the torus $\mathbf{T}^d$. We define the integer valued p-cochain on $\mathbf{T}^d_N$

$$(f_{N,b})_{i_1\cdots i_p}(\mathbf{m}) = [N^b f_{i_1\cdots i_p}(2\pi R N^{-1}\mathbf{m})], \tag{1.11}$$

where $N = N_2 - N_1 + 1$, b is a strictly positive integer and $[r]$ is the integer part of the real number r. In order to define the continuum limit we need to know how the constant β in the Villain energy function (1.1) depends on the lattice spacing parameter $a = 2\pi R N^{-1}$. B.K. Driver requires that being multiplied by $\beta(a)$ the scalar product (1.5) of two lattice approximations (1.6) of the smooth differential 2–forms (electromagnetic field strength is the differential 2–form) tends as $a \to 0$ to the usual scalar product of these smooth differential 2–forms multiplied by the constant g^{-2}. This requirement implies that $\beta(a) = g^{-2}a^{d-4}$. For the dimensions $d = 3, 4$ we get $\beta(a) = (ag^2)^{-1}$ and $\beta(a) = g^{-2}$, respectively. Due to the Theorem 4.1 from [4] for the dimensions $d > 4$ this scaling implies that the continuum limit on the current sector of $U(1)$ gauge lattice models is degenerate. We choose the non–standard scaling, when $\beta(a) = g_1^{-2}a^{-d-2b}$, where b is a strictly positive integer introduced above. We require that being multiplied by $(\beta(N))^{-1}$ the scalar product (1.5) of two lattice approximations (1.11) of the smooth differential p–forms tends as $N \to \infty$ to the usual scalar product of these smooth differential p–forms multiplied by the constant g^2. Let the function $f(\mathbf{x})$ on the torus $\mathbf{T}^d$ be equal to one. By the definition (1.11) the 0-cochain $f_{N,b}(\mathbf{m}) \equiv N^b$. The definition (1.5) implies that $(f_{N,b}, f_{N,b}) = N^{d+2b}$. Choose $\beta(N)$ such that $\beta(N)^{-1}(f_{N,b}, f_{N,b}) = g^2(2\pi R)^d$, where $(2\pi R)^d$ is the volume of the torus $\mathbf{T}^d$ and $g > 0$, i.e. $\beta(N) = g^{-2}(2\pi R)^{-d}N^{d+2b}$. It seems that this geometrical definition of the continuum limit may be useful also for the $U(1)$ gauge models including the interaction with the Fermions. In the next sections it will be proved that for any real smooth differential 2-form ϕ on the torus $\mathbf{T}^d$, $d = 3, 4$,

$$\lim_{N\to\infty} \langle \exp[i(\partial\phi_{N,b}, \theta)]\rangle_{\mathbf{T}^d_N, \beta(N)} = \exp[-g^2(d^*\phi, G(d^*\phi))/2] \tag{1.12}$$

where d^* is the adjoint operator of the differential operator d, the operator G is the Green operator for the Laplace-Beltrami operator on the differential 1-forms on the torus $\mathbf{T}^d$ and the inner product of the differential 1-forms on the torus is similar to the inner product (1.5).

When $\phi = dj$ the right-hand side of the equality (1.12) is a torus analogue of the right-hand sides of the equalities (1.8) and (1.10) for $\alpha = 1$. Due to $(d^*)^2 = 0$ the substitution $\phi = d^*\psi$ into the right-hand side of the equality (1.12) yields 1 and we obtain a torus generalization of the equalities (1.9). It is important to note that the right-hand side of the equality (1.12) coincides with the correlation function of $\mathbf{R}$-gauge electrodynamics on the torus [6]. As the torus radius R tends to infinity the limit of (1.12) gives the correlation function of $\mathbf{R}$-gauge Euclidean electrodynamics [6]. We studied the continuum limit (1.12) of the correlation functions of the free $U(1)$ gauge model. We believe that the limit (1.12) may be applied for the study of the correlation functions of the $U(1)$ gauge model which includes the interaction with the fermions.

The remaining section is devoted to the proof of the equality (1.12) for $d = 3$. In order to prove the relation (1.12) for $d = 4$ the four dimensional versions of the below results are obtained in the paper [7] .

2 Three Dimensional Torus

The p-cochains with the coefficients in the Abelian group $G = \mathbf{Z}, \mathbf{R}, U(1) = \mathbf{R}/2\pi\mathbf{Z}$ satisfying the periodic boundary conditions (1.2) form the Abelian group $C^p(\mathbf{T}^d_N, G)$, where $N = N_2 - N_1 + 1$. In order to simplify the situation we assume that $N_1 = 0$, $N_2 = N-1$. Now the definition (1.7) for the correlation function may be rewritten in the form

$$\langle \exp[i(\jmath,\theta)]\rangle_{\mathbf{T}^d_N,\beta} = Z^{-1}_{\mathbf{T}^d_N} \int_{C^1(\mathbf{T}^d_N,U(1))} \exp[i(\jmath,\theta) - \beta(h_\beta(\partial^*\theta),1)]d\theta, \qquad (2.1)$$

where a 1-cochain $\jmath \in C^1(\mathbf{T}^d_N, \mathbf{Z})$ and $\exp[i(\jmath,\theta)]$ is a character of the compact group $C^1(\mathbf{T}^d_N, U(1))$. The Villain energy function $h_\beta(\theta)$ is given by (1.1). Here $d\theta$ is the normalized Haar measure on the compact group $C^1(\mathbf{T}^d_N, U(1))$ and $Z_{\mathbf{T}^d_N}$ is the normalization constant.

By [5, Lemma 1] the correlation function (2.1) isn't zero only for the boundaries $\jmath = \partial\phi$, where $\phi \in C^2(\mathbf{T}^d_N, \mathbf{Z})$, and

$$\langle \exp[i(\partial\phi,\theta)]\rangle_{\mathbf{T}^d_N,\beta} = Z^{-1}_{\mathbf{T}^d_N} \int_{B^2(\mathbf{T}^d_N,U(1))} \exp[i(\phi,\psi) - \beta(h_\beta(\psi),1)]d\psi, \qquad (2.2)$$

where the group of coboundaries $B^2(\mathbf{T}^d_N, U(1))$ is the image of the homomorphism $\partial^* : C^1(\mathbf{T}^d_N, U(1)) \to C^2(\mathbf{T}^d_N, U(1))$ and $d\psi$ is the normalized Haar measure on the compact group $B^2(\mathbf{T}^d_N, U(1))$.

It is easy to compute the Fourier transform of the function (1.1)

$$1/2\pi \int_0^{2\pi} \exp[in\theta - \beta h_\beta(\theta)]d\theta = c_\beta (2\pi\beta)^{-1/2} \exp[-(2\beta)^{-1} n^2]. \qquad (2.3)$$

By using the Fourier transform on the group $B^2(\mathbf{T}^d_N, U(1))$, due to the formula (2.3) and [5, Proposition 1] we obtain

$$\langle \exp[i(\partial\phi,\theta)]\rangle_{\mathbf{T}^d_N,\beta} = Z^{-1}_{\mathbf{T}^d_N} c^g_\beta (2\pi\beta)^{-g/2} \times \qquad (2.4)$$
$$\sum_{\mathbf{m}\in\mathbf{Z}^g} \exp\Big[-1/2\beta(\phi + \sum_{i=1}^g m_i z_i, \phi + \sum_{i=1}^g m_i z_i)\Big],$$

where $z_1, ..., z_g$ form a basis of the group of 2-cycles $Z_2(\mathbf{T}^d_N, \mathbf{Z})$ which is the kernel of the homomorphism $\partial : C^2(\mathbf{T}^d_N, \mathbf{Z}) \to C^1(\mathbf{T}^d_N, \mathbf{Z})$. The symmetric $g \times g$ matrix $\Omega_{ij} = (z_i, z_j)$ is positively definite and invertible. Let us introduce the dual basis $\bar{z}_i = \sum_{j=1}^g \Omega^{-1}_{ij} z_j$. For every $i = 1, ..., g$ the 2-cochain $\bar{z}_i \in Z_2(\mathbf{T}^d_N, \mathbf{R})$ has the following properties: $(\bar{z}_i, z_j) = \delta_{ij}$ and $(\bar{z}_i, \bar{z}_j) = \Omega^{-1}_{ij}$. Let $\bar{Z}_2(\mathbf{T}^d_N, \mathbf{Z})$ be a free group with the basis $\bar{z}_1, ..., \bar{z}_g$. The group $\bar{Z}_2(\mathbf{T}^d_N, \mathbf{Z})$ may be defined also as the maximal subgroup of $Z_2(\mathbf{T}^d_N, \mathbf{R})$ so that for any elements $z \in Z_2(\mathbf{T}^d_N, \mathbf{Z})$ and $\bar{z} \in \bar{Z}_2(\mathbf{T}^d_N, \mathbf{Z})$ the inner product $(z, \bar{z})$ is an integer.

Applying the Poisson summation formula

$$\sum_n f(n) = \sum_n \int dx f(x) \exp[2\pi i n x], \qquad (2.5)$$

we can rewrite the relation (2.4) as

$$\langle \exp[i(\partial\phi,\theta)]\rangle_{\mathbf{T}^d_N,\beta} = Z^{-1}_{\mathbf{T}^d_N} c^g_\beta [\det \Omega]^{-1/2} \times$$
$$W_{\mathbf{T}^d_N,\beta}(\partial\phi)\Theta((\phi,\bar{\mathbf{z}})|2\pi i\beta\Omega^{-1}), \tag{2.6}$$

where

$$W_{\mathbf{T}^d_N,\beta}(\partial\phi) = \exp\big[-1/2\beta[(\phi,\phi) - \sum_{i=1}^{g}(\phi,z_i)(\phi,\bar{z}_i)]\big] \tag{2.7}$$

is the correlation function calculated for **R**-gauge electrodynamics on the lattice $\mathbf{T}^d_N$ in the paper [8] and the Riemann θ-function

$$\Theta(\mathbf{y}|\omega) = \sum_{\mathbf{m}\in\mathbf{Z}^g} \exp\big[i\pi \sum_{j,k=1}^{g} m_i\omega_{ij}m_j + 2\pi i\sum_{j=1}^{g} m_j y_j\big] \tag{2.8}$$

depends on the vector $\mathbf{y} \in \mathbf{C}^g$ and on the symmetric $g \times g$ matrix ω with the positively definite imaginary part. In our case $\omega = 2\pi i\beta\Omega^{-1}$.

Taking the trivial 2-cochain $\phi = 0$ we obtain

$$Z_{\mathbf{T}^d_N} = c^g_\beta[\det \Omega]^{-1/2}\Theta(\mathbf{0}|2\pi i\beta\Omega^{-1}). \tag{2.9}$$

The substitution of the equality (2.9) into the right-hand side of the relation (2.6) gives

$$\langle \exp[i(\partial\phi,\theta)]\rangle_{\mathbf{T}^d_N,\beta} = W_{\mathbf{T}^d_N,\beta}(\partial\phi)\frac{\Theta((\phi,\bar{\mathbf{z}})|2\pi i\beta\Omega^{-1})}{\Theta(\mathbf{0}|2\pi i\beta\Omega^{-1})} \tag{2.10}$$

This formula was obtained in the paper [8] for the Wilson energy function but only in the weak-coupling region when the inverse temperature approaches infinity.

The definition of the group $\bar{Z}_2(\mathbf{T}^d_N,\mathbf{Z})$ and the definition (2.8) imply that

$$\Theta((\phi,\bar{\mathbf{z}})|2\pi i\beta\Omega^{-1}) = \sum_{\bar{z}\in\bar{Z}_2(\mathbf{T}^d_N,\mathbf{Z})} \exp[-2\pi^2\beta(\bar{z},\bar{z}) + 2\pi i(\phi,\bar{z})] \tag{2.11}$$

To obtain the equality (2.10) we used the Fourier transform on the group $U(1)$. Our space $\mathbf{T}^d$ is a product of groups $U(1)$ and a lattice approximation $\mathbf{T}^d_N = \mathbf{Z}^d_N$ where a group $\mathbf{Z}_N = \mathbf{Z}/N\mathbf{Z}$. In order to study the groups $Z_2(\mathbf{T}^d_N,\mathbf{Z})$ and $\bar{Z}_2(\mathbf{T}^d_N,\mathbf{Z})$ we use the Fourier transform on the group $\mathbf{Z}^d_N$. The Fourier transform of a p-cochain $f \in C^p(\mathbf{T}^d_N,\mathbf{R})$ is defined by

$$f^{\sim}_{i_1\cdots i_p}(\mathbf{l}) = \sum_{\mathbf{m}\in\Lambda} \exp\left[2\pi i N^{-1}\sum_{k=1}^{d} l_k m_k\right] f_{i_1\cdots i_p}(\mathbf{m}) \tag{2.12}$$

where $\mathbf{l} \in \Lambda$ and a cube $\Lambda = \{\mathbf{m} \in \mathbf{Z}^d : 0 \le m_i \le N-1, i = 1, ..., d\}$. We denote the group of functions (2.12) by $C^p(\mathbf{T}_N^d, \mathbf{R})^\sim$.

The following relation

$$N^{-1} \sum_{k=0}^{N-1} \exp[2\pi i N^{-1} km] = \delta_{m,0} \tag{2.13}$$

holds for any integer $-N+1 \le m \le N-1$. The relations (1.5) and (2.13) imply

$$(f, g) = N^{-d} \sum_{i_1 < \cdots < i_p} \sum_{\mathbf{l} \in \Lambda} \overline{f^\sim_{i_1 \cdots i_p}(\mathbf{l})} g^\sim_{i_1 \cdots i_p}(\mathbf{l}) \tag{2.14}$$

The right-hand side of the equality (2.14) we denote by $(f^\sim, g^\sim)$.

Applying the Fourier transform (2.12) we can rewrite the equalities (1.3) and (1.4) as

$$(\tilde{\partial} f^\sim)_{i_1 \cdots i_{p-1}}(\mathbf{l}) = \sum_{i_0=1}^{d} (\exp[2\pi i N^{-1} l_{i_0}] - 1) f^\sim_{i_0 i_1 \cdots i_{p-1}}(\mathbf{l}) \tag{2.15}$$

$$(\tilde{\partial}^* f^\sim)_{i_1 \cdots i_{p+1}}(\mathbf{l}) = \sum_{k=1}^{p+1} (-1)^{k+1} (\exp[-2\pi i N^{-1} l_{i_k}] - 1) f^\sim_{i_1 \cdots \widehat{i_k} \cdots i_{p+1}}(\mathbf{l}) \tag{2.16}$$

Therefore a lattice Laplace-Beltrami operator is given by

$$((\tilde{\partial}^* \tilde{\partial} + \tilde{\partial} \tilde{\partial}^*) f^\sim)_{i_1 \cdots i_p}(\mathbf{l}) = \sum_{k=1}^{d} |\exp[2\pi i N^{-1} l_k] - 1|^2 f^\sim_{i_1 \cdots i_p}(\mathbf{l}) \tag{2.17}$$

A p-cochain f is said to be harmonic if the expression (2.17) equals zero.

Lemma 2.1 *Any harmonic p-cochain $f \in C^p(\mathbf{T}_N^d, \mathbf{R})$ is constant.*

Proof. Taking the inner product (2.14) of the functions $f^\sim_{i_1 \cdots i_p}(\mathbf{l})$ and (2.17) we get that p-cochain $f_{i_1 \cdots i_p}(\mathbf{m})$ is harmonic iff for any $k = 1, ..., d$

$$(\exp[2\pi i N^{-1} l_k] - 1) f^\sim_{i_1 \cdots i_p}(\mathbf{l}) = 0 \tag{2.18}$$

Hence $f^\sim_{i_1 \cdots i_p}(\mathbf{l})$ isn't zero only for $l_k = 0$, $k = 1, ..., d$. By using the inverse Fourier transform we have that the p-cochain $f_{i_1 \cdots i_p}(\mathbf{m})$ is constant.

Let P be an orthogonal projector on the subspace $Z_2(\mathbf{T}_N^d, \mathbf{R})$ of the linear space $C^2(\mathbf{T}_N^d, \mathbf{R})$. Lemma 2.1 and the equalities (2.15) - (2.17) imply that

$$(Pf)^\sim_{i_1 i_2}(\mathbf{l}) = f^\sim_{i_1 i_2}(\mathbf{0}) \prod_{k=1}^{d} \delta_{l_k,0} +$$

$$\left(\tilde{\partial} \left(\sum_{k=1}^{d} |\exp[2\pi i N^{-1} l_k] - 1|^2\right)^{-1} \tilde{\partial}^* f^\sim\right)_{i_1 i_2}(\mathbf{l}) \tag{2.19}$$

Due to equalities (2.15) and (2.16) we may consider the second term in the right-hand side of (2.19) to be equal zero at $\mathbf{l} = \mathbf{0}$. We denote the right-hand side of the relation (2.19) by $(\tilde{P}f^{\sim})_{i_1 i_2}(\mathbf{l})$.

Proposition 2.2 *The Fourier transform (2.12) of every 2-cycle $z \in Z_2(\mathbf{T}_N^3, \mathbf{Z})$ has the form : for any permutation i_1, i_2, i_3 of the numbers 1, 2, 3*

$$z^{\sim}_{i_1 i_2}(\mathbf{l}) = a^{\sim}_{i_1 i_2}(\mathbf{l}) + (\exp[2\pi i N^{-1} l_{i_3}] - 1) \times \\ [b^{\sim}_{i_2 i_3}(\mathbf{l}) - b^{\sim}_{i_1 i_3}(\mathbf{l}) + c^{\sim}_{i_1 i_2 i_3}(\mathbf{l})] \tag{2.20}$$

where a 2-cochain $a_{j_1 j_2}(\mathbf{m})$ is independent of the variables m_{j_1}, m_{j_2}, a 2-cochain $b_{j_1 j_2}(\mathbf{m})$ depends on the variables m_{j_1}, m_{j_2} only and it is equal to zero if one of these variables equals $N-1$, a 3-cochain $c_{i_1 i_2 i_3}(\mathbf{m})$ equals zero if one of the variables m_1, m_2, m_3 is equal to $N-1$. The above cochains determine the 2-cycle given by (2.20) uniquely.

Proof. By using the formula for the sum of geometric progression we get

$$z^{\sim}_{i_1 i_2}(\mathbf{l}) = z^{\sim}_{i_1 i_2}(\mathbf{l}) \mid_{l_{i_3}=0} + (\exp[2\pi i N^{-1} l_{i_3}] - 1)(z^{\sim})'_{i_1 i_2; i_3}(\mathbf{l}) \tag{2.21}$$

where

$$(z^{\sim})'_{i_1 i_2; i_3}(\mathbf{l}) = \sum_{m_{i_1}, m_{i_2}=0}^{N-1} \sum_{m_{i_3}=1}^{N-1} \sum_{m'_{i_3}=0}^{m_{i_3}-1} \exp[2\pi i N^{-1}[\sum_{k=1.2} l_{i_k} m_{i_k} + l_{i_3} m'_{i_3}]] z_{i_1 i_2}(\mathbf{m}) \tag{2.22}$$

Since a function $z^{\sim}_{i_1 i_2}(\mathbf{l})$ is antisymmetric under the permutation of the indices i_1, i_2 the substitution $l_{i_3} = 0$ into two equations $(\tilde{\partial} z^{\sim})_{i_1}(\mathbf{l}) = 0$ and $(\tilde{\partial} z^{\sim})_{i_2}(\mathbf{l}) = 0$ provides two equations $(\exp[2\pi i N^{-1} l_{i_k}] - 1) z^{\sim}_{i_1 i_2}(\mathbf{l}) \mid_{l_{i_3}} = 0$, $k = 1, 2$. Hence a function $z^{\sim}_{i_1 i_2}(\mathbf{l}) \mid_{l_{i_3}=0}$ is not equal to zero only at $l_{i_1} = 0$, $l_{i_2} = 0$ and by the relation (2.13) we have

$$z^{\sim}_{i_1 i_2}(\mathbf{l}) \mid_{l_{i_3}=0} = N^2 \delta_{l_{i_1},0} \delta_{l_{i_2},0} a_{i_1 i_2} \tag{2.23}$$

where a constant $a_{i_j i_2}$ is antisymmetric under a permutation of the indices i_1, i_2.

The equation $(\tilde{\partial} z^{\sim})_{i_1}(\mathbf{l}) = 0$ and the equalities (2.21), (2.23) imply that

$$\left[\prod_{k=2,3} (\exp[2\pi i N^{-1} l_{i_k}] - 1) \right] [(z^{\sim})'_{i_1 i_2; i_3}(\mathbf{l}) + (z^{\sim})'_{i_1 i_3; i_2}(\mathbf{l})] = 0 \tag{2.24}$$

Let us introduce the function

$$b^{\sim}_{i_2 i_3}(\mathbf{l}) = \delta_{l_{i_1},0} \sum_{l'_{i_1}=0}^{N-1} \exp[-2\pi i N^{-1}(N-1) l'_{i_1}](z^{\sim})'_{i_1 i_2; i_3}(\mathbf{l}) \mid_{l_{i_1}=l'_{i_1}} - \\ \delta_{l_{i_1},0} \delta_{l_{i_2},0} \sum_{l'_{i_1}, l'_{i_2}=0}^{N-1} \exp[-2\pi i N^{-1}(N-1)(l'_{i_1} + l'_{i_2})](z^{\sim})'_{i_1 i_2; i_3}(\mathbf{l}') \mid_{l'_{i_3}=l_{i_3}} \tag{2.25}$$

By definition the function (2.25) satisfies the equation (2.24) and therefore it satisfies the equation $b^{\sim}_{i_2 i_3}(\mathbf{l}) + b^{\sim}_{i_3 i_2}(\mathbf{l}) = N^2 \delta_{l_{i_1},0}[\delta_{l_{i_2},0} f_3(l_{i_3}) + \delta_{l_{i_3},0} f_2(l_{i_2})]$. The definitions (2.22) and (2.25) imply that the Fourier expansion of the left-hand side of this equation does not contain the components with $m_{i_2} = N-1$ or $m_{i_3} = N-1$. It is easy to show now that $b^{\sim}_{i_2 i_3}(\mathbf{l}) + b^{\sim}_{i_3 i_2}(\mathbf{l}) = 0$. Hence the function $b^{\sim}_{i_2 i_3}(\mathbf{l})$ is antisymmetric under a permutation of indices i_2, i_3. By the definitions (2.22), (2.25) and the relation (2.13) all components in the Fourier expansion of the function $b^{\sim}_{i_2 i_3}(\mathbf{l})$ are integers. Then the function $b^{\sim}_{i_2 i_3}(\mathbf{l})$ is the Fourier transform (2.12) of some 2-cochain $b_{i_2 i_3}(\mathbf{m}) \in C^2(\mathbf{T}^3_N, \mathbf{Z})$. Due to equality (2.25) a cochain $b_{i_2 i_3}(\mathbf{m})$ depends on the variables m_{i_2}, m_{i_3} only and it is equal to zero if one of these variables equals $N-1$.

We define the function $c^{\sim}_{i_1 i_2 i_3}(\mathbf{l})$ by the following equality

$$(z^{\sim})'_{i_1 i_2; i_3}(\mathbf{l}) = c^{\sim}_{i_1 i_2 i_3}(\mathbf{l}) + b^{\sim}_{i_2 i_3}(\mathbf{l}) - b^{\sim}_{i_1 i_3}(\mathbf{l}) +$$
$$\delta_{l_{i_1},0}\delta_{l_{i_2},0} \sum_{l'_{i_1}, l'_{i_2}=0}^{N-1} \exp[-2\pi i N^{-1}(N-1)(l'_{i_1} + l'_{i_2})](z^{\sim})'_{i_1 i_2; i_3}(\mathbf{l}') \,|_{l'_{i_3}=l_{i_3}} \quad (2.26)$$

A function $c^{\sim}_{i_1 i_2 i_3}(\mathbf{l})$ is obviously antisymmetric under a permutation of indices i_1, i_2. By definitions (2.22), (2.25) and (2.26) its Fourier expansion does not contain the components with $m_k = N-1$, where k is one of the numbers $1, 2, 3$ and the remaining components are integers. It is easy now to verify that a function $c^{\sim}_{i_1 i_2 i_3}(\mathbf{l})$ satisfies the equation (2.24) and therefore it is antisymmetric under a permutation of indices i_2, i_3. Hence it is antisymmetric under a permutation of all indices i_1, i_2, i_3. Then a function $c^{\sim}_{i_1 i_2 i_3}(\mathbf{l})$ is the Fourier transform (2.12) of some 3-cochain $c_{i_1 i_2 i_3}(\mathbf{m}) \in C^3(\mathbf{T}^3_N, \mathbf{Z})$ which equals zero if one of the variables m_1, m_2, m_3 is equal to $N-1$.

Now the equalities (2.21), (2.23) and (2.26) imply the equality (2.20) where a function $a^{\sim}_{i_1 i_2}(\mathbf{l})$ is the Fourier transform (2.12) of some 2-cochain $a_{i_1 i_2}(\mathbf{m}) \in C^2(\mathbf{T}^3_N, \mathbf{Z})$ independent of the variables m_{i_1}, m_{i_2}. By definitions the cochains in the expansion (2.20) determine the 2-cycle defined by (2.20) uniquely.

Let S_n be a symmetric group, i.e. a group of all permutations of the numbers $1, ..., n$. For any permutation $\sigma \in S_3$ and for any point $\mathbf{m} \in \mathbf{T}^3_N$ we introduce three 2-cochains from $C^2(\mathbf{T}^3_N, \mathbf{Z})$ by defining their Fourier transforms

$$(\hat{a}[\sigma(1), \sigma(2); m_{\sigma(3)}])^{\sim}_{i_1 i_2}(\mathbf{l}) = (\delta_{i_1,\sigma(1)}\delta_{i_2,\sigma(2)} - \delta_{i_1,\sigma(2)}\delta_{i_2,\sigma(1)}) \times$$
$$\exp[2\pi i N^{-1}((N-1)(l_{\sigma(1)} + l_{\sigma(2)}) + m_{\sigma(3)} l_{\sigma(3)})] \quad (2.27)$$

$$(\hat{b}[\sigma(2), \sigma(3); m_{\sigma(2)}, m_{\sigma(3)}])^{\sim}_{i_1 i_2}(\mathbf{l}) = (\delta_{i_1,\sigma(1)}\delta_{i_2,\sigma(2)} - \delta_{i_1,\sigma(2)}\delta_{i_2,\sigma(1)}) \times$$
$$\left[\prod_{k=2,3} (\exp[2\pi i N^{-1} m_{\sigma(k)} l_{\sigma(k)}] - \exp[2\pi i N^{-1}(N-1) l_{\sigma(k)}])\right] \times$$
$$(\exp[-2\pi i N^{-1} l_{\sigma(3)}] - 1)^{-1} \exp[2\pi i N^{-1}((N-1) l_{\sigma(1)}] \quad (2.28)$$

$$(\hat{c}[\sigma(1),\sigma(2),\sigma(3);\mathbf{m}])^{\sim}_{i_1 i_2}(\mathbf{l}) = (\delta_{i_1,\sigma(1)}\delta_{i_2,\sigma(2)} - \delta_{i_1,\sigma(2)}\delta_{i_2,\sigma(1)}) \times \left[\prod_{k=1,2,3}(\exp[2\pi i N^{-1} m_{\sigma(k)} l_{\sigma(k)}] - \exp[2\pi i N^{-1}(N-1) l_{\sigma(k)}])\right] (\exp[-2\pi i N^{-1} l_{\sigma(3)}] - 1)^{-1} \tag{2.29}$$

The inner product (2.14) of the 2-cochains given by their Fourier transforms (2.20) and (2.27) is equal to $a_{\sigma(1)\sigma(2)}(m_{\sigma(3)})$. It is antisymmetric under a permutation of the indices $\sigma(1), \sigma(2)$. The function (2.27) has the same property. Hence the independent functions (2.27) are related to three permutations $\sigma \in S_3$ satisfying the condition $\sigma(1) < \sigma(2)$. The inner product (2.14) of the 2-cochains given by their Fourier transforms (2.20) and (2.28) is equal to $b_{\sigma(2)\sigma(3)}(\mathbf{m})$. Since it is antisymmetric under a permutation of the indices $\sigma(2), \sigma(3)$ the independent projections on the subspace $Z_2(\mathbf{T}^3_N, \mathbf{R})$ of the 2-cochains given by the Fourier transforms (2.28) correspond to three permutations $\sigma \in S_3$ satisfying the condition $\sigma(2) < \sigma(3)$. The inner product (2.14) of the 2-cochains given by their Fourier transforms (2.20) and (2.29) is equal to $c_{\sigma(1)\sigma(2)\sigma(3)}(\mathbf{m})$. Since it is antisymmetric under a permutation of all indices $\sigma(1), \sigma(2), \sigma(3)$ the only projection on the subspace $Z_2(\mathbf{T}^3_N, \mathbf{R})$ of the 2-cochain given by the Fourier transform (2.29) corresponding to the identity permutation $\sigma \in S_3$ is independent. Thus we have proved the following

Proposition 2.3. *Every element $\bar{z} \in \bar{Z}_2(\mathbf{T}^3_N, \mathbf{Z})$ has the following form*

$$\bar{z}_{i_1 i_2}(\mathbf{m}) = (P\hat{z})_{i_1 i_2}(\mathbf{m}), \tag{2.30}$$

$$\begin{aligned}\hat{z}_{i_1 i_2}(\mathbf{m}) = & \sum_{\sigma \in S_3; \sigma(1)<\sigma(2)} \sum_{k_{\sigma(3)}=0}^{N-1} \bar{a}_{\sigma(1)\sigma(2)}(k_{\sigma(3)})(\hat{a}[\sigma(1),\sigma(2);k_{\sigma(3)}])_{i_1 i_2}(\mathbf{m}) + \\ & \sum_{\sigma \in S_3; \sigma(2)<\sigma(3)} \sum_{k_{\sigma(2)},k_{\sigma(3)}=0}^{N-2} \bar{b}_{\sigma(2)\sigma(3)}(k_{\sigma(2)},k_{\sigma(3)})(\hat{b}[\sigma(2),\sigma(3);k_{\sigma(2)},k_{\sigma(3)}])_{i_1 i_2}(\mathbf{m}) \\ & + \sum_{k_1,k_2,k_3=0}^{N-2} \bar{c}_{123}(\mathbf{k})(\hat{c}[1,2,3;\mathbf{k}])_{i_1 i_2}(\mathbf{m})\end{aligned} \tag{2.31}$$

where P is the projector (2.19), the 2-cochains $(\hat{a}[\sigma(1),\ \sigma(2);\ k_{\sigma(3)}])_{i_1 i_2}$ $(\mathbf{m})$, $(\hat{b}[\sigma(2),\ \sigma(3);\ k_{\sigma(2)},\ k_{\sigma(3)}])_{i_1 i_2}$ $(\mathbf{m})$ and $(\hat{c}[1,2,3;\ \mathbf{k}])_{i_1 i_2}$ $(\mathbf{m})$ are defined by their Fourier transforms (2.27), (2.28) and (2.29). The integer valued functions $\bar{a}_{\sigma(1)\sigma(2)}$ $(k_{\sigma(3)})$, $\bar{b}_{\sigma(2)\sigma(3)}(k_{\sigma(2)}, k_{\sigma(3)})$ and $\bar{c}_{123}$ $(\mathbf{k})$ in the equalities (2.30), (2.31) are independent and they determine the element (2.30) uniquely.

As explained above for the continuum limit $\mathbf{T}^3_N \to \mathbf{T}^3$ of the correlation function (2.10) we need to choose the special sequence (1.11) of the 2-cochains $\phi_N \in C^2(\mathbf{T}^3_N, \mathbf{Z})$ and the inverse temperature $\beta = \beta_0 N^{3+2b}$, where $\beta_0^{-1} = g^2(2\pi R)^3 > 0$ and b is a strictly positive integer.

Proposition 2.4. *Let a θ-function $\Theta((\phi,\bar{\mathbf{z}})|2\pi i\beta\Omega^{-1})$ be given by the equality (2.11). Then for any sequence $\phi_N \in C^2(\mathbf{T}^3_N,\mathbf{Z})$ and for any numbers $\beta_0>0$, $\gamma>3$*

$$\lim_{N\to\infty}\Theta((\phi_N,\bar{\mathbf{z}})|2\pi i\beta_0 N^{\gamma}\Omega^{-1})=1 \tag{2.32}$$

Proof. It follows from the equalities (2.16), (2.27) - (2.29) and (2.31) that: for $0\le m_1,m_2,m_3\le N-2$

$$(\partial^*\hat{z})_{123}(\mathbf{m})=\bar{c}_{123}(\mathbf{m}), \tag{2.33}$$

for $0\le m_1,\widehat{m_j},m_3\le N-2$, $m_j=N-1$, $j=1,2,3$

$$(\partial^*\hat{z})_{123}(\mathbf{m})=(-1)^{j+1}\bar{b}_{1\hat{j}3}(m_1,\widehat{m_j},m_3)-\sum_{m'_j=0}^{N-2}\bar{c}_{123}(\mathbf{m})\,|_{m_j=m'_j} \tag{2.34}$$

for $0\le m_j\le N-2$, $m_1,\widehat{m_j},m_3=N-1$, $j=1,2,3$

$$\begin{aligned}(\partial^*\hat{z})_{123}(\mathbf{m})=&(-1)^{j+1}(\bar{a}_{1\hat{j}3}(m_j+1)-\bar{a}_{1\hat{j}3}(m_j))-\\ &\sum_{\sigma\in S_3;j=\sigma(2)<\sigma(3)}\mathrm{sgn}\sigma\sum_{m'_{\sigma(3)}=0}^{N-2}\bar{b}_{j\sigma(3)}(m_j,m'_{\sigma(3)})-\\ &\sum_{\sigma\in S_3;\sigma(2)<\sigma(3)=j}\mathrm{sgn}\sigma\sum_{m'_{\sigma(2)}=0}^{N-2}\bar{b}_{\sigma(2)j}(m'_{\sigma(2)},m_j)+\\ &\sum_{m'_k=0;k\neq j}^{N-2}\bar{c}_{123}(\mathbf{m}')\,|_{m'_j=m_j}\end{aligned} \tag{2.35}$$

where $\mathrm{sgn}\sigma$ is a parity of permutation.

Due to equalities (2.27) - (2.29) for $1\le i_1,i_2\le 3$ we get

$$(\hat{z})^{\sim}_{i_1i_2}(\mathbf{0})=\sum_{k=0}^{N-1}\bar{a}_{i_1i_2}(k) \tag{2.36}$$

By definition the terms in the right-hand side of the relation (2.19) are orthogonal to each other. Now the equalities (2.15), (2.16), (2.36) and the obvious estimation for any integers l_k, $k=1,2,3$,

$$\left(\sum_{k=1}^{3}|1-\exp[2\pi iN^{-1}l_k]|^2\right)^{-1}\ge 1/6 \tag{2.37}$$

imply the following estimation

$$|\Theta((\phi,\bar{\mathbf{z}})|2\pi i\beta\Omega^{-1})-1| \leq \sum_{P\hat{z}\in\bar{Z}_2(\mathbf{T}^3_N,\mathbf{Z})} \exp\Big[-2\pi^2\beta N^{-3}\cdot \tag{2.38}$$

$$\sum_{\sigma\in S_3;\sigma(1)<\sigma(2)} \Big(\sum_{m=0}^{N-1}\bar{a}_{\sigma(1)\sigma(2)}(m)\Big)^2 - \pi^2\beta/3 \sum_{m_1,m_2,m_3=0}^{N-1} ((\partial^*\hat{z})_{123}(\mathbf{m}))^2\Big]-1$$

Since

$$\bar{a}_{\sigma(1)\sigma(2)}(m) = \bar{a}_{\sigma(1)\sigma(2)}(0) + \sum_{k=0}^{m-1}(\bar{a}_{\sigma(1)\sigma(2)}(k+1) - \bar{a}_{\sigma(1)\sigma(2)}(k)) \tag{2.39}$$

it is possible to consider $\bar{a}_{\sigma(1),\sigma(2)}(0)$ and the right-hand sides of the equalities (2.33) - (2.35) as the summation variables in the sum (2.38). It follows from the equality (2.39) that

$$\sum_{m=0}^{N-1}\bar{a}_{\sigma(1)\sigma(2)}(m) = N\bar{a}_{\sigma(1)\sigma(2)}(0) + \sum_{m=0}^{N-2}(N-m-1)(\bar{a}_{\sigma(1)\sigma(2)}(m+1) - \bar{a}_{\sigma(1)\sigma(2)}(m)) \tag{2.40}$$

Extending the summation over the integer variables $\bar{a}_{\sigma(1)\sigma(2)}(0)$ in the sum (2.38) into the summation over $\bar{a}_{\sigma(1)\sigma(2)}(0) \in N^{-1}\mathbf{Z}$ in view of (2.40) we get the extended sum (2.38) where the independent summation variables are the right-hand sides of the equalities (2.33) - (2.36). Now if we leave in the second exponent (2.38) the components (2.33) - (2.35) only we obtain the obvious estimation for this extended sum and therefore for the left-hand side of the inequality (2.38)

$$|\Theta((\phi,\bar{\mathbf{z}})|2\pi i\beta\Omega^{-1})-1| \leq (\Theta(0|2\pi i\beta N^{-3}))^3(\Theta(0|i\pi\beta/3))^{N^3-1}-1 \tag{2.41}$$

where N^3-1 is the total number of the component (2.33) - (2.35), i.e. the total number of the generators of the group of 2-boundaries on $\mathbf{T}^3_N$.

Since for any strictly positive integer n the following estimation $n^2 > n$ holds, the definition (2.8) of the one dimensional θ-function implies that for any $t>0$

$$1 < \Theta(0|it) < 1 + 2(e^{\pi t}-1)^{-1} \tag{2.42}$$

By using this estimation we have

$$\begin{aligned} 0 &< (\Theta(0|i\pi\beta_0 N^{\gamma}/3))^{N^3-1} - 1 \\ &< \sum_{k=1}^{N^3-1}\frac{(N^3-1)\cdots(N^3-k)}{k!}2^k(\exp[\pi^2\beta_0 N^{\gamma}/3]-1)^{-k} \end{aligned} \tag{2.43}$$

The estimations (2.41) - (2.43) imply the relation (2.32).

In order to compute the continuum limit of the correlation function (2.10) it is necessary to calculate the continuum limit of the correlation function (2.7). Let us consider again the d-dimensional torus, $d > 2$. Due to [9, Sect. 22, Proposition 1] for any 2-cochain $\phi \in C^2(\mathbf{T}_{N^d}, \mathbf{Z})$

$$(\phi, \phi) - \sum_{i=1}^{g} (\phi, z_i)(\phi, \bar{z}_i) = (\phi, Q\phi) \tag{2.44}$$

where Q is the orthogonal projector on the subspace of the 2-coboundaries $B^2(\mathbf{T}_N^d, \mathbf{R})$. Lemma 2.1, the relations (2.14) - (2.17) and the relation $\exp[2\pi i N^{-1} (N-l)m] = \exp[-2\pi i N^{-1} lm]$ for any integers l, m imply that

$$(\phi, Q\phi) = \sum_{\mu=1}^{d} N^{-d} \sum_{l_1,\ldots,l_d=-(N-1)/2; l_1^2+\cdots+l_d^2\neq 0}^{(N-1)/2} \Big(\sum_{k=1}^{d} |\exp[2\pi i N^{-1} l_k] - 1|^2\Big)^{-1} \times$$
$$|\sum_{\lambda=1}^{d} (\exp[2\pi i N^{-1} l_\lambda] - 1)\phi^{\sim}_{\lambda\mu}(\mathbf{l})|^2 \tag{2.45}$$

Here we assume N to be odd. Let a 2-cochain $\phi_{N,b} \in C^2(\mathbf{T}_N^d, \mathbf{Z})$ be constructed from the coefficients $\phi_{i_1 i_2}(\theta)$ of a smooth differential 2-form on the torus $\mathbf{T}^d$ by means of the definition (1.11) for some strictly positive integer b. ¿From the definitions (1.11) and (2.12) we get

$$\lim_{N\to\infty} N^{-d-b} (\phi_{N,b})^{\sim}_{i_1 i_2}(\mathbf{l}) = \phi^{\sim}_{i_1 i_2}(\mathbf{l}) =$$
$$(2\pi R)^{-d} \int_0^{2\pi R} d\theta_1 \cdots \int_0^{2\pi R} d\theta_d \exp\big[iR^{-1} \sum_{k=1}^{d} l_k \theta_k\big] \phi_{i_1 i_2}(\theta) \tag{2.46}$$

Hence the limit (2.46) is a square summable function of the variable $\mathbf{l} \in \mathbf{Z}^d$. Now the relations (2.45) and (2.46) imply that

$$\lim_{N\to\infty} g^2 (2\pi R)^d N^{-d-2b} (\phi_{N,b}, Q\phi_{N,b}) =$$
$$g^2 (2\pi R)^{-d} \sum_{\mu=1}^{d} \sum_{l_1,\ldots,l_d=-\infty; l_1^2+\cdots+l_d^2\neq 0}^{\infty} R^2 (l_1^2 + \cdots + l_d^2)^{-1} |(d^*\phi)^{\sim}_{\mu}(\mathbf{l})|^2 \tag{2.47}$$

where d^* is the adjoint operator of the differential operator d

$$(d^*\phi)_\mu(\theta) = -\sum_{\lambda=1}^{d} \frac{\partial}{\partial\theta_\lambda} \phi_{\lambda\mu}(\theta) \tag{2.48}$$

Now it follows from the relations (2.7), (2.44) and (2.47) that

$$\lim_{N\to\infty} W_{\mathbf{T}^d_N, g^{-2}(2\pi R)^{-d}N^{d+2b}}(\partial\phi_{N,b}) = \tag{2.49}$$

$$\exp\left[-(g^2/2)(2\pi R)^{-d}\sum_{\mu=1}^{d}\sum_{l_1,\ldots,l_d=-\infty;l_1^2+\cdots+l_d^2\neq 0}^{\infty} R^2(l_1^2+\cdots+l_d^2)^{-1}|(d^*\phi)^{\sim}_{\mu}(1)|^2\right]$$

It is interesting to note that the right-hand side of this relation is a correlation function of the **R**-gauge electrodynamics on a torus [6]. The equalities (2.10), (2.32) and (2.49) imply the equality (1.12) for $d = 3$.

If the differential 2-form ϕ has a compact support independent of the radius R of a torus then by using the equalities (2.46) and (2.49) it is easy to prove that

$$\lim_{R\to\infty}\lim_{N\to\infty} W_{\mathbf{T}^d_N, g^{-2}(2\pi R)^{-d}N^{d+2b}}(\partial\phi_{N,b}) =$$

$$\exp\left[-(g^2/2)(2\pi)^{-d}\int_{\mathbf{R}^d} d^dp(p_1^2+\cdots+p_d^2)^{-1}|(d^*\phi)^{\sim}_{\mu}(\mathbf{p})|^2\right] \tag{2.50}$$

where the operator d^* is defined by the same equality (2.48) and a function $f^{\sim}_{\mu}(\mathbf{p})$ is an usual Fourier transform of a function $f_{\mu}(\mathbf{x})$ on the Euclidean space $\mathbf{R}^d$. The right-hand side of the equality (2.50) is a correlation function of the **R**-gauge Euclidean electrodynamics [6].

For $d = 2$ the relation (1.12) is also fulfilled but now the right-hand side of (1.12) has the continuum form of (2.7) for $g = 1$. When the torus radius tends to infinity this correlation function of type (2.7) converges to the trivial correlation function of the **R**-gauge Euclidean two dimensional electrodynamics [6].

References

1. Wilson, K.G. : Confinement of quarks. Phys. Rev. D**10**, 2445-2459 (1974)
2. Villain, J. : Theory of one- and two-dimensional magnets with an easy magnetization plane. 2. The planar, classical, two-dimensional magnet. J. Phys. **36**, 581-590 (1975)
3. Gross, L. : Convergence of $U(1)_3$ lattice gauge theory to its continuum limit. Commun. Math. Phys. **92**, 137-162 (1983)
4. Driver, B.K. : Convergence of $U(1)_4$ lattice gauge theory to its continuum limit. Commun. Math. Phys. **110**, 479-501 (1987)
5. Zinoviev, Yu.M. : Duality in the Abelian gauge lattice theories. Theor. Math. Phys. **43**, 481-490 (1980)
6. Zinoviev, Yu.M. : **R**-gauge theories. Theor. Math. Phys. **50**, 135-143 (1982)
7. Zinoviev, Yu.M. : $U(1)$ Gauge Theory on a Torus. Commun. Math. Phys. (to appear)
8. Zinoviev, Yu.M. : Lattice **R**-gauge theories. Theor. Math. Phys. **49**, No.2 (1981)
9. de Rham, G. : Variétés différentiables. Paris : Hermann 1955

A Low Temperature Expansion and "Spin Wave Picture" for Classical N-Vector Models

Tadeusz Balaban

Rutgers University
Mathematics Department
New Brunswick, N.J. 08903

Abstract
We obtain a convergent multi-scale expansion for a class of low temperature classical vector spin models, of the type of lattice "$\lambda|\phi|^4$" field theory, in dimensions $d \geq 3$. With the help of this expansion we prove main statements of the so called "spin wave picture", like the existence of a continuum of phases parametrized by vectors of the unit sphere in the space of spins ϕ, and the existence of Goldstone bosons, i.e. free massless decay of truncated transversal two-point correlation functions.

We consider a class of low temperature lattice spin models for spin variables $\phi(x) \in \mathbb{R}^N, x \in \mathbb{Z}^d, N \geq 2, d \geq 3$. They are defined by the actions

$$A(\phi) = \frac{1}{2} \sum_{<x,x'>} |\phi(x') - \phi(x)|^2 + \sum_x \left(\frac{\lambda}{8} |\phi(x)|^4 - \frac{\mu}{2} |\phi(x)|^2 - h \cdot \phi(x) \right), \quad (1)$$

where $\lambda > 0, \mu > 0, h \in \mathbb{R}^N$. The effective potential for the above action is given by

$$V(\phi) = \frac{\lambda}{8} |\phi|^4 - \frac{\mu}{2} |\phi|^2 - h \cdot \phi = \frac{\lambda}{8} \left(|\phi|^2 - \frac{2\mu}{\lambda} \right)^2 - \nu h_o \cdot \phi - \frac{\mu^2}{2\lambda}, \quad (2)$$

where $h = \nu h_o, \nu \geq 0, |h_o| = 1$. We are interested particularly in the case $\nu = 0$, when the action is invariant with respect to the full orthogonal group $O(N)$ acting on the spin configurations, and the effective potential has minima at all points of the sphere $|\phi|^2 = \frac{2\mu}{\lambda}$. According to the standard picture at low enough temperature to each minimum there corresponds a pure phase with one-point correlation function proportional to the minimum. This pure phase can be constructed either by choosing proper boundary conditions for finite volume models, and then taking a thermodynamic limit, or by choosing a non-zero external field h in the direction of the minimum, taking the thermodynamic limit, and then the limit $\nu = |h| \to 0$. We follow here the second way, so we have to consider the actions (1) with $h \neq 0$, or $\nu > 0$. Then the effective potential (2) has exactly one minimum at a vector proportional to h_o. It is convenient to normalize it in such a way that this minimum is exactly equal to h_o. This can be achieved simply by scaling the variables ϕ. Denoting the rescaled coefficients in the same way we obtain the effective potential of the form

$$V(\phi) = \frac{\lambda}{8}\left(|\phi|^2 - 1 + \frac{2\nu}{\lambda}\right)^2 - \nu h_o \cdot \phi - \frac{\lambda}{8}\left(1 - \frac{2\nu}{\lambda}\right)^2$$

$$= \frac{\lambda}{8}\left(|\phi|^2 - 1\right)^2 + \frac{\nu}{2}|\phi - h_o|^2 - \frac{\lambda}{8}\left(1 + \frac{4\nu}{\lambda}\right) . \qquad (3)$$

We take finite volume models defined on tori $T = \{x \in \mathbb{Z}^d : -L_\mu \leq x_\mu < L_\mu, \mu = 1, ..., d\}$ with periodic boundary conditions. The corresponding probability measure is defined by

$$d\mu(\phi) = \rho(\phi)d\phi, \rho(\phi) = \exp[-\beta A(\phi) - E], \qquad (4)$$

where $d\phi$ is the Lebesgue measure on the space of all configurations ϕ defined on the torus $T, \beta > 0$ is a parameter proportional to the inverse temperature $\beta = \frac{1}{kT}, E$ is a normulization constant, $E = \log Z, Z = \int d\phi \exp[-\beta A(\phi)]$ is the partition function. The action $A(\phi)$ can be written in the following form

$$A(\phi) = \frac{1}{2}\|\partial\phi\|^2 + \frac{\lambda}{8}\||\phi|^2 - 1\|^2 + \frac{\nu}{2}\|\phi - h\|^2, \qquad (5)$$

where the norms are L^2–norms on the unit lattice tori T. We consider this class of models for β sufficiently large, λ not too small. For simplicity we assume that $\lambda \geq 1, \nu > 0$ and h in the unit sphere of $\mathbb{R}^N, h \in S^{N-1}$. The method we describe here is uniform in λ for $\lambda \in [1, +\infty]$, and the case $\lambda = +\infty$ is understood as the limit

$$\lim_{\lambda \to +\infty} \exp[-\beta A(\phi) - E_\lambda]$$

$$= \exp[-\beta\left(\frac{1}{2}\|\partial\phi\|^2 + \frac{\nu}{2}\|\phi - h\|^2\right) - E_\infty] \prod_{x \in T} \delta\left(|\phi(x)|^2 - 1\right),$$

which is the classical Heisenberg model, or its N-component generalization. Finally, to construct and discuss also correlation functions we extend the density $\rho(\phi)$ in (4) to a density of a generating functional for these functions, which simply means that to the exponent in (4) we add the linear function $\langle g, \phi \rangle$, where g is defined on T and has values in $\mathbb{C}^N$.

The action (5) has some important symmetries. The first two terms are invariant with respect to transformations of the orthogonal group $O(N)$. The third term breaks this symmetry, it is invariant with respect to the subgroups of $O(N)$ leaving the vector h invariant. This can be formulated also as an invariance with respect to transformations of $O(N)$ acting simultaneously on both variables ϕ, h. The action is also invariant with respect to Euclidean transformations of the lattice T. Let us write the action $A(\phi)$ more explicitly as $A(\phi; h, \lambda, \nu)$. The symmetries can be found as the equalities

$$A(R\phi; Rh, \lambda, \nu) = A(\phi; h, \lambda, \nu) \text{ for all } R \in O(N)$$

$$A(r\phi; h, \lambda, \nu) = A(\phi; h, \lambda, \nu) , \tag{6}$$

for all Euclidean transformations r of the lattice T, where $(r\phi)(x) = \phi(rx)$, $x \in T$. In particular the Euclidean symmetric includes all translations of the torus T.

There are quite a number of rigorous mathematical results concerning the above and related models. We do not intend to make a survey of these results, but let us mention the papers [5,6,7,11,12,17] in which some fundamental aspects of the above picture have been proved, like the existence of non-zero magnetization, the domination of the two-point correlation function by the inverse of the Laplace operator, etc. These results have been obtained by elegant and simple methods, like infrared bounds, chessboard bounds, correlation inequalities, etc. We develop here a different approach based on expansion methods, which yield more precise and complete results. Actually our fundamental goal is not to get these more precise results, but to construct a convergent low temperature expansion for this class of models, an analog of such expansions for the Ising model, or for the considered ones with $N = 1$ constructed in [15]. A reason for this is that a sufficiently flexible and robust expansion could be applied to more complicated models, like models with disorders, stochastic or quantum models, in the way the low temperature Peierls expansion for the Ising model has been applied to Euclidean quantum field theory with "$\lambda\phi^4 - \frac{\mu}{2}\phi^2$" interaction in [15]. It should be clear also that in our case we have to construct a multi-scale expansion, because of massless modes, or slow decay properties of correlation functions. We do it using renormalization group ideas and techniques, in particular the ideas and techniques developed for lattice scalar and gauge field models in [3,4,13]. Nevertheless, before going into a description of these methods, let us formulate a theorem which summarizes the most important features of the above mentioned heuristic picture, the so called "spin wave picture".

Theorem 1 *The model defined by (4), (5) has exactly one thermodynamic limit if β is large enough and $\nu > 0$. Denote the corresponding expectation value by $\langle\cdot\rangle_\nu$, although it depends obviously on the parameters β, h, λ also. The one-point correlation function satisfies*

$$\langle\phi(x)\rangle_\nu = m(\beta,\nu)h, |m(\beta,\nu) - 1| < C\beta^{-1+\rho_1}, \tag{7}$$

where $\rho_1 > 0$ may be chosen to be arbitrarily small, and $C > 0$ is a constant independent of β, ν. The many-point truncated correlation functions have an exponential decay with a decay rate proportional to $\sqrt{\nu}$, for example the two-point truncated correlation function $\langle\phi(x)\phi(x')\rangle_\nu^T = \langle\phi(x)\phi(x')\rangle_\nu - \langle\phi(x)\rangle_\nu\langle\phi(x')\rangle_\nu$ satisfies

$$\langle\phi(x)\phi(x')\rangle_\nu^T = G_{tr}(x-x';\beta,\nu)(I - h\otimes h) + G_{\text{long}}(x-x';\beta,\nu)h\otimes h,$$

$$|G_{tr}(x-x';\beta,\nu)|, |G_{\text{long}}(x-x';\beta,\nu)| C\frac{1}{|x-x'|^{d-2}}\exp(-c\sqrt{\nu}|x-x'|) \tag{8}$$

with an absolute positive constant c. The states $\langle\cdot\rangle_\nu$ have a limit as $\nu \to 0+$, $\langle\cdot\rangle = \lim_{\nu\to 0+}\langle\cdot\rangle_\nu$, and the limit is a pure state with many-point truncated correlation *functions having power law decays. More precisely, the one-point function $\langle\phi(x)\rangle$ satisfies (7) with $\nu = 0$, and the truncated two-point function $\langle\phi(x)\phi(x')\rangle^T$ satisfies the equality in (8) with $\nu = 0$, and the functions G_{tr}, G_{long} satisfying*

$$G_{tr}(x - x'; \beta, 0) = \frac{1}{\beta\gamma_\infty(\beta)} m^2(\beta, 0)\Delta^{-1}(x - x') + R(x - x'; \beta)$$

where $|\gamma_\infty(\beta) - 1| < C\beta^{-1}$;

$$|R(x - x'; \beta)|, |G_{\text{long}}(x - x'; \beta, 0)| < C\frac{1}{|x - x'|^{d-2+\rho}}. \tag{9}$$

This theorem is a consequence of the multi-scale low temperature expansion, and is proved in a separate paper [2] written together with Michael O'Carrol. It gives only an example of results which can be proved with the help of the expansion; in fact much more detailed properties of the correlation functions can be obtained. In particular the property (9) can be generalized to expansions with terms ordered by increasing rates of their power law decays, and remainders with arbitrarily large rates. Unfortunately such expansions can be constructed only by a lengthy and cumbersome, although quite straightforward, process and they do not seem to have interesting applications, so we do not discuss them here, or in other papers.

Let us describe now some basic features of the "block-spin" renormalization group approach we apply to the above models. We will be very brief because it has been described already several times, so we refer the reader to the previous papers for more details, in particular to [3,4]. We apply a sequence of renormalization transformations with linear averaging operations and Gaussian densities. Let us recall the most important definitions. A renormalization transformation applied to a density $\rho(\phi; \beta)$ depending on the parameter β is defined by the formulas

$$(T\rho)(\psi) = \int d\phi t(\psi, \phi; \beta, a)\rho(\phi; \beta),$$

$$t(\psi, \phi; \beta, a) = \prod_{y \in T_L^{(1)}} t(y; \psi, \phi; \beta, a),$$

$$t(y; \psi, \phi; \beta, a) = \exp\left[-\frac{1}{2}\beta a L^{d-2}|\psi(y) - (Q\phi)(y)|^2 + \frac{N}{2}\log\frac{\beta a L^{d-2}}{2\pi}\right],$$

$$(Q\phi)(y) = \sum_{x \in B(y)} L^{-d}\phi(x), \tag{10}$$

where L is a positive integer, $B(y)$ is the L-block with center at $y \in T_L^{(1)}, T_L^{(1)}$ denotes the new L-lattice of centers of blocks, and ψ is a new spin configuration defined on the new lattice. Let us make a few comments on this definition. It has the usual normalization property

$$\int d\psi (T\rho)(\psi) = \int d\phi \rho(\phi), \tag{11}$$

which is the basic property underlying the whole renormalization group ideology. There is a new parameter a introduced in (10), and we take $a \approx 1$, e.g. $\frac{1}{2} < a < \frac{3}{2}$. The definition also depends explicitly on β. The transformation T determines one step in the renormalization group procedure. We apply it to the density $\rho(\phi)$ in (4) and we obtain a new density $\rho_1(\psi)$, which is a function of new spin configurations ψ on the L-lattice $T_L^{(1)}$. Next we rescale the lattice $T_L^{(1)}$ to the unit lattice $T_1^{(1)}$, rescaling properly the configurations ψ also. Then we iterate this step many times. How many times we have to iterate depends crucially on the scaling transformations, so we discuss them now in detail.

Let us begin with the remark that the scaling transformations are not uniquely defined by a given model, they can be chosen in many different ways. Usually the choice depends on a parameterization of the model, and on properties we want to display. In our case important properties are determined by the effective potential $V(\phi)$ given by (3). For $\nu > 0$ it has a minimum at $\phi = h$, and for $\nu = 0$ a set of minima is the unit sphere $|\phi| = 1$. We want to preserve this property by the scaling transformation, so the variable ϕ does not scale, or it scales with the trivial factor 1. Consider a scaling of the unit lattice T by a positive number ϵ, so we obtain the ϵ-lattice $T_\epsilon = \{\epsilon x : x \in T\}$. Thus we take a rescaled configuration $\phi(\epsilon x) = \phi(x)$. Actually we should distinguish between the two configurations defined on the lattice with different scales, but because of this scaling property we use the same notation for configurations defined on rescaled lattices. Using this property we have

$$\beta A(\phi; h, \lambda, \nu) = \beta\epsilon^{2-d}\left[\frac{1}{2}\sum_{\langle x,x'\rangle \subset T_\epsilon} \epsilon^d |(\partial^\epsilon \phi)(\langle x,x'\rangle)|^2 + \right.$$

$$\left. + \sum_{x\in T_\epsilon}\left(\frac{\lambda\epsilon^{-2}}{8}(|\phi(x)|^2 - 1)^2 + \frac{\nu\epsilon^{-2}}{2}|\phi(x) - h|^2\right)\right] =$$

$$= \beta\epsilon^{2-d} A^\epsilon(\phi; h, \lambda\epsilon^{-2}, \nu\epsilon^{-2}) = \beta^\epsilon A^\epsilon(\phi; h, \lambda^\epsilon, \nu^\epsilon), \tag{12}$$

where A^ϵ denotes the action with all the expressions, like derivatives, norms, etc., defined on the ϵ-lattice, and $\beta^\epsilon, \lambda^\epsilon, \nu^\epsilon$ are the corresponding coefficients calculated for this scale. From (12) we obtain the scaling laws for the coefficients

$$\beta^\epsilon = \beta\epsilon^{2-d}, \lambda^\epsilon = \lambda\epsilon^{-2}, \nu^\epsilon = \nu\epsilon^{-2} \tag{13}$$

These we call the canonical scalings. Renormalizations connected with the renormalization group approach introduce some additional rescalings of spin configurations and the coefficients, but for $d > 2$ the canonical scalings dominate an asymptotic behavior as $\epsilon \to 0$. They decrease the temperature, strengthen an effect of a barrier well around the sphere $|\phi| = 1$ by increasing λ, and they increase the external field ν. Our approach to the low temperature problem for this model is based on these scaling properties. In particular they determine the number of renormalization steps we have to perform. This is based on the observation that if $\nu \approx 1$ in the model (4), (5), then we can prove easily the existence of a unique thermodynamic phase with exponential decay properties by a fairly straightforward cluster expansion, uniformly in β, λ in the considered region. In each renormalization step we scale by the factor L^{-1}, so after n steps we scale by the factor L^{-2n}, and for an effective parameter ν_n we have $\nu_n \approx \nu L^{2n}$. Thus we can end the procedure if $\nu L^{2n} \approx 1$. For a positive ν this determines the number if steps n almost uniquely. Obviously $n \to \infty$ if $\nu \to 0+$.

Let us denote the one step scaling transformation, i.e. the scaling corresponding to the factor L^{-1}, by S. The one step renormalization transformation is the composition ST. Actually we write the transformations with superscripts indicating to what densities they are applied, because T depends on the effective parameter β, which changes with each step, so successive densities depend on different parameters β. Let us denote the density in (4), (5) by ρ_o, with the parameters in it having the subscript "0"also. The first transformations S, T applied to it have the superscript "(0)". They yield the density ρ_1, which can be expressed in terms of some new parameters $\beta_1, \lambda_1, \nu_1$ in a way which will be described later. Then we apply the next transformations S, T with the superscript "(1)", and so on. Applying the renormalization and scaling transformations k times we obtain a density ρ_k for which we have

$$\rho_k = S^{(k-1)}T^{(k-1)}\rho_{k-1} = \prod_{j=1}^{k} S^{(k-j)}T^{(k-j)}\rho_o = (ST)_k\rho_o. \tag{14}$$

This density depends on new spin configurations defined on the unit lattice $T_1^{(k)}$, which is obtained by replacing k-blocks of the lattice $T_\eta, \eta = L^{-k}$, i.e blocks containing L^{kd} points of this lattice, by their centers. It depends also on new, renormalized parameters $\beta_k, \lambda_k, \nu_k, a_k$ where a_k is determined by the parameter a in the definition (10). Our basic goal is to give a sufficiently detailed and precise description of the densities ρ_k, and in particular of the new parameters. It will be an inductive description, as usual in renormalization group methods, see [3,8,13].

To understand various issues connected with such a description it is best to discuss briefly the first renormalization transformation $T^{(0)}$ applied to ρ_o, in which they appear in their clearest form. It is given by the integral

$$(T^{(0)}\rho_o)(\psi) = \int d\phi \exp[-\beta_o\{\frac{1}{2}aL^{-2}\|\psi - Q\phi\|_L^2 + \frac{1}{2}\|\partial\phi\|^2 +$$

$$+\frac{\lambda_o}{8}\||\phi|^2 - 1\|^2 + \frac{\nu_o}{2}\|\phi - h\|^2\} + \langle g, \phi\rangle - E_o'], \tag{15}$$

where the normalization constant of $T^{(0)}$ is included into the constant E_o'. This integral is called a fluctuation integral, and the expression in the curly brackets is its effective action. We denote it by $A_1(\psi, \phi; h, aL^{-2}, \lambda_o, \nu_o)$. We are interested in low temperature properties of the model, so β_o may be arbitrarily large and it is natural to apply a saddle point method to calculate the integral. This method requires finding all critical points of the function $A_1(\psi, \phi)$, which is multiplied by β_o in (15). The problem of finding the critical points on the whole space of configurations ϕ is a highly singular one, it is easy to construct examples of configurations ψ with many critical points, in fact with quite singular sets of critical points. Also, expanding the function $A_1(\psi, \phi)$ around some of the critical points we obtain functions with no good positivity properties, so the resulting integrals are also quite singular and impossible to treat by known methods. This is a manifestation of a very well know problem in renormalization group methods, the so called "large field problem". Every renormalization group approach deals with this problem in its own specific way, for example, see [3,4,8,9,10,13]. For "block-spin" approaches it is typical to introduce restrictions on basic variables, in our case on the spin configurations ψ, ϕ. The restrictions are naturally connected with positive terms in the effective action $A_1(\psi, \phi)$, for example they may be defined by

$$|\psi(y) - (Q\phi)(y)| < \beta_o^{-\frac{1}{2}} p(\beta_o), |(\partial\phi)(b)| < \beta_o^{-\frac{1}{2}} p(\beta_o),$$

$$\big||\phi(x)|^2 - 1 < \lambda_o^{-\frac{1}{2}} \beta_o^{-\frac{1}{2}} p(\beta_o), |\phi(x) - h| < \nu_o^{-\frac{1}{2}} \beta_o^{-\frac{1}{2}} p(\beta_o),$$

$$p(\beta_o) = (A_o + \log\beta_o)^{p_o} \ , \tag{16}$$

where A_o is a large positive constant and p_o is an even integer.

These restrictions are used in the following way. Denote by P subsets of the set of blocks of the lattice T, the set of bonds of the lattice T, and the lattice T, and denote by $\mathcal{X}_P$ the characteristic function of the domain of configurations ψ, ϕ satisfying the inequalities (16) on blocks, bonds and points of the subset P. We substitute the decomposition $1 = \sum_P \mathcal{X}_P \mathcal{X}^c_{P^c}$ into the integral (15), where the meaning of the symbol $\mathcal{X}^c_{P^c}$ should be obvious. A set P^c determines a subset X of the lattice T, which is a union of blocks, bonds and points belonging to P^c. The integral (15) is written as a sum over P of integrals restricted by the characteristic functions $\mathcal{X}_P \mathcal{X}^c_{P^c}$. Each term determines its set X, which is roughly speaking a "large field domain". The part $\exp[-\beta_o A_1]$ of the density ρ_o restricted to this domain can be bounded by $\exp[-\frac{1}{8}p^2(\beta_o)|X|]$, and the integral of this density restricted to X can be estimated also by such a factor with the constant $\frac{1}{8}$ replaced by another absolute positive constant. This factor is sufficiently small to control the combinatorics of the decomposition and the other bounds, at least for this and the next few renormalized steps. This solves the easiest part of "the large field problem". A difficult part appears if we perform enough steps without creating new large field domains sufficiently close to X. We will discuss very briefly the "large field problems" at the end of this paper. Now consider the action and the integral restricted to a complement of the set X. It is a "small field domain", and the restricted integral can be calculated by the saddle point method mentioned above. The action A_1 restricted to this domain has exactly one critical point, which is its minimum, and the expansion around the minimum has all the necessary properties, so that the integral can be analyzed by a simplified version of a cluster expansion. This is an easier part of "the small field problem", although it becomes technically quite involved in subsequent steps. A more difficult part appears again when we perform many steps and we have to obtain uniform bounds for the resulting effective actions. We have to renormalize these actions, and to solve renormalization group equations for the "running" parameters. We discuss now these "small field problems" in some detail, because they form the most important part of the whole problem, and their solution decides the properties described in Theorem 1. It is sometimes said that the small field picture is a perturbation picture, which is true, but only on some general, ideological level. On a methodological level our approach does not make any use of perturbation series, in fact it is difficult even to make contact with a standard "spin wave" type series for most of the renormalization steps. This contact becomes clearer only after the last step.

Let us discuss the small field contributions to the densities (14). For simplicity and clarity we consider the case when a whole lattice at each step is the small field domain. After this it will be quite simple to make the necessary modifications in general cases. We will discuss it in the last part of this paper. Thus instead of the densities given by the formula (14) we consider here the small field densities denoted also by ρ_k and given by the formula

$$\rho_k = \mathcal{X}_k S^{(k-1)} T^{(k-1)} \mathcal{X}^{(k-1)} \cdot \ldots \cdot S^{(0)} T^{(0)} \mathcal{X}^{(0)} \rho_o, \tag{17}$$

where the characteristic functions $\mathcal{X}_k, \mathcal{X}^{(j)}, j = k-1, ..., 0$, yield restrictions to proper small field spaces. The restrictions are basically of the form (16), only for technical reasons are they modified and formulated in a more elaborate way. We refer the interested reader to the paper [1] for a detailed discussion of these restrictions. Now we will give a relatively precise description of the densities ρ_k, and of their properties and bounds. A fundamental role in this description is played by critical points of effective actions, which are simple generalizations of the action $A_1(\psi, \phi; h, aL^{-2}, \lambda_o, \nu_o)$ in (15). These critical points turn out to be minima of the corresponding actions, so we consider appropriate variational problems. We rescale the lattice T to the scale $\eta = L^{-k}$, so we get the lattice T_η. Spin configurations ϕ are defined on T_η, and new spin configurations ψ are defined on the lattice $T_1^{(k)}$, which is a unit lattice of centers of k-blocks, i.e. blocks containing L^{kd} sites of T_η. We define

$$A_k(\psi, \phi; h, a, \lambda, \nu) = \frac{a}{2}\|\psi - Q_k\phi\|_1^2 + \frac{1}{2}\|\partial^\eta \phi\|^2 + \frac{\lambda}{8}\||\phi|^2 - 1\|^2 + \\ + \frac{\nu}{2}\|\phi - h\|^2, \tag{18}$$

where the norms are defined on the η-lattice, except the first one which is defined on the unit lattice. The coefficients are as before, i.e. $\frac{1}{2} < a < \frac{3}{2}, \lambda \geq 1, 0 < \nu \leq 1$. We are interested in the variational problem

$$\inf_\phi A_k(\psi, \phi; h, a, \lambda, \nu). \tag{19}$$

As was explained before, we have to restrict the function A_k to a proper small field domain. Such domains are defined by conditions which are simple generalizations of (16). We define

$$\Phi_k(\epsilon; \lambda, \nu) = \{(\psi, \phi, h) : \psi \text{ is defined on } T_1^{(k)} \text{ and has values in } \mathbb{R}^N,$$

$$\phi \text{ is defined on } T_\eta \text{ and has values in } \mathbb{R}^N, h \in S^{N-1},$$

$$|\psi - Q\phi| < \epsilon, |\partial^\eta \phi| < \epsilon, |\Delta^\eta \phi| < \epsilon, |\frac{\lambda}{2}(|\phi|^2 - 1)| < \epsilon, |\phi - h| < \frac{\epsilon}{\sqrt{\nu}}\}. \qquad (20)$$

This is a real space, and we extend it to a complex space $\Phi_k^c(\epsilon;\lambda,\nu)$ of configurations $(\psi+\psi', \phi+\phi', h+h')$ requiring that $(\psi,\phi,h) \in \Phi_k(\epsilon;\lambda,\nu), \psi', \phi', h'$ have values in $\mathbb{C}^N$ and the sums $\psi+\psi', \phi+\phi', h+h'$ satisfy the inequalities in (20). The complex spaces are important to formulate analyticity properties of various functions, but these are connected with more technical aspects of the procedure, which we want to avoid, so we make use mainly of the real spaces (20). The conditions in the definition (20) imply some conditions on configurations ψ, which can be described in the following definition

$$\Psi_k(\delta;\nu) = \{(\psi,h) : \psi \text{ is defined on } T_1^{(k)} \text{ and has values in } \mathbb{R}^N,$$

$$h \in S^{N-1}, |\partial^1 \psi| < \delta, \big||\psi|^2 - 1\big| < \delta, \nu(1 - \psi_o \cdot h) < \delta^2 \text{ where} \psi = \frac{\psi}{|\psi|}\}. \qquad (21)$$

It is easy to see that if $(\psi,\phi,h) \in \Phi_k(\epsilon;\lambda,\nu)$, then $(\psi,h) \in \Psi_k(3\epsilon;\nu)$. With the help of the above spaces we can formulate one of the maim technical results in our approach, a result concerning the variational problem (19).

Theorem 2 *There exist positive constants $c_o, c_1, K_1, K_1 c_1 \leq c_o$, such that if $\epsilon \leq c_1$ and $(\psi,h) \in \Psi_k(\epsilon;\nu)$, then the variational problem (19) has a solution belonging to the space $\Phi_k(K_1\epsilon;\lambda,\nu)$, and this solution is unique in the space $\Phi_k(c_o;\lambda,\nu)$. It is also the only critical point of the function (18) in the last space.*

This theorem covers only a small part of theory of the problem (19). In this theory other properties of solutions are proved. The most important is the existence of localization expansions which play a crucial technical role, and which will be briefly discussed in a moment. Let us denote the solution by $\phi_k(\psi;h,a,\lambda,\nu)$ or simpler by $\phi_k(\psi)$, or ϕ_k. It is a function defined on the space $\Psi_k(c_1;\nu)$, and has an analytic extension onto a complexification of this space defined in a similar way as was the complexification of the space (20). The solutions for different k's are connected by a transformation of corresponding spin variables. More precisely, for a pair of natural numbers $j,k,j \leq k$, there exists a function $\psi_k^{(j)}(\psi)$ defined on spin configurations on the lattice $T^{(k)}$, whose values are spin configurations on the lattice $T^{(j)}$, such that

$$\phi_j(\psi_k^{(j)}(\psi)) = \phi_k(\psi), \qquad (22)$$

where the coefficients in the variational problems (19) determining the functions ϕ_j, ϕ_k are connected by the scaling laws (13). The functions $\psi_k^{(j)}$ can be defined also as solutions of variational problems similar to the problem (19), see [1]. They play an equally important role in our renormalization groups approach as the solutions ϕ_k do, but they can be easily expressed in terms of ϕ_k, so there is no need for a separate theory. Now we are ready to formulate an inductive description of the densities ρ_k given by (17). It has the form of seven inductive assumptions, which we write now together with some comments. The first inductive assumption is about a general form of the k-th density. We assume that there exist constants $\beta_k, a_k, \lambda_k, \nu_k, E_k$, and functions $\mathcal{E}_k(\psi, h), \mathcal{F}_k(\psi, h, g)$ such that

$$\rho_k(\psi) = \mathcal{X}_k \exp\left[\mathcal{A}_k + \mathcal{F}_k(\psi, h, g)\right],$$
$$\mathcal{A}_k = -\beta_k A_k(\psi, \phi_k; h, a_k, \lambda_k, \nu_k) + \mathcal{E}_k(\psi, h) - E_k, \tag{H.1}$$

where ψ is a new spin variable on the unit lattice $T_1^{(k)}$, and the minimal configuration ϕ_k is determined by ψ, h and the coefficients a_k, λ_k, ν_k. The constants $\beta_k, a_k, \lambda_k, \nu_k$ are determined by a sequence of inductive renormalization equations, which will be discussed later, now let us mention only that their leading asymptotic behavior is determined by the scaling laws (13), i.e. we have approximately

$$\beta_k \approx \beta L^{k(d-2)}, a_k \approx a\frac{1-L^{-2}}{1-L^{-2k}}, \lambda_k \approx \lambda L^{2k}, \nu_k \approx \nu L^{2k}. \tag{23}$$

The function $\mathcal{X}_k$ is a characteristic function of a domain in a space of spin configurations ψ, defined as the set of all ψ such that $(\psi, \phi_k, h) \in \Phi_k(\beta_k^{-\frac{1}{2}} p(\beta_k); \lambda_k, \nu_k)$. This domain is approximately the domain (21) with $\delta = \beta_k^{-\frac{1}{2}} p(\beta_k), \nu = \nu_k$, in fact it is contained between two such domains defined with some additional constants multiplying this δ. Notice then that the expressions in ψ occurring in (21) can be bounded by const. $\beta^{-\frac{1}{2}+\alpha}\eta^{\frac{d-2}{2}-\alpha}$, where $\eta = L^{-k}$ and α can be arbitrarily small.

The second assumption describes a structure of the functions $\mathcal{E}_k(\psi, h)$. There exist functions $\mathcal{E}^{(j)}(y; \psi_j, h), y \in T^{(j)}$ and ψ_j is a spin variable on the lattice $T^{(j)}$, such that

$$\mathcal{E}_k(\psi, h) = \sum_{j=1}^{k} \mathcal{E}^{(j)}\left(\psi_k^{(j)}(\psi, h), h\right), \text{ where}$$
$$\mathcal{E}^{(j)}(\psi_j, h) = \sum_{y \in T^{(j)}} \mathcal{E}^{(j)}(y; \psi_j, h). \tag{H.2}$$

The functions $\mathcal{E}^{(j)}(y; \psi_j, h)$ have analytic extensions onto a complex domain which is defined as the set of all complex ψ, h such that $(\psi_j, \phi_j(\psi_j, h), h) \in \Phi_j^c(\epsilon_j; \lambda_j, \nu_j)$, where $\epsilon_j = \epsilon_o L^{-j(\frac{d-2}{2}-\gamma)}, \epsilon_o$ is a sufficiently small positive constant, and γ is a positive constant, $\gamma < \gamma_o = \min\{\frac{d-2}{2}, 1\}$, which otherwise can be chosen arbitrarily and fixed for the whole procedure. The transformation $\psi_k^{(j)}$ maps the complex domain defined for k into the complex domain for j, so the function $\mathcal{E}_k(\psi, h)$ has an analytic extension onto the domain for k.

Let us make a remark about the above domains. The domain of the characteristic function $\mathcal{X}_k$ is in some sense an optimal domain for bounds and uniformity of the whole procedure, and it is characterized by the constant $\beta^{-\frac{1}{2}+\alpha}\eta^{\frac{d-2}{2}-\alpha}, \alpha$ arbitrarily small, mentioned above. The domain of the function $\mathcal{E}_k$ is characterized by the constant $\epsilon_o\eta^{\frac{d-2}{2}-\gamma}$, whre γ can be a fixed positive number, e.g. for

$d = 3$ we can choose it arbitrarily close to $\frac{d-2}{2}$. This means that this domain can be much larger than the previous one, in fact we have an improvement characterized by the scaling factor $L^{k(\gamma-\alpha)}$. This plays an important role in the proof of the asymptotic behavior of the correlation functions.

The next inductive assumption plays a crucial technical role in the renormalization group methods. It states that the functions $\mathcal{E}^{(j)}(y;\psi_j,h)$ have localization expansions of the form

$$\mathcal{E}^{(j)}(y;\psi_j,h) = \sum_{X\in\mathcal{D}_j:y\in X} \mathcal{E}^{(j)}(y,X;\psi_j,h), \tag{H.3}$$

where the sum is over subsets of $T_\xi, \xi = L^{-j}$, which are connected unions of cubes of some fixed large size M. The functions $\mathcal{E}^{(j)}(y,X;\psi_j,h)$ depend on (ψ_j,h) restricted to X, and satisfy the bounds

$$|\mathcal{E}^{(j)}(y,X;\psi_j,h)| < E_o \exp\left(-\kappa d_j(X)\right),$$

where E_o, κ are some positive constants, and $d_j(X)$ measures a linear size of the set X, see [1] for a definition. The equality and the bounds hold for the analytically extended functions also. The constant κ is chosen sufficiently large, so that we have a bound

$$\sum_{X\in\mathcal{D}_j:y\in X} \exp(-\kappa d_j(X)) \leq K_o.$$

uniformly in j, T and κ, with an absolute constant K_o dependent on d only. Then we get the bound $|\mathcal{E}^{(j)}(y;\psi_j,h)| < E_oK_o$, so

$$|\mathcal{E}^{(j)}(\psi_j,h)| < E_oK_o|T_1^{(j)}| = E_oK_oL^{-jd}|T_1|, \text{ and}$$

$$|\mathcal{E}_k(\psi,h)| < \sum_{j=1}^{k} E_oK_oL^{-jd}|T_1| < (L^d-1)^{-1}E_oK_o|T_1|.$$

This is a very bad bound, we would expect that after k steps we can bound the effective action by a constant times $|T_1^{(k)}| = |T_\eta| = \eta^d|T_1|$. Such an improvement is connected with a renormalization of the effective action after each step, and is a result of some additional renormalization conditions satisfied by the actions. We will formulate them after the next inductive assumption.

The functions $\mathcal{E}^{(j)}(y;\psi_j,h)$ have symmetry properties which we have described already in (6) for the original action. We assume that

$$\mathcal{E}^{(j)}(ry;r\psi_j,h) = \mathcal{E}^{(j)}(y;\psi_j,h)$$

for Euclidean transformations r of the lattice $T^{(j)}$ into itself, and that

$$\mathcal{E}^{(j)}(y;R\psi_j,Rh) = \mathcal{E}^{(j)}(y;\psi_j,h) \text{ for } R\in O(N). \tag{H.4}$$

The last property holds also for the localized functions $\mathcal{E}^{(j)}(y,X;\psi_j,h)$ introduced in (H.3). The first one implies that for constant configurations $\psi_j, \psi_j(y) =$

ψ for $y \in T^{(j)}, \psi \in \mathbb{R}^N$, the function $\mathcal{E}^{(j)}(y;\psi,h)$ does not depend on y, because $\mathcal{E}^{(j)}(y+z;\psi,h) = \mathcal{E}^{(j)}(y;\psi,h)$. This allows us to define an effective potential for the action $\mathcal{E}^{(j)}$ by the equality

$$\mathcal{V}^{(j)}(\psi,h) = \mathcal{E}^{(j)}(y;\psi,h) \text{ for any } y \in T^{(j)}. \tag{24}$$

It is convenient to use the effective potential to formulate the renormalization conditions. We assume that

$$\mathcal{V}^{(j)}(e_1,e_1) = 0, \left(\frac{\partial}{\partial h}\mathcal{V}^{(j)}\right)(e_1,e_1)\cdot e_1 = 0, e_1 = (1,0,...,0),$$

$$\left(\frac{\partial}{\partial \psi}\mathcal{V}^{(j)}\right)(e_1,e_1)\cdot e_1 = \sum_{y'\in T^{(j)}} \left(\frac{\partial}{\partial \psi(y')}\mathcal{E}^{(j)}\right)(y;,e_1,e_1)\cdot e_1 = 0,$$

$$\sum_{x\in\mathbb{Z}^d} tr\left[\lim_{T_1^{(j)}\to\mathbb{Z}^d}\left(\frac{\partial^2}{\partial\psi_j(x)\partial\psi_j(0)}\mathcal{E}^{(j)}\right)(e_1,e_1)\cdot(e_2\otimes e_2)\right]|x|^2 = 0, \tag{H.5}$$

for all $j \leq k$. These conditions allow us to improve essentially the bound of the effective action $\mathcal{E}_k$. In a more technical renormalization group language they imply that $\mathcal{E}^{(j)}$ contributes only "irrelevant terms" to the action. A precise formulation of this statement is in the following theorem.

Theorem 3 *If a function $\mathcal{E}_k(\psi,h)$ satisfies the assumptions (H.2) - (H.5), then there exist positive constants K_2, α_2 such that*

$$|\mathcal{E}^{(j)}(\psi_k^{(j)}(\psi,h),h)| < K_2(L^j\eta)^{d+\alpha_2}E_o|T_1^{(j)}| = K_2(L^j\eta)^{\alpha_2}E_o|T_1^{(k)}|, \tag{25}$$

$$|\mathcal{E}_k(\psi,h)| < \sum_{j=1}^{k}(L^j\eta)^{\alpha_2}K_2E_o|T_1^{(k)}| < \frac{1}{1-L^{-\alpha_2}}K_2E_o|T_1^{(k)}|, \tag{26}$$

on the domain of analyticity of the function $\mathcal{E}_k(\psi,h)$ described in (H.2).

Notice that the first inequality in (25) is the crucial one, the remaining are its simple consequencies. There are two important mechanisms for the appearance of the scaling factor $(L^j\eta)^{d+\alpha_2}$: the first is caused by the renormalization conditions (H.5), which remove relevant and marginal terms; the second by the fact that there is the function $\psi_k^{(j)}(\psi,h)$ substituted instead of ψ_j into the function $\mathcal{E}^{(j)}(\psi_j,h)$, which transforms the domain of analyticity of $\mathcal{E}_k(\psi,h)$ into a domain smaller than the analyticity domain of $\mathcal{E}^{(j)}(\psi_j,h)$, with better regularity properties and bounds. A proof of the theorem, along the guidelines sketched above, is given in [1].

A next inductive assumption clarifies what we mean by the asymptotic behavior (23). The precise form of this behavior depends on a renormalization procedure, which is not unique. We describe here a particularly simple one, we assume that

$$\beta_k = \beta L^{k(d-2)}\gamma_k, a_k = a\frac{1-L^{-2}}{1-L^{-2k}}, \lambda_k = \lambda L^{2k}, \nu_k = \nu L^{2k}\delta_k, \tag{H.6}$$

where the constants γ_k, δ_k are arbitrarily close to 1 if β is sufficiently large.

It will follow from renormalization groups equations, which will be formulated later, that we can strengthen the above assumption, requiring that there exist limits $\lim_{k\to\infty}\gamma_k = \gamma_\infty, \lim_{k\to\infty}\delta_k = \delta_\infty$, but it is not important here.

Our final assumption concerns the second term in the exponential in (H.1). This term determines a generating functional for truncated correlation functions, and here we describe the minimal assumptions which can be reproduced by the renormalization transformations, and which are sufficient to control convergence as $T \longrightarrow \mathbb{Z}^d$ and $\nu \longrightarrow 0+$, or $k \longrightarrow \infty$. To prove Theorem 1 in full we would need a more precise description, which is given in the paper [2]. Now we assume that

$$\mathcal{F}_k(\psi, h, g) = \langle g, \phi_k\rangle_1 + \sum_{j=1}^{k} \beta_{j-1}^{-\frac{1}{2}} \mathcal{F}^{(j)}(\psi_k^{(j)}(\psi, h), h, g),$$

$$\mathcal{F}^{(j)}(\psi_j, h, g) = \langle g, \mathcal{M}^{(j)}(\psi_j, h, g)\rangle_1, \tag{H.7}$$

and the functions $\mathcal{M}^{(j)}(x; \psi_j, h, g)$ can be analytically extended in (ψ, h) onto the same complex domain as the functions $\mathcal{E}^{(j)}$ in (H.2), and in g onto the complex domain $\|g\|_\ell^{\frac{1}{\ell}} < 1$; they have symmetry properties as in (H.4), and they have localization expansions as in (H.3), with localized terms satisfying the bounds

$$|\mathcal{M}^{(j)}(x, X; \psi_j, h, g)| < M_o L^{-j(\gamma-\alpha_1)} \exp(-\kappa d_j(X)),$$

where M_o is a positive constant and α_1 is a sufficiently small positive constant.

The above bounds and (H.6) imply in particular that the sum over j in (H.7) can be bounded by $4M_oK_o\beta^{-\frac{1}{2}}$, so it is small for β large. We have finished the description of the inductive assumptions. One of our main results is that actually the small field densities ρ_k defined by (17) satisfy the inductive hypotheses for properly chosen constant. We formulate it in the theorem below.

Theorem 4 *Consider the model given by (4), (5) for $d \geq 3, N \geq 2, \beta > 0, 1 \leq \lambda \leq +\infty, 0 < \nu \leq 1$, and define the sequence of densities ρ_k applying the small field renormalization transformations (17). For β sufficiently large, $\gamma \in (0, \gamma_o)$, there exist constants $E_o, \kappa, \epsilon_o, M_o, a_1$, and the coefficients $\beta_k, a_k, \lambda_k, \nu_k$ such, that the densities ρ_k satisfy the inductive hypotheses (H.1) - (H.7), as long as $\nu_k \leq 1$.*

This theorem is a consequence of another, more general theorem on one step renormalization transformation applied to densities satisfying the seven inductive assumptions. To formulate it we introduce the following definition

$$\begin{aligned}\mathcal{R}_k(E_o, \kappa, \epsilon_o, \gamma, M_o, \alpha_1, B_o) = \{ &\text{ the space of densities of the form } (H.1),\\ &\text{with the coefficients } \beta_k, a_k, \lambda_k, \nu_k \text{ satisfying (H.6) and } \beta_k > B_o, \nu_k \leq 1,\\ &\text{the functions } \mathcal{E}_k \text{ satisfying } (H.2) - (H.5), \text{ and the functions } \mathcal{F}_k\\ &\text{satisfying } (H.7)\}.\end{aligned} \tag{27}$$

Let us stress that in the above definition we treat the coefficients $\beta_k, a_k, \lambda_k, \nu_k$, the functions $\mathcal{E}_k$, and the functions $\mathcal{F}_k$ as independent variables. Notice that there are still some little details which are left uprecised by the definition, for example restrictions on the constants γ_k, δ_k in (H.6). It is easy to add such restrictions using another small constant in the definition, for example $|\gamma_k - 1| < \alpha_2, |\delta_k - 1| < \alpha_2$, but we did not want to overburden it. There are other quite technical aspects of the inductive assumptions which we have not mentioned, but they all give only slight modifications of some details without changing essential properties. Notice also that the functions $\mathcal{E}_k$ satisfying the assumptions (H.2) - (H.5), except the bounds in (H.3), form a vector space, and the bounds determine a ball in this space with a properly defined norm. The same is true for the functions $\mathcal{F}_k$. We can now formulate the fundamental theorem.

Theorem 5 *There exist constants $E_o, \kappa, \epsilon_o, M_o, \alpha_1, B_o$ independent of k, κ and B_o sufficiently large, ϵ_o, α_1 sufficiently small, such that*

$$\mathcal{X}_{k+1} S^{(k)} T^{(k)} \mathcal{X}^{(k)} : \mathcal{R}_k(E_o, \kappa, \epsilon_o, \gamma, M_o, \alpha_1, B_o) \longrightarrow$$

$$\longrightarrow \mathcal{R}_{k+1}(E_o, \kappa, \epsilon_o, \gamma, M_o, \alpha_1, B_o) \tag{28}$$

The value of the above transformation on a density ρ_k is a density ρ_{k+1} with coefficients $\beta_{k+1}, a_{k+1}, \lambda_{k+1}, \nu_{k+1}$ determined by the renormalization group equations

$$\beta_{k+1} = \beta_k L^{d-2} - b_{k+1}, a_{k+1} = \frac{a a_k}{a L^{-2} + a_k}, \lambda_{k+1} = \lambda_k L^2,$$

$$\nu_{k+1} = \frac{\beta_k L^{d-2}}{\beta_{k+1} + c_{k+1}} \nu_k L^2 = \frac{\beta_{k+1} + b_{k+1}}{\beta_{k+1} + c_{k+1}} \nu_k L^2, \tag{29}$$

where the constants b_{k+1}, c_{k+1} are determined by $\beta_k, a_k, \lambda_k, \nu_k$ and $\mathcal{E}_k$, and they are small compared with β_{k+1}, for example $|\beta^1_{k+1} b_{k+1}|, |\beta^{-1}_{k+1} c_{k+1}| < \alpha_3 L^{-(k+1)2\gamma}$ with α_3 arbitrarily small for B_o large enough.

This theorem can be given a much more precise form, with a more detailed description of how various elements of ρ_{k+1} depend on ρ_k, but the above formulation gives a basic idea of this approach, and is sufficient for us to conclude Theorem 4 from it. Let us notice that the transformation $\mathcal{X}_{k+1} S^{(k)} T^{(k)} \mathcal{X}^{(k)}$ depends on the density to which it is applied, more exactly it depends on the constant β_k in the assumption (H.1) on the density, so actually it is a quite complicated non-linear transformation, despite its appearance. We have not defined yet the characteristic function $\mathcal{X}^{(k)}$. We will make now a few remarks on the proof of the theorem, and a definition of $\mathcal{X}^{(k)}$ will be given in one of the remarks. A first part of the proof is to "calculate" the integral in

$$(\mathcal{X}_{k+1} T^{(k)} \mathcal{X}^{(k)} \rho_k)(\theta, h) = \mathcal{X}_{k+1} \int d\psi \mathcal{X}^{(k)} \mathcal{X}_k \exp[-\beta_k\{\frac{1}{2} a L^{-2} \|\theta - Q\psi\|^2_L +$$

$$+ A_k(\psi, \phi_k; h, a_k \lambda_k, \nu_k)\} + \mathcal{E}_k(\psi, h) - E'_k + \mathcal{F}_k(\psi, h, g)], \tag{30}$$

where θ is a new spin variable defined on the lattice $T_L^{(k+1)}$, and the normalization constant of the transformation $T^{(k)}$ is included into the constant E'_k. This integral has the same basic properties as the integral in (15), although it is obviously much more complicated from a technical point of view. The constant β_k may be arbitrarily large, so we apply the saddle point method. We want to find critical points of the function in the curly brackets. It turns out that on the considered small field domain there is exactly one critical point, and it is a solution of the variational problem

$$\inf_{\psi}\{\frac{1}{2}aL^{-2}\|\theta - Q\psi\|_L^2 + A_k(\psi, \phi_k; h, a_k, \lambda_k, \nu_k)\}. \tag{31}$$

This solution is equal to the function $\psi_{k+1}^{(k)}(\theta, h)$, which is one of the family of functions $\psi_k^{(j)}$ we have discussed before. They satisfy the equalities (22), in particular we have $\phi_k(\psi_{k+1}^{(k)}(\theta, h), h) = \phi_{k+1}(\theta, h)$. We have mentioned after (22) that the functions $\psi_k^{(j)}$ may be defined as solutions of variational problems, the problem (31) is one of them. In fact they all have this general form with proper modifications, like replacing k by $j, Q\psi$ by $Q_{k-j}\psi$, etc. Now we take the solution $\psi_{k+1}^{(k)}(\theta, h)$ and we make the change of variables

$$\psi = \psi_{k+1}^{(k)}(\theta, h) + \beta^{-\frac{1}{2}}\psi'. \tag{32}$$

The variable ψ' is called a fluctuation variable. We expand the function in the exponent of the underintegral expression in the variable ψ', and we obtain an integral in ψ', which is called a fluctuation integral, and which has the following form. The exponent of the underintegral exponential function is a sum of terms, the first term is a value of the exponent in (30) at the minimal configuration $\psi_{k+1}^{(k)}(\theta, h)$, the second term is a quadratic form in ψ', which is bounded from below and above by positive constants uniformly in k, and which has exponential decay properties. The remaining terms are small because they have factors which are powers of $\beta_k^{-\frac{1}{2}}$, and they have also good exponential decay properties. Thus the fluctuation integral can be "calculated", or represented, by a form of a cluster expansion. Of course all the above mentioned issues become technically quite involved and cannot be discussed here. Let us come back only to the question of the characteristic function $\mathcal{X}^{(k)}$. It may be defined simply by restrictions $|\psi'| < p_1(\beta_k) = (A_1 + \log \beta_k)^{p_1}$ where A_1, p_1 satisfy some restrictions with respect to A_o, p_o in (16). A second part of the proof, after the simple scaling transformation $S^{(k)}$, concerns a renormalization of the obtained new action. This new action satisfies all the inductive hypotheses, except the renormalization conditions (H.5). We can rearrange the action by changing the coefficients of the main part A_{k+1} according to the equations (29) with two unknown parameters b_{k+1}, c_{k+1}. This change yields some new terms which we include into $\mathcal{E}_{k+1}$. It can be shown that it is possible to choose the parameters in such a way that the conditions (H.5) are satisfied, and this completes the proof. For all the details see [1].

A last remark we want to make is about the last step in the procedure. As we have explained before, we end the procedure if the external field ν_k is approximately equal to $1, \nu_k \approx 1$. The last step is simply the integration of the density ρ_k with respect to ψ. This integral has the form (30), but without the term connected with the renormalization transformation in the exponent, i.e. with $a = 0$. The corresponding variational problem (31) still has a unique solution in the small field domain, because $\nu_k \approx 1$, and the solution is given by a constant configuration equal to h. We make the change of variables $\psi = h + \beta_k^{-\frac{1}{2}}\psi'$, and the rest of the analysis is the same as in the first part of a general case. We obtain a new action equal to $\mathcal{F}_{k+1}(h,h,g)$, which is a generating functional of truncated correlation functions: all the other expressions are equal to 0. This functional still satisfies (H.7), so now we can use it to prove various properties of these functions. We can prove most of the statements of Theorem 1 for the corresponding small field correlation functions, and all of them if we make a more precise inductive description of $\mathcal{F}_k$, see [2].

Let us finally make a few remarks about the "large field problem". In the scope of this paper it is impossible to give a discussion comparable to the one given above for the "small field problem". The "large fields" do not change any essential properties of the model, and actually the changes in the renormalization group procedure they cause are quite straightforward and ideologically simple, but they involve long and very technical inductive descriptions. We will make only some general remarks on these issues. It has been explained in the discussion of the first step that we decompose the integral (15) into a sum of integrals corresponding to the decomposition $1 = \sum_P \mathcal{X}_P \mathcal{X}^c_{P^c}$. A term in the decomposition determines the set X, to which we add a layer of large cubes of a proper size and denote it by Ω_1^c. This is a "large field domain". We leave the integration on Ω_1^c, and perform the integral in (15) over the small field domain Ω_1. It has the same feature as the whole integral (15), except that we have to consider boundary conditions on the boundary of Ω_1. In particular, critical points of the function $A_1^{B.C.}(\Omega_1;\psi,\phi)$ play a crucial role in the analysis. We obtain a new partial density, which can be written basically as a product of two factors: one is a contribution from Ω_1^c, and another is a contribution from Ω_1, which can be written in the form (H.1) restricted to Ω_1 by boundary conditions. The whole new density can be written as a sum of the partial densities over Ω_1. We repeat this procedure at each step. After k steps we obtain an effective density of the form

$$\rho_k(\psi) = \sum_{\Omega_k} \rho_k(\Omega_k^c;\psi,h,g)\mathcal{X}_k(\Omega_k)\exp\left[-\beta_k A_k^{B.C.}(\Omega_k;\psi,\phi_k;h,a_k,\lambda_k,\nu_k)+ \right.$$

$$\left. +\mathcal{E}_k(\Omega_k;\psi,h) - E_k(\Omega_k) + \mathcal{F}_k(\Omega_k;\psi,h,g)\right], \tag{33}$$

where the sum is over connected subdomains of the lattice T, the "small field domains". All the expressions in the exponential are as before, they satisfy all the properties (H.2) - (H.7), only the sums of the localization expansions in (H.3), (H.7) are restricted to $X \subset \Omega_k$. The "large field" factors $\rho_k(\Omega_k^c)$ are not localized to the domains Ω_k^c, unfortunately they have boundary terms which creep into the whole domain Ω_k. They are very small, so that the factors can be bounded by $\exp[-K|\Omega_k^c|_\eta]$ with K sufficiently large, but they complicate immensely the inductive description. The exponential bound controls the sum over Ω_k, so that we can get the desired uniform bound for the whole density, as the one implied by Theorem 3. The main result can be formulated as the following theorem.

Theorem 6 *The renormalization transformation $S^{(k)}T^{(k)}$ preserves the form and the bounds of the densities (33), with the same equations for the coefficients $\beta_{k+1}, a_{k+1}.\lambda_{k+1}, \nu_{k+1}$, and similar restrictions on the remaining constants.*

This is actually only an idea of a theorem, because we have not described the densities (33) in sufficient detail, but it captures the main features of the method. Let us remark that the proof of this theorem follows the steps described before. We introduce a decomposition $1 = \sum_{P_k} \mathcal{X}_{P_k} \mathcal{X}^c_{P^c_k}$ connected with proper spin variables, and we define new small field domains Ω_{k+1}. The integral (30) is restricted to the domain Ω_{k+1} now, but otherwise we analyze it in the same way, in particular we consider the variational problem (31) restricted to Ω_{k+1}. All the remaining steps in the small field domains are the same as before. This gives us the action (33) for Ω_{k+1}, satisfying all the required conditions, except possibly one. With each step the exponential bounds on the large field factors $\rho_k(\Omega_k^c)$ may become worse, because we use a part of them to control various sums and some terms in the exponentials coming from old small field domains. This actually depends on a size of Ω_k^c, or rather its connected components, the bounds become worse for components of some minimal size. It turns out that we can exponentiate a sum over all such components of the corresponding factors, and we can obtain instead a small field contribution in (33) with some additional terms in the exponent, which we add to the effective action. This is a second difficult technical point in the procedure, but it is quite universal and has been discussed before for other problems. We have overstepped already the scopes of this paper, so we finish and refer the interested reader to the papers mentioned above.

References

[1] Balaban, T. A low temperature expansion in classical N-vector models preprint, to appear in Commun. Math Phys.

[2] Balaban, T., O'Carroll, M. preprint 1994

[3] Balaban, T., Imbrie, J., Jaffee, J. Commun. Math. Phys. (a) Vol. **97** (1985) 299; (b) Vol **114** (1988) 257

[4] Balaban, T. Commun. Math. Phys. (a) Vol. **85** (1982) 603; (b) Vol. **86** (1982) 555; (c) Vol **89** (1983) 571; (d) Vol. **109** (1987) 249; (e) Vol. **116** (1988) 1

[5] Bleher, P.M., Major, P. (a) Commun. Math. Phys., Vol. **125** (1989) 43; (b) Ann. Inst. Henri Poincaré, Phys. Theor. Vol. **49** (1988) 1

[6] Bricmont, J., Fontaine, J., Lebowitz, J., Spencer, T. Commun. Math. Phys. (a) Vol. **78** (1980) 281; (b) Vol **78** (1980) 363

[7] Bricmont, J., Fontaine, J., Lebowitz, J., Lieb, E., Spencer, T. Commun. Math. Phys. Vol. 78 , 595 1981
[8] Brydges, D., Yau, H. Commun. Math. Phys. Vol. 129 , 351 1990
[9] Brydges, D., Dimock, J., Hurd, T. Weak Perturbations of Gaussian Measures preprint (a set of lectures)
[10] Feldman, J., Magnen, J., Rivasseau, V., Sénéor, R. Commun. Math. Phys. Vol. 109 , 437 1987
[11] Fröhlich, J., Simon, B., Spencer, T. Commun. Math. Phys. Vol. 50 , 79 1976
[12] Fröhlich, J., Pfister, C.-E. Commun. Math. Phys. Vol. 89 , 30 1983
[13] Gawedzki, K., Kupiainen, A., Commun. Math. Phys. Vol. 99 , 197 1985
[14] Glimm, J., Jaffe, A., Quantum Physics: A Functional Integral Point of View Springer, New York 1989
[15] Glimm, J., Jaffe, A., Spencer, T. Annals of Physics Vol. 101 , 610, 631 1976
[16] Negele, J.W., Orland, H. Quantum Many-Particle Systems Addison-Wesley Publishing Company 1988
[17] Schor, R., O'Carroll, M. Common. Math. Phys. (a) Vol. 138, 487 (1991); (b) J. Stat. Phys. Vol. 64, 163 (1991)
[18] Zinn-Justin, J., Quantum Field Theory and Critical Phenomena Oxford Science Publications 1989

Renormalization Group Approach to Zero Temperature Bose Condensation

Giuseppe Benfatto

Dipartimento di Matematica
Università di Roma "Tor Vergata"
Via della Ricerca Scientifica, 00133 Roma, Italy

1 Introduction

In this lecture, I will present some recent results [2] about the renormalization group approach to the problem of Bose condensation at zero temperature for a three dimensional system of bosons, interacting with a repulsive short range potential.

Before starting, I have to acknowledge the contribution of many other people in the different stages of this work. In fact, I was introduced to this problem by G. Gallavotti and J. F. Perez two years ago, but at the beginning we were only able to understand the difficulties, without finding a solution, so that we gave up the project. I started again to work on the subject during a visit to the University of Sao Paulo, where I had many useful discussions with J. F. Perez; these discussions allowed to clarify the problem, that I could really understand only recently. In this last stage the comments and the suggestions of G. Gallavotti had again an important role, as well as the criticism of C. Castellani, C. Di Castro and M. Grilli, which allowed me to find and correct an important mistake in my calculations.

The ideas that I shall discuss in this lecture are not yet sufficient for a complete rigorous treatment of the Bose condensation problem, but they are formulated in a way, that makes reasonable the conjecture that such a treatment is possible. The aim of my talk will be to give convincing arguments that the usual picture of three dimensional Bose condensation (see, for example, [1], [16]) can be explained in terms of a renormalization group flow, showing asymptotic freedom and anomalous behaviour of the two-point correlation function $S(x-y)$, at least order by order in the running coupling constants.

As we shall see, this anomalous behaviour is related to the fact the Fourier transform of $S(x)$ has a singularity of the type $(k_0^2 + v^2\mathbf{k}^2)^{-1}$, to compare with the singularity $(-ik_0+\mathbf{k}^2/2m)^{-1}$ of the free Bose gas. This anomaly explains, according to Landau's criterion [12], the superfluid properties of the system, whose spectrum is expressed, for small momenta, in terms of collective excitations with speed v.

The main missing point, in order to get a rigorous proof, is a suitable cluster expansion, similar to the one used to analyze the infrared ϕ_4^4 problem [8]; this would solve the *large field problem.* However, I do not expect this is a trivial generalization of known techniques, since in the Bose gas problem one has to use complex Gaussian measures, instead of positive Gaussian measures, and this would introduce new technical problems.

There is of course a huge physical literature on the subject of Bose condensation at zero or small temperature. As far as I know, the more convincing results about the superfluid behaviour at zero temperature are contained in [14] and [17] (see also references therein), where the authors use arguments similar to the ones that I will explain (C. Castellani made me aware of this point). However, at my knowledge, there is no treatment of the problem explicitly based on renormalization group arguments.

In the next section I will give the relevant definitions and I will discuss the so called *Bogoliubov approximation* [5], which is the starting point of the usual picture of Bose condensation. In §3 I will introduce the renormalization group transformation and the associated effective potentials and running coupling constants. Finally, in §4 I will discuss the flow equations for the running coupling constants, truncated to second order, showing that they can solved; the solution implies that the theory is asymptotically free with a superfluid two-point correlation.

2 The Bogoliubov Approximation

The problem is the following. Let:

$$H = \sum_{i=1}^{N} \left(-\frac{\Delta_{\mathbf{x}_i}}{2m} - \mu \right) + \lambda \sum_{i<j} v(\mathbf{x}_i - \mathbf{x}_j) \tag{1}$$

be the hamiltonian describing a system of N bosons in $\mathbf{R}^3$, enclosed in a periodic box of side size L, interacting with a pair potential $\lambda v(\mathbf{x}-\mathbf{y})$ which is supposed C^∞ and with short range p_0^{-1}. In fact we shall suppose that the interaction is repulsive in the sense that $v(0) > 0$, $v(\mathbf{x}) \geq 0$ and $\lambda \geq 0$.

Let $\varphi^{\pm}_{\mathbf{x}}$ be the creation and the annihilation operators for the bosons and

$$\varphi^\sigma_x = e^{Ht} \varphi^\sigma_{\mathbf{x}} e^{-Ht} \quad , \quad \sigma = \pm, x = (t, \mathbf{x}) \tag{2}$$

Define, for $\beta > t_i \geq 0$, $i = 1, \ldots, n$, with $t_i \neq t_j$ for $i \neq j$:

$$S_{\sigma_1 \cdots \sigma_n}(x_1, \ldots, x_n) = \lim_{\beta\to\infty} \lim_{L\to\infty} \frac{\mathrm{Tr}\left[e^{-\beta H} \varphi^{\sigma_{\pi(1)}}_{x_{\pi(1)}} \cdots \varphi^{\sigma_{\pi(n)}}_{x_{\pi(n)}}\right]}{\mathrm{Tr}\; e^{-\beta H}} \tag{3}$$

where $\sigma_i = \pm 1$ and π is the permutation of $(1, \ldots, n)$, such that $t_{\pi(1)} > t_{\pi(2)} > \ldots > t_{\pi(n)}$. In particular consider:

$$S(x) = S_{-+}(x, 0) \tag{4}$$

The functions (3), (4) describe the properties of the *ground state* of the above bosons system (essentially *by definition*) in the grand canonical ensemble with chemical potential μ.

The case $\lambda = 0$ is trivial and one finds that, if $\mu < 0$, the Schwinger functions at finite β and L (the r.h.s. of (3)) can be calculated by the Wick rule (see for example [15]) with propagator:

$$S_{\beta,L}(x) = L^{-d} \sum_{\mathbf{k}} e^{-i\mathbf{k}\cdot\mathbf{x}} e^{-\varepsilon(\mathbf{k})t} \Big(\frac{\vartheta(t>0)}{1-e^{-\beta\varepsilon(\mathbf{k})}} + \frac{\vartheta(t\le 0)e^{-\beta\varepsilon(\mathbf{k})}}{1-e^{-\beta\varepsilon(\mathbf{k})}} \Big) \tag{5}$$

where

$$\varepsilon(\mathbf{k}) = \frac{\mathbf{k}^2}{2m} - \mu \tag{6}$$

Note that it is not possible to take $\mu = 0$ in (5), since in this case the term in the sum with $k_0 = |\mathbf{k}| = 0$ involves a division by zero. However the interesting phenomena just occur when $\mu = 0$ and we shall therefore have to deal also with such case. For this reason we take μ as a function of β, L, which goes to 0 as $L, \beta \to \infty$, in such a way that the number of particles in the *condensate* (the state $\mathbf{k} = 0$) is fixed, that is:

$$e^{\beta\mu}(1-e^{\beta\mu})^{-1} = L^d \rho \tag{7}$$

where ρ is the condensate density. One finds, if $x_0 = t$:

$$S(x) = S_\rho(x) \equiv \rho + \frac{1}{(2\pi)^3} \int d\mathbf{k} e^{-i\mathbf{k}\cdot\mathbf{x}} e^{-\frac{\mathbf{k}^2}{2m}x_0} \vartheta(x_0 > 0) \tag{8}$$

so that:

$$S_0(x) = \frac{1}{(2\pi)^4} \int dk \, \frac{e^{-ikx}}{-ik_0 + \frac{\mathbf{k}^2}{2m}} \tag{9}$$

where $k = (k_0, \mathbf{k})$.

The Schwinger functions generated by the above limiting procedure and by the Wick rule will describe the ground state of a system of non interacting bosons with density ρ. They describe a *Bose condensed state.*

Let us now consider the case $\lambda > 0$. As it is well known [15], if β and L are finite, the Schwinger functions can be expressed as functional integrals in the following way:

$$S_{\sigma_1\ldots\sigma_n}(x_1\ldots x_n) = \int \varphi_{x_1}^{\sigma_1} \cdots \varphi_{x_n}^{\sigma_n} \frac{e^{-V(\varphi)}P(d\varphi)}{\int e^{-V(\varphi)}P(d\varphi)} \tag{10}$$

where $P(d\varphi)$ is a complex Gaussian measure, such that the fields φ_x^-, $\varphi_x^+ = (\varphi_x^-)^*$ have covariance:

$$\begin{aligned} &\int \varphi_x^- \varphi_y^+ P(d\varphi) = S_{\beta,L}(x-y) \\ &\int \varphi_x^- \varphi_y^- P(d\varphi) = \int \varphi_x^+ \varphi_y^+ P(d\varphi) = 0 \end{aligned} \tag{11}$$

and, if $x = (x_0, \mathbf{x})$, $y = (y_0, \mathbf{y})$, $\Lambda = [-\frac{1}{2}\beta, \frac{1}{2}\beta] \times [-\frac{1}{2}L, \frac{1}{2}L]^3$:

$$V(\varphi) = \lambda \int_\Lambda v(\mathbf{x} - \mathbf{y})\delta(x_0 - y_0)\varphi_x^+ \varphi_x^- \varphi_y^+ \varphi_y^- \, dx\, dy \tag{12}$$

Note that the fields $\varphi_x^\pm$ satisfy periodic boundary conditions in Λ.

In order to study the limit $L, \beta \to \infty$, it is convenient to add to $V(\varphi)$ a term proportional to $\int \varphi_x^+ \varphi_x^- \, dx$, by changing the definition of μ, and then fix μ so that (7) is satisfied. The limiting theory will be still described by the functional integrals (10), where $P(d\varphi)$ is the formal integration with propagator (8) and

$$V(\varphi) = \lambda \int_\Lambda v(\mathbf{x} - \mathbf{y})\delta(x_0 - y_0)\varphi_x^+ \varphi_x^- \varphi_y^+ \varphi_y^- \, dx\, dy + \nu \int_\Lambda \varphi_x^+ \varphi_x^- \, dx \tag{13}$$

ν has the role of a control parameter that should be fixed so that the limiting theory is meaningful as a perturbation of the free theory with propagator (8). If this program is successful, it is natural to say that there is Bose condensation at $T = 0$ with given density ρ and chemical potential $-\nu$. One could also fix the physical mass of the particles by adding to $V(\varphi)$ a term $\alpha \int \varphi_x^+ (-\Delta_{\mathbf{x}}/2m) \varphi_x^- \, dx$, but we shall not do that, because we do not expect that the choice of α is critical.

The form of the covariance (8) shows that the field $\varphi_x^\pm$ can be represented as:

$$\varphi_x^\pm = \xi^\pm + \psi_x^\pm \tag{14}$$

with $\xi^\pm$ being variables independent from $\psi_x^\pm$ and with covariance $\langle \xi^- \xi^+ \rangle = \rho$, while the fields $\psi_x^\pm$ have covariance $\langle \psi_x^- \psi_y^+ \rangle = S_0(x - y)$.

The integration with respect to $\xi^\pm$ can be thought as a Gaussian integral by writing $\xi^\pm = \xi_1 \pm i\xi_2$ and:

$$P(d\xi) = e^{-\frac{\xi_1^2 + \xi_2^2}{\rho}} \frac{d\xi_1 \, d\xi_2}{\pi \rho} \tag{15}$$

Hence, if we define:

$$W^{(-\infty)}(\xi) = -\frac{1}{\Lambda} \log \int e^{-V(\xi + \psi)} P(d\psi) \tag{16}$$

we see that the computation of $\langle \xi^+ \xi^- \rangle$ in presence of interaction will lead to the integral:

$$\rho = \int \frac{d\xi_1 d\xi_2}{2\pi\rho} (\xi_1^2 + \xi_2^2)\, e^{-(\xi_1^2 + \xi_2^2)/\rho}\, e^{-(W^{(-\infty)}(\xi) - C^{(-\infty)})|\Lambda|} \tag{17}$$

where $C^{(-\infty)}$ is a normalization constant and the equality to ρ of the above integral is just the requirement that the condensate density should be ρ. Therefore equality (17) can hold if and only if the function $W^\infty(\xi)$, which is a function of the product $\xi^+\xi^-$, by symmetry considerations, reaches its minimum at $\xi^+\xi^- = \rho$. And in this case $\xi^+\xi^-$ will be a sure random variable, provided the minimum is not degenerate.

This implies that, in order to get a condensate with density ρ, one has to choose ν so that the free energy (16) has a minimum in

$$\xi^+ = \xi^- = \sqrt{\rho} > 0 \tag{18}$$

These considerations are a heuristic justification of the so called *Bogoliubov approximation*, very usually used in the literature, consisting in replacing the fields $\xi^\pm$ by a real positive constant external field and by choosing its value so that the ground state energy is minimum.

At my knowledge, there is only one rigorous general result in this direction, found by Ginibre in 1967, [7]. Ginibre proved that, at any finite temperature and any fixed chemical potential, the finite volume pressure of the Bose system in the Bogoliubov approximation has a supremum (as a function of ξ), whose thermodynamical limit coincides with the thermodynamical limit of the real pressure. However, Ginibre was not able to prove a similar result for the correlation functions.

In any case, I will suppose that the Bogoliubov approximation is correct in the limit $\beta, L \to \infty$ and that the equation relating ν and ρ is invertible; this implies that, in order to study Bose condensation at $T = 0$, one has to consider the measure

$$\frac{1}{\mathcal{N}} e^{-V_\rho(\psi)} P(d\psi) \tag{19}$$

where $\mathcal{N}$ is a normalization constant, $V_\rho(\psi)$ is obtained from (13) trough the substitution (14), $\xi^\pm$ are positive constants satisfying (18) and ν has to be chosen so that the free energy is minimum for the fixed value of $\xi^\pm$.

The Schwinger functions of the field ψ are defined by an expression similar to (10) and we shall use the symbol $\tilde{S}$ to denote them. Their perturbation expansion is obtained in the usual way in terms of the propagator $S_0(x-y)$. Note that the measure (19) does not preserve the free measure property $\tilde{S}_{--}(x-y) = \tilde{S}_{++}(x-y) = 0$.

An old perturbative argument [9] allows to show (see also [1]) that the free energy is minimum if the following formal equation is satisfied:

$$\Sigma_{-+}(0) = \Sigma_{++}(0) \tag{20}$$

where $\Sigma_{\sigma_1\sigma_2}(k)$ is the Fourier transform of the sum of all one particle irreducible graphs (connected graphs which can not become disconnected by cutting one leg) with two external lines $\psi_x^{\sigma_1}$, $\psi_y^{\sigma_2}$.

The formal proof of (20) is very simple. It starts from the observation that:

$$W^{(-\infty)}(\xi) = \sum_{s=0}^{\infty} w_s \rho^s \tag{21}$$

where w_s is the sum of all irreducible graphs with $2s$ zero momentum ξ external legs (s of type $+$ and s of type $-$), constructed by the interaction (13) and the propagator $S_{\beta,L} - \rho$ (whose limit for $\Lambda \to \infty$ is given by (9)); the irreducibility condition follows from the fact that the propagator is zero at $k = 0$. Hence the free energy is stationary (and it turns out that the stationary point is unique and a minimum, at least for λ small enough) if the following equation is satisfied:

$$\sum_{s=1}^{\infty} s w_s \rho^{s-1} = 0 \tag{22}$$

Let us now consider $\Sigma_{-+}(0)$. It is clear that, given a graph contributing to w_s, there are s^2 graphs contributing to $\Sigma_{-+}(0)$ and having the same value; they are obtained by substituting in all possible ways an external leg ξ^- with ψ^- and an external leg ξ^+ with ψ^+, so that:

$$\Sigma_{-+}(0) = \sum_{s=1}^{\infty} s^2 w_s \rho^{s-1} \tag{23}$$

A similar argument can be used for $\Sigma_{++}(0)$ or $\Sigma_{--}(0)$, which are equal by symmetry reasons. Given a graph contributing to w_s, there are $s(s-1)/2$ equal contributions to $\Sigma_{++}(0)$, obtained by substituting two external legs ξ^+ with two ψ^+. Moreover there is a factor 2, coming from the possibility of interchanging the two external ψ^+ fields in the calculation of $\Sigma_{++}(0)$; hence:

$$\Sigma_{++}(0) = \sum_{s=1}^{\infty} s(s-1) w_s \rho^{s-1} \tag{24}$$

The condition (20) immediately follows from (22), (23) and (24). Note also that we could repeat the previous arguments in presence of an ultraviolet cut off, that we shall indeed introduce in the following section.

The *renormalization condition* (20) has here the same role of the condition which determines the critical temperature in Statistical Mechanics; for example, in the case of the φ_4^4 infrared model, a similar condition allows to define the stable manifold of the trivial fixed point in the renormalization group flow [8]. Therefore it is natural to use it, instead of the free energy minimum condition, as the condition fixing the chemical potential, in a renormalization group analysis of the measure (19).

3 The Effective Potentials

The problem we want to study is an infrared problem; therefore we shall consider a simplified model by substituting $S_0(x)$ with

$$g_{\leq 0}(x) = \frac{1}{(2\pi)^4} \int dk\, t_0(k) \frac{e^{-ikx}}{-ik_0 + \frac{\mathbf{k}^2}{2m}} \tag{25}$$

where $t_0(k)$ is a smooth function, which imposes an ultraviolet cutoff on scale p_0 (the scale of the potential). We shall choose $t_0(k)$ as a regularization of the characteristic function of the set $\{k_0^2 + \frac{\mathbf{k}^2}{2m}\frac{p_0^2}{2m} \le (\frac{p_0^2}{2m})^2\}$, such that, if γ is a fixed positive number greater than 1 and $q_0 = p_0^2/2m$:

$$t_0(k) = \theta(k_0^2 + \frac{\mathbf{k}^2}{2m}\frac{p_0^2}{2m}) \quad , \quad \theta(\lambda) = \begin{cases} 1 & \text{if } |\lambda| \le \frac{2\gamma^2}{\gamma^2+1}q_0^2 \\ 0 & \text{if } |\lambda| \ge \frac{2}{\gamma^2+1}q_0^2 \end{cases} \tag{26}$$

The particular choice of $\theta(\lambda)$ is not relevant; it is done only to simplify some calculations in the following.

Note that the assumed presence of the ultraviolet cut off on the scale of the interaction potential is reasonable only if $\rho p_0^{-3} \ll 1$, *i.e.* only if there is in mean much less than one particle in a cube with side equal to the range p_0^{-1} of the potential.

Hence we have to study the measure:

$$\frac{1}{\mathcal{N}} e^{-V_\rho(\psi)} P^{(\le 0)}(d\psi) \tag{27}$$

where $P^{(\le 0)}(d\psi)$ is the measure with covariance (25). We shall do that by a multiscale analysis, in the form presented in [6] and applied to a "similar" infrared problem, the one dimensional Fermi gas, in [4], [3].

At first, one could try to define a family of effective potentials based on a multiscale decomposition of the covariance (25). However, this approach does not work, as a consequence of the well known fact that the functions $\tilde{S}_{\sigma_1\sigma_2}(x-y)$ are expected to have a large distance behaviour very different from the behaviour of the *free propagator* (25). In fact, by simple (very formal) perturbative arguments, using the normalization condition (20), one can prove [1] that the Fourier transform of $\tilde{S}_{\sigma_1\sigma_2}(x-y)$ behaves for $k \to 0$ as

$$\frac{A}{k_0^2 + c^2\mathbf{k}^2} \tag{28}$$

where A and c are suitable constants (c has the physical interpretation of sound waves velocity in the condensate). The behaviour (28) is considered typical of superfluids.

Hence we are faced with a model with an *anomalous* behaviour, as in the $d = 1$ Fermi gas. The anomaly is now of a different nature, but we shall see that we can extend in a very natural way the renormalization group techniques of [4] to cover also this case.

We start writing the potential V_ρ in the following way:

$$V_\rho = |\Lambda|(\nu\rho + \lambda\hat{v}(0)\rho^2) + \mathcal{L}V_\rho + \mathcal{R}V_\rho \tag{29}$$

where the *local part* $\mathcal{L}V_\rho$ is obtained by substituting $\psi_y^\pm$ with $\psi_x^\pm$ in the expression of $V_\rho(\psi)$, after extracting the constant term, so that:

$$\begin{aligned}\mathcal{L}V_\rho(\psi) = \lambda\hat{v}(0)\int_\Lambda (\psi_x^+\psi_x^-)^2\,dx + 2\lambda\hat{v}(0)\sqrt{\rho}\int_\Lambda \psi_x^+\psi_x^-(\psi_x^+ + \psi_x^-)\,dx + \\ +(2\lambda\hat{v}(0)\rho + \nu)\int_\Lambda \psi_x^+\psi_x^-\,dx + \lambda\hat{v}(0)\rho\int_\Lambda (\psi_x^+ + \psi_x^-)^2\,dx\end{aligned} \tag{30}$$

We now observe that, if the "remainder" $\mathcal{R}V_\rho$ and the local terms of order greater than two were not present in (30) (so obtaining the exactly soluble *Bogoliubov model*), then the condition (20) could be imposed, by choosing $\nu = -2\lambda\rho\hat{v}(0)$, that is by putting equal to zero the coefficient of $\psi^+\psi^-$ in (30). This observation implies that the "natural" way to study the measure (27) is to change the free measure, by absorbing in it the local terms of V_ρ, quadratic in the field, which remain after imposing the condition $\nu^0 \equiv \nu + 2\lambda\rho\hat{v}(0) = 0$. We shall denote $P_B(d\psi)$ the new measure and we shall call it the *renormalized free measure*.

By using the results of Appendix 1 with $a = \lambda\rho\hat{v}(0)$, $b_0 = b = c = 0$, we see that the renormalized free measure has indeed the behaviour (28) for $k \to 0$, with $c^2 = c_B^2 \equiv 2\lambda\rho\hat{v}(0)/m$.

Therefore the measure (27) should be written:

$$\frac{1}{\mathcal{N}} e^{-\tilde{V}_0(\psi)} P_B(d\psi) \tag{31}$$

where

$$P_B(d\psi) \equiv P^{(\leq 0)}(d\psi) e^{-\lambda\hat{v}(0)\rho \int_\Lambda (\psi_x^+ + \psi_x^-)^2\, dx} \tag{32}$$

and, if we neglect non local terms:

$$\tilde{V}_0(\psi) = \lambda\hat{v}(0) \int_\Lambda (\psi_x^+ \psi_x^-)^2\, dx + \\ + 2\lambda\hat{v}(0)\sqrt{\rho} \int_\Lambda \psi_x^+ \psi_x^- (\psi_x^+ + \psi_x^-)\, dx + \nu^0 \int_\Lambda \psi_x^+ \psi_x^-\, dx \tag{33}$$

and the parameter ν^0 has to be determined so that the condition (20) is satisfied.

It is convenient, before proceeding, to change the basic fields and to perform a rescaling (amounting at fixing $\rho = 1$); we set:

$$\chi_x^\pm = \frac{1}{\sqrt{2\rho}} (\psi_x^+ \pm \psi_x^-), \qquad \psi^\pm = \sqrt{\frac{\rho}{2}} (\chi_x^+ \pm \chi_x^-) \tag{34}$$

$$\varepsilon = \lambda\hat{v}(0)\rho \frac{2m}{p_0^2} \tag{35}$$

so that $\tilde{V}_0$ becomes a function of χ:

$$\tilde{V}_0(\chi) = \frac{p_0^2 \rho}{2m} \Big(\frac{\varepsilon}{4} \int_\Lambda ((\chi_x^-)^4 - 2(\chi_x^-)^2(\chi_x^+)^2 + (\chi_x^+)^4) dx + \\ + \varepsilon\sqrt{2} \int_\Lambda ((\chi_x^+)^3 - (\chi_x^-)^2 \chi_x^+)\, dx + \frac{\nu^0}{2} \frac{2m}{p_0^2} \int_\Lambda ((\chi_x^+)^2 - (\chi_x^-)^2)\, dx \Big) \tag{36}$$

and the covariance of the fields $\chi^\pm$ in the distribution (32) is, in the limit $\Lambda \to \infty$:

$$\tilde{C}^{\sigma_1\sigma_2}(x-y) = \int P_B(d\chi) \chi_x^{\sigma_1} \chi_y^{\sigma_2} = \frac{1}{(2\pi)^4} \int dk e^{-ik(x-y)} t_0(k) \tilde{G}_0^{-1}(k)_{\sigma_1\sigma_2} \tag{37}$$

where the matrix $\tilde{G}_0(k)$ is defined by (see appendix 1):

$$\tilde{G}_0(k) = \rho \begin{pmatrix} \frac{\mathbf{k}^2}{2m} + 4\varepsilon \frac{p_0^2}{2m} t_0(k) & ik_0 \\ -ik_0 & -\frac{\mathbf{k}^2}{2m} \end{pmatrix} \tag{38}$$

where the first row and column correspond to $\sigma = +$ and the second row and column correspond to $\sigma = -$.

Note that $\tilde{C}^{--}(k)$ behaves, for $k \to 0$, as $1/k^2$ (like in φ_4^4 theory), while $\tilde{C}^{-+}(k)$ and $\tilde{C}^{++}(k)$ are less singular. In a sense, the field χ^+ behaves like the derivative of the field χ^-.

In order to study the measure (31), we make the scale decomposition

$$\tilde{C}^{\sigma_1\sigma_2}(x) = \sum_{h=-\infty}^{0} \tilde{C}_h^{\sigma_1\sigma_2}(x) \tag{39}$$

where $\tilde{C}_h^{\sigma_1\sigma_2}(x)$ is obtained from (37) by substituting $t_0(k)$ with

$$\tilde{T}_h(k) = t_0(\gamma^{-h}k) - t_0(\gamma^{-h+1}k) \tag{40}$$

so that we have the scaling relations:

$$\tilde{C}_h^{\sigma_1\sigma_2}(x) = \gamma^{(3+\frac{\sigma_1+\sigma_2}{2})h} \tilde{g}_h^{\sigma_1\sigma_2}(\gamma^h x) \tag{41}$$

with

$$\tilde{g}_h^{\sigma_1\sigma_2}(x) = \frac{1}{(2\pi)^4} \int dk e^{-ikx} \tilde{T}_0(k) \tilde{G}_h^{-1}(k)_{\sigma_1\sigma_2} \tag{42}$$

and

$$\tilde{G}_h(k) = \rho \begin{pmatrix} \frac{\gamma^{2h}\mathbf{k}^2}{2m} + 4\varepsilon \frac{p_0^2}{2m} t_0(\gamma^h k) & ik_0 \\ -ik_0 & -\frac{\mathbf{k}^2}{2m} \end{pmatrix} \tag{43}$$

Note that $\tilde{g}_h^{\sigma_1\sigma_2}(x)$ are essentially independent of h, for $h \to -\infty$.

Let us now define the *effective potentials* $\tilde{V}_h(\chi)$ in the usual way [6]:

$$e^{-\tilde{V}_h(\chi)} = \int \tilde{P}_B^{(h+1)}(d\chi^{(h+1)}) \dots \tilde{P}_B^{(0)}(d\chi^{(0)}) e^{-\tilde{V}_0(\chi+\chi^{(h+1)}+\dots+\chi^{(0)})} =$$

$$= \int \tilde{P}_B^{(h+1)}(d\chi^{(h+1)}) e^{-\tilde{V}_{h+1}(\chi+\chi^{(h+1)})} \tag{44}$$

where $\tilde{P}_B^{(h)}(d\chi)$ denotes the measure with covariance $\tilde{C}_h$.

The scaling relations (41) imply that we can define dimensionless fields $\bar{\chi}^{\pm}$ by the relations:

$$\chi_x^- = \gamma^h \bar{\chi}_{\gamma^h x}^- \quad , \quad \chi_x^+ = \gamma^{2h} \bar{\chi}_{\gamma^h x}^+ \tag{45}$$

and this allows us to analyze the relevant local part $\mathcal{L}\tilde{V}_h(\chi)$ of the effective potentials $\tilde{V}_h(\chi)$.

$\mathcal{L}\tilde{V}_h(\chi)$ is in principle a linear combination of all the local terms relevant or marginal. However only a few of them are different from zero, thanks to the symmetries of the original potential (13). Let us consider first the local terms not involving derivatives of the fields, that is terms of the form

$$F_{m_1 m_2} = \int dx\, \chi_x^{-m_1} \chi_x^{+m_2} \tag{46}$$

By using (45), it is easy to show that $F_{m_1 m_2}$ is not irrelevant only if $m_1+m_2-4 \leq 0$. The terms with $m_1+m_2=1$ are absent, since the fields $\chi^\pm$ have no $k=0$ Fourier component. If $m_1+m_2=2$, we have two relevant terms, F_{20} and F_{11}, and one marginal term, F_{02}. If $m_1+m_2 \geq 3$, we have one relevant term, F_{30}, and two marginal terms, F_{40} and F_{31}. However, F_{11} and F_{30} are absent, as a consequence of the particular structure of the potential (13), which implies that the local monomials of order 2 and 3 in the fields $\psi^\pm$ *must* appear in the following combinations:

$$\begin{aligned}
(\psi_x^+)^3 + (\psi_x^-)^3 &= 2\frac{\rho^{3/2}}{2^{3/2}}((\chi_x^+)^3 + 3\chi^+(\chi^-)^2) \\
\psi_x^+\psi_x^-(\psi_x^+ + \psi_x^-) &= 2\frac{\rho^{3/2}}{2^{3/2}}((\chi_x^+)^3 - (\chi_x^+)(\chi_x^+)^2) \\
\psi_x^+\psi_x^- &= \frac{\rho}{2}((\chi_x^+)^2 - (\chi_x^-)^2) \\
(\psi_x^+)^2 + (\psi_x^-)^2 &= 2\frac{\rho}{2}((\chi_x^+)^2 + (\chi_x^-)^2)
\end{aligned} \tag{47}$$

As regards the local terms involving derivatives of the field, we have three marginal terms, that is:

$$\begin{aligned}
D_{tt} &= -(\frac{2m}{p_0^2})^2 \int_\Lambda (\partial_{x_0}\chi_x^-)^2\, dx \quad , \quad D_{ss} = -p_0^{-2} \int_\Lambda (\partial_{\mathbf{x}}\chi_x^-)^2\, dx \\
D_t &= -\frac{2m}{p_0^2} \int_\Lambda \chi_x^+ \partial_{x_0}\chi_x^-\, dx
\end{aligned} \tag{48}$$

We did not include in the list the local term $-\int_\Lambda dx \chi_x^+ \partial_{\mathbf{x}} \chi_x^- = (1/\Lambda)\sum_k \chi_k^+ i\mathbf{k}\, \chi_k^-$, since it is identically zero; in fact the fields χ_k^σ are even functions of $\mathbf{k}$. Finally, the local terms $\int_\Lambda dx \chi_x^- \partial \chi_x^-$ and $\int_\Lambda dx (\chi_x^-)^2 \partial \chi_x^-$ are absent, since they are integrals of total derivatives and the fields satisfy periodic boundary conditions.

We can now define the relevant part of the effective potentials in terms of a *localization operator*, defined in the usual way [6]. For example, if (m,n) is equal to $(4,0)$, $(2,1)$ or $(0,2)$, we define:

$$\mathcal{L}\chi_{x_1}^- \cdots \chi_{x_m}^- \chi_{y_1}^+ \cdots \chi_{y_n}^+ = (\chi_{x_1}^-)^m (\chi_{x_1}^+)^n \tag{49}$$

and a similar definition is used for the other marginal terms. The monomials of positive dimension are localized trough a suitable Taylor expansion:

$$\begin{aligned}
\mathcal{L}\chi_x^-\chi_y^- &= (\chi_x^-)^2 + \sum_{i=0}^{3}(y_i - x_i)\chi_x^-\partial_i\chi_x^- + \frac{1}{2}\sum_{i,j=0}^{3}(y_i - x_i)(y_j - x_j)\chi_x^-\partial_i\partial_j\chi_x^- \\
\mathcal{L}\chi_x^+\chi_y^- &= \chi_x^+\chi_x^- + \sum_{i=0}^{3}(y_i - x_i)\chi_x^+\partial_i\chi_x^-
\end{aligned} \tag{50}$$

Note that, according to the remark preceding (47) and rotation invariance, many terms in the r.h.s. of (50) cancel out in the full effective potential.

The previous discussion implies that $\mathcal{L}\tilde{V}_h(\chi)$ is of the form

$$\mathcal{L}\tilde{V}^{(h)}(\chi) = \frac{p_0^2\rho}{2m}(\tilde{\lambda}_h F_{40} + \tilde{\mu}_h F_{21} + \gamma^{2h}\tilde{\nu}_h(F_{02} - F_{20}) + \\ + 2\tilde{z}_h F_{02} + 2\tilde{\zeta}_h D_{tt} + 2\tilde{\alpha}_h D_{ss} + 2\tilde{d}_h D_t) \tag{51}$$

where the factors 2 have no special meaning and the dimension fixing factor $\frac{p_0^2\rho}{2m}$ is introduced to keep track of the dimensions of the various quantities (its physical dimension is that of an action density, in space time).

As usual, we can hope to have a perturbative control of the model only if we can show that all the running constants appearing in (51) stay small for $h \to -\infty$, for a suitable choice of ν_0, such that:

$$|\tilde{\nu}_h| \leq \bar{\nu} \quad , \quad \text{for } h \to -\infty \tag{52}$$

This condition is equivalent to the renormalization condition (20), as it is easy to see at least if the theory is asymptotically free. In fact, in this case, for $h \to -\infty$, the effective potential is of the second order in the fields $\psi^{\pm}$, which appear only in the combination $(\psi_x^+ + \psi_x^-)^2 = 2\rho F_{02}$, up to terms containing field derivatives, which do not influence (20). A simple calculation allows to prove that this structure of the effective potential implies the renormalization condition.

For $h = 0$ we have:

$$\tilde{\lambda}_0 = \frac{\varepsilon}{4}, \quad \tilde{\mu}_0 = -\varepsilon\sqrt{2}, \quad \tilde{\nu}_0 = \frac{\nu^0}{2}\frac{2m}{p_0^2} \tag{53}$$

and

$$\tilde{z}_0 = \tilde{\zeta}_0 = \tilde{\alpha}_0 = \tilde{d}_0 = 0 \tag{54}$$

However $\tilde{z}_h$, $\tilde{\zeta}_h$, $\tilde{d}_h$ and $\tilde{\alpha}_h$ will be different from zero for $h < 0$ and they could grow as $h \to -\infty$.

This problem can be solved by the same strategy used in passing from the representation (27) of the measure to the representation (31). We define iteratively a new family of effective potentials $V_h(\chi)$ in the following way.

Given $V_0(\chi) = \tilde{V}_0(\chi)$, we define $\tilde{V}_{-1}(\chi)$ as before, so that:

$$\frac{1}{\mathcal{N}}\int P_B(d\chi)e^{-V_0(\chi)} = \frac{1}{\mathcal{N}}\int \tilde{P}_B^{(\leq -1)}(d\chi)\tilde{P}_B^{(0)}(d\chi^{(0)})e^{-V_0(\chi+\chi^{(0)})} = \\ = \frac{1}{\mathcal{N}}\int \tilde{P}_B^{(\leq -1)}(d\chi)e^{-\tilde{V}_{-1}(\chi)} \tag{55}$$

where $\tilde{P}_B^{(\leq -1)}(d\chi)$ and $\tilde{P}_B^{(0)}(d\chi)$ denote the measures with covariance $\sum_{h=-\infty}^{-1}\tilde{C}_h^{\sigma_1\sigma_2}$ and $\tilde{C}_0^{\sigma_1\sigma_2}$, respectively.

We then define $V_{-1}(\chi)$ by absorbing the terms proportional to $\tilde{z}_{-1}$, $\tilde{\zeta}_{-1}$, $\tilde{d}_{-1}$ and $\tilde{\alpha}_{-1}$ in the measure $\tilde{P}_B^{(\leq -1)}(d\chi)$, so that

$$\frac{1}{\mathcal{N}}\int \tilde{P}_B^{(\leq -1)}(d\chi)e^{-\tilde{V}_{-1}(\chi)} = \frac{1}{\mathcal{N}}\int P_B^{(\leq -1)}(d\chi)e^{-V_{-1}(\chi)} \tag{56}$$

We can iterate this procedure, so defining a family of measures $P_B^{(\leq h)}(d\chi)$ and a family of effective potentials $V_h(\chi)$, such that:

$$\mathcal{L}V_h(\chi) = \frac{p_0^2\rho}{2m}\left[\lambda_h F_{40} + \mu_h F_{21} + \gamma^{2h}\nu_h(F_{02} - F_{20})\right] \tag{57}$$

and the covariance $C_{\leq h}^{\sigma_1\sigma_2}$ of $P_B^{(\leq h)}(d\chi)$ is of the form (as follows from the calculations in appendix 1):

$$C_{\leq h}^{\sigma_1\sigma_2}(x-y) = \frac{1}{(2\pi)^4}\int dk e^{-ik(x-y)}\, t_0(\gamma^{-h}k)\, G_{\leq h}^{-1}(k)_{\sigma_1\sigma_2} \tag{58}$$

where

$$G_{\leq h}(k) = \rho\begin{pmatrix} \frac{\mathbf{k}^2}{2m} + 4\frac{p_0^2}{2m}\bar{Z}_h(k) & ik_0\bar{E}_h(k) \\ -ik_0\bar{E}_h(k) & -(\frac{\mathbf{k}^2}{2m} + 4\bar{B}_h(k)\frac{2m}{p_0^2}k_0^2 + 4\bar{A}_h(k)\frac{\mathbf{k}^2}{2m}) \end{pmatrix} \tag{59}$$

Note that $V_h(\chi)$ can be easily obtained from the potential $\bar{V}_h(\chi)$ defined by:

$$e^{-\bar{V}_{h-1}(\chi)} = \frac{1}{\mathcal{N}}\int \bar{P}_B^{(h)}(d\chi^{(0)})e^{-V_h(\chi+\chi^{(0)})} \tag{60}$$

where $\bar{P}_B^{(h)}(d\chi)$ is a single scale covariance obtained from (58) by substituting $t_0(\gamma^{-h}k)$ with

$$T_0(\gamma^{-h}k) = t_0(\gamma^{-h}k) - t_0(\gamma^{-h+1}k) \tag{61}$$

$V_h(\chi)$ is calculated by the equation:

$$V_h(\chi) = \bar{V}_h(\chi) - \frac{p_0^2\rho}{2m}\left[2z_h F_{02} + 2\zeta_h D_{tt} + 2\alpha_h D_{ss} + 2d_h D_t\right] \tag{62}$$

if the local part of $\bar{V}_h$ is written in the form

$$\begin{aligned}\mathcal{L}\bar{V}_h(\chi) = \frac{p_0^2\rho}{2m}[\quad &\lambda_h F_{40} + \mu_h F_{21} + \gamma^{2h}\nu_h(F_{02} - F_{20}) + \\ &+ 2z_h F_{02} + 2\zeta_h D_{tt} + 2\alpha_h D_{ss} + 2d_h D_t]\end{aligned} \tag{63}$$

Note that the choice (26) of $t_0(k)$ implies that

$$T_h(k)T_{h'}(k) = 0 \qquad \text{if} |h-h'| > 1 \tag{64}$$

The functions $\bar{A}_h(k)$, $\bar{B}_h(k)$, $\bar{Z}_h(k)$ and $\bar{E}_h(k)$ satisfy, for $h \leq -1$, some recursion relations, which can be easily obtained using (62) and the relation (116) of appendix 1, implying that

$$G_{\leq h}(k) = G_{\leq h+1} + 2\Delta_h(k)t_0(\gamma^{-h}k) \tag{65}$$

where:

$$\Delta_h(k) = \rho \begin{pmatrix} 2z_h \frac{p_0^2}{2m} & id_h k_0 \\ -id_h k_0 & -2\frac{2m}{p_0^2} k_0^2 \zeta_h - 2\alpha_h \frac{\mathbf{k}^2}{2m} \end{pmatrix} \tag{66}$$

It follows that:

$$\begin{aligned} \bar{Z}_h(k) &= \bar{Z}_{h+1}(k) + z_h t_0(\gamma^{-h} k) \\ \bar{A}_h(k) &= \bar{A}_{h+1}(k) + \alpha_h t_0(\gamma^{-h} k) \\ \bar{B}_h(k) &= \bar{B}_{h+1}(k) + \zeta_h t_0(\gamma^{-h} k) \\ \bar{E}_h(k) &= \bar{E}_{h+1}(k) + 2d_h t_0(\gamma^{-h} k) \end{aligned} \tag{67}$$

with initial conditions:

$$\bar{Z}_0(k) = \varepsilon t_0(k) \quad , \quad \bar{A}_0(k) = \bar{B}_0(k) = 0 \quad , \quad \bar{E}_0(k) = 1 \tag{68}$$

The covariance $\bar{C}_h^{\sigma_1\sigma_2}(x)$ associated with the measure $\bar{P}_B^{(h)}(d\chi)$ satisfies the following scaling relations:

$$\bar{C}_h^{\sigma_1\sigma_2}(x) = \gamma^{(3+\frac{\sigma_1+\sigma_2}{2})h} g_h^{\sigma_1\sigma_2}(\gamma^h x) \tag{69}$$

where

$$g_h^{\sigma_1\sigma_2}(x) = \frac{1}{(2\pi)^4} \int dk \frac{e^{-ikx}\, T_0(k)\, p_h^{\sigma_1\sigma_2}(k)}{\rho \mathcal{D}_h(k)} \tag{70}$$

$$\mathcal{D}_h(k) = b_h(k) k_0^2 + a_h(k) \frac{v_0^2}{4} \mathbf{k}^2 + \gamma^{2h} [\frac{\mathbf{k}^4}{4m^2}(1 + 4A_h(k)) + 4B_h(k) \frac{k_0^2 \mathbf{k}^2}{p_0^2}] \tag{71}$$

$$\begin{aligned} a_h(k) &= 4Z_h(k) + 16 Z_h(k) A_h(k) \\ b_h(k) &= E_h(k)^2 + 16 Z_h(k) B_h(k) \end{aligned} \tag{72}$$

$$\begin{aligned} &p_h^{-+}(k) = p_h^{+-}(-k) = -ik_0 E_h(k) \quad , \quad p_h^{--}(k) = -\gamma^{2h} \frac{\mathbf{k}^2}{2m} - 4Z_h(k) \frac{p_0^2}{2m} \\ &p_h^{++}(k) = \frac{\mathbf{k}^2}{2m} + 4[B_h(k) k_0^2 \frac{2m}{p_0^2} + A_h(k) \frac{\mathbf{k}^2}{2m}] \quad , \quad v_0 = \frac{p_0}{m} \end{aligned} \tag{73}$$

By using (67), (68) and (26), it is easy to see that, on the support of $T_0(k)$:

$$\begin{aligned} Z_h(k) &= \bar{Z}_h(\gamma^h k) = \sum_{h'=h}^{0} z_{h'} t_0(\gamma^{-(h'-h)} k) = \sum_{h'=h+1}^{0} z_{h'} + z_h t_0(k) \\ A_h(k) &= \bar{A}_h(\gamma^h k) = \sum_{h'=h}^{0} \alpha_{h'} t_0(\gamma^{-(h'-h)} k) = \sum_{h'=h+1}^{0} \alpha_{h'} + \alpha_h t_0(k) \\ B_h(k) &= \bar{B}_h(\gamma^h k) = \sum_{h'=h}^{0} \zeta_{h'} t_0(\gamma^{-(h'-h)} k) = \sum_{h'=h+1}^{0} \zeta_{h'} + \zeta_h t_0(k) \\ E_h(k) &= \bar{E}_h(\gamma^h k) = 1 + 2\sum_{h'=h}^{0} d_{h'} t_0(\gamma^{-(h'-h)} k) = 1 + 2\sum_{h'=h+1}^{0} d_{h'} + 2d_h t_0(k) \\ &z_0 = \varepsilon \quad , \quad \alpha_0 = \zeta_0 = d_0 = 0 \end{aligned} \tag{74}$$

Note that there are some relations between λ_h, μ_h and $Z_h(0)$, which can be defined in another equivalent way by the following steps.
a) We first integrate in a single step the fluctuations up to scale $h+1$, without any free measure renormalization and by using the original representation (12) of the potential. The result of this operation can be written in the form:

$$\tilde{U}_h(\varphi) = \sum_{s=1}^{\infty} \int d\underline{x} d\underline{y} W_{h,s}(\underline{x}, \underline{y}) \varphi^+_{x_1} \cdots \varphi^+_{x_s} \varphi^-_{y_1} \cdots \varphi^-_{y_s} \tag{75}$$

where $\underline{x} = (x_1, \ldots, x_s)$, $\underline{y} = (y_1, \ldots, y_s)$. $\tilde{U}_h(\varphi)$ satisfies the identity:

$$\int P(d\psi) e^{-V(\varphi)} = \frac{1}{\mathcal{N}} \int P_0^{(\leq h)}(d\psi) e^{-\tilde{U}_h(\varphi)} \tag{76}$$

where $P_0^{(\leq h)}(d\psi)$ is the measure with propagator:

$$C_0^{(\leq h)}(x) = \frac{1}{(2\pi)^4} \int dk\, t_0(\gamma^{-h} k) \frac{e^{-ikx}}{-ik_0 + \mathbf{k}^2} \tag{77}$$

b) We insert the representation (14) of $\varphi^{\pm}$ in (75), we collect the terms containing the monomials $(\prod_{i=1}^{r_1} \psi^+_{x_i})(\prod_{j=1}^{r_2} \psi^-_{y_j})$, with $2 \leq r_1 + r_2 \leq 4$, and we localize them, by a Taylor expansion of the fields of order 0, if $r_1 + r_2 > 2$, and of order 2, if $r_1 + r_2 = 2$. We obtain various local terms; those which do not contain any derivative of a field are of the form

$$\lambda_h^{r_1, r_2} \int dx (\psi^+_x)^{r_1} (\psi^-_x)^{r_2} \tag{78}$$

with

$$\lambda_h^{r_1, r_2} = \sum_{s=1}^{\infty} \binom{s}{r_1} \binom{s}{r_2} \bar{W}_{h,s} \rho^{s - \frac{r_1 + r_2}{2}} \quad , \quad \bar{W}_{h,s} = \lim_{|\Lambda| \to \infty} \frac{1}{\Lambda} \int d\underline{x} d\underline{y} W_{h,s}(\underline{x}, \underline{y}) \tag{79}$$

c) We insert the representation (34) in (78) and we collect the terms proportional to F_{40}, F_{21}, $F_{02} - F_{20}$, F_{02}, D_{tt}, D_{ss} and D_t; let us call $\tilde{\lambda}_h$, $\tilde{\mu}_h$, $\tilde{\nu}_h$, $2\tilde{Z}_h$, $2\tilde{B}_h$, $2\tilde{A}_h$ and $(\tilde{E}_h - 1)$ their respective coefficients. By using (79), it is easy to prove that:

$$\begin{aligned}
\tilde{\lambda}_h &= \frac{1}{4} \sum_{s=2}^{\infty} \bar{W}_{h,s} \rho^s \sum_{r=0}^{4} \binom{s}{r} \binom{s}{4-r} (-1)^r \\
\tilde{\mu}_h &= -\frac{1}{\sqrt{2}} \sum_{s=2}^{\infty} s(s-1) \bar{W}_{h,s} \rho^s \\
\tilde{Z}_h &= \frac{1}{2} \sum_{s=2}^{\infty} s(s-1) \bar{W}_{h,s} \rho^s
\end{aligned} \tag{80}$$

Note that there is no contribution with $s = 1$ in the equation for $\tilde{Z}_h$, because they could come only from the monomials $\psi^+_x \psi^-_y$, which only contribute to $F_{02} - F_{20}$, see (47).

d) We absorb in the free measure the local terms proportional to F_{02}, D_{tt}, D_{ss} and D_t. If we call $U_h(\chi)$ the remaining part of $\tilde{U}_h(\psi)$, thought as a function of χ, and $P_{0,B}^{(\leq h)}(\chi)$ the renormalized free measure, we can write:

$$\int P^{(\leq 0)}(d\psi)e^{-V(\varphi)} = \frac{1}{\mathcal{N}}\int P_{0,B}^{(\leq h)}(d\chi)e^{-U_h(\chi)} \tag{81}$$

However, this identity and the property that the local terms proportional to F_{02}, D_{tt}, D_{ss} and D_t are equal to zero uniquely determine the effective potential $V_h(\chi)$ and the measure $P_B^{(\leq h)}(d\chi)$. Hence $\tilde{\lambda}_h = (p_0^2\rho/2m)\lambda_h$, $\tilde{\mu}_h = (p_0^2\rho/2m)\mu_h$ and (by using the results of appendix 1) $\tilde{Z}_h t_0(\gamma^{-h}k) = (p_0^2\rho/2m)Z_h(k)$.

The last observation and (80) imply the *exact* identity:

$$Z_h(0) = -\frac{\mu_h}{\sqrt{2}} \quad , \quad \forall h \leq 0 \tag{82}$$

which will play a very important role in the discussion of the following section.

4 The Beta Function

We want to study perturbatively the flow of the *running couplings* λ_h, μ_h and ν_h (the *beta function* of our problem) and of the *renormalization functions* $Z_h(k)$, $A_h(k)$, $B_h(k)$ and $E_h(k)$, by keeping only the leading terms in the expansion of r_{h-1} in terms of $\{r_{h'}, h' \geq h\}$, if $r_h \equiv \{\lambda_h, \mu_h, \nu_h, Z_h, A_h, B_h, E_h\}$.

In order to understand which are the leading terms, we shall consider only the Feynman graphs calculated by using the single scale propagator (69); this corresponds to the approximation of neglecting the irrelevant part of $V_h(\chi)$ in calculating the relevant part of $V_{h-1}(\chi)$. In terms of the tree expansion described in [6], this approximation implies that we consider only trees with one tree vertex on scale h. The tree expansion allows to prove also that one obtains in this way the same bounds that one would obtain by summing over all trees with the same number of λ-vertices, μ-vertices and ν-vertices on different scales.

In order to perform this analysis, we need bounds on the rescaled propagators (70). If we define a scale h_0 so that:

$$v_0^2 z_0 = \gamma^{2h_0}\frac{p_0^2}{4m^2} \tag{83}$$

then we have to distinguish the case $h \geq h_0$ from the case $h \leq h_0$. In the first case, if $T_0(k) \neq 0$, $\gamma^{2h}\mathbf{k}^4$ is dominating over $a_h\mathbf{k}^2$, at least if $|a_h\mathbf{k}^2| \leq cz_0$, that we shall prove is true for λ small enough. In the second case $a_h\mathbf{k}^2$ is dominating over $\gamma^{2h}\mathbf{k}^4$.

If $\gamma^{2h} \geq Z_h(0)$, which essentially means $h \geq h_0$, and

$$|A_h| \leq \eta \quad , \quad \gamma^{2h}|B_h| \leq \eta \quad , \quad |E_h - 1| \leq \eta \tag{84}$$

where η is some number sufficiently small, one can show that there exists a fast decaying function $f(x)$ such that:

$$|g_h^{\sigma_1\sigma_2}(x)| \leq \gamma^{-h\frac{\sigma_1+\sigma_2}{2}} f(\gamma^h x_0, \mathbf{x}) \tag{85}$$

Let us now consider a graph contributing to the flow equation for λ_h, μ_h or $Z_h(0)$ and denote n_σ the number of external lines of type χ^σ, n_λ the number of λ-vertices, n_μ the number of μ-vertices. In order to estimate the contribution of this graph, we observe that the bound (85) implies that each internal half-line of type χ^σ gives a contribution $\gamma^{-h\sigma/2}$, while each one of the $(n_\lambda + n_\mu - 1)$ integrations that one has to perform to evaluate the local part of the graph gives a contribution proportional to γ^{-h}. Hence, up to a constant, the graph can be bounded by:

$$\begin{aligned} &|\lambda_h|^{n_\lambda} |\mu_h|^{n_\mu} \gamma^{h[\frac{1}{2}(4n_\lambda+2n_\mu-n_-)-\frac{1}{2}(n_\mu-n_+)-(n_\lambda+n_\mu-1)]} = \\ &= |\lambda_h|^{n_\lambda} |\mu_h|^{n_\mu} \gamma^{h[n_\lambda-\frac{1}{2}n_\mu+\frac{1}{2}(n_+-n_-+2)]} \end{aligned} \tag{86}$$

If we denote N_L the number of independent loops of the graph (equal to the number of propagators minus $(n_\lambda + n_\mu - 1)$) and we use the identity (82), we can write the bound (86) in the form:

$$\lambda_h^{n_\lambda} (Z_h^2 \gamma^{-2h})^{n_\mu/2} \gamma^{h(N_L+n_+)} \leq \lambda_h^{n_\lambda} Z_h^{n_\mu/2} \gamma^{h(N_L+n_+)} \tag{87}$$

Note that the contribution of a graph to the beta function is obtained by calculating its value at zero momentum of the external lines. Therefore no graph with $N_L = 0$ can contribute, because the single scale propagator vanishes at zero momentum. Moreover, (53) and (68) imply that $\lambda_0 = Z_0/4$, so that, by using (87), one can easily show that, if λ_0 is small enough,:

$$Z_h \leq 5\lambda_h \tag{88}$$

In fact this inequality follows by induction from the remark that, if it is true for $h' \geq h$, then (87) implies that the leading contributions to the beta function are those with $N_L = 1$. Hence one expects that there exists a constant c such that:

$$\begin{aligned} |\lambda_{h-1} - \lambda_h| &\leq c\lambda_h^2 \gamma^h \\ |Z_{h-1} - Z_h| = \frac{|\mu_{h-1} - \mu_h|}{\sqrt{2}} &\leq \frac{c}{\sqrt{2}} \lambda_h^{3/2} \gamma^{2h} \end{aligned} \tag{89}$$

which implies (88), together with

$$0 < Z_h \leq 2Z_0 \quad , \quad \text{if } 2h \geq h_0 \tag{90}$$

if λ_0 is small enough.

Note that all the previous bounds rest on the hypothesis (84); hence we have to check that (84) is consistent with (88) and (90). This can be easily done, by noticing that the graphs contributing to B_{h-1} and A_{h-1} are the graphs with $n_- = 2$ and $n_+ = 0$ and that their contribution is of the form $\int dx x_0^2 W_G(x)$ and $\int dx \mathbf{x}^2 W_G(x)$, respectively. Hence, we can use (87) for bounding $A_{h-1} - A_h$, while we have to multiply that bound by a factor γ^{-2h}, in the case of $B_{h-1} - B_h$; by using also (88), we find that:

$$|B_{h-1} - B_h| \le c\lambda_h \gamma^{-h} \quad , \quad |A_{h-1} - A_h| \le c\lambda_h \gamma^{h} \tag{91}$$

In a similar way one can prove that

$$|E_{h-1} - E_h| = 2|d_h| \le c\lambda_h \gamma^h \tag{92}$$

Hence (84) are satisfied, if λ_0 is small enough.

Remark. In order to make rigorous the previous considerations, one can not really use a complete perturbative argument, because the formal series representing the beta function are at most asymptotic series. In fact, by applying the technique of reference [6], one can prove only $n!$ bounds on the sum of all graphs with n point vertices, like in the Fermi gas problem, see [4].

The previous discussion shows that, in order to study the beta function, it is sufficient to consider carefully only the region $h < h_0$. The latter is the region where, if, for some constants c, c_1:

$$\frac{1}{2} \le 1 + 4A_h \le c, \qquad cZ_h \le E_h^2 + 16B_h Z_h \le c_1 Z_h, \quad 0 < Z_h < 2Z_0 \equiv 2\varepsilon \tag{93}$$

the rescaled propagators $g_h^{--}(x)$ and $g_h^{-+}(x)$ are essentially independent of h, that is they can be bounded by a rapidly decaying function $f(x)$, uniformly in A_h, B_h, Z_h, E_h verifying (93) (see appendix 2). On the contrary, the propagator $g_h^{++}(x)$ can be bounded by $f(x)/Z_h$.

The previous properties of the rescaled propagators and the identity (82) imply that, in the bound of a generic graph contributing to the beta function, two μ_h vertices are essentially equivalent to one λ_h vertex. A further simple analysis allows to prove that, if one wants to keep in each flow equation only the leading terms, then one has to consider only one loop graphs without ν_h vertices. Therefore, we shall study the flow equations, by keeping only these contributions; the properties of the corresponding solutions will be used to justify the approximation.

In this approximation ν_h does not appear in the flow equations of λ_h and μ_h; hence the growth of ν_h is controlled in a simple way by the right choice of ν_0, that is the right value of the chemical potential, if one can control the flow of λ_h and μ_h, which only depend on the renormalization functions.

It turns out that the leading terms of the beta function depend in a smooth way on the cutoff function $t_0(k)$, as one expects, because these terms determine the asymptotic behaviour of the model, which is independent of the choice of $t_0(k)$. Therefore we shall approximate the smooth cutoff function $t_0(k)$, whenever this is possible, by the characteristic function of the set $\{k_0^2 + \mathbf{k}^2 \le 1\}$.

This approximation simplifies the discussion of the beta function, because it implies that the supports of the Fourier transforms of the single scale covariances $\bar{C}_h^{\sigma_1\sigma_2}(x)$ are disjoint. This will guarantee that the leading terms in the beta function for the running couplings on scale $(h-1)$ will depend only on the running couplings on scale h; in fact we have to consider only graphs with up to four point vertices, one loop and all external momenta equal to 0, so that all the propagators belonging to the loop must have the same momentum. In terms of the tree expansion described in [6] this means that we must consider only trees with one tree vertex on scale h.

Note that this approximation is apparently not allowed everywhere; in fact the constants ζ_h and α_h (see (63)) depend on $t_0'(k)^2$, hence they are divergent when the regularization of $t_0(k)$ goes to zero. However there is a cancellation of these contributions, as well as of all contributions containing derivatives of $t_0(k)$, in the calculation of the leading terms, as explained in appendix 2. Therefore also in this case we can take the same approximation for the cutoff function.

There is another simplification following from the choice of $t_0(k)$ described before: $Z_h(k)$, $A_h(k)$ and $B_h(k)$ are independent of k, for $k^2 \leq \gamma^h$, that is for the values of k in the support of the Fourier transform of $C^{\sigma_1\sigma_2}_{\leq h}(x)$. This immediately follows from (67). Hence, also Z_h, A_h and B_h are constants.

There are very "few" graphs with one loop; therefore the calculation is straightforward: but quite long. Here we report the result, while the details are exposed in appendix 2. Neglecting terms of order γ^{2h} (which essentially come from the corrections to the scaling of the propagators, which become quickly scale independent), one finds after using (82) to eliminate μ_h from the row result (see appendix 2):

$$\begin{aligned}
\lambda_{h-1} &= \lambda_h - 36\beta_{2,h}(4Z_h)^2\big(\lambda_h - \frac{Z_h}{6}\big)^2 \\
Z_{h-1} &= Z_h - \frac{1}{4}\beta_{2,h}(4Z_h)^3 Z_h \\
E_{h-1} &= E_h - \frac{1}{4}\beta_{2,h}(4Z_h)^3 E_h \\
A_{h-1} &= A_h \\
B_{h-1} &= B_h + \frac{1}{16}\beta_{2,h}(4Z_h)^2 E_h^2 \\
\nu_{h-1} &= \gamma^2\left(\nu_h + \beta_{1,h}(24 Z_h\lambda_h - 4Z_h^2)\right)
\end{aligned} \tag{94}$$

where:

$$\begin{aligned}
\beta_{2,h} &= \frac{\log\gamma}{8\pi^2\sqrt{a_h^3 b_h}}\frac{p_0^3}{\rho}, \quad \beta_{1,h} = \frac{1-\gamma^{-2}}{8\pi^2(\sqrt{a_h b_h}+a_h)}\frac{p_0^3}{\rho} \\
a_h &= 4Z_h(1+4A_h), \quad b_h = (E_h^2 + 16 Z_h B_h)
\end{aligned} \tag{95}$$

It is very easy to analyze this flow, under the conditions (93), implying that $\beta_{2,h}$ is of order Z_h^{-2}. In fact, this observation is sufficient to prove that $Z_h = O(1/|h|)$ for $h \to -\infty$. But this property has to be true also for E_h, since, by (94):

$$\frac{E_{h-1}-E_h}{E_h} = \frac{Z_{h-1}-Z_h}{Z_h} \tag{96}$$

It is now very easy to check that

$$B_h \xrightarrow[h\to-\infty]{} B_{-\infty} > 0 \tag{97}$$

while A_h stays constant (indeed a small constant, roughly equal to its value on scale h_0).

Finally, if we define:

$$c_0 = \frac{p_0^3 \log\gamma}{16\pi^2\rho\sqrt{B_{-\infty}(1+4A_{-\infty})^3}} \tag{98}$$

we see that the first two equations in (94) can be written, for $h \to -\infty$, in the form:

$$\begin{aligned} \lambda_{h-1} &= \lambda_h - 36 c_0 Z_h^2 \left(\frac{\lambda_h}{Z_h} - \frac{1}{6}\right)^2 \\ Z_{h-1} &= Z_h - c_0 Z_h^2 \end{aligned} \tag{99}$$

The discussion of the above equations is elementary and, starting from initial data $Z_0 = \varepsilon, \lambda_0 = Z_0/4$ (or any other close to them), the result is that, if ν_0 is chosen so that ν_h is bounded uniformly in h, then, asymptotically:

$$Z_h = \bar{c}|h|^{-1}, \quad \lambda_h = \frac{1}{4} Z_h, \quad \nu_h = O(\lambda_h) \tag{100}$$

if $\bar{c}$ is a suitable constant (ε independent).

Note that these results are consistent with (93), which can be then proved inductively, together with (100).

At this point it is very easy to check that all the neglected terms in the beta function are at least of order $1/|h|^3$. Hence they can not change in a substantial way the asymptotic properties of the flow (up to convergence problems, see the remark above); only the values of $A_{-\infty}$, $B_{-\infty}$ and $\bar{c}$ depend on them, and $A_{-\infty}$ has to be a small number (of order ε). Note that this last observation is important, in order to be sure that the first condition in (93) is preserved, since we do not have a control on the sign of $A_{-\infty}$.

The main consequence of the previous discussion is that, for $k \to 0$ (that is for $h \to -\infty$), the model is Gaussian (asymptotic freedom) and the pair Schwinger function of the fields $\psi^{\pm}$ behaves as:

$$\tilde{S}_{-+}(k) = -\tilde{S}_{--}(k) = -\tilde{S}_{++}(k) \simeq \frac{q_0}{8\rho B_{-\infty}} \frac{1}{k_0^2 + c^2\mathbf{k}^2} \tag{101}$$

where the sound speed c is given by:

$$c^2 = \frac{(1+4A_{-\infty})v_0^2}{16 B_{-\infty}} \equiv c_B^2[1 + O(\sqrt{\varepsilon})] \tag{102}$$

where $c_B^2 = \varepsilon v_0^2$ is the square sound speed in the Bogoliubov model. In fact, the first bound in (91) implies that B_{h_0} is of order $\sqrt{\varepsilon}$, (92) implies that $1 - E_{h_0}$

is of order ε and (89) implies that $Z_{h_0} = \varepsilon[1 + O(\sqrt{\varepsilon})]$; moreover, by (96), for $h < h_0$:

$$Z_h \simeq \frac{Z_{h_0}}{E_{h_0}} E_h = \varepsilon E_h [1 + O(\sqrt{\varepsilon})] \tag{103}$$

and, by (94):

$$\frac{B_{h-1} - B_h}{E_{h-1} - E_h} = -\frac{E_h}{16 Z_h} = -\frac{1}{16\varepsilon}[1 + O(\sqrt{\varepsilon})] \tag{104}$$

implying that $B_{-\infty}$ is of order $1/\varepsilon$, since $E_{-\infty} = 0$, so that:

$$B_{-\infty} = \frac{1}{16\varepsilon}[1 + O(\sqrt{\varepsilon})] \tag{105}$$

Note Added in Proof . During a discussion with J. Magnen, we corrected an algebraic error in the first line of an earlier version of equations (80). In the current corrected version, equations (80) imply, besides (82), another exact identity (following, as the other, from the gauge invariance of the model, which is here used through the representation (75) of the effective potential on scale h). This identity is

$$\lambda_h = \frac{1}{4} Z_h$$

whose use would simplify the discussion of Section 4.

Note that the role of this new identity is in any case marginal, because, as we have shown, the surface $\{\lambda_h = \frac{1}{4} Z_h\}$ is locally attracting for the beta function. As discussed at the end of Appendix 2, this is not the case for the identity (82).

Appendix 1: Propagators for the Bose Gas

Let $P^{(t)}(d\psi)$ be the formal complex Gaussian measure with covariance

$$g^{(t)}(x) = \frac{1}{(2\pi)^4} \int dk \, e^{-ikx} \frac{t(k)}{-ik_0 + \frac{\mathbf{k}^2}{2m}} \tag{106}$$

where $x = (x_0, \mathbf{x})$, $k = (k_0, \mathbf{k})$ and $t(k)$ is a positive cutoff function.

We consider the fields $\chi_x^{\pm} = \frac{1}{\sqrt{2\rho}}(\psi_x^+ \pm \psi_x^-)$. Their propagator has the form:

$$\langle \chi_x^{\sigma} \chi_y^{\sigma'} \rangle = \frac{1}{(2\pi)^4} \int e^{-ikx} t(k) G^{-1}(k)_{\sigma\sigma'} \tag{107}$$

where the matrix G, which we call the propagator matrix, is:

$$G = \rho \begin{pmatrix} \frac{\mathbf{k}^2}{2m} & ik_0 \\ -ik_0 & -\frac{\mathbf{k}^2}{2m} \end{pmatrix} \tag{108}$$

Therefore the formal Gaussian measure:

$$P^{(t)}(d\psi)e^{-\rho\sum_{\sigma\sigma'}\int\chi^{\sigma}_{\sigma k}\Delta_{\sigma\sigma'}\chi^{\sigma'}_{-\sigma' k}dk} \tag{109}$$

has a propagator matrix $G'\equiv G+2\rho\Delta t(k)$. In particular, if Δ is associated with the quadratic form:

$$\int dx[2a(\chi^+_x)^2-2\sum_{i=0}^{3}b_i(\partial_i\chi^-_x)^2-2c\chi^+_x\partial_0\chi^-_x] \tag{110}$$

with $a, b_0, b=b_1=b_2=b_3, c$ not negative real numbers, that is:

$$\Delta=\begin{pmatrix}2a & ick_0\\ -ick_0 & -2(b_0k_0^2+b\mathbf{k}^2)\end{pmatrix} \tag{111}$$

then it is:

$$G'=\rho\begin{pmatrix}\frac{\mathbf{k}^2}{2m}+4at(k) & ik_0[1+2ct(k)]\\ -ik_0[1+2ct(k)] & -\frac{\mathbf{k}^2}{2m}-4(bk_0^2+b\mathbf{k}^2)t(k)\end{pmatrix} \tag{112}$$

And the $\psi^\pm$ fields propagators with respect to the measure $P^{(t)}_{ab}(d\psi)$ with propagator matrix G' can be immediately checked to be:

$$\int P^{(t)}_{a,b}(d\psi)\psi^-_x\psi^+_y=\frac{1}{(2\pi)^4}\int dk\, e^{-ik(x-y)}\,. \tag{113}$$

$$\cdot\frac{ik_0[1+2ct(k)]+\frac{\mathbf{k}^2}{2m}+2d^+(k)t(k)}{[1+2ct(k)]^2k_0^2+\frac{\mathbf{k}^4}{4m^2}+4d^+(k)\frac{\mathbf{k}^2}{2m}t(k)+16a(\sum_{i=0}^3 b_ik_i^2)t(k)^2}t(k)$$

and:

$$\int P^{(t)}_{a,b}(d\psi)\psi^+_x\psi^+_y=\int P^{(t)}_{a,b}(d\psi)\psi^-_x\psi^-_y=\frac{1}{(2\pi)^4}\int dk\, e^{-ik(x-y)}\,.$$

$$\cdot\frac{-2d^-(k)t(k)}{[1+2ct(k)]^2k_0^2+\frac{\mathbf{k}^4}{4m^2}+4d^+(k)\frac{\mathbf{k}^2}{2m}t(k)+16a(\sum_{i=0}^3 b_ik_i^2)t(k)^2}t(k) \tag{114}$$

where $d^\pm(k)=a\pm(b_0k_0^2+b\mathbf{k}^2)$. Note that (114) can be derived also by performing a Bogoliubov transformation.

The previous calculation can be immediately generalized, in the sense that, if the measure $P^{(t)}_G(d\psi)$ has propagator matrix G and cutoff function $t(k)$, then the measure

$$P^{(t)}_G(d\psi)e^{-\rho\sum_{\sigma\sigma'}\int\chi^{\sigma}_{\sigma k}\Delta_{\sigma\sigma'}\chi^{\sigma'}_{-\sigma' k}dk} \tag{115}$$

has propagator matrix

$$G'=G+2\rho t(k)\Delta \tag{116}$$

Appendix 2: The Second Order Beta Function

In this appendix we want to prove (94). We first note that all the terms in the r.h.s. of (94) are obtained by applying the localization operator to:

$$V_h^{\leq 4} \equiv \mathcal{E}_h(V_h) - \frac{1}{2!}\mathcal{E}_h^T(V_h^2) + \frac{1}{3!}\mathcal{E}_h^T(V_h^3) - \frac{1}{4!}\mathcal{E}_h^T(V_h^4) \tag{117}$$

where $\mathcal{E}_h$ and $\mathcal{E}_h^T$ denote, respectively, the expectation and the truncated expectation with respect to the measure describing the fluctuations on scale h, whose propagator is given by (69)-(73).

Some other remarks are important.

1) In order to calculate the beta function one has to evaluate some Feynman graphs at zero momentum of the external lines. Therefore only terms without internal lines or terms with at least one loop can contribute, since the single scale propagator vanishes at zero momentum.

2) The previous remark also implies that, in the graphs with only one loop, all internal lines must carry the same momentum. Hence, if we suitably choose the cutoff function $t_0(k)$, the internal lines of the loop may only have propagators of scale h or $h+1$; in fact at least one propagator must be of scale h and the supports of the Fourier transforms of the propagators of scale h and $h' \geq h$ are disjoint if $h' > h+1$. This implies that the trees (see [6] for the definitions), that one has to consider in evaluating the contribution of such graphs, are the trees with only one vertex of scale h and at most three endpoints and the trees with one vertex on scale h and one non trivial vertex on scale $h+1$, corresponding to some irrelevant contribution.

3) In the calculation of the graphs with only one loop, the subgraph associated with the tree vertex of scale $h+1$ has no loop. Therefore in this tree vertex the $\mathcal{R}$ operation coincides with the identity.

4) Since we are interested only in the leading orders, we can neglect in the rescaled propagators (70) the terms proportional to γ^{2h} and the dependence on k of Z_h, A_h, B_h and E_h (see (74)). For the same reason we can approximate Z_{h+1}, A_{h+1}, B_{h+1} and E_{h+1} by Z_h, A_h, B_h and E_h in the expression of $g_{h+1}^{\sigma_1\sigma_2}(x)$. Finally we can neglect the difference between λ_{h+1}, μ_{h+1} and λ_h, μ_h in the endpoints of the trees involving a tree vertex on scale $h+1$.

The previous remarks imply that the leading terms in the beta function can be obtained by the following steps:

a) Evaluate the graphs with one loop and propagator given by the sum of the single scale propagators of scale h and $h+1$, approximated as explained in remark 4), that is:

$$\gamma^{(3+\frac{1}{2}(\sigma_1+\sigma_2))h} g_{1,h}^{\sigma_1\sigma_2}(\gamma^h x) \tag{118}$$

where

$$g_{1,h}^{\sigma_1\sigma_2}(x) = \frac{1}{(2\pi)^4}\int dk\, e^{-ikx}\frac{T_1(k)\,\bar{p}_h^{\sigma_1\sigma_2}(k)}{\rho\mathcal{D}_h(k)} \tag{119}$$

Here $\mathcal{D}_h(k)$ and $\bar{p}_h^{\sigma_1\sigma_2}$ denote the expressions (71) and (73), modified as explained in remark 4) above, that is:

$$\mathcal{D}_h(k) = b_h k_0^2 + a_h \frac{v_0^2}{4} \mathbf{k}^2 \qquad (120)$$
$$a_h = 4Z_h(1+4A_h) \quad , \quad b_h = E_h^2 + 16 Z_h B_h$$

$$\bar{p}_h^{--}(k) = -4q_0 Z_h \quad , \quad \bar{p}_h^{-+}(k) = \bar{p}_h^{+-}(-k) = -ik_0 E_h \qquad (121)$$
$$\bar{p}_h^{++}(k) = \frac{16 Z_h B_h k_0^2 + a_h \mathbf{k}^2 \frac{v_0^2}{4}}{4q_0 Z_h} \quad , \quad q_0 = \frac{p_0^2}{2m}, \quad v_0 = \frac{p_0}{m}$$

and

$$T_1(k) = T_0(k) + T_0(\gamma^{-1}k) = t_0(\gamma^{-1}k) - t_0(\gamma k) \qquad (122)$$

b) Evaluate the same graphs with propagator of scale $h+1$, again approximated as in remark 4). This propagator is obtained from (119) by substituting $T_1(k)$ with

$$T_2(k) = T_0(\gamma^{-1}k) = t_0(\gamma^{-1}k) - t_0(k) \qquad (123)$$

We shall denote $g_{2,h}^{\sigma_1\sigma_2}$ the corresponding rescaled propagator.

c) Subtract the values found in b) from the values found in a) and add the trivial graphs without any internal line.

d) Approximate in the result the cutoff function $t_0(k)$ by the characteristic function of the set $\{k_0^2 + \frac{\mathbf{k}^2}{2m}\frac{p_0^2}{2m} \le (\frac{p_0^2}{2m})^2\}$. Note that this approximation is everywhere equivalent to calculating the graphs with all propagators on the single scale h, except in the case of the beta function for B_{h-1}, A_{h-1} and E_{h-1}, which involve derivatives with respect to the loop momentum. Hence, except in this case, we have to calculate the graphs by using the propagator obtained from (119) by substituting $T_1(k)$ with $T_0(k)$; we shall denote again $g_h^{\sigma_1\sigma_2}$ this rescaled propagator.

The trivial graphs give the linear terms in the r.h.s. of (94), except the term linear in λ_h appearing in the equation for ν_{h-1}. This term is obtained by contracting two χ^- fields in the λ_h vertex; one gets:

$$-\binom{4}{2}\frac{p_0^2}{2m}\lambda_h \gamma^{2h} 4q_0 Z_h \beta_{1,h} F_{20} \qquad (124)$$

where F_{20} is defined as in (46) and

$$\beta_{1,h} = \int \frac{dk}{(2\pi)^4}\frac{T_0(k)}{\mathcal{D}_h(k)} = \frac{1-\gamma^{-2}}{8\pi^2(\sqrt{a_h b_h} + a_h)} q_0^2 \left(\frac{2}{v_0}\right)^3 \qquad (125)$$

The quadratic terms in the equations for λ_h and μ_h are associated with the graphs drawn in Fig. 1, where the heavy lines represent the χ^- fields and the dotted ones represent χ^+.

Note that the coefficients in front of the different graphs, here and in the following figures, indicate how many times the graph appears, if one expands the powers of the potential in the r.h.s. of (117) in terms of the different monomials of the field, whose sum gives the potential, and consider the different possibilities of connecting different point vertices, giving rise to the same graph. In order to get the right contribution to the beta function, one has to consider also the coefficients of the expectations in (117) and the combinatorial factors which count the different possibilities of choosing the external lines in the different vertices of the graph and the different possibilities of contracting the internal lines.

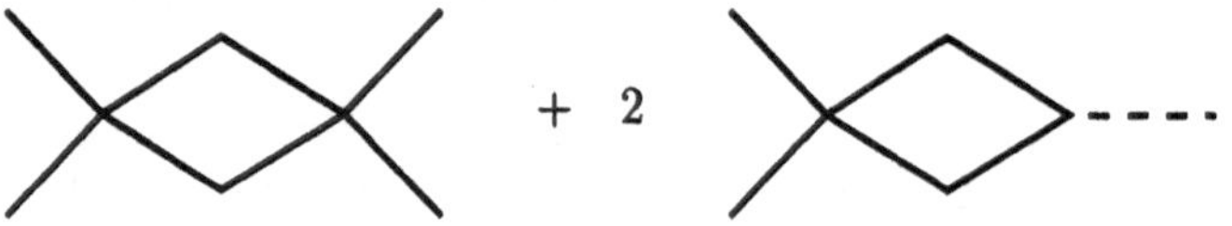

Fig. 1

The contribution to $\mathcal{L}V_{h-1}(\chi)$ coming from the graphs in Fig. 1 is:

$$-\frac{1}{2}\left[\binom{4}{2}^2 2\lambda_h^2(4q_0Z_h)^2\beta_{2,h}F_{40}+2\binom{4}{2}2\lambda_h\mu_h(4q_0Z_h)^2\beta_{2,h}F_{21}\right]\left(\frac{p_0^2}{2m}\right)^2 \quad (126)$$

where

$$\beta_{2,h}=\int\frac{dk}{(2\pi)^4}\frac{T_0(k)}{\mathcal{D}_h(k)^2}=\frac{\log\gamma}{8\pi^2\sqrt{a_h^3b_h}}\left(\frac{2}{v_0}\right)^3 \quad (127)$$

The cubic terms in the equations for λ_h and μ_h are associated with the graphs drawn in Fig. 2.

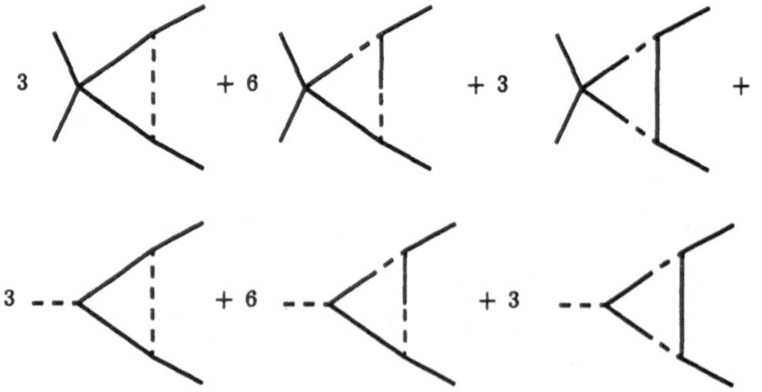

Fig. 2

The contribution to $\mathcal{L}V_{h-1}(\chi)$ coming from the graphs in Fig. 2 is:

$$\frac{1}{6}3\cdot 8\left[\binom{4}{2}\lambda_h\mu_h^2F_{40}+\mu_h^3F_{21}\right]\left(\frac{p_0^2}{2m}\right)^3\beta_{3,h} \quad (128)$$

where:

$$\beta_{3,h} = \rho^3 \int \frac{dk}{(2\pi)^4} \Big[g_h^{--}(k)^2 g_h^{++}(k) + 2 g_h^{--}(k) g_h^{-+}(k)^2 + \\ + g_h^{--}(k) g_h^{-+}(k) g_h^{+-}(k) \Big] = \\ = \rho^3 \int \frac{dk}{(2\pi)^4} \Big[g_h^{--}(k)^2 g_h^{++}(k) + g_h^{--}(k) g_h^{-+}(k)^2 \Big] = 4 q_0 Z_h \beta_{2,h} \tag{129}$$

The last equality follows from the identity:

$$\rho^2 [g_h^{--}(k) g_h^{++}(k) + g_h^{-+}(k)^2] = -\frac{T_0(k)^2}{\mathcal{D}_h(k)} \tag{130}$$

and from the observation that $T_0(k)^2 = T_0(k)$, in the approximation of item d) above. We also used the fact that $g_h^{-+}(k) = -g_h^{+-}(k) = -g_h^{-+}(-k)$. Hence (128) can be written as:

$$4 \Big[\binom{4}{2} \lambda_h \mu_h^2 F_{40} + \mu_h^3 F_{21} \Big] 4 q_0 Z_h \left(\frac{p_0^2}{2m} \right)^3 \beta_{2,h} \tag{131}$$

The quartic term in the equation for λ_h is associated with the graphs drawn in Fig. 3.

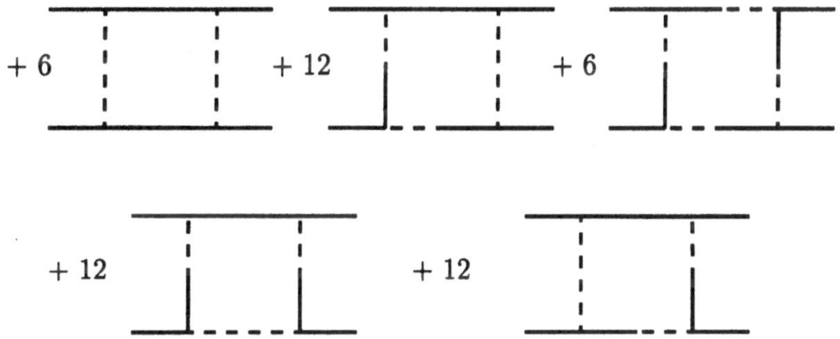

Fig. 3

The contribution to $\mathcal{L}V_{h-1}(\chi)$ coming from the graphs in Fig. 3 is:

$$-\frac{1}{4!} 2^4 \cdot 6 \left(\frac{p_0^2}{2m} \right)^4 \mu_h^4 \beta_{4,h} F_{40} \tag{132}$$

where:

$$\beta_{4,h} = \rho^4 \int \frac{dk}{(2\pi)^4} \Big[g_h^{--}(k)^2 g_h^{++}(k)^2 + 2 g_h^{--}(k) g_h^{-+}(k)^2 g_h^{++}(k) + \\ + g_h^{-+}(k)^4 - 2 g_h^{--}(k) g_h^{-+}(k)^2 g_h^{++}(k) + 2 g_h^{--}(k) g_h^{-+}(k)^2 g_h^{++}(k) \Big] = \\ = \rho^4 \int \frac{dk}{(2\pi)^4} \Big[g_h^{--}(k) g_h^{++}(k) + g_h^{-+}(k)^2 \Big]^2 = \beta_{2,h} \tag{133}$$

where we used again the identity (130).

The leading contribution in the flow equation for Z_h is the quadratic term associated with the graph in the first line of Fig. 4, whose contribution to the local part is:

$$-\frac{1}{2}\,2\left(\frac{p_0^2}{2m}\right)^2 \mu_h^2 (4q_0 Z_h)^2 \beta_{2,h} F_{02} \tag{134}$$

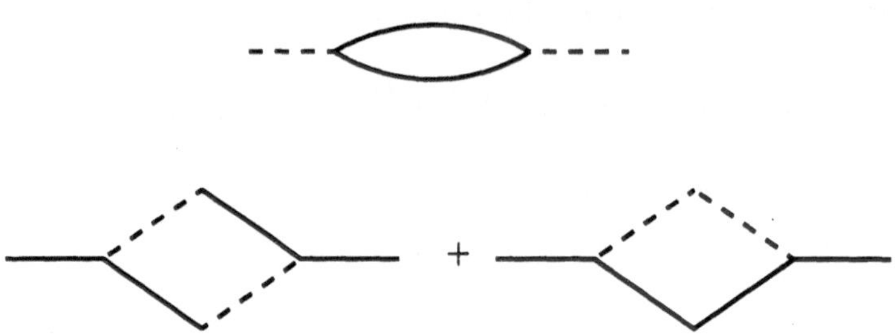

Fig. 4

and no operators involving field derivatives arise because this is a marginal operator.

The contribution to the local part of the last two graphs of Fig. 4 is:

$$\begin{aligned}&-\frac{1}{2}2^2\left(\frac{p_0^2\rho}{2m}\right)^2 \mu_h^2\Big\{\gamma^{2h}F_{20}\int\frac{dk}{(2\pi)^4}\Big[g_h^{--}(k)g_h^{++}(k)+g_h^{-+}(k)^2\Big]-\\&-\frac{1}{2}[\beta_{t,h}^{(1)}-\beta_{t,h}^{(2)}]q_0^2 D_{tt}-\frac{1}{6}[\beta_{s,h}^{(1)}-\beta_{s,h}^{(2)}]4m^2q_0^2D_{ss}\Big\}\end{aligned} \tag{135}$$

where D_{tt} and D_{ss} are defined as in (48) and

$$\beta_{t,h}^{(i)} = \int dx\; x_0^2\Big[g_{i,h}^{-+}(x)^2 - g_{i,h}^{--}(x)g_{i,h}^{++}(x)\Big] \tag{136}$$

$$\beta_{s,h}^{(i)} = \int dx\; \mathbf{x}^2\Big[g_{i,h}^{-+}(x)^2 - g_{i,h}^{--}(x)g_{i,h}^{++}(x)\Big] \tag{137}$$

$$g_{i,h}^{\sigma_1\sigma_2}(x) = \frac{1}{(2\pi)^4}\int dk \frac{e^{-ikx}\,T_i(k)\,\bar{p}_h^{\sigma_1\sigma_2}(k)}{\rho\mathcal{D}_h(k)} \tag{138}$$

We have:

$$\begin{aligned}\rho^2\beta_{t,h}^{(i)} &= \int\frac{dk}{(2\pi)^4}\Big\{E_h^2\Big[\frac{\partial}{\partial k_0}\frac{k_0T_i}{\mathcal{D}_h}\Big]^2+\Big[\frac{\partial}{\partial k_0}\frac{T_i}{\mathcal{D}_h}\Big]\frac{\partial}{\partial k_0}\Big[\Big(1-\frac{E_h^2k_0^2}{\mathcal{D}_h}\Big)T_i\Big]\Big\}\\&=\int\frac{dk}{(2\pi)^4}\Big\{E_h^2\frac{T_i^2}{\mathcal{D}_h^2}+\Big(\frac{\partial}{\partial k_0}\frac{T_i}{\mathcal{D}_h}\Big)\frac{\partial T_i}{\partial k_0}\Big\}\end{aligned} \tag{139}$$

It is easy to see that the terms containing the derivatives of T_i give the same contribution to $\beta_{t,h}^{(1)}$ and $\beta_{t,h}^{(2)}$, so that, by doing in the remaining term the approximation of item d) above, we get:

$$\rho^2[\beta_{t,h}^{(1)} - \beta_{t,h}^{(2)}] = \beta_{2,h}E_h^2 \tag{140}$$

We have also:

$$\rho^2\beta^{(i)}_{s,h} = \sum_{j=1}^{3}\int \frac{dk}{(2\pi)^4}\Big\{E_h^2\Big[\frac{\partial}{\partial k_j}\frac{k_0 T_i}{\mathcal{D}_h}\Big]^2 + \Big[\frac{\partial}{\partial k_j}\frac{T_i}{\mathcal{D}_h}\Big]\frac{\partial}{\partial k_j}\Big[\Big(1-\frac{E_h^2k_0^2}{\mathcal{D}_h}\Big)T_i\Big]\Big\} =$$

$$= \sum_{j=1}^{3}\int \frac{dk}{(2\pi)^4}\Big(\frac{\partial}{\partial k_j}\frac{T_i}{\mathcal{D}_h}\Big)\frac{\partial T_i}{\partial k_j} \tag{141}$$

and one can see again that the terms containing the derivative of T_i give the same contribution to $\beta^{(1)}_{s,h}$ and $\beta^{(2)}_{s,h}$, so that:

$$\beta^{(1)}_{s,h} - \beta^{(2)}_{s,h} = 0 \tag{142}$$

By summarizing, the contribution to the local part of the last two graphs of Fig. 4 is:

$$2\beta_{1,h}\left(\frac{p_0^2}{2m}\right)^2 \mu_h^2\gamma^{2h}F_{20} + \beta_{2,h}E_h^2\left(\frac{p_0^2}{2m}\right)^2 \mu_h^2 q_0^2 D_{tt} \tag{143}$$

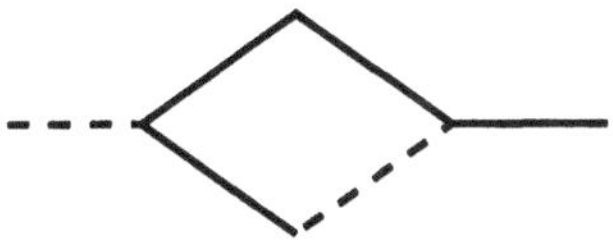

Fig. 5

The leading contribution in the flow equation for E_h is associated with the graph drawn in Fig. 5, whose local part is calculated by taking the first order Taylor expansion of the external χ^- field. The term of order zero, which would contribute to the relevant term F_{11}, cancels out for parity reasons, as well as the term of order one containing the spatial derivatives, so that the local part is equal to:

$$-\frac{1}{2}\,2\,2\,2\left(\frac{p_0^2}{2m}\right)^2 \mu_h^2(\beta^{(1)}_{5,h} - \beta^{(2)}_{5,h})q_0(-D_t) \tag{144}$$

where D_t is defined as in (48) and, if we use also the definition (138):

$$\beta^{(i)}_{5,h} = \rho^2\int dx\; x_0 g^{--}_{i,h}(x)g^{+-}_{i,h}(x) \tag{145}$$

We have:

$$\beta^{(i)}_{5,h} = 4q_0Z_hE_h\int \frac{dk}{(2\pi)^4}\frac{k_0T_i}{\mathcal{D}_h}\Big[\frac{\partial}{\partial k_0}\frac{T_i}{\mathcal{D}_h}\Big] \tag{146}$$

and we can see, as before, that the terms containing the derivatives of T_i give the same contribution to $\beta^{(1)}_{5,h}$ and $\beta^{(2)}_{5,h}$, so that, in the usual approximation:

$$\beta^{(1)}_{5,h} - \beta^{(2)}_{5,h} = -8q_0Z_hE_hb_h\int \frac{dk}{(2\pi)^4}\frac{k_0^2T_0(k)}{\mathcal{D}_h^3} = -2q_0Z_hE_h\beta_{2,h} \tag{147}$$

where the last equality follows from an explicit calculation and $\beta_{2,h}$ is exactly the same function of a_h and b_h defined in (127). Hence the local part of the graph drawn in Fig. 5 can be written as:

$$-8\left(\frac{p_0^2}{2m}\right)^2 q_0^2 \mu_h^2 Z_h E_h \beta_{2,h} D_t \tag{148}$$

The flow equations immediately follow from (124), (126), (131), (132), (134), (143) and (148):

$$\begin{aligned}
\lambda_{h-1} &= \lambda_h - 36(4Z_h)^2\beta_{2,h}\left[\lambda_h - \frac{\mu_h^2}{12Z_h}\right]^2 \\
\mu_{h-1} &= \mu_h - 12(4Z_h)^2\beta_{2,h}\mu_h\left[\lambda_h - \frac{\mu_h^2}{12Z_h}\right] \\
Z_{h-1} &= Z_h - 8Z_h^2\beta_{2,h}\mu_h^2 \\
A_{h-1} &= A_h \\
B_{h-1} &= B_h + \frac{1}{2}\beta_{2,h}E_h^2\mu_h^2 \\
E_{h-1} &= E_h - 8Z_hE_h\beta_{2,h}\mu_h^2 \\
\nu_{h-1} &= \gamma^2\left[\nu_h + 24Z_h\beta_{1,h}\lambda_h - 2\beta_{1,h}\mu_h^2\right]
\end{aligned} \tag{149}$$

where:

$$\beta_{2,h} = \frac{\log\gamma}{8\pi^2\sqrt{a_h^3 b_h}}\frac{p_0^3}{\rho}, \quad \beta_{1,h} = \frac{1-\gamma^{-2}}{8\pi^2(\sqrt{a_h b_h}+a_h)}\frac{p_0^3}{\rho} \tag{150}$$

with:

$$a_h = 4Z_h(1+4A_h) \quad , \quad b_h = E_h^2 + 16Z_hB_h \tag{151}$$

and the (94) follow immediately from the above by replacing μ_h via (82).

One may wonder whether it might be that the equation for μ_h is compatible with the others. Of course, if (82) is valid, it must be; this can also be seen directly (and in fact it is this remark that would make one conjecture the exact relation (82), if one did not know it). However the compatibility is true only to the order of the calculation that we are performing, *i.e.* within the neglected corrections: if we did not know *a priori* the validity of (82) we could not guarantee that the corrections and the uncertainty on the initial data would not spoil the relation (82) and turn the flow equation into an unmanageable relation of little use.

References

1. Abrikosov, A.A., Gorkov, L.P., Dzyaloshinski, I.E.: *Methods of quantum field theory in Statistical Physics*, Dover Publications, 1963.
2. Benfatto, G.: *Renormalization group approach to zero temperature Bose condensation problem*, work in progress, 1994.

3. Benfatto, G., Gallavotti, G., Procacci, A., Scoppola, B.: *Beta function and Schwinger functions for many fermion systems in one dimension. Anomaly of the Fermi surface*, Communications in Mathematical Physics, **160**, 93–172, 1994.
4. Benfatto, G., Gallavotti, G.: *Perturbation theory of the Fermi surface in a quantum liquid. A general quasi particle formalism and one dimensional systems*, Journal of Statistical Physics **59**, 541–664, 1990.
5. Bogoliubov, N. N.: Journal of Physics (USSR) **II**, 23,, 1947.
6. Gallavotti, G.: *Renormalization theory and ultraviolet stability via renormalization group methods*, Reviews of Modern Physics, **57**, 471–569, 1985.
7. Ginibre, J.: *On the asymptotic exactness of the Bogoliubov approximation for many boson systems*, Communications in Mathematical Physics **8**, 26–51, 1968.
8. Gawedzky, K., Kupiainen, A.: *Massless lattice ϕ_4^4 theory: rigorous control of a renormalizable asymptotically free model*, Communications in Mathematical Physics, **99**, 197–252, 1985.
9. Hugenholtz, N.M., Pines, D.: *Ground-state energy and excitation spectrum of a system of interacting bosons*, Physical Review **116**, 489, 1959.
10. Jimbo, M, Miwa, T., Mori, Y., Sato, M.: *Density matrix of an impenetrable gas and the fifth Painlevé transcendent*, Physica **D1**, 80–158, 1980.
11. Kennedy, T., Lieb, E., Shastry: *The XY model has long range order for all spins and all dimensions greater than one*, Physical Review Letters, **61**, 2582–2584, 1988.
12. Landau, L.D., Lifshits, E.M.: *Statistical Physics*, Pergamon Press, 1968.
13. Lieb, E., Lininger, W.: *Exact analysis of an interacting Bose gas. I. The general solution and the ground state*, Physical Review **130**, 1605–1624, 1963.
14. Nepomnyashchii, Yu.A., Nepomnyashchii, A.A.: *Infrared divergence in field theory of a Bose system with a condensate*, Soviet Phys. JETP **48**, 493-501, 1978.
15. Negele, J.W., Orland, H.: *Quantum many-particle systems*, Addison-Wesley, 1987.
16. Popov, V.N.: *Functional integration in Quantum Field Theory and Statistical Mechanics*, Reidel, Dordricht, 1983.
17. Popov, V.N., Seredniakov, A.V.: *Low-frequency asymptotic form of the self-energy parts of a superfluid Bose system at $T = 0$*, Soviet Phys. JETP **50**, 193–195, 1979.

Random and Interacting Surfaces

François Dunlop

Département de Physique
Université de Cergy-Pontoise, 95806 Cergy-Pontoise, France
&
Centre de Physique Théorique *(CNRS – UPR14)*
Ecole Polytechnique, 91128 Palaiseau, France
dunlop@orphee.polytechnique.fr

Abstract

We review the statistical mechanics of two-dimensional interfaces separating phases in $\mathbb{Z}^3$, with particular attention to constructive aspects, understood here as the construction of convergent expansions for the free energy and correlation functions. These will take different forms depending on whether the interface is rigid, like in the three-dimesional Ising model at low temperature, or rough like in the capillary waves model. The corresponding techniques may be associated respectively to the Pirogov-Sinaï theory or to perturbations of a massless gaussian field. The unifying feature is the occurrence of scaled expansions, which may allow to control the neighborhood of a phase transition. Some open problems are presented and discussed.

Contents

1 Introduction

The rigorous study of two-dimensional interfaces and random surfaces in statistical mechanics really started with Dobrushin's 1972 paper [D72] "Gibbs State Describing Coexistence of Phases for a Three-Dimensional Ising Model". The difficulty associated to the degeneracy of the Ising Hamiltonian with respect to arbitrary integer translations perpendicular to the interface was overcomed by the wall-and-ceiling decomposition of the interface. The Ising model with Dobrushin ± boundary conditions is given by

$$\Lambda = [-\frac{L}{2}, \frac{L}{2}]^3 \cap \mathbb{Z}^3, \qquad \Lambda^c = \mathbb{Z}^3 \setminus \Lambda$$

$$\Lambda^c_+ = \Lambda^c \cap (\mathbb{Z}^2 \times \mathbb{Z}^+), \qquad \Lambda^c_- = \Lambda^c \setminus \Lambda^c_+$$

$$\underline{\sigma}^\Lambda \in \{1, -1\}^{|\Lambda|}, \qquad \sigma_x = 1 \quad \text{if} \quad x \in \Lambda^c_+, \qquad \sigma_x = -1 \quad \text{if} \quad x \in \Lambda^c_-$$

$$H(\underline{\sigma}^\Lambda | \underline{\sigma}^{\Lambda^c}) = -J \sum_{\substack{|x-y|=1 \\ x,y \in \Lambda}} \left(\sigma_x \sigma_y - 1\right) - J \sum_{\substack{|x-y|=1 \\ x \in \Lambda, y \in \Lambda^c}} \left(\sigma_x \sigma_y - 1\right)$$

and the corresponding finite volume Gibbs measure by

$$\mu_\Lambda(\underline{\sigma}^\Lambda | \underline{\sigma}^{\Lambda^c}) = (Z^\pm_\Lambda)^{-1} \exp(-\beta H(\underline{\sigma}^\Lambda | \underline{\sigma}^{\Lambda^c}))$$

Defining a plaquette as a unit square drawn on the dual lattice $\mathbb{Z}^3 + (1/2, 1/2, 1/2)$ separating a (+)spin and a (−)spin, and defining contours as maximal connected sets of plaquettes gives the usual one-to-one correspondence between spin configurations and compatible families of contours. The ± boundary conditions give rise to one potentially infinite contour which is called the interface, while finite contours may be called bubbles. Plaquettes belonging to the interface are divided into vertical plaquettes, used as bricks contributing to walls, and horizontal plaquettes which are bricks if they overhang another plaquette and element of ceiling otherwise. A wall is then defined as a maximal connected set of bricks, and

$$\mu_\Lambda(\underline{\sigma}^\Lambda | \underline{\sigma}^{\Lambda^c}) = (Z^\pm_\Lambda)^{-1} e^{-2\beta J (L+1)^2} \prod_{\substack{\text{bubbles} \\ \gamma}} e^{-2\beta J |\gamma|} \prod_{\substack{\text{walls} \\ \omega}} e^{-2\beta J |\omega|} \tag{1.1}$$

where $|\gamma|$ and $|\omega|$ are the area or number of plaquettes of a bubble or wall. That (1.1) can be exponentiated into a convergent polymer expansion is not obvious because of the hard core interaction between bubbles and ceilings, considering that ceilings do not get any damping factor. The remarkable fact is that ceilings are costless, as inherited from translation invariance, and yet that the hard core interaction between bubbles and ceilings can be factored out, leaving a convergent expansion over walls, bubbles and generated polymers.

The gas of walls and bubbles is dilute for large β ; in particular the probability that a point in the interface be surrounded by a wall is $\mathcal{O}(\exp(-8\beta J))$, which means that the interface is rigid with a constant height $-1/2$ except for scarse excitations. This proves the existence of non-translation Gibbs states, and gives a convergent low-temperature expansion for the surface tension $\sigma^{\pm}$, defined as the excess free energy per unit area of the $\pm$ state compared to the + or − states,

$$\sigma^{\pm}(\beta) = -\beta^{-1} \lim_{\Lambda \nearrow \mathbb{Z}^3} \frac{1}{L^2} \log \frac{Z_\Lambda^{\pm}}{\left(Z_\Lambda^{+} Z_\Lambda^{-}\right)^{1/2}}$$

giving at lowest orders

$$\sigma^{\pm}(\beta) = 2J - 2\beta^{-1} e^{-8\beta J} + \mathcal{O}(e^{-12\beta J}) \tag{1.2}$$

where the ceiling-bubble interaction enters in $\mathcal{O}(e^{-12\beta J})$.

Various improvements have been obtained since 1972, notably the extension to Potts and more general lattice models in the Pirogov-Sinaï family [BLP79, BLPO79, BKL83, LMR86, HKZ88, MMRS91, M94].

As the temperature is raised, the interface fluctuations increase and diverge at the roughening transition temperature T_R, which is believed to satisfy $T_R < T_C$. Van Beijeren [B75] proved $T_R > T_C^{(2)}$ and Fröhlich and Spencer [FS81] established the roughening transition as a Kosterlitz-Thouless transition, within the Solid-On-Solid model defined from the Ising model by suppressing all bubbles and overhangs. In principle, integrating the Ising model below T_C over bubbles and overhangs would produce an SOS model with exponentially decaying interactions, but the resulting estimates would need to be sharpened for this line of approach to be really conclusive [W83].

The roughening temperature T_R should also be the temperature above which the step free energy vanishes. This is defined from boundary conditions allowing one infinite step in a rigid interface : for $x = (x_1, x_2, x_3) \in \Lambda^c$, $\sigma_x = +1$ if $x_3 > 0$ or if $x_3 = 0$ and $x_1 \cos\phi + x_2 \sin\phi \geq 0$ and $\sigma_x = -1$ otherwise. Taking $0 \leq \phi \leq \pi/4$ for definiteness and denoting the corresponding partition function by $Z_\Lambda^{step}(\phi)$, the step tension is defined as the excess free energy per unit length compared to the $\pm$ state,

$$\tau^{step}(\phi) = -\beta^{-1} \lim_{\Lambda \nearrow \mathbb{Z}^3} \frac{\cos\phi}{L} \log \frac{Z_\Lambda^{step}(\phi)}{Z_\Lambda^{\pm}}$$

Inequalities involving the step tension have been obtained by Bricmont, Fontaine and Lebowitz [BFL82], and by Bricmont, El Mellouki and Fröhlich [BEF86]. A convergent low-temperature expansion has been given recently by Miracle-Sole [M94] :

$$\begin{aligned}\tau^{step}(\phi) = 2J\big(|\cos\phi| + |\sin\phi|\big) & -\beta^{-1}\Big\{\big(|\cos\phi| + |\sin\phi|\big)\log\big(|\cos\phi| + |\sin\phi|\big) \\ & -|\cos\phi|\log|\cos\phi| - |\sin\phi|\log|\sin\phi| + e^{-2\beta J}\tilde{\tau}_\phi(e^{-2\beta J})\Big\}\end{aligned} \tag{1.3}$$

where $\tilde{\tau}_\phi(z)$ is an analytic function of its argument in a neighborhood of the origin. These results for the step free energy imply, in principle, estimates of the size of facets of an Ising crystal, once the Wulff construction is established within statistical mechanics.

Let us now go over to interacting surfaces and give an outline of the next sections. First "Wetting", which can be modelled by an Ising model in a half space $\mathbb{Z}^2 \times \mathbb{Z}_+$ with Hamiltonian

$$H(\underline{\sigma}^{\Lambda}|\underline{\sigma}^{\Lambda^c}) = -J \sum_{\substack{|x-y|=1 \\ x,y\in\Lambda}} \left(\sigma_x\sigma_y - 1\right) - J \sum_{\substack{|x-y|=1 \\ x\in\Lambda, y\in\Lambda^c}} \left(\sigma_x\sigma_y - 1\right) + h \sum_{x\in W} (\sigma_x + 1)$$

where $\Lambda \subset \mathbb{Z}^2 \times \mathbb{Z}_+$, $\Lambda^c = (\mathbb{Z}^2 \times \mathbb{Z}_+) \setminus \Lambda$, and $W = \Lambda \cap (\mathbb{Z}^2 \times \{0\})$. The surface field $h > 0$ favours the $(-)$ phase and the boundary conditions $\sigma_x = 1$ if $x \in \Lambda^c$ give a $(+)$ bulk phase.

Defining contours and interface as before, one speaks of complete wetting if there is an infinite interface, and of partial wetting otherwise. Clearly complete wetting is obtained whatever $\beta < \beta_C$ if $h \geq J$. For $h < J$, partial wetting has been proven by Fröhlich and Pfister [FP87] in the region $h < \sigma^{\pm}(\beta)/2$, using correlation inequalities. Pfister and Penrose [PP88] have given a low-temperature expansion, proving analyticity and partial wetting for $h < J - \mathcal{O}(\beta^{-1})$. The essential open problem here is the existence of the wetting transition or equivalently of a delocalised interface for $h < J$ and some $T < T_C$. A reasonable goal for constructivists would be to establish the wetting transition line near $T = 0$.

The understanding of the wetting transition requires the understanding and estimation of entropic repulsion, for which important results have been obtained within Solid-On-Solid models by Bricmont, El Mellouki and Fröhlich [BEF86] : Consider an interface above a wall, without overhangs, given by a positive height function $\Phi \in \mathbb{Z}_+^{\Lambda}$ for $\Lambda \subset \mathbb{Z}^2$, with Hamiltonian

$$H(\Phi^{\Lambda}) = J \sum_{\substack{|x-y|=1 \\ x,y\in\Lambda}} \left|\Phi(x) - \Phi(y)\right|^{\alpha} + J \sum_{\substack{|x-y|=1 \\ x\in\Lambda, y\in\Lambda^c}} \left|\Phi(x) - \Phi(y)\right|^{\alpha}$$

with boundary conditions $\Phi(x) = 0$ for $x \in \Lambda^c$. Denote the corresponding finite volume Gibbs state by $\langle\, . \,\rangle_{\alpha}$ and let $\alpha = 1$ or $\alpha = 2$. Then for large β

$$\frac{1}{|\Lambda|} \sum_{x\in\Lambda} \langle\Phi(x)\rangle_{\alpha} \approx ((\beta J)^{-1} \log|\Lambda|)^{1/\alpha} \tag{1.5}$$

This result is natural in terms of spikes from the mean height (1.5), the damping factor being just compensated by the entropic factor L^2. By comparison, when $\Phi \in \mathbb{R}_+^{\Lambda}$ and $\alpha = 2$, denoting the corresponding finite volume Gibbs state by $\langle\, . \,\rangle_0$, Bricmont, El Mellouki and Fröhlich prove

$$\frac{1}{|\Lambda|} \sum_{x\in\Lambda} \langle\Phi(x)\rangle_0 > cst.(\beta J)^{-1/2} \log|\Lambda| \tag{1.6}$$

which isn't obvious from the capillary waves picture, which suggests $(\log|\Lambda|)^{1/2}$ instead.

In the next three sections, we shall describe in more detail some results and open problems which we feel belong to the constructive approach, with scaled polymer or contour expansions :

- layering
- critical wetting
- Wulff shape

Unfortunately, we shall not discuss nor review the following topics :

- exactly solved models
- one-dimensional interfaces
- tubes, foams, statistics of genus of random surfaces
- interfaces in quenched random media
- dynamics of random surfaces

2 Layering

Competition between entropic repulsion pushing the interface away from the wall and a bulk external field pushing the interface towards the wall, in the case of a rigid interface, leads to layering transitions : for large β and small external field h, the interface should be localised at an integer height of order $\frac{1}{4\beta J}\log((\beta h)^{-1})$. This integer height k increases as $h \searrow 0$; at values of h corresponding to jumps from k to $k+1$, the two corresponding Gibbs states are expected to coexist.

The first rigorous results for this first order phase transition were obtained for the semi-infinite Ising model by Fröhlich and Pfister [FP87] who used correlation inequalities and chessboard estimates. More precise results, showing in particular coexistence of phases, have been obtained recently by Dinaburg and Mazel [DM94] for the Solid-On-Solid model, using a low-temperature expansion and the Pirogov-Sinaï theory. We give below their theorem and some hints on the expansion, which has some new interesting features, in particular the use of cylinders instead of the more standard walls, and a resummation of "elementary" cylinders into the activity of large contours.

For $\Lambda \subset \mathbb{Z}^2$, the SOS Hamiltonian for the interface height $\Phi^\Lambda \in \mathbb{Z}_+^\Lambda$ with boundary condition $k \in \mathbb{Z}_+$ is given by

$$H(\Phi^\Lambda|k^{\Lambda^c}) = J \sum_{\substack{|x-y|=1\\ x,y\in\Lambda}} \Big|\Phi(x)-\Phi(y)\Big| + h\sum_{x\in\Lambda}\Phi(x) + J \sum_{\substack{|x-y|=1\\ x\in\Lambda, y\in\Lambda^c}} \Big|\Phi(x)-k\Big|$$

and the corresponding finite volume Gibbs state by

$$\Xi(\Lambda^{(k)})^{-1}\exp(-\beta H(\Phi^\Lambda|k^{\Lambda^c}))$$

as a probability measure on $(\mathbb{Z}_+)^\Lambda$.

Theorem (Dinaburg and Mazel) : *There exist a constant β_0 and a sequence of continuous functions $h_k^*(\beta)$, $k=1,2,...$ with $h_k^*(\beta) \searrow 0$ as $k \nearrow \infty$ such that for any $\beta > \beta_0$ the following hold true:*

(i) if $h_k^*(\beta) < h < h_{k-1}^*(\beta)$, then the model posseses a unique $\mathbb{Z}^2$-periodic Gibbs state generated by the boundary condition k.

(ii) if $h = h_k^*(\beta)$, then the set of $\mathbb{Z}^2$-periodic extremal Gibbs states contains precisely two elements generated by the boundary conditions k and $k+1$.

The proof is based on an expansion over cylinders, of which we now attempt to give a flavour. A cylinder $\gamma = (\tilde{\gamma}, E, I)$ is a cylinder in the usual sense, truncated between an external height $E \in \mathbb{Z}^+$ and an internal height $I \in \mathbb{Z}^+$. The basis $\tilde{\gamma}$ of the cylinder is a connected set of bonds drawn on $\tilde{\mathbb{Z}}^2 = \mathbb{Z}^2 + (1/2, 1/2)$. Both signs of $S(\gamma) = \operatorname{sign}(I(\gamma) - E(\gamma))$ will occur. Compatible families of cylinders are defined so as to ensure one-to-one correspondence with interface configurations. A notable feature is that compatible cylinders should not annihilate their effects: if the bases share some bonds, $\tilde{\gamma}' \cap \tilde{\gamma}'' \neq \emptyset$, then the two cylinders must have the same sign, $S(\gamma') = S(\gamma'')$, which means that both cylinders contribute to raising the surface, or to lowering the surface. Cylinders are partially ordered by inclusion of the projection of their interiors, denoted $\bar{\gamma}$, and $E(\gamma) = I(\gamma')$ whenever $\bar{\gamma} \subset \bar{\gamma}'$.

External (maximal) cylinders contributing to $\Xi(\Lambda^{(k)})$ must have $E(\gamma) = k$. The set of families of compatible cylinders contributing to $\Xi(\Lambda^{(k)})$ is denoted $\Lambda^{(k)}$. Then

$$\Xi(\Lambda^{(k)}) = e^{-\beta h k |\Lambda|} \sum_{\{\gamma_i\} \in \Lambda^{(k)}} \prod_i \omega(\gamma_i) \tag{2.1}$$

with, for each $\gamma = (\tilde{\gamma}, E, I)$,

$$\omega(\gamma) = e^{-\beta J |I - E| |\tilde{\gamma}| - \beta h (I - E) |\bar{\gamma}|}$$

The compatibility conditions introduce a non-local interaction between nested cylinders, which does not allow to take directly the logarithm of (2.1) as in Dobrushin's expansion over walls [D72]. Some resummation is necessary, and will be achieved by a moderate use of the phase changing trick,

$$\Xi(\Lambda^{(k)}) = e^{-\beta h k |\Lambda|} \sum_{\{\gamma_i\}^{ext} \in \Lambda^{(k)}} \prod_i \Big(\omega(\gamma_i) e^{\beta h I(\gamma_i) |\bar{\gamma}_i|} . \cdot \frac{\Xi(\bar{\gamma}_i^{(I(\gamma_i), S(\gamma_i))})}{\Xi(\bar{\gamma}_i^{(k, S(\gamma_i))})} \Xi(\bar{\gamma}_i^{(k, S(\gamma_i))}) \Big)$$

where the additional superscript $S(\gamma_i)$ in the configuration ensembles $\bar{\gamma}_i^{(I(\gamma_i), S(\gamma_i))}$ and $\bar{\gamma}_i^{(k, S(\gamma_i))}$ means that compatible cylinders inside γ_i must also be compatible with γ_i itself, which involves $S(\gamma_i)$. Taking now

$$\tilde{\omega}(\gamma) = e^{-\beta J |I - E| |\tilde{\gamma}|} \frac{\Xi(\bar{\gamma}^{(I,S)})}{\Xi(\bar{\gamma}^{(E,S)})}$$

leads to

$$\Xi(\Lambda^{(k,S)}) = e^{-\beta h k |\Lambda|} \sum_{[\gamma_i] \in \Lambda^{(k,S)}} \prod_i \tilde{\omega}(\gamma_i) \tag{2.2}$$

where $[\gamma_i]$ denotes a weakly compatible family, meaning that the conditions involving $I(\gamma_i)$ and $E(\gamma_i)$ have been dropped, except those involving only $S(\gamma_i)$.

The advantage is that the weak compatibility conditions are local ; the disadvantage is that $\tilde{\omega}(\gamma_i)$ is not known.

The renormalised weights $\tilde{\omega}(\gamma_i)$ can be controlled when γ_i is small enough, depending upon $E(\gamma_i)$. Dinaburg and Mazel call elementary and denote $\epsilon = (\tilde{\epsilon}, E, I)$ those cylinders obeying

$$\text{diam } \tilde{\epsilon} < 100 \min(E, -[\frac{1}{4\beta J} \log \beta h]) \tag{2.3}$$

and keep the notation $\gamma = (\tilde{\gamma}, E, I)$ for the non-elementary ones. Note that (2.3) is natural in terms of equilibrium between spikes responsible of entropic repulsion, cf. (1.5), and the field h.

Then $\{\gamma_i; \epsilon_j]$ denotes a family of cylinders with compatibility conditions which are "weak" between two ϵ's or one ϵ and one γ, and strong between two γ's, and such that for any pair (γ_i, ϵ_j) not separated by a γ_k, one has $E_j = E_i$ if $\bar{\epsilon}_j \cap \bar{\gamma}_i \neq \emptyset$ and $E_j = I_i$ if $\bar{\epsilon}_j \subset \bar{\gamma}_i$ or $\bar{\gamma}_i \subset \bar{\epsilon}_j$. Note that an elementary cylinder can sometimes be larger than a non-elementary one.

The expansion over non-elementary cylinders with bare weights and elementary ones with renormalised weights now takes the form

$$\Xi(\Lambda^{(k)}) = e^{-\beta hk|\Lambda|} \sum_{\{\gamma_i;\epsilon_j]\in\Lambda^{(k)}} \prod_i \omega(\gamma_i) \prod_j \tilde{\omega}(\epsilon_j) \tag{2.4}$$

Non-elementary cylinders still have a non-local interaction ; naturally the next step is to group interacting non-elementary cylinders into clusters which Dinaburg and Mazel call contours. This is done as follows : given $\{\gamma_i; \epsilon_j]$, consider the interface configuration obtained from the family $\{\gamma_i\}$ of non-elementary cylinders. Each connected component of $\{x \in \mathbb{Z}^2; \Phi(x) \neq k\}$ defines the support of a contour Γ_l . Each contour can be written as $\Gamma = \{\gamma^{ext}, \gamma_i, \gamma^{int,j}\}$ with γ^{ext} the outer boundary of SuppΓ, $\gamma^{int,j}$ the inner boundaries of SuppΓ and γ_i all other non-elementary cylinders lying inside SuppΓ. The height function of an interface defined from Γ alone is constant on the connected components of $\Lambda \setminus (\cup\tilde{\gamma}_i)$, i.e.

$$\text{Supp}_e\Gamma = \bar{\gamma}^{ext} \setminus (\bigcup_{\gamma_i \neq \gamma^{ext}} \bar{\gamma}_i)$$

and each

$$\text{Supp}_i\Gamma = \bar{\gamma}_i \setminus (\bigcup_{\bar{\gamma}_j \subset \bar{\gamma}_i} \bar{\gamma}_j)$$

Elementary cylinders can be resummed independently in Supp$_e\Gamma$ and each Supp$_i\Gamma$ giving elementary partition functions

$$Z(V^{I,S}) = \sum_{[\epsilon_j]\in V^{I,S}} \prod \tilde{\omega}(\epsilon_j) \tag{2.5}$$

where V is one of Supp$_e\Gamma$ or Supp$_i\Gamma$ and I, S the corresponding heights and heredited signs. These elementary partition are shown to be fine, in the sense that taking the logarithm of (2.5) gives a convergent expansion, an extensive free energy. The proof is by induction over V.

Introducing (2.5) into (2.4) yields

$$\Xi(\Lambda^{(k)}) = e^{-\beta hk|\Lambda|} \sum_{\{\Gamma_l\}\in\Lambda^{(k)}} Z((\Lambda \setminus \cup \mathrm{Supp}\Gamma_l)^{(k)}) \prod_l \omega(\Gamma_l) Z((\mathrm{Supp}\Gamma_l)^{(k)})$$

with

$$\omega(\Gamma) = Z\big((\mathrm{Supp}\Gamma)^{E(\gamma^{ext}),S(\gamma^{ext})}\big)^{-1} \quad \omega(\gamma^{ext}) Z\big((\mathrm{Supp_e}\Gamma)^{I(\gamma^{ext}),S(\gamma^{ext})}\big). \\ \cdot \prod_j \omega(\gamma^{int,j}) \prod_i \omega(\gamma_i) Z\big((\mathrm{Supp_i}\Gamma)^{I(\gamma_i),S(\gamma_i)}\big)$$

The expansion for $\Xi(\Lambda^{(k)})$ can also be written

$$\Xi(\Lambda^{(k)}) = e^{-\beta hk|\Lambda|} \sum_{[\Gamma_l;\epsilon_m]\in\Lambda^{(k)}} \prod_l \omega(\Gamma_l) \prod_m \tilde{\omega}(\epsilon_m) \tag{2.6}$$

where the compatibility conditions are local by construction, with $E(\Gamma_l) = E(\epsilon_m) = k$ for all l and m. Dinaburg and Mazel then proceed to estimate the weights of monomers $\omega(\Gamma)$ and the sums over the resulting polymers, and prove the cluster expansion of (2.6) in the Kotecký-Preiss fashion [KP86].

3 Critical Wetting

The wetting transition is expected to be of second order in the case of short range forces and rough interfaces. The interface taken alone, without an attracting wall, should have large fluctuations as described by capillary waves. With an attracting wall, starting from a partially wet situation, the thickness of the partial wetting layer then grows continuously to infinity as the temperature is raised up to the transition temperature T_W, and one is interested in the corresponding critical exponents.

In the three-dimensional Ising model, this picture should hold when the wetting transition temperature T_W is higher than the roughening transition temperature T_R, i.e. when the boundary field is sufficiently small.

In order to find an effective potential for the interaction of the interface with the wall, one starts from a mean-field approximation parallel to the wall, so that the magnetisation, or the appropriate bulk order parameter, varies only perpendicular to the wall and is of the form

$$m(z) = m^* \tanh(\frac{z-\Phi}{2\xi})$$

where Φ is the distance from the interface to the wall, m^* is the magnetisation of the + phase and ξ is the bulk correlation length. Near the wetting transition, ξ remains finite whereas $\Phi \nearrow \infty$; therefore, assuming only short range forces, the contribution to the free energy due to the wall will depend upon Φ only through the tail of the *tanh* function. Expanding to second order in $\exp(-\Phi/\xi)$ gives an effective potential

$$-a\exp(-\frac{\Phi}{\xi})+b\exp(-\frac{2\Phi}{\xi}))$$

One has partial wetting when $a>0$ and complete wetting otherwise. Generically $a=a(T)\approx(T_W-T)$ near the wetting transition, while b is positive and remains of order one.

One then allows $\Phi=\Phi(x)$ to vary slowly with x, and expands to second order in $\nabla\Phi$ to obtain a Hamiltonian of the form

$$\frac{\sigma}{2}(\nabla\Phi)^2-a\exp(-\frac{\Phi}{\xi})+b\exp(-\frac{2\Phi}{\xi})$$

Rescaling the parameters and fixing b arbitrarily gives

$$\frac{1}{kT}H(\Phi)=\frac{1}{2}(\nabla\Phi)^2-\epsilon\exp(-\alpha\Phi)+\frac{1}{2}\exp(-2\alpha\Phi)\equiv\frac{1}{2}(\nabla\Phi)^2+V(\Phi) \quad (3.1)$$

with

$$\alpha=\left(\frac{kT}{\sigma\xi^2}\right)^{1/2}=\mathcal{O}(1) \quad \text{and} \quad \epsilon\approx(T_W-T) \quad \text{as} \quad T\nearrow T_W$$

The following questions for the one-point correlation function and two-point truncated correlation arise when $\epsilon\searrow 0$:

$$\langle\Phi(x)\rangle\sim ...?...\log\epsilon^{-1}$$

$$\langle\Phi(x);\Phi(y)\rangle\sim ...?...\exp(-m|x-y|) \quad \text{as} \quad |x-y|\nearrow+\infty$$

$$m\sim\epsilon^{-\nu} \qquad \nu=?$$

where $\langle.\rangle$ denotes the average in a Gibbs measure associated to (3.1). In mean-field theory, one expands around the minimum,

$$V(\Phi)\equiv-\epsilon\exp(-\alpha\Phi)+\frac{1}{2}\exp(-2\alpha\Phi)\simeq-\frac{\epsilon^2}{2}+\frac{\epsilon^2\alpha^2}{2}(\Phi-\alpha^{-1}\log\epsilon^{-1})^2+...$$

so that $\nu_{MF}=1$. We shall see that this value can hold only if α goes to zero together with ϵ, which we don't want to assume here.

A constructive formulation could start as follows : for a large square $\Lambda\subset\mathbb{Z}^2$, consider the measure

$$\exp\Big(-\sum_{x\in\Lambda}V(\Phi_0(x)+\zeta_0)+\frac{1}{2}m_0^2\sum_{x\in\Lambda}\Phi_0^2(x)\Big)d\mu_{m_0}(\Phi_0) \quad (3.2)$$

where $d\mu_{m_0}(\Phi_0)$ is the infinite volume Gaussian measure of inverse covariance $-\Delta+m_0^2$. Then one can see that the field $\Phi(x)=\Phi_0(x)+\zeta_0$ is governed by Hamiltonian (3.1) with boundary conditions given by $d\mu_{m_0}(\Phi_0)$. The translation ζ_0 will be chosen later as the location of the minimum of the renormalised potential, and will differ significantly from $\zeta_0^{MF}=\alpha^{-1}\log\epsilon^{-1}$.

The Gaussian measure $d\mu_{m_0}(\Phi_0)$ is then decomposed into momentum slices according to

$$\frac{1}{p^2+m_0^2} = \frac{1}{p^2+m_0^2}\{(1-e^{-M^2p^2}) + .. + (e^{-M^{2i}p^2} - e^{-M^{2(i+1)}p^2}) + .. + e^{-M^{2n}p^2}\}$$

$$\Phi_0 \quad = \quad \Phi^0 \quad + \quad \ldots \quad + \quad \Phi^i \quad + \quad \ldots \quad + \quad \Phi_n$$

where $M > 1$ is a fixed constant, e.g. 2, so that the field Φ^i has a support in momentum space localised around $M^{-i} > |p| > M^{-i-1}$ for $i = 0, 1, ..., n-1$ while Φ_n is supported around $|p| < M^{-n}$.

Following a renormalisation group scheme, one would like to integrate over the short distance fluctuating fields Φ^i for $i = 0, 1, ..., n-1$ to obtain an effective potential for Φ_n, whose length scale should be the true correlation length. One step in this direction is to write the exponential terms in the potential $V(\Phi)$ as their expectation value in the uncoupled measure, times Wick ordered exponentials :

$$e^{-\alpha\Phi^i} = \langle e^{-\alpha\Phi^i}\rangle_{C^i} : e^{-\alpha\Phi^i} :_{C^i} = e^{\frac{1}{2}\alpha^2 C^i} : e^{-\alpha\Phi^i} :_{C^i}$$

Then

$$\begin{aligned} V(\Phi) &= V_n(\Phi^0, ..., \Phi^{n-1}, \Phi_n) \\ &= -\epsilon e^{\frac{1}{2}\alpha^2(C^0+...+C^{n-1})} : e^{-\alpha(\Phi^0 + ... + \Phi^{n-1})} : e^{-\alpha(\Phi_n + \zeta_0)} \\ &\quad + \frac{1}{2} e^{2\alpha^2(C^0+...+C^{n-1})} : e^{-2\alpha(\Phi^0 + ... + \Phi^{n-1})} : e^{-2\alpha(\Phi_n + \zeta_0)} \end{aligned}$$

Let us now choose ζ_0 such that the effective potential for Φ_n is approximately centred, i.e. such that

$$\begin{aligned} V_n(\Phi^0, ..., \Phi^{n-1}, \Phi_n) = -\tfrac{1}{2}\epsilon_0^2 + \tfrac{1}{2}\epsilon_0^2\Big\{1 - 2 : e^{-\alpha(\Phi^0 + ... + \Phi^{n-1})} : e^{-\alpha\Phi_n} \\ + : e^{-2\alpha(\Phi^0 + ... + \Phi^{n-1})} : e^{-2\alpha\Phi_n}\Big\} \end{aligned}$$

This is achieved with

$$\begin{aligned} e^{-\alpha\zeta_0} &= \epsilon e^{-\frac{3}{2}\alpha^2(C^0+...+C^{n-1})} \\ \epsilon_0^2 &= \epsilon^2 e^{-\alpha^2(C^0+...+C^{n-1})} \end{aligned}$$

In order that the effective potential V_n give a correct prediction at the mean field level, the number of "renormalisation group steps" n is chosen so that the ultra-violet cutoff M^{-n} equal the mean field mass as given by the curvature of the effective potential :

$$M^{-n} = \epsilon_0\alpha = \epsilon\alpha e^{-\frac{1}{2}\alpha^2(C^0+...+C^{n-1})}$$

If all goes well, this will give the physical mass m as

$$m \simeq \left(\epsilon\alpha\right)^{\frac{1}{1-\frac{\alpha^2}{4\pi}}} \approx \left(T_W - T\right)^{\frac{1}{1-\frac{\alpha^2}{4\pi}}} \tag{3.3}$$

as predicted [BHL83, LKZ83]. These authors also wanted to take into account the restriction $\Phi > 0$, which led them to conjecture the validity of (3.3) only for $\alpha^2/4\pi < 1/2$ instead of a possible $\alpha^2/4\pi < 1$. Estimated values of $\alpha^2/4\pi$ appropriate to the Ising model have decreased from around 0.8 at the beginning of the story to around 0.25 more recently [GKL90]. A more thorough discussion can be found in [PEB91, B94].

A constructive proof would require a small parameter, and α in front of the field is the natural candidate; one could aim at proving (3.3) for $\alpha \ll 1$ fixed and $\epsilon \searrow 0$.

There is one aspect not addressed by heuristic renormalisation group analyses and which turns out to be essential when looking for a mathematical proof : the potential $V(\Phi)$ goes to a constant when $\Phi \nearrow +\infty$, and only boundary conditions seem to prevent the field from escaping to infinity. The stable phase lies near the minimum of the potential, but there is a metastable phase at large fields where the potential is just $\mathcal{O}(\epsilon^2)$ higher and there is more entropy available.

A scaled expansion over large versus small fields has been devised for the case of some analogous but symmetric potentials [DMRR92] to show that expectations of the fields remained bounded in the thermodynamic limit. This made extensive use also of correlation imequalities. Combined large versus small field expansions and cluster expansions were then given for various potentials [DMR92, DMR94, R93, L94], but apply only in cases where the mean-field exponent is exact. The latest of these papers [L94] is also closest to the model discussed above :

Theorem (Lemberger) : *Let $d\mu_C$ be the Gaussian measure of covariance*

$$\hat{C}(p) = \frac{e^{-p^2/m^2}}{p^2 + m^2}$$

Let $\Lambda \subset \mathbb{R}^2$ and

$$\bar{V}_\Lambda(\Phi) = \int_\Lambda dx \, \frac{1}{2}\epsilon^2 \left(1 - e^{-\alpha\Phi(x)}\right)^2 - \frac{1}{2}m^2 \int_\Lambda dx \, \Phi(x)^2$$

Let $\langle . \rangle_\Lambda$ denote expectation values in the probability measure proportional to

$$\exp(-\bar{V}_\Lambda(\Phi)) d\mu_C$$

Then for α small

$$\lim_{\Lambda \nearrow \mathbb{R}^2} \left| \langle \Phi(x)\Phi(y) \rangle_\Lambda - \langle \Phi(x) \rangle_\Lambda \langle \Phi(y) \rangle_\Lambda \right| < K e^{-m'|x-y|}$$

with $m' = \mathcal{O}(m)$.

Among the ingredients in the proof, let us only point out to the definition of the large versus small field expansion : the plane is paved by squares Δ of area $|\Delta| = m^{-2}$. A "largeness" function

$$g(\Phi) = \alpha^2 \Phi^6 + e^{-4\alpha\Phi} \tag{3.4}$$

allows to define large field squares from a smooth version of

$$\frac{1}{|\Delta|}\int_{\Delta} g(\Phi(x))\, dx > 1 \tag{3.5}$$

which implies

$$\int_{\Delta} V(\Phi(x))dx > C\alpha^{-2/3} \tag{3.6}$$

or

$$\int_{\Delta} (\nabla\Phi(x))^2 dx > C\alpha^{-2/3} \tag{3.7}$$

Any reader who is fond of Sobolev inequalities should go and see in [L94] how (3.4–3.7) work and fit into the cluster expansion... Large field squares yield a small factor in the measure (3.2) either from the interaction if (3.6) holds or from the massless Gaussian part of the measure if (3.7) holds.

4 Wulff Shape

The Wulff shape is the solution of the following classical variational problem : given a function $\sigma(\hat{n})$ from S^2 (unit vectors in $\mathbb{R}^3$) to $\mathbb{R}_+$ (surface tension), find a closed surface Σ^V minimising

$$F = \int_{\Sigma} \sigma(\hat{n})dS$$

under the constraint that the measure V of the enclosed volume is fixed. The problem originates in metallurgy and cristallography when asking for the shape of crystals which indeed are not spheres and show interesting features [RW84] . In thermodynamics or statistical physics, one considers two coexisting phases, e.g. liquid and solid metal or + and − phases of the Ising model below T_C and asks for the equilibrium shape of a drop of one phase surrounded by the other phase, under the constraint of fixed volume of the drop. One expects to recover the classical variational problem in the limit of large volumes, when the fluctuation amplitude of the interface will be much smaller than the size of the drop. Note that the asymptotics associated to the thermodynamic limit is studied, but that the problem makes sense only for finite volumes. The subject can be considered as a study of a metastable phase, which is reasonable also form a more practical point of view : the fixed volume constraint often means that the shape relaxes to equilibrium much faster than the volume measure changes.

A complete mathematical proof of the Wulff shape for the two-dimensional Ising model at low temperatures has been given by Dobrushin, Kotecky and Shlosman [DKS92]. In the corresponding Solid-On-Solid approximation, the interface is a kind of random walk enclosing a given area, and the proof of the Wulff shape was given in [DDR89].

As we restrict our attention to two-dimensional surfaces or interfaces in three-dimensional space, we shall present in this section only some conjectures, which may be appealing to constructivists...

The starting point is again a Solid-On-Solid model of a rough interface to which we add the volume constraint. For a large connected $\Lambda \subset \mathbb{Z}^2$, and a volume parameter $V \approx |\Lambda|^{3/2}$, we wish to describe a sessile drop of basis Λ and volume V by the following measure :

$$d\mu_\Lambda^V = (Z_\Lambda^V)^{-1} \exp\Big(-\sum_{x\in\Lambda} U\big(\nabla\Phi(x)\big)\Big)\delta\Big(\sum_{x\in\Lambda}\Phi(x) - V\Big) \prod_{y\in\partial\Lambda}\delta(\Phi(y)) \prod_{x\in\Lambda} d\Phi(x) \tag{4.1}$$

where Z_Λ^V normalises the probability, $\partial\Lambda$ is the set of $x \in \Lambda$ which are neighbours of points not in Λ, and the interaction $U(\nabla\Phi(x))$ can be chosen as

$$U\big(\nabla\Phi(x)\big) = \frac{1}{2}\sum_{|y-x|=1}\big(\Phi(y)-\Phi(x)\big)^2 + \frac{\lambda}{4}\Big(\sum_{|y-x|=1}\big(\Phi(y)-\Phi(x)\big)^2\Big)^2 \tag{4.2}$$

with λ small. A bounded perturbation of the Gaussian could also be considered instead of $(\nabla\Phi)^4$.

The basis Λ of the drop can also be random with a statistical weight $\exp(-a|\Lambda|)$ associated to a surface tension between the drop phase and the substrate. Because of translation invariance, we require the basis Λ to be centred, meaning that its center of gravity $c(\Lambda)$ lies within distance one from the origin, $|c(\Lambda)| < 1$. The grand canonical partition function then reads

$$\Xi^V = \sum_{\substack{\Lambda \text{ connected} \\ |c(\Lambda)|<1}} Z_\Lambda^V e^{-a|\Lambda|}$$

Denoting expectations in this grand canonical ensemble by $\langle . \rangle^V$, with the convention that $\Phi(x) = 0$ outside Λ, we can formulate the following conjectures for the asymptotics as $V \nearrow \infty$:

$$\langle\Phi(x)\rangle^V = \text{Wulff shape} + \mathcal{O}((\log V)^{1/2}) \tag{4.3}$$

$$\log \Xi^V = \int_{\Sigma^V} \sigma(\hat{n})dS + \mathcal{O}(V^{1/3}) \tag{4.4}$$

where Σ^V is the Wulff shape enclosing a volume V, also called Winterbottom shape in the case of sessile drops. Partial wetting should be ensured by taking a not too small. The surface tension $\sigma(\hat{n})$ is defined in the usual way, without any boundary wall or volume constraint, as we now recall :

$$\sigma(\hat{n}) = \lim_{\Lambda\nearrow\mathbb{Z}^2} -\frac{n_3}{|\Lambda|}\log\int \exp\Big(-\sum_{x\in\Lambda} U\big(\nabla\Phi(x)\big)\Big) \prod_{y\in\partial\Lambda}\delta(n_3\Phi(y) + \hat{n}\cdot y)\prod_{x\in\Lambda} d\Phi(x) \tag{4.5}$$

where n_3 is the vertical component of $\hat{n}$. This is the free energy per unit area of an interface bounded by a plane curve orthogonal to $\hat{n}$. For the three-dimensional Ising model or two-dimensional discrete Solid-On-Solid models, a proof of the existence of (4.5) can be found in [MMR92], who also prove for these models that the function $f(x) = |x|\sigma(x/|x|)$ is convex on $\mathbb{R}^3$.

It is convenient to change ensemble by introducing a slope chemical potential $\gamma(\hat{n})$ such that

$$\sigma(\hat{n}) \stackrel{?}{=} \lim_{\Lambda \nearrow \mathbb{Z}^2} -\frac{n_3}{|\Lambda|} \log \int \exp\Big(-\sum_{x\in\Lambda} U(\nabla\Phi(x)) + \sum_{x\in\Lambda} \gamma(\hat{n}) \cdot (\nabla\Phi(x) - \langle\nabla\Phi(x)\rangle)\Big) \, \delta(\Phi(0)) \prod_{x\in\Lambda} d\Phi(x) \qquad (4.6)$$

where $\langle.\rangle$ denotes expectations in the same measure of which the integral is evaluated. The existence of $\gamma(\hat{n})$ would first need to be proven for the present model.

The reason why (4.6) should be equal to (4.5) is that the boundary condition imposed in (4.5) is satisfied in average in (4.6), and conversely that the slope potential terms in (4.6) would sum up to zero with the fixed boundary conditions in (4.5). The difference is therefore due only to fluctuations at the boundary, which should not contribute to the extensive part of the free energy.

The change of ensemble from fixed plane boundary curve to constant slope potential may be extended to deal with the volume constraint, which would be replaced by a varying slope potential, as follows : suppose that measure (4.1) is multiplied by

$$\exp\Big(\sum_{x\in\Lambda} \nabla\Psi(x) \cdot \big(\nabla\Phi(x) - \langle\nabla\Phi(x)\rangle\big)\Big) \qquad (4.7)$$

where $\langle.\rangle$ again denotes expectations in the same measure in which the integral is evaluated. The slope potential is now the gradient of a function, which is further required to satisfy

$$\Delta\Psi(x) = \text{constant}$$

Integrating by parts over $x \in \Lambda$ and using the constraints in the measure (4.1) shows that (4.7) is then identically equal to one. Let us now make the assumption, which can be valid only for suitable Λ, that there exists $\Psi(x)$ and Φ_0 such that the constraints are satisfied in average in the measure

$$d\mu_\Lambda^\Psi = (Z_\Lambda^\Psi)^{-1} \exp\Big(-\sum_{x\in\Lambda} U\big(\nabla\Phi(x)\big) + \sum_{x\in\Lambda} \nabla\Psi(x) \cdot \big(\nabla\Phi(x) - \langle\nabla\Phi(x)\rangle\big)\Big). \, .\delta(\Phi(0) - \Phi_0) \prod_{x\in\Lambda} d\Phi(x) \qquad (4.8)$$

Let us denote $\langle.\rangle_\Lambda^V$ and $\langle.\rangle_\Lambda^\Psi$ respectively the expectations in (4.1) and (4.8). We expect

$$\Big|\langle\Phi(x)\rangle_\Lambda^V - \langle\Phi(x)\rangle_\Lambda^\Psi\Big| \stackrel{?}{=} \mathcal{O}((\log V)^{1/2})$$

The measure (4.8) corresponds to a massless perturbed Gaussian field with a varying slope potential, which one may hope to control with a multi-scale cluster expansion. One would obtain the extensivity of the free energy, which with the varying slope potential should yield a variant of (4.4), implying also a weak form of (4.3).

Let us now turn to the questions about the basis Λ, and the existence or not of $\Psi(x)$ with $\Delta\Psi(x) =$ constant and Φ_0 such that the constraints are satisfied in average in (4.8). One can clearly satisfy the average volume condition together with a vanishing of the average of $\Phi(y)$ averaged over the boundary $\partial\Lambda$. But we conjecture that if Λ is not a section of a Wulff shape, then typically

$$\left|\langle\Phi(y)\rangle_\Lambda^\Psi\right| \approx V^{1/3} \qquad \text{for} \qquad y \in \partial\Lambda$$

The problem is simpler when $\sigma(\hat{n})$ has cylindrical symmetry, as should be the case with (4.2), at least if $\mathbb{Z}^2$ is replaced by $\mathbb{R}^2$ with a rotation invariant short distance cutoff. Then one can aim at proving (4.3–4.4) with a fixed large disk for Λ.

If the basis Λ is random, then even the Gaussian case deserves interest. Indeed let

$$F_\Lambda^0 = \min_{\substack{\Phi \\ \int_\Lambda \Phi dx = V \\ \Phi|_{\partial\Lambda}=0}} \int_\Lambda (\nabla\Phi)^2 dx = V^2 \min_{\substack{\Phi \\ \Phi|_{\partial\Lambda}=0}} \frac{\int_\Lambda (\nabla\Phi)^2 dx}{\left(\int_\Lambda \Phi dx\right)^2} \equiv 4V^2 P_\Lambda^{-1} \tag{4.9}$$

P_Λ is called the torsional rigidity of Λ. The grand canonical partition function then looks like

$$\Xi_0^V = \sum_{\substack{\Lambda \text{ connected} \\ |c(\Lambda)|<1}} e^{-4V^2 P_\Lambda^{-1} - a|\Lambda|}$$

The available isoperimetric inequalities for the torsional rigidity may not not be enough to get the desired results, namely that there is a circle from which $\partial\Lambda$ remains within distance $\mathcal{O}((\log V)^{1/2})$ with probability going to one as $V \nearrow \infty$.

The basic isoperimetric inequality for the torsional rigidity is due to Pólya and Saint Venant and reads

$$P_\Lambda \leq \frac{|\Lambda|^2}{2\pi}$$

with equality only for a disk.

References

[PEB91] A.O. Parry, R. Evans and K. Binder : *Critical Amplitude Ratios for Critical Wetting in Three Dimensions : Observation of Non-Classical Behaviour in the Ising Model*, Phys. Rev. **B43**, 11535 (1991).

[B94] K. Binder : *Monte-Carlo Simulations of Wetting Transitions*, in Proceedings of the East-West Surface Science Workshop "Thin Films and Phase Transitions at Surfaces", Febr. 1994, Pamporowa, Bulgaria.

[B75] H. van Beijeren : *Interface Sharpness in the Ising System*, Commun. Math. Phys **40**, 1 (1975).

[BEF86] J. Bricmont, A. El Mellouki and J. Fröhlich : *Random Surfaces in Statistical Mechanics : Roughening, Rounding, Wetting,...* J. Stat. Phys. **42**, 743–798 (1986).

[BFL82] J. Bricmont, J.R. Fontaine and J.L. Lebowitz : *Surface Tension, Percolation and Roughening*, J. Stat. Phys. **29**, 193 (1982).

[BHL83] E. Brézin, B. I. Halperin and S. Leibler : *Critical Wetting in Three Dimensions*, Phys. Rev. Lett. **50** 1387 (1983).

[BKL83] J. Bricmont, K. Kuroda and J.L. Lebowitz : *Surface Tension and Phase coexistence for General Lattice Systems*, J. Stat. Phys. **33**, 59–75 (1983).

[BLP79] J. Bricmont, J.L. Lebowitz and C.E. Pfister : *Non-Translation Invariant Gibbs States with Coexisting Phases III : Analyticity properties*, Commun. Math. Phys **69**, 267 (1979).

[BLPO79] J. Bricmont,, J.L. Lebowitz, C.E. Pfister and E. Olivieri : *Existence of Sharp Interface for Widom-Rowlinson Type Lattice Models in Three Dimensions*, Commun. Math. Phys **66**, 1–20 (1979).

[D72] R. L. Dobrushin : *Gibbs State describing Coexistence of Phases for a Three-dimensional Ising Model*, Theory Prob. Appl. **17**, 582–600 (1972).

[DDR89] J. De Coninck, F. Dunlop, V. Rivasseau : *On the Microscopic Validity of the Wulff Construction and of the Generalized Young Equation*, Commun. Math. Phys. **121**, 401-419 (1989).

[DKS92] R. L. Dobrushin, R. Kotecký and S. B. Shlosman: *Wulff Construction : a Global Shape from Local Interactions*, Am. Math. Soc., Providence (1992).

[DM94] E. I. Dinaburg and A. E. Mazel : *Layering Transition in SOS Model with External Magnetic Field*, J. Stat. Phys. **74** 533–563 (1994).

[DMRR92] F. Dunlop, J. Magnen, V. Rivasseau, Ph. Roche : *Pinning of an Interface by a Weak Potential*, J. Stat. Phys. **66**, 71-98 (1992).

[DMR92] F. Dunlop, J. Magnen and V. Rivasseau : *Mass Generation for an Interface in the Mean Field Regime*, Ann. Inst. Henri Poincaré **57**, 333 (1992).

[DMR94] F. Dunlop, J. Magnen, V. Rivasseau : *Mass Generation for an Interface in the Mean Field Regime : Addendum*, Ann. Inst. Henri Poincaré, **61** (1994).

[FP87] J. Fröhlich and C.-E. Pfister : *The Wetting and Layering Transitions in the Half-Infinite Ising Model*, Europhys. Lett. **3**, 845–852 (1987) ; *Semi-Infinite Ising Model II. The Wetting and Layering Transitions*, Commun. Math. Phys **112**, 51–74 (1987).

[FS81] J. Fröhlich and T. Spencer : *The Kosterlitz-Thouless Transition in Two-Dimensional Abelian Spin Systems and the Coulomb Gas*, Commun. Math. Phys. **81**, 527–602 (1981).

[GKL90] G. Gomper, D.M. Kroll and R. Lipowski, Phys. Rev. **B42**, 961 (1990).

[HKZ88] P. Holický, R. Kotecký and M. Zahradnik : *Rigid Interfaces for Lattice Models at Low Temperatures*, J. Stat. Phys. **50**, 755–812 (1988).

[KP86] R. Kotecký and D. Preiss : *Cluster Expansion for Abstract Polymer Model*, Commun. Math. Phys **103**, 491–498 (1986).

[L94] P. Lemberger : *Large Field Versus Small Field Expansions and Sobolev Inequalities*, preprint (1994).

[LKZ83] R. Lipowsky, D. M. Kroll and R. K. P. Zia : *Effective Field Theory for Interface Delocalisation Transitions*, Phys. Rev. **B27** 4499 (1983).

[LMR86] L. Laanait, A. Messager and J. Ruiz : *Phases Coexistence and Surface Tensions for the Potts Model*, Commun. Math. Phys. **105** 572 (1986).

[M94] S. Miracle-Sole : *Surface Tension, Step Free Energy and Facets in the Equilibrium Crystal*, preprint.

[MMR92] A. Messager, S. Miracle-Sole and J. Ruiz : *Convexity Properties of the Surface Tension and Equilibrium crystals*, J. Stat. Phys. **67**, 449–470 (1992).

[MMRS91] A. Messager, S. Miracle-Sole, J. Ruiz and S. Shlosman : *Interfaces in the Potts Model. (II) Antonov's rule and rigidity of the order-disorder Interface*, Commun. Math. Phys **140**, 275 (1991).

[PP88] C.-E. Pfister and O. Penrose : *Analyticity Properties of the Surface Free Energy of the Ising model*, Commun. Math. Phys **115**, 691–699 (1988).

[R93] V. Rivasseau : *Cluster Expansions with Small/Large Field Conditions*, Vancouver conference (1993)

[RW84] C. Rottman and M. Wortis : *Statistical Mechanics of Equilibrium Crystal Shapes : Interfacial Phase Diagrams and Phase Transitions*, Phys. Rep. **103**, 59–79 (1984).

[W83] C.E. Wayne : *Surface Models with Non-Local Potentials : Upper Bounds*, Commun. Math. Phys **90**, 293–315 (1983).

Fermi Liquids in Two–Space Dimensions

Joel Feldman[1*],
Detlef Lehmann[2],
Horst Knörrer[2],
Eugene Trubowitz[2]

1 Department of Mathematics
University of British Columbia
Vancouver, B.C. CANADA V6T 1Z2
2 Mathematik
ETH-Zentrum
CH-8092 Zürich
SWITZERLAND

1 Introduction

In this review, we consider a many-body system which is somewhat unusual in that the Fermi surface survives the turning on of all sufficiently weak short range interactions. The system lives in $d = 2$ space dimensions and consists of a gas of fermions with prescribed, strictly positive, density, together with a crystal lattice of <u>magnetic</u> ions. The fermions interact with each other through a two-body potential. The lattice provides periodic scalar and vector background potentials. As well, the ions oscillate, generating phonons and then the fermions interact with the phonons.

To start, turn off the fermion-fermion and fermion-phonon interactions. Then we have a gas of independent fermions, each with Hamiltonian

$$H_0 = \tfrac{1}{2m}\left(i\nabla + \mathbf{a}(\mathbf{x})\right)^2 + U(\mathbf{x})$$

The vector and scalar potentials $\mathbf{a}$, U are periodic with respect to some lattice Γ in $\mathbb{R}^2$. We use the convention that bold face characters are two component vectors. Because the Hamiltonian commutes with lattice translations it is possible to simultaneously diagonalize the Hamiltonian and the generators of lattice translations. Call the eigenvalues and eigenvectors $\varepsilon_\nu(\mathbf{k})$ and $\phi_{\nu,\mathbf{k}}(\mathbf{x})$ respectively. They obey

$$\begin{aligned} H_0\phi_{\nu,\mathbf{k}}(\mathbf{x}) &= \varepsilon_\nu(\mathbf{k})\phi_{\nu,\mathbf{k}}(\mathbf{x}) \\ \phi_{\nu,\mathbf{k}}(\mathbf{x}+\gamma) &= e^{i<\mathbf{k},\gamma>}\phi_{\nu,\mathbf{k}}(\mathbf{x}) \qquad \text{for all } \gamma \in \Gamma \end{aligned} \tag{I.1}$$

The crystal momentum $\mathbf{k}$ runs over $\mathbb{R}^2/\Gamma^\#$ where

$$\Gamma^\# = \{\, b \in \mathbb{R}^2 \mid\ < b,\gamma > \in 2\pi\mathbb{Z} \text{ for all } \gamma \in \Gamma \,\}$$

---

* Research supported in part by the Natural Sciences and Engineering Research Council of Canada

is the dual lattice to Γ. The band index $\nu \in \mathbb{N}$ just labels the eigenvalues for boundary condition $\mathbf{k}$ in increasing order.

In the grand canonical ensemble, the Hamiltonian H is replaced by $H - \mu N$ where N is the number operator and the chemical potential μ is used to control the density of the gas. At very low temperature, which is the physically interesting domain, only those pairs $\nu, \mathbf{k}$ for which $\varepsilon_\nu(\mathbf{k}) \approx \mu$ are important. To keep things as simple as possible, we assume that $\varepsilon_\nu(\mathbf{k}) \approx \mu$ only for one value ν_0 of ν and we put on a fixed ultraviolet cutoff so that we consider only those crystal momenta for which $|\varepsilon_{\nu_0}(\mathbf{k}) - \mu|$ is smaller than some fixed small constant.

Precisely, we denote $e(\mathbf{k}) = \varepsilon_{\nu_0}(\mathbf{k}) - \mu$ and make the following assumptions.

Hypothesis I: *The dispersion relation* $e(\mathbf{k})$ *is a real-valued, real analytic function on a compact subset* B *of* $\mathbb{R}^d$. *For all points* $\mathbf{p} \in \mathrm{B}$,

$$\nabla e(\mathbf{p}) \neq 0$$

Hypothesis II: *The Fermi curve*

$$\mathrm{F} = \{ \mathbf{p} \in \mathrm{B} \mid e(\mathbf{p}) = 0 \}$$

for e is a simple closed curve, whose curvature is bounded away from zero.

Hypothesis III: *For all* $\mathbf{q} \in \mathbb{R}^d$,

$$-\mathrm{F} + \mathbf{q} \neq \mathrm{F}$$

By definition,

$$-\mathrm{F} + \mathbf{q} = \{ \mathbf{p} \in \mathbb{R}^2 \mid -\mathbf{p} + \mathbf{q} \in \mathrm{B} \text{ and } e(-\mathbf{p} + \mathbf{q}) = 0 \}$$

It is Hypothesis III that makes this class of models somewhat unusual and permits the system to remain a Fermi liquid when the interaction is turned on. If $\mathbf{a} = 0$ then, taking the complex conjugate of (I.1), we see that $\varepsilon_\nu(-\mathbf{k}) = \varepsilon_\nu(\mathbf{k})$ so that Hypothesis III is violated for $\mathbf{q} = 0$. Hence the presence of a nonzero vector potential is essential.

In order to have simple sounding hypotheses, we have made them much stronger than necessary. One model that violates these hypotheses, not only for technical reasons but because it exhibits different physics, is the Hubbard model at half filling. Its Fermi surface looks like

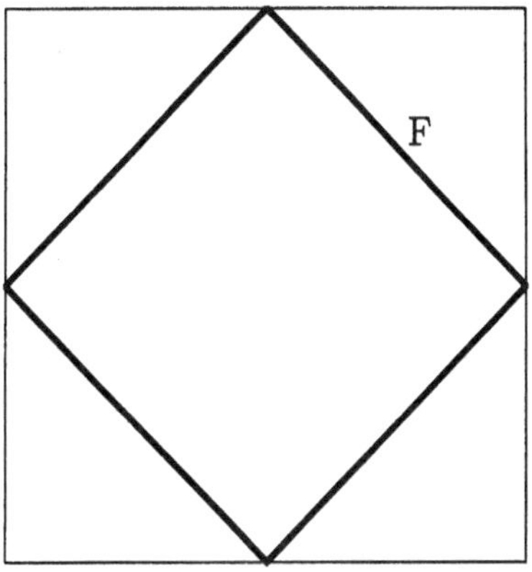

This Fermi curve is not smooth, violating Hypothesis I, has zero curvature almost everywhere, violating Hypothesis II and is reflection invariant so that F = −F, violating Hypothesis III with $\mathbf{q} = 0$.

The interacting models are formally characterized by the Euclidean Green's functions

$$\left\langle \prod_{i=1}^{n} \psi_{p_i}\bar{\psi}_{q_i} \right\rangle = \frac{\int \left(\prod_{i=1}^{n} \psi_{p_i}\bar{\psi}_{q_i}\right) e^{\mathcal{A}(\psi,\bar{\psi})} \prod_{k,\sigma} d\psi_{k,\sigma}\, d\bar{\psi}_{k,\sigma}}{\int e^{\mathcal{A}(\psi,\bar{\psi})} \prod_{k,\sigma} d\psi_{k,\sigma}\, d\bar{\psi}_{k,\sigma}} \tag{I.2a}$$

The action

$$\mathcal{A}(\psi,\bar{\psi}) = -\int đk\, (ik_0 - e(\mathbf{k}))\bar{\psi}_k\psi_k - \int đk\, \varepsilon(\lambda,\mathbf{k})\bar{\psi}_k\psi_k - \mathcal{V}(\psi,\bar{\psi}) \tag{I.2b}$$

We now take some time to explain this formula. The fermion fields are vectors

$$\psi_k = \begin{pmatrix} \psi_{k,\uparrow} \\ \psi_{k,\downarrow} \end{pmatrix} \qquad \bar{\psi}_k = (\,\bar{\psi}_{k,\uparrow} \quad \bar{\psi}_{k,\downarrow}\,)$$

whose components $\psi_{k,\sigma}, \bar{\psi}_{k,\sigma}$, $k = (k_0,\mathbf{k}) \in \mathcal{B} = (-1,1) \times \mathrm{B}$, $\sigma \in \{\uparrow,\downarrow\}$, are generators of an infinite dimensional Grassmann algebra over $\mathbb{C}$. That is, the fields anticommute with each other.

$$\overset{(-)}{\psi}_{k,\sigma}\overset{(-)}{\psi}_{p,\tau} = -\overset{(-)}{\psi}_{p,\tau}\overset{(-)}{\psi}_{k,\sigma}$$

We have deliberately chosen $\bar{\psi}$ to be a row vector and ψ to be a column vector so that

$$\bar{\psi}_k\psi_p = \bar{\psi}_{k,\uparrow}\psi_{p,\uparrow} + \bar{\psi}_{k,\downarrow}\psi_{p,\downarrow} \qquad \psi_k\bar{\psi}_p = \begin{pmatrix} \psi_{k,\uparrow}\bar{\psi}_{p,\uparrow} & \psi_{k,\uparrow}\bar{\psi}_{p,\downarrow} \\ \psi_{k,\downarrow}\bar{\psi}_{p,\uparrow} & \psi_{k,\downarrow}\bar{\psi}_{p,\downarrow} \end{pmatrix}$$

In the argument $k = (k_0,\mathbf{k})$, the last d components $\mathbf{k}$ are to be thought of as a crystal momentum and the first component k_0 as the dual variable to an imaginary time. Hence the $\sqrt{-1}$ in $ik_0 - e(\mathbf{k})$. For convenience only, we have put an ultraviolet cutoff on k_0 as well as on $\mathbf{k}$. In the full model k_0 runs over $\mathbb{R}$ and

$\mathbf{k}$ is replaced by $(\nu, \mathbf{k})$ with ν summed over $\mathbb{N}$ and $\mathbf{k}$ integrated over $\mathbb{R}^d/\Gamma^\#$. The relationship between the position space field $\psi(\xi)$, with $\xi = (t, \mathbf{x})$ running over (imaginary) time×space, and the momentum space field ψ_k is given, in our single band approximation, by

$$\psi_k = \int d\xi\ e^{-ik_0t}\phi_{\nu_0,\mathbf{k}}(\mathbf{x})\psi(\xi)$$

$$\psi(\xi) = \int đk\ e^{ik_0t}\overline{\phi_{\nu_0,\mathbf{k}}(\mathbf{x})}\psi_k \tag{I.3}$$

where

$$đk = \frac{dk_0}{2\pi}\ đ\mathbf{k} = \frac{d^{d+1}k}{(2\pi)^{d+1}}$$

The general spin independent form of the interaction is

$$\mathcal{V}(\psi, \bar\psi) = \frac{\lambda}{2}\int \prod_{i=1}^{4} đk_i\ (2\pi)^{d+1}\delta(k_1+k_2-k_3-k_4)\ \bar\psi_{k_1}\psi_{k_3}\ \langle k_1, k_2|\mathrm{V}|k_3, k_4\rangle\ \bar\psi_{k_2}\psi_{k_4} \tag{I.4}$$

Spin independence is imposed purely for notational convenience. It plays no role. The delta function δ is that for $\mathbb{R}^d/\Gamma^\#$ and imposes the appropriate conservation of crystal momentum for the present setting. The function $\langle k_1, k_2|\mathrm{V}|k_3, k_4\rangle$ implements the fermion-fermion and fermion-phonon interaction. Its precise value does not concern us. We just assume

Hypothesis IV *The interaction is short range. That is* $\langle k_1, k_2|\mathrm{V}|k_3, k_4\rangle \in C^\infty$.

The net coefficient $e(\mathbf{k}) - \mathcal{E}(\lambda, \mathbf{k})$ of $\bar\psi_k\psi_k$ in $\mathcal{A}$ has been deliberately split into two parts, with $\mathcal{E}(\lambda, \mathbf{k})$ chosen to satisfy an explicit renormalization condition. This is called renormalization of the dispersion relation. It is done to ensure that $\langle \prod_{i=1}^n \psi_{p_i}\bar\psi_{q_i}\rangle$ is C^∞ in λ at $\lambda = 0$. Define the proper self energy $\Sigma(p)$ for the action $\mathcal{A}$ by the equation

$$\Big(ip_0 - e(\mathbf{p}) - \Sigma(p)\Big)^{-1} (2\pi)^{d+1}\delta(p-q) = \frac{\int \psi_p\bar\psi_q\ \ e^{\mathcal{A}(\psi)} \prod d\psi_{k,\sigma}d\bar\psi_{k,\sigma}}{\int\ e^{\mathcal{A}(\psi)} \prod d\psi_{k,\sigma}d\bar\psi_{k,\sigma}}$$

The counterterm $\mathcal{E}(\lambda, \mathbf{k})$ is chosen so that

$$\Sigma(0, \mathbf{p})\big|_{\mathbf{p}\in\mathbf{F}} = 0$$

To give a rigorous definition of (I.2) one must introduce cutoffs and then take the limit in which the cutoffs are removed. To impose an infrared cutoff in the spatial directions one may put the system in a finite periodic box $\mathbb{R}^d/L\Gamma$. To impose an infrared cutoff in the zero direction one may make the inverse temperature $\beta < \infty$. Then momenta $k = (k_0, \mathbf{k})$ are restricted to lie on the lattice

$$k_0 \in \tfrac{\pi}{\beta}(2\mathbb{Z}+1)$$

$$\mathbf{k} \in \tfrac{1}{L}\Gamma^\#$$

The ultraviolet cutoffs further restrict $|k_0| \leq 1$, $|e(\mathbf{k})| \leq 1$. Then the Grassmann algebra becomes finite dimensional and (I.2b) with the integral symbol reinterpreted as

$$\int dk\ f(k) = \frac{1}{\beta} \sum_{\substack{k_0 \in \frac{\pi}{\beta}(2\mathbb{Z}+1) \\ |k_0| \leq 1}} \frac{1}{L^2} \sum_{\substack{\mathbf{k} \in \frac{1}{L}\Gamma^\# \\ |e(\mathbf{k})| \leq 1}} f(k)$$

is a well-defined element of that algebra.

Theorem. *Let $d = 2$ and Hypotheses I-IV be satisfied. There is an $r > 0$ and a dispersion relation counterterm $\mathcal{E}(\lambda, \mathbf{k})$, such that the limits*

$$\lim_{\beta, L \to \infty} \frac{\int \prod_{i=1}^{n} \psi_{p_i} \bar{\psi}_{q_i}\ e^{\mathcal{A}(\psi,\bar{\psi})} \prod d\psi_{k,\sigma} d\bar{\psi}_{k,\sigma}}{\int e^{\mathcal{A}(\psi,\bar{\psi})} \prod d\psi_{k,\sigma} d\bar{\psi}_{k,\sigma}} \tag{I.5}$$

exist in the sense of distributions and are independent of the order in which the limits are taken. The counterterm and the limit are both analytic functions of the coupling constant λ for $|\lambda| < r$. Furthermore, there is a jump in the average occupation number $n_{\mathbf{k}}$ at the Fermi curve. Precisely, if

$$n_{\mathbf{k}} = \lim_{x_0 \searrow 0} \int dk_0\, e^{ik_0 x_0} \Big(ik_0 - e(\mathbf{k}) - \Sigma(k_0, \mathbf{k}) \Big)^{-1}$$

then

$$\lim_{\varepsilon \searrow 0} n_{\mathbf{p} - \varepsilon \nu_{\mathbf{p}}} - n_{\mathbf{p} + \varepsilon \nu_{\mathbf{p}}} = \Big(1 + i \tfrac{\partial}{\partial k_0} \Sigma(0, \mathbf{p}) \Big)^{-1}$$

$$\geq 1 - O(\lambda)$$

for all $\mathbf{p}$ on the Fermi curve $\mathbf{F}$. Here, $\nu_{\mathbf{p}}$ is the outward pointing unit normal to $\mathbf{F}$ at $\mathbf{p}$. In other words, the infinite volume system is a Fermi liquid.

Our main goal here is to explain why this Theorem is true, though the complete proof [FKLT1] is too long to include. There are two main aspects to that proof: the control of four legged Feynman diagrams and the control of high orders of perturbation theory. The first aspect is discussed in section **3** while the second is discussed in section **2**.

2 Analyticity of Greens Functions

In this section we give an outline of the main ideas which are necessary for controlling large orders of perturbation theory. Roughly speaking, what we want to show is the following:

Theorem. *"The sum of all graphs that contribute to (I.5) and do not contain nontrivial four legged subgraphs is analytic."*

This theorem is true no matter whether you have $e(\mathbf{k}) = \frac{\mathbf{k}^2}{2m} - \mu$ where the Fermi curve is a circle or a dispersion relation $e(\mathbf{k})$ obeying the hypothesises of the first section, where the Fermi curve is not perfectly round. In particular, Theorem II.1 below is also the starting point for a rigorous construction of the theory of BCS-superconductivity. It means that the physical behaviour of the model is completely determined by the four legged subgraphs. In the case of a dispersion relation of section 1, four legged graphs are summable making the whole Greens functions analytic, whereas $e(\mathbf{k}) = \frac{\mathbf{k}^2}{2m} - \mu$ produces logarithmic divergences which drive the renormalization group flow to a superconducting fixed point. To prove the above theorem, one has to do two things.

First, one has to control the magnitude of each graph. When four legged subgraphs are removed, this can be done by power counting and renormalization of two legged subgraphs. Second, one has to control the number of graphs, since, if one expands the exponential in (I.5), there is a $\frac{1}{n!}$ from that expansion but one gets $(2n)! \approx \text{const}^{\,n} n!^2$ Feynman graphs after the evaluation of the functional integral with $4n$ fields. Thus, one is left with $\sum_{n=0}^{\infty} \text{const}^{\,n} n!\, \lambda^n$ which has radius of convergence zero. That is, we are not allowed to expand completely to Feynman graphs. Instead of this, one has to use the antisymmetry of the fermionic integral to exploit some cancellations of the $(2n)!$ graphs.

The strategy will be to decompose the propagator into scales and integrate out one scale at a time, removing four legged kernels by hand after each integration. Then, after renormalization of the two legged kernels, the result has summable power counting. To control the number of graphs, one Taylor expands the fields at each scale to sufficiently large order and uses the fact that the functional integral vanishes if two fields are equal.

Let us first give a precise mathematical formulation of what we have proven. Since we want to give an inductive proof which uses only a single scale expansion, instead of considering (I.5) we start with the generating functional for the connected amputated Greens functions given by

$$\mathcal{G}(\phi, \bar{\phi}) = \log \frac{1}{Z} \int e^{-\mathcal{V}(\psi+\phi, \bar{\psi}+\bar{\phi})} d\mu_C(\psi, \bar{\psi}) \,,$$

This is convenient for an inductive proof. Here $d\mu_C$ is the Grassmann Gaussian measure with covariance

$$C(x, x') = C(x - x') = \int \frac{d^3k}{(2\pi)^3}\, e^{i\mathbf{k}(\mathbf{x}-\mathbf{x}') - ik_0(\tau-\tau')} \frac{1}{ik_0 - e(\mathbf{k})} \,,$$

where $e(\mathbf{k})$ is either the dispersion relation of section 1 or, in this section, $\frac{\mathbf{k}^2}{2m} - \mu$. The coordinate and momentum space variables are

$$x = (x_0, \mathbf{x}) = (x_0, x_1, x_2) \in \mathbb{R}^3, \quad k = (k_0, \mathbf{k}) = (k_0, k_1, k_2) \in \mathbb{R}^3 \,.$$

For simplicity, we do everything in infinite volume and at zero temperature. A more careful treatment, which starts at positive temperature and in finite volume will be given in [FKLT1].

First let us introduce scales $j = 0, -1, -2, \cdots$ which select shells of thickness M^j around the Fermi curve $k_0 = 0,\ e(\mathbf{k}) = 0$ which is the singular locus of $C(k)$.

$$C(k) = \frac{1}{ik_0 - e(\mathbf{k})} = \sum_{j=-\infty}^{0} \frac{f_j(k)}{ik_0 - e(\mathbf{k})} + \frac{h(k)}{ik_0 - e(\mathbf{k})} = \sum_{j=-\infty}^{0} C^j(k) + C^{>0}(k),$$

where the f_j are smooth functions with support on $\{M^j \leq |ik_0 - e(\mathbf{k})| \leq M^2 M^j\}$ and the ultraviolet part h has support on $\{1 \leq |ik_0 - e(\mathbf{k})|\}$. Here, $M \geq 2$ is a constant which eventually has to be chosen sufficiently large.

We consider only the infrared part of the model and introduce an infrared cuttoff at scale $r > -\infty$. Furthermore, to renormalize two legged kernels, we introduce a counterterm

$$\delta e_r(\mathbf{k}, \lambda) = \sum_{j=r}^{-1} \delta e_r^j(\mathbf{k}, \lambda) = \sum_{j=r}^{-1} \sum_{l=1}^{\infty} \delta e_{r,l}^j(\mathbf{k})\, \lambda^l\,,$$

$$\delta \mathcal{V}_r(\psi) = \int \frac{d^3k}{(2\pi)^3} \delta e_r(\mathbf{k}, \lambda)\, \bar{\psi}(k) \psi(k)$$

which is an analytic function of λ and will be determined below. If the Fermi curve is a circle, δe_r is independent of $\mathbf{k}$, but in general, it will depend on $\mathbf{k}$. So we consider

$$\mathcal{G}^r(\phi) = \log \frac{1}{Z_{r+1}} \int e^{(-\mathcal{V}+\delta\mathcal{V}_r)(\sum_{j=r+1}^{0} \psi^j + \phi)} \prod_{j=r+1}^{0} d\mu_{C^j}(\psi^j)\,.$$

Then $\mathcal{G}^r$ can be computed inductively $(r + 1 \leq j \leq 0)$

$$\mathcal{G}^0 = -\mathcal{V} + \delta\mathcal{V}_r\,, \qquad \mathcal{G}^{j-1}(\psi^{\leq j-1} + \phi) = \log \frac{1}{\tilde{Z}_j} \int e^{\mathcal{G}^j(\psi^{\leq j} + \phi)} d\mu_{C^j}(\psi^j)$$

where

$$\tilde{Z}_j = \int e^{\mathcal{G}^j(\psi^j)} d\mu_{C^j}(\psi^j)$$

The $2q$-point functions G_q^{j-1} at scale $j - 1$ are defined by

$$\mathcal{G}^{j-1}(\phi) = \sum_{q=1}^{\infty} \mathcal{G}_q^{j-1}(\phi)$$

$$= \sum_{q=1}^{\infty} \int d\xi_1 \cdots d\xi_{2q}\, G_q^{j-1}(\xi_1, \cdots, \xi_{2q})\, \phi(\xi_1) \cdots \phi(\xi_{2q})$$

where $G_q^{j-1}(\xi_1, \cdots, \xi_{2q})$ is some antisymmetric kernel and $\xi = (x, \sigma, b)$ belongs to $\mathbb{R}^3 \times \{\uparrow, \downarrow\} \times \{0, 1\}$,

$$\phi(\xi) = \begin{cases} \phi(x,\sigma) & \text{if } b = 0 \\ \bar{\phi}(x,\sigma) & \text{if } b = 1 \end{cases} \quad \text{and} \quad \int d\xi = \sum_{b\in\{0,1\}} \sum_{\sigma\in\{\uparrow,\downarrow\}} \int d^3x .$$

Define an operator Q_4 which projects out nontrivial four legged subgraphs, that is four legged subgraphs of order at least $O(\lambda^2)$, by

$$Q_4 \sum_{q=1}^{\infty} \mathcal{G}_q^{j-1}(\phi) = \sum_{\substack{q=1 \\ q\neq 2}}^{\infty} \mathcal{G}_q^{j-1}(\phi) + \lambda \tfrac{d}{d\lambda} \mathcal{G}_2^{j-1}(\phi)\big|_{\lambda=0}$$

Theorem II.1 *Define the effective potential without four legged subgraphs* $\mathcal{W}^j$ *inductively by* $\mathcal{W}^0 = -\mathcal{V} + \delta\mathcal{V}_r$ *and for* $r + 1 \leq j \leq 0$

$$\mathcal{W}^{j-1}(\psi^{\leq j-1} + \phi) = Q_4 \log \frac{1}{Y_j} \int e^{\mathcal{W}^j(\psi^{\leq j} + \phi)} d\mu_{C^j}(\psi^j) ,$$

$$Y_j = \int e^{\mathcal{W}^j(\psi^j)} d\mu_{C^j}(\psi^j) .$$

Then there is an $\varepsilon > 0$ *which is independent of the infrared cuttoff* r *and a function* $\delta e_r(\mathbf{k}, \lambda)$ *which is analytic in* λ *for* $|\lambda| < \varepsilon$ *such that* $\mathcal{W}^r$ *is analytic for* $|\lambda| < \varepsilon$. *Furthermore, for all test functions* $f_1, \cdots, f_{2q}$,

$$\int \prod_{i=1}^{2q} d\xi_i \, |f_i(\xi_i)| \, |W_q^r(\xi_1, \cdots, \xi_{2q})| \leq |\lambda|^{\frac{q}{2}} \prod_{i=1}^{2q} \left(\|f_i\|_{L^1} + \|f_i\|_{L^\infty} \right) .$$

Remark. *$\mathcal{W}^r$ is not really the sum of all graphs (with propagators $C^{\geq r}$) without four legged subgraphs. Rather, $\mathcal{W}^r$ may be expressed as a sum of labelled graphs, with each line of each graph labelled by a fixed scale. Then only those four legged subgraphs for which the maximal scale of the external legs is strictly less than the minimal scale of the internal lines are forbidden.*

In the rest of this section, we sketch the proof of Theorem II.1, stating without proof the main Lemmata. Details will be given in [FKLT1]. Also see [FMRT1].

Let us first take a look at the power counting of the graphs. In coordinate space, we use the following norms. Fix test functions $f_k \in L^1 \cap L^\infty$, $1 \leq k \leq 2q$. Let $G_q = G_q(\xi_1, \cdots, \xi_{2q})$ be some kernel. Then

$$|\!|\!| G_q |\!|\!|_\emptyset = \sup_i \sup_{\xi_i} \int \prod_{\substack{k=1 \\ k\neq i}}^{2q} d\xi_k \, |G_q(\xi_1, \cdots, \xi_{2q})|$$

and for $S \subset \{1,\cdots,2q\}$, $S \neq \emptyset$

$$\|G_q\|_S = \int \prod_{i=1}^{2q} d\xi_i \prod_{k\in S} |f_k(\xi_k)|\, |G_q(\xi_1,\cdots,\xi_{2q})| \,.$$

Lemma II.2 (Power Counting) *Let G_q be a connected amputated graph with $2q$ external legs built up from generalized, $2q_v$ legged vertices or subgraphs I_{q_v} with $\|I_{q_v}\|_\emptyset < \infty$. Suppose each line of the graph has a covariance C^j with*

$$\|C^j\|_{L^\infty} \leq c\, M^j \,, \qquad \|C^j\|_{L^1} \leq c\, M^{-j} \,.$$

Then there are the following bounds

$$\|G_q\|_\emptyset \leq c^{\sum_v q_v - q} \prod_{v\in V} \left(\|I_{q_v}\|_\emptyset M^{(q_v-2)j}\right) M^{-(q-2)j} \,,$$

$$\|G_q\|_S \leq c^{\sum_v q_v - q} \prod_{v\in V_{int}} \left(\|I_{q_v}\|_\emptyset M^{(q_v-2)j}\right) \times$$

$$\prod_{v\in V_{ext}} \left(\|I_{q_v}\|_{S_v} M^{\frac{1}{2}(2q_v-|S_v|)j}\right) M^{-\frac{1}{2}(2q-|S|)j} \,.$$

Here a vertex is called external, if at least one of its legs is integrated against a test function.

There is an analogous bound in momentum space. Then the L^∞ and L^1 norms reverse roles. In momentum space, one easily verifies that $C^j(k) = \frac{f^{(j)}(k)}{ik_0 - e(\mathbf{k})}$ obeys $|C^j(k)| \leq c\, M^{-j}$ and $\int d^3k\, |C^j(k)| \leq c\, M^j$, so that one obtains the above bound. In coordinate space, one has to work harder (see [FT1], lemma V.2,3).

The reason we have formulated the above lemma in coordinate space is, that the whole expansion for the generating functional is done in coordinate space since the fields have to be Taylor expanded in coordinate space. Then, to get small factors from the Taylor expansion, the covariance has to be decomposed further into pieces $C^{j,\ell}$ and the power counting lemma will be applied to graphs, or more precisely, to kernels which are sums of graphs where each line has covariance $C^{j,\ell}$. The estimates are done in coordinate space and one effectively obtains the above bound.

If one iteratively applies Lemma II.1 for all scales one gets a similar bound with $M^{(q_v-2)j}$ replaced by $M^{(q_v-2)(j-i_v)}$ where $i_v > j$ is the scale of the generalized vertex v. This scale must be summed over, which yields

$$\sum_{i_v=j+1}^{0} M^{(q_v-2)(j-i_v)} \sim \begin{cases} M^{-(q_v-2)} \leq M^{-\frac{1}{3}q_v} \leq 1 & \text{if } q_v \geq 3 \\ |j| & \text{if } q_v = 2 \\ M^{|j|} & \text{if } q_v = 1 \end{cases}$$

Vertices with at least six external legs are summable. In fact, they produce a small factor $M^{-\frac{1}{3}q_v}$ which can be used to control the q_v-sums coming with each vertex. Four legged vertices give a $|j| = |\log M^j| = |\log e(\mathbf{k})|$. In the next section it is shown that this logarithm is really there if the Fermi curve is a circle, but it is absent if the dispersion relation $e(\mathbf{k})$ satisfies hypotheses I-III of section I. Finally, a two legged vertex gives the exploding factor $M^{|j|}$, but this factor can be eliminated by renormalization of the two legged kernels. As we mentioned earlier, we will give an inductive proof which uses only a single scale expansion. We will not write down the scale sums explicitly. They are replaced by a suitable induction hypothesis on V_q^{j-1}, see (II.16,17) below. We hope that this makes the proof clearer.

When renormalization is performed, the scale j can no longer be used as the induction index, because the definition of the counterterm at scale j involves the sum of all scales below j. In that case all scales are treated simultaneously and the induction is on "iteration step", corresponding to the depth of the tree in the Gallavotti Nicolo tree expansion. This was also the method used in [FT1,2]. The corresponding formalism is presented in Lemma II.5 below.

We start with an easier case, in which we remove both two and nontrivial four legged subgraphs. That is, we replace the Q_4 in Theorem II.1 with $Q_{2,4}$ defined by

$$Q_{2,4}\sum_{q=1}^{\infty}\mathcal{G}_q^{j-1}(\phi) = \lambda\tfrac{d}{d\lambda}\mathcal{G}_2^{j-1}(\phi)\big|_{\lambda=0} + \sum_{q=3}^{\infty}\mathcal{G}_q^{j-1}(\phi)\,.$$

Then one does not have to renormalize and the induction is on scales. We want to show that $\mathcal{W}^{j-1}$ defined inductively by $\mathcal{W}^0 = -\lambda\mathcal{V}$ and

$$\mathcal{W}^{j-1}(\psi^{\le j-1}+\phi) = Q_{2,4}\log\frac{1}{Y_j}\int e^{\mathcal{W}^j(\psi^{\le j}+\phi)}d\mu_{C^j}(\psi^j)$$

$$Y_j = \int e^{\mathcal{W}^j(\psi^j)}d\mu_{C^j}(\psi^j)$$

is analytic for all sufficiently small λ, independent of $j > -\infty$.

First write

$$\mathcal{W}^{j-1} = \sum_{i=j-1}^{0}\mathcal{V}^i$$

where

$$\mathcal{V}^0(\psi^{\le j-1}+\phi) = -\mathcal{V}(\psi^{\le j-1}+\phi)$$

and for $j \le 0$

$$\mathcal{V}^{j-1}(\psi^{\le j-1}+\phi) = Q_{2,4}\log\frac{1}{Y_j}\int e^{\mathcal{W}^j(\psi^{\le j}+\phi)-\mathcal{W}^j(\psi^{\le j-1}+\phi)}d\mu_{C^j}(\psi^j)$$

$$= Q_{2,4}\left\{\log\frac{1}{Y_j}\int e^{\mathcal{W}^j(\psi^{\le j}+\phi)}d\mu_{C^j}(\psi^j) - \mathcal{W}^j(\psi^{\le j-1}+\phi)\right\}$$

Since $\mathcal{W}^j(\psi^{\leq j-1}+\phi)$ is subtracted in the exponential, $\mathcal{V}^{j-1}$ must contain at least one contraction at scale $j-1$. This is not the case for $\mathcal{W}^{j-1}$. From a technical point of view, the $\mathcal{V}^{j-1}$'s are the basic objects. In particular, the induction hypothesis is stated in terms of them.

Let $P_{\leq n}$ be the operator which projects onto contributions up to n^{th} order in λ. That is $P_{\leq n}\sum_{k=0}^{\infty}a_k\lambda^k=\sum_{k=0}^{n}a_k\lambda^k$ and let V_q^{j-1}, $j\leq 0$, be given by

$$\mathcal{V}^{j-1}(\phi)=\sum_{q=3}^{\infty}\mathcal{V}_q^{j-1}(\phi)=\sum_{q=3}^{\infty}\int d\xi_1\cdots d\xi_{2q}\,V_q^{j-1}(\xi_1,\cdots,\xi_{2q})\,\phi(\xi_1)\cdots\phi(\xi_{2q})\,.$$

There will be an expansion

$$P_{\leq n}V_q^{j-1}(\eta_1,\cdots,\eta_{2q})=\sum_{\gamma\in\mathcal{P}_{n,q}}P_{\leq n}K_\gamma^{j-1}(W_{q_v}^j)(\eta_1,\cdots,\eta_{2q})$$

where the sum is over all paths γ joining the root to some other extremal vertex of a tree $\mathcal{P}_{n,q}$, called the expansion or partial integration tree. This tree is **not** the tree of the Gallavotti-Nicolo Tree Expansion. The expansion tree contains a description of the entire expansion. Each fork corresponds to a substep of the expansion process and each branch leaving the fork corresponds to a possible outcome for that substep.

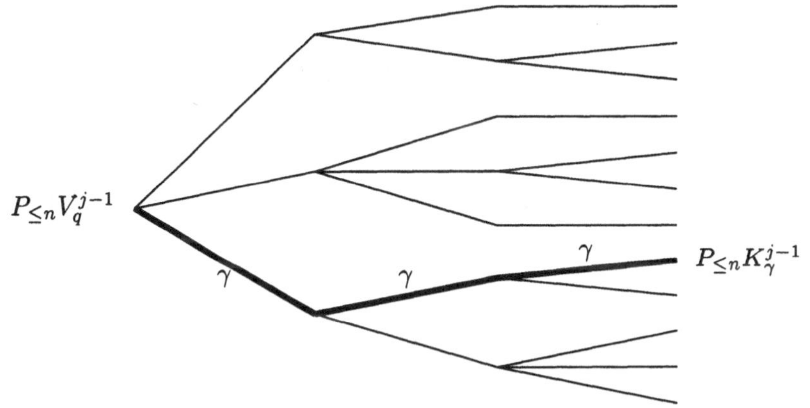

To give a bound on $P_{\leq n}V_q^{j-1}$ one has to bound the kernels $K_\gamma^{j-1}(W_{q_v}^j)$ and to control the sum $\sum_{\gamma\in\mathcal{P}_{n,q}}$.

The sum is controlled by

Lemma II.3 (Combinatorial Tree Lemma) *Let $\mathcal{T}$ be a tree, $w_\ell>0$ a weight factor assigned to the line ℓ and K_γ a real number assigned to the end of the path $\gamma\in\mathcal{T}$. Then*

$$\sum_{\gamma\in\mathcal{T}}|K_\gamma|\leq\sup_{\gamma\in\mathcal{T}}\left\{\prod_{f\in\gamma}b(f)\prod_{\ell\in\gamma}w_\ell\,|K_\gamma|\right\}$$

where $b(f)=\sum_{\ell\in f}\frac{1}{w_\ell}$ is a generalized branching number for the fork f.

In the case of $\mathcal{P}_{n,q}$ we can choose the w_ℓ's so that

$$\prod_{f\in\gamma} b(f) \prod_{\ell\in\gamma} w_\ell \leq \text{const}^{\sum_{v\in V(\gamma)} q_v}$$

where $\sum_{v\in V(\gamma)} q_v$ is half the number of legs of the vertices of which the kernel K_γ is made of. Since power counting gives a factor $M^{-\frac{1}{3}q_v}$ coming with each vertex, the above factor can be controlled by choosing M sufficiently large. This is a general rule for the expansion: you can allow all kinds of sums as long as the branching numbers or generalized branching numbers times the weight factors is bounded by const$^{\sum_v q_v}$. Factorials like $\prod_v q_v!$ are not allowed. Weight factors are necessary when there are infinite sums. In our model, this will only be the case for the q_v-sums. One may take 2^{q_v} as a weight factor.

Before we write down what the kernels are, let us first explain how the expansion is generated. We start with the fundamental theorem of calculus $f(1) = f(0) + \int_0^1 d\varepsilon \frac{d}{d\varepsilon} f(\varepsilon)$:

$$\mathcal{V}^{j-1}(\psi^{\leq j-1}) = Q_{2,4} \log \frac{1}{Y_j} \int e^{\mathcal{W}^j(\psi^j+\varepsilon\psi^{\leq j-1})-\mathcal{W}^j(\varepsilon\psi^{\leq j-1})} d\mu_{C^j}(\psi^j)\Big|_{\varepsilon=0}^{\varepsilon=1}$$

$$= Q_{2,4} \sum_{q=3}^{\infty} \sum_{\substack{J\subset\{1,\cdots,2q\}\\ J\neq\emptyset}} \int \prod_{i=1}^{2q} d\xi_i \, W_q^j(\underline{\xi}) \prod_{k\in J^c} \psi^{\leq j-1}(\xi_k) \times$$

$$\int_0^1 d\varepsilon \frac{d}{d\varepsilon} \left\{\varepsilon^{|J^c|}\right\} \frac{\int \prod_{i\in J} \psi^j(\xi_i) \, e^{\mathcal{W}^j(\psi^j+\varepsilon\psi^{\leq j-1})-\mathcal{W}^j(\varepsilon\psi^{\leq j-1})} d\mu_{C^j}(\psi^j)}{\int e^{\mathcal{W}^j(\psi^j+\varepsilon\psi^{\leq j-1})-\mathcal{W}^j(\varepsilon\psi^{\leq j-1})} d\mu_{C^j}(\psi^j)}. \tag{II.1}$$

The J-sum comes from multplying out the $2q$ brackets $\Big(\psi^j(\xi_i)+\varepsilon\psi^{\leq j-1}(\xi_i)\Big)$. The condition $J \neq \emptyset$ ensures that there is at least one contraction. Together with the q-sum, they yield the first branching of the expansion tree.

Now we do integration by parts. That is, we eliminate all the fields $\prod_{i\in J} \psi^j(\xi_i)$ in the numerator of (II.1) by repeatedly applying the formula

$$\int \psi(\xi) \, F(\psi) \, d\mu_C = \int d\eta \, C(\xi,\eta) \int \frac{\delta F(\psi)}{\delta\psi(\eta)} \, d\mu_C \tag{II.2}$$

where, if $\xi = (x, \sigma, b)$ and $\eta = (y, \tau, c)$,

$$C(\xi,\eta) = \delta_{b,1-c}\delta_{\sigma,\tau} \begin{cases} C(x,y) & \text{if } b = 0 \\ -C(y,x) & \text{if } b = 1 \end{cases} \quad \text{and} \quad \int d\eta = \sum_{c\in\{0,1\}} \sum_{\tau\in\{\uparrow,\downarrow\}} \int d^3y \, .$$

The result is formalized in the following general

Lemma II.4 (Integration by Parts) *Let C be some covariance and*

$$\mathcal{W}(\psi) = \sum_{q=1}^{\infty} \int d\eta_1 \cdots d\eta_{2q}\, W_q(\eta_1, \cdots, \eta_{2q})\, \psi(\eta_1) \cdots \psi(\eta_{2q})$$

Then one has

$$\int \psi(\xi_1)\psi(\xi_2)\cdots\psi(\xi_p)\, e^{\mathcal{W}(\psi+\phi)} d\mu_C(\psi) =$$

$$\sum_{A\subset\{1,\cdots,p\}} \sum_{\substack{q_i=1 \\ \text{if } i\in A}}^{\infty} \sum_{\substack{J_i\subset\{1,\cdots,2q_i\} \\ \text{if } i\in A}} \sum_{\substack{l_i\in J_i \\ \text{if } i\in A}} \sum_{U\subset W(A,\underline{J},\underline{l})} \operatorname{sign}(\underline{J},\underline{l},U)$$

$$\int \prod_{i\in A} \left\{ \prod_{k=1}^{2q_i} d\eta_k^i\, W_{q_i}(\eta_1^i, \cdots, \eta_{2q_i}^i) \times C(\xi_i, \eta_{l_i}^i) \prod_{k\in J_i^c} \phi(\eta_k^i) \right\}$$

$$\int \prod_{i=1}^{p} \chi_{{}_{A,U,i}}(\psi)\, d\mu_C'(\psi', \psi) \int \prod_{i\in A} \prod_{(i,k)\in U^c} \psi(\eta_k^i)\, e^{\mathcal{W}(\psi+\phi)} d\mu_C(\psi)$$

where

$$\chi_{{}_{A,U,i}}(\psi) = \begin{cases} \psi'(\xi_i) & \text{if } i \notin A \\ \prod_{(i,k)\in U} \psi(\eta_k^i) & \text{if } i \in A, \end{cases}$$

and

$$\int \prod_{i=1}^{p} \chi_{{}_{A,U,i}}(\psi)\, d\mu_C'(\psi', \psi) = \det \begin{bmatrix} \text{same matrix as for the} \\ \text{functional integral without primes,} \\ \text{but with } C(\eta_l^i, \eta_m^l) \text{ replaced by 0} \\ \text{and } C(\xi_i, \eta_k^j) \text{ replaced by 0 if } i<j \end{bmatrix}.$$

Furthermore, if the tree $\mathcal{P}$ is defined by

$$\sum_{\gamma\in\mathcal{P}} = \sum_{A\subset\{1,\cdots,p\}} \sum_{\substack{q_i=1 \\ if\ i\in A}}^{\infty} \sum_{\substack{J_i\subset\{1,\cdots,2q_i\} \\ if\ i\in A}} \sum_{\substack{l_i\in J_i \\ if\ i\in A}} \sum_{U\subset W(A,\underline{J},\underline{l})}$$

with weight factors $w_{q_i} = 2^{q_i}$ for the q_i-sums and all other weights being one, then

$$\prod_{f\in\gamma} b(f) \prod_{\ell\in\gamma} w_\ell \le \text{const}^{\; p+\sum_{v\in V(\gamma)} q_v}$$

for all $\gamma \in \mathcal{P}$. Finally $W(A, \underline{J}, \underline{l}) = \cup_{i\in A} \cup_{k\in J_i\setminus\{l_i\}} \{(i,k)\}$, $J_i^c = \{1, \cdots, 2q_i\} \setminus J_i$,, $U^c = W(A, \underline{J}, \underline{l}) \setminus U$ and $\operatorname{sign}(\underline{J}, \underline{l}, U) \in \{1, -1\}$.

Let us briefly explain how the five sums in the lemma arise. The first sum tells you which fields of the p 'downstairs' fields $\psi(\xi_1), \cdots, \psi(\xi_p)$ differentiate the exponential when you apply (II.2). Each time one hits the exponential, say with the field $\psi(\xi_i)$, the sum

$$\sum_{q_i=1}^{\infty} \int d\eta_1^i \cdots d\eta_{2q_i}^i\, W_{q_i}(\eta_1^i, \cdots, \eta_{2q_i}^i)\, (\phi+\psi)(\eta_1^i) \cdots (\phi+\psi)(\eta_{2q_i}^i) \qquad \text{(II.3)}$$

is brought down from the exponential. This explains the second sums in the lemma. The third sums come from multiplying out the $(\phi+\psi)(\eta_k^i)$ brackets and the fourth sums are there because of the derivative in $\frac{\delta}{\delta\psi}e^{\mathcal{W}} = \frac{\delta\mathcal{W}}{\delta\psi}e^{\mathcal{W}}$. Finally, the last sum tells you which of the new downstairs fields in (II.3) are differentiated by some later $\psi(\xi)$ fields which do not hit the exponential. That is, $U \subset W(A, \underline{J}, \underline{l})$, is the set of η-fields (the new downstairs fields) which are contracted to some ξ-fields. The sum of all possible contractions between them is given by the "primed integral"

$$\int \prod_{i=1}^{p} \chi_{_{A,U,i}}(\psi)\, d\mu'_C(\psi',\psi) = \det \begin{bmatrix} \text{same matrix as for the} \\ \text{functional integral without primes,} \\ \text{but with } C(\eta_l^i,\eta_m^l) \text{ replaced by 0} \\ \text{and } C(\xi_i,\eta_k^j) \text{ replaced by 0 if } i<j \end{bmatrix} . \tag{II.4}$$

We used a prime on $d\mu'_C$ to indicate that this is not the sum of all contractions given by the usual functional integral. It is restricted to contractions initiated by ξ-fields so that (η,η)-contractions are forbidden. Furthermore, an η-field which is generated by, say, $\psi(\xi_p)$ hitting the exponential, cannot contract to $\psi(\xi_1)$, $\cdots$, $\psi(\xi_{p-1})$, since, again by construction, we begin the integration by parts procedure with $\psi(\xi_1)$ and at that time the η-field has not yet been produced. Therefore the primed ξ-fields in (II.4) can only contract to η-fields sitting on the left of the ξ-field, or to an arbitrary other primed ξ-field. A rigorous proof of the Integration by Parts Lemma will be given in [FKLT1].

Let us return to (II.1). Eliminate the fields $\prod_{i\in J}\psi^j(\xi_i)$ in the numerator of (II.1),

$$\int \prod_{i\in J} \psi^j(\xi_i)\, e^{\mathcal{W}^j(\psi^j+\varepsilon\psi^{\le j-1})-\mathcal{W}^j(\varepsilon\psi^{\le j-1})} d\mu_{C^j}(\psi^j)\,, \tag{II.5}$$

by doing one round of partial integration. That is, apply the above lemma one time. This produces a big sum of terms. There are terms where U^c, which labels the fields in the new functional integral, is empty. Then the remaining functional integral is just

$$\int e^{\mathcal{W}^j(\psi^j+\varepsilon\psi^{\le j-1})-\mathcal{W}^j(\varepsilon\psi^{\le j-1})} d\mu_{C^j}(\psi^j)$$

and cancels against the denominator in (II.1). On the other hand, when U^c is not the empty set, apply the integration by parts lemma again. Repeat as necessary.

Consider a term where, after n steps of partial integration, the functional integral has not cancelled. Then, in each step there must have been a field which hit the exponential. Since each $\mathcal{W}^j$ comes at least with one λ and (II.1) already has one $\mathcal{W}_q^j$ downstairs, the term must be of order at least λ^{n+1}. Thus, to isolate all contributions up to n^{th} order, it suffices to do n rounds of partial integration discarding all terms for which the functional integral has not cancelled. So we may write

$$P_{\le n} V_q^{j-1}(\eta_1,\cdots,\eta_{2q}) = \sum_{\gamma\in\mathcal{P}_{n,q}} P_{\le n} K_\gamma^{j-1}(W_{q_v}^j)(\eta_1,\cdots,\eta_{2q}) \tag{II.6}$$

where the expansion tree $\mathcal{P}_{n,q}$ is obtained by iterating the partial integration tree $\mathcal{P}$ of Lemma II.2 n times and removing all paths which lead to contributions in which functional integral has not cancelled or the number of external fields is not equal to $2q$. The kernels are given by

$$K_\gamma^{j-1}(W_{q_v}^j)(\eta_1,\cdots,\eta_{2q}) \equiv K_\gamma^{j-1}(W_{q_v}^j)(\underline{\eta}^{ext})$$

$$= \int d\underline{\eta}^{int} \prod_{v\in V_\gamma} W_{q_v}^j(\underline{\eta}^v) \prod_{l\in T_\gamma} C^j(\eta_{i_1(l)},\eta_{i_2(l)}) \prod_{r=1}^{n} I_r'(\underline{\eta}) \tag{II.7}$$

where V_γ is the set of all vertices of K_γ^{j-1} and T_γ is a spanning tree for K_γ^{j-1}. The primed integral produced at the r'th step of partial integration is abbreviated by $I_r'(\underline{\eta})$ in (II.7).

As mentioned above, the sum over the expansion tree is estimated with the combinatorial tree lemma and causes no problem. The only place where factorials may arise are the primed integrals. In fact, full expansion of the determinant in (II.4) for all primed integrals in (II.7) generates all Feynman graphs. Our expansion is designed to produce all contributions up to n^{th} order in such a way that potential factorials are isolated and can be controlled.

To eliminate the factorials produced by the primed integrals, one has to do two things. First, one introduces sectors, that is, one decomposes the shells around the Fermi curve into smaller pieces as shown in the following figure. The decomposition for a nonspherical Fermi curve is analogous.

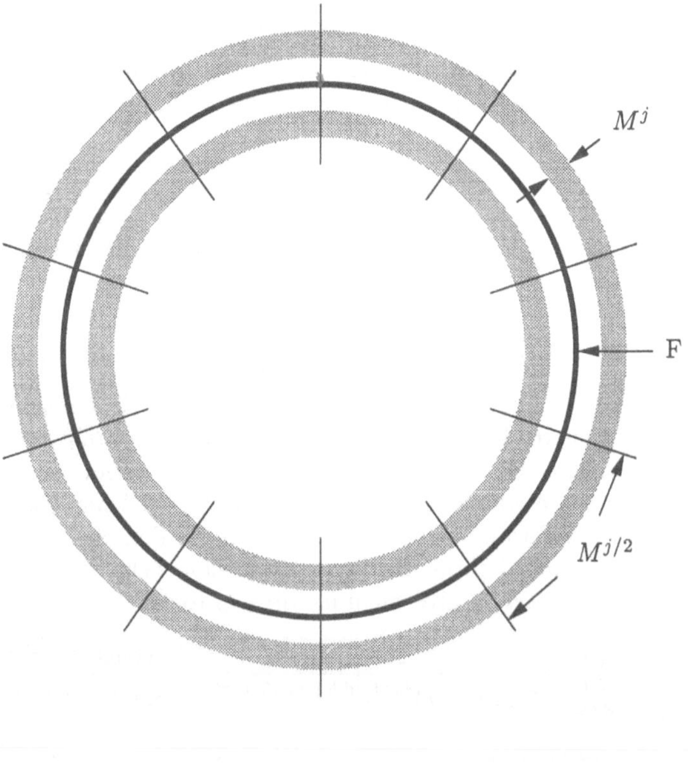

Second, one Taylor expands the fields (in coordinate space) at the beginning of each step of partial integration.

We now explain why. One starts with a functional integral like (II.5). Suppose that two fields in (II.5) are equal. Then the functional integral vanishes. If you were to integrate by parts with these fields, you would produce terms whose sum is zero but which are assigned to different paths in the expansion tree. If one applies the combinatorial tree lemma, one does not see the cancellations between these terms.

The reason for sectors is the following: Each covariance C^j comes with a decay factor $(1+M^j|x|)^{-N}$ where N can be chosen arbitrary large. That is, one has summable decay between boxes of length M^{-j}. Using this decay, one can bound, as in the classical cluster expansion, the primed integral by a power counting factor times a product of "local" factorials,

$$\prod_{\substack{\Delta\in\mathcal{D}_j\\ b\in\{0,1\}\\ \sigma\in\{\uparrow,\downarrow\}}} \pi(\underline{\xi},\Delta,b,\sigma)!$$

where $\underline{\xi}$ is the set of coordinates of the downstairs fields in (II.5) and $\pi(\underline{\xi},\Delta,b,\sigma)$ is the number of those fields which have $x_i\in\Delta$, $b_i=b$ and $\sigma_i=\sigma$. Here $\mathcal{D}_j$ is a set of space time boxes of length M^{-j} which cover $\mathbb{R}^3$.

For a given set of coordinates $\underline{\xi}$ and a given field $\psi(\xi_i)$, divide $\Delta\ni x_i$ into $\pi(\underline{\xi},\Delta,b_i,\sigma_i)$ subboxes d of length

$$\frac{M^{-j}}{\pi(\underline{\xi},\Delta,b_i,\sigma_i)^{\frac{1}{3}}}.$$

Let c_i be the center of the subbox $d\subset\Delta$ to which ξ_i belongs, that is, for which $x_i\in d$. The Taylor expansion produces powers of $((x_i-c_i)\cdot\nabla)$. Now, by construction,

$$|x_i-c_i|\le\sqrt{3}\frac{M^{-j}}{\pi(\underline{\xi},\Delta,b_i,\sigma_i)^{\frac{1}{3}}}. \tag{II.8}$$

Thus, the Taylor expansion will generate a net small factor if each derivative in coordinate space gives an M^j. This would be the case if the covariance had a singularity at a single point in momentum space, for example if $C^j(x)=\int d^{d+1}k\, e^{ikx}\frac{f_j(k)}{|k|^{\frac{d+1}{2}}}$. However, in our model the singularity is on the Fermi curve $k_0=0$, $e(\mathbf{k})=0$ and, in C^j, $\mathbf{k}$ is only localized in a shell of thickness M^j around the Fermi curve. Each coordinate space derivative brings down a factor of k from e^{ikx}, which is of order one rather than M^j. Alternatively, one may say that the phase space volume $M^{-3j}\times M^{2j}=M^{-j}$ is too big. Therefore the spatial momentum has to be localized further. This is done by introducing sectors ℓ. Then, if the order t of the the Taylor expansion is choosen sufficiently large, the operators $(x_i-c_i)\cdot\nabla$ give small factors (up to constants)

$$\prod_{d\in\Delta}\frac{1}{\pi(\underline{\xi};\Delta,b,\sigma,\ell)^{\frac{1}{3}(\pi(\underline{\xi};d,b,\sigma,\ell)-3^t)}}\le\frac{\mathrm{const}^{\,\pi(\underline{\xi};\Delta,b,\sigma,\ell)}}{\pi(\underline{\xi};\Delta,b,\sigma,\ell)^{2\pi(\underline{\xi};\Delta,b,\sigma,\ell)}}$$

which kill the local factorials produced by the primed integrals. The number $\pi(\underline{\xi}; d, b, \sigma, \ell) - 3^t$ arises as follows. Expand each field

$$\psi^{j,\ell}(\xi_k) = \psi^{j,\ell}(c) + (x_k - c) \cdot \nabla\psi^{j,\ell}(c) + \cdots + \frac{1}{(t-1)!}\big((x_k - c)\cdot\nabla\big)^{t-1}\psi^{j,\ell}(c)$$

$$+ \frac{1}{(t-1)!}\int_0^1 dw\,(1-w)^{t-1}\big((x_k - c)\cdot\nabla\big)^t\psi^{j,\ell}(c + w(x_k - c))$$

$$\equiv \sum_{s=0}^{t} T^s_{\underline{\xi}}\psi^{j,\ell}(\xi_k)\,, \tag{II.9}$$

Since the functional integral (II.5) vanishes if two fields are equal, at most $1 + 3 + 3^2 + \cdots + 3^{t-1} \leq 3^t$ fields can fail to be Taylor remainders. So at least $\pi(\underline{\xi}; d, b, \sigma, \ell) - 3^t$ fields must be Taylor remainders.

The most natural thing would be to break up the Fermi curve into sectors of length M^j. However, there is a further subtlety which forces one to use sectors of length $M^{\frac{j}{2}}$ instead of M^j. Of course, one has to control the sector sums associated with the decomposition of the covariance and fields. Since the primed integrals are to be estimated without full evaluation, one no longer has Feynman graphs. There no longer are momentum loops and one can use conservation of momentum only at each generalized vertex to control the sector sums. To be more precise on this point: if one computes the primed integral, one obtains for each contraction of, say, ψ^{j,ℓ_1} and ψ^{j,ℓ_2} a Kronecker delta δ_{ℓ_1,ℓ_2}, that mimics conservation of momentum in a line. However, to have access to all these delta's one must fully expand the primed integrals and then the crucial bound (II.14) below fails.

If $\zeta_\ell(\mathbf{k})$ has support on a sector $\ell \in \Sigma_j$ of length $M^{\frac{j}{2}}$ and

$$C^j(k) = \frac{f_j(k)}{ik_0 - e(\mathbf{k})} = \sum_{\ell\in\Sigma_j}\frac{f_j(k)\zeta_\ell(\mathbf{k})}{ik_0 - e(\mathbf{k})} = \sum_{\ell\in\Sigma_j} C^{j,\ell}(k) \tag{II.10}$$

is the corresponding decomposition of the covariance, then one can prove that

$$\left|e^{i\mathbf{q}_\ell(\mathbf{x}-\mathbf{x}')}D_0^{n_0}D_{\|}^{n_1}D_{\perp}^{n_2}\left(e^{-i\mathbf{q}_\ell(\mathbf{x}-\mathbf{x}')}C^{j,\ell}(x,x')\right)\right|$$

$$\leq \text{const } M^{(\frac{3}{2}+n_0+n_1+\frac{n_2}{2})j}\rho^{j,\ell}(x,x')^{-N}\,, \tag{II.11}$$

where the decay factor is given by

$$\rho^{j,\ell}(x,x') = 1 + M^j|x_0 - x_0'| + M^j|\mathbf{x}_{\|} - \mathbf{x}_{\|}'| + M^{\frac{j}{2}}|\mathbf{x}_\perp - \mathbf{x}_\perp'|\,,$$

and

$$D_0 = \frac{\partial}{\partial x_0},\quad D_{\|} = \frac{\mathbf{q}_\ell}{|\mathbf{q}_\ell|}\cdot\nabla_{\mathbf{x}_1},\quad D_\perp = \hat{\pi}_\ell\cdot\nabla_{\mathbf{x}_1}$$

where $\mathbf{q}_\ell$ denotes center of the sector ℓ, $\hat{\pi}_\ell$ is some unit vector perpendicular to $\mathbf{q}_\ell$ and $\mathbf{x} = (\mathbf{x},\mathbf{q}_\ell)\mathbf{q}_\ell + (\mathbf{x},\pi_\ell)\pi_\ell = \mathbf{x}_{\|} + \mathbf{x}_\perp$. A derivative in the π_ℓ direction only gives an $M^{\frac{j}{2}}$ but the decay rate in that direction is also $M^{-\frac{j}{2}}$ instead of M^{-j}.

Thus the phase space volume is still one. $M^{\frac{j}{2}}$-sectors are the largest ones for which this can be achieved.

So we introduce the decomposition (II.10) and write the fields $\psi^j = \sum_{\ell\in\Sigma_j}\psi^{j,\ell}$ as a sum of independent variables. At the beginning of each step of partial integration, we Taylor expand the downstairs fields as in (II.9), where the expansion point $c = c(\xi, \xi_k)$ is determined by the number of fields which are localized in the box $\Delta \ni x_k$. Observe that, because of the anisotropic decay of the covariance (II.11), one has to use boxes of size $M^{-j}\times M^{-j}\times M^{-\frac{j}{2}}$ in the x_0-, $\mathbf{q}_\ell$- and π_ℓ-direction.

The sums coming with the Taylor expansion add new branchs to the expansion tree, which is still denoted by $\mathcal{P}_{n,q}$, while the sector sums are left explicit rather than bounded by Lemma II.3. The result of the integration by parts expansion with sectors and Taylor expansion is

$$P_{\leq n}V^{j-1}_{q,\leq n}(\psi^{\leq j-1}) = \sum_{\gamma\in\mathcal{P}_{n,q}} P_{\leq n}\mathcal{K}^{j-1}_{\gamma}(\psi^{\leq j-1}) \tag{II.12}$$

where

$$\mathcal{K}^{j-1}_\gamma(\psi^{\leq j-1}) = \prod_{v\in V}\Bigg(\sum_{j_v=j}^{-1}\sum_{\ell^v_k\in\Sigma_{j_v+1}}\sum_{m^v_k\in\Sigma_j(\ell^v_k)}\int\prod_{k=1}^{2q_v} d\eta^v_k\, V^{j_v}_{q_v}(\underline{\eta}^v;\underline{\ell}^v)\,\delta_{q_v}(\underline{m}_v)\Bigg)\times$$

$$\prod_{l\in T}\delta_{m_{i_1(l)},m_{i_2(l)}}T^{s_l}_{\underline{\eta}}C^{j,m_{i_1(l)}}(\eta_{i_1(l)},\eta_{i_2(l)})\prod_{r=1}^{n}I'_r(\underline{\eta};\underline{m})\prod_{\substack{\text{external}\\ \text{legs}}}\psi^{\leq j-1,m_k}(\eta_k)$$

$$= \sum_{m_1,\cdots,m_{2q}\in\Sigma_j}\int\prod_{k=1}^{2q}d\eta_k\,K^{j-1}_\gamma(\underline{\eta};\underline{m})\,\psi^{\leq j-1,m_1}(\eta_1)\cdots\psi^{\leq j-1,m_{2q}}(\eta_{2q}). \tag{II.13}$$

Here the first sector sums $\sum_{\ell^v_k}$ are over the set Σ_{j_v+1} of sectors of scale j_v+1 where the vertex $V^{j_v}_{q_v}$ was created. The second sector sum decomposed each sector ℓ^v_k into $M^{-\frac{j-(j_v+1)}{2}}$ sectors m^v_k of length $M^{\frac{j}{2}}$. Furthermore, the sector delta function

$$\delta_q(m_1,\cdots,m_{2q}) = \begin{cases} 1, & \text{if } \int\prod_{i=1}^{2q}\Big(d^{d+1}k_i\,f_j(k_i)\zeta_{m_i}(\mathbf{k}_i)\Big)\delta(k_1+\cdots+k_{2q})>0\\ 0, & \text{if } \int\prod_{i=1}^{2q}\Big(d^{d+1}k_i\,f_j(k_i)\zeta_{m_i}(\mathbf{k}_i)\Big)\delta(k_1+\cdots+k_{2q})=0\end{cases}$$

ensures conservation of momentum at each vertex.

The effect of the Taylor expansion is that

$$\sup_{\underline{\eta},\underline{m}}\left\{\prod_{r=1}^{n}|I'_r(\underline{\eta};\underline{m})|\right\} \leq \Big(\underbrace{\text{const}_M\,M^{\frac{3}{2}j}}_{=\|C^{j,m}\|_{L^\infty}}\Big)^{\frac{1}{2}\left(\substack{\text{number of fields in}\\ \text{the primed integrals}}\right)}. \tag{II.14}$$

One gets the same bound if one estimates a single diagram.

Because there are sector sums on the external legs, we introduce new norms which are suitable for an inductive treatment.

$$\|K_\gamma\|_{\emptyset,\Sigma_j} = \sup_i \sup_{\eta_i, m_i} \sum_{\substack{m_k\in\Sigma_j \\ k\neq i}} \int \prod_{\substack{k=1 \\ k\neq i}}^{2q} d\eta_k \, |K_\gamma(\underline{\eta};\underline{m})| \, ,$$

$$\|K_\gamma\|_{S,\Sigma_j} = \sum_{\substack{m_k\in\Sigma_j \\ k\in S^c}} \int \prod_{k=1}^{2q} d\eta_k \prod_{i\in S} |f_i(\eta_i)| \, |K_\gamma(\underline{\eta};\underline{m})| \, .$$

In particular

$$\|K_\gamma\|_{\{1,\cdots,2q\},\Sigma_j} = \|K_\gamma\|_{\{1,\cdots,2q\}} = \int \prod_{k=1}^{2q} d\eta_k \, |f_k(\eta_k)| \, |K_\gamma(\underline{\eta})| \, .$$

Furthermore, if $K_\gamma = \sum_{k=1}^{\infty} \lambda^k K_{\gamma,k}$, we define

$$\|K_\gamma\|_{\emptyset,\Sigma_j,\leq n} = \sum_{k=1}^{n} |\lambda|^k \|K_{\gamma,k}\|_{\emptyset,\Sigma_j} \, .$$

Then one obtains the following bound

$$\|K_\gamma^{j-1}\|_{\emptyset,\Sigma_j,\leq n} \leq \prod_{v\in V} \left(M^{\frac{j}{2}(q_v-2)} \sum_{j_v=j}^{-1} M^{\frac{j_v+1}{2}(2q_v-3)} \|V^{j_v}_{q_v,k_v}\|_{\emptyset,\Sigma_{j_v+1},\leq n} \right) \times$$

$$c_1^{\sum_v q_v - q} M^{-\frac{j}{2}(q-2)} M^{-\frac{j}{2}(2q-3)} \, . \tag{II.15}$$

Let us briefly explain how the different powers of M^j arise. The primed integrals give the factors

$$\left(M^{\frac{3}{2}j}\right)^{\frac{1}{2}\left(\substack{\text{number of fields in} \\ \text{the primed integrals}}\right)} = \left(M^{\frac{3}{2}j}\right)^{\frac{1}{2}\left(\substack{\text{number of fields} \\ \text{not on the tree}}\right)} = M^{\frac{3}{2}j(\sum_v q_v - q - \sum_v 1 + 1)} \, .$$

On the tree T in (II.13), the L^1-norm of $C^{j,m}$ has to be taken which give the factors

$$\left(M^{\frac{3}{2}j} M^{-\frac{5}{2}j}\right)^{\sum_v 1 - 1} = M^{-j(\sum_v 1 - 1)} \, .$$

The sector sums $\sum_{\ell_k^v\in\Sigma_{j_v+1}}$ are contained in the norms $\|V^{j_v}_{q_v,k_v}\|_{\emptyset,\Sigma_{j_v+1}}$. The sector sums $\sum_{m_k^v\in\Sigma_j(\ell_k^v)}$ are estimated by Lemma II.3' of [FMRT1]. This lemma covers only the spherical case, but there is an analogous version for the $e(\mathbf{k})$'s specified in section I. It says that, if one fixes the sector of one leg of the vertex v, the number of independent sector sums for the remaining legs is at most $2q_v - 3$. By conservation of momentum, one can get rid of two sector sums at each generalized vertex. This gives the factor

$$\prod_v M^{-\frac{j-(j_v+1)}{2}(2q_v-3)}$$

since for all but one vertex one sector is fixed by the Kronecker delta's $\delta_{m_{i_1(l)},m_{i_2(l)}}$ on the tree T and for the last vertex one sector is fixed by the definition of the norm $\| \cdot \|_{\emptyset,\Sigma_j}$.

Altogether, one obtains

$$M^{\frac{3}{2}j(\sum_v q_v - q - \sum_v 1+1)} M^{-j(\sum_v 1-1)} M^{-\sum_v (2q_v-3)\frac{j-(j_v+1)}{2}} =$$
$$\prod_v \left(M^{\frac{j}{2}(q_v-2)} M^{\frac{j_v+1}{2}(2q_v-3)} \right) M^{-\frac{j}{2}(q-2)} M^{-\frac{j}{2}(2q-3)}$$

which coincides with the power counting of (II.15).

Now we apply the combinatorial tree lemma to obtain

$$\begin{aligned}
\| V_q^{j-1} \|_{\emptyset,\Sigma_j,\leq n} &\leq \sum_{\gamma\in\mathcal{P}_{n,q}} \| K_\gamma^{j-1} \|_{\emptyset,\Sigma_j,\leq n} \\
&\leq \sup_{\gamma\in\mathcal{P}_{n,q}} \left\{ \prod_{f\in\gamma} b(f) \prod_{\ell\in\gamma} w_\ell \, \| K_\gamma^{j-1} \|_{\emptyset,\Sigma_j,\leq n} \right\}
\end{aligned}$$

$$\leq \sup_{\gamma\in\mathcal{P}_{n,q}} \left\{ c_2^{\sum_v q_v} c_1^{\sum_v q_v - q} \prod_{v\in V} \left(M^{\frac{j}{2}(q_v-2)} \sum_{j_v=j}^{-1} M^{\frac{j_v+1}{2}(2q_v-3)} \| V_{q_v}^{j_v} \|_{\emptyset,\Sigma_{j_v+1},\leq n} \right) \times \right.$$
$$\left. M^{-\frac{j}{2}(q-2)} M^{-\frac{j}{2}(2q-3)} \right\}. \qquad \text{(II.16)}$$

The constant c_1 comes from covariance estimates and depends on M but c_2 is a pure combinatorial constant. If some external legs in $S \subset \{1,\cdots,2q\}$ are integrated against test functions, one gets the following bound

$$\| V_q^{j-1} \|_{S,\Sigma_j,\leq n} \leq$$

$$\sup_{\gamma\in\mathcal{P}_{n,q}} \left\{ c_2^{\sum_v q_v - |S|} c_1^{\sum_v q_v - q} \prod_{v\in V_{int}} \left(M^{\frac{j}{2}(q_v-2)} \sum_{j_v=j}^{-1} M^{\frac{j_v+1}{2}(2q_v-3)} \| V_{q_v}^{j_v} \|_{\emptyset,\Sigma_{j_v+1},\leq n} \right) \right.$$
$$\times \prod_{v\in V_{ext}} \left(M^{\frac{j}{2}(q_v-\frac{1}{2}|S_v|)} \sum_{j_v=j}^{-1} M^{\frac{j_v+1}{2}(2q_v-|S_v|)} \| V_{q_v}^{j_v} \|_{S_v,\Sigma_{j_v+1},\leq n} \right)$$
$$\left. M^{-\frac{j}{2}(q-\frac{1}{2}|S|)} M^{-\frac{j}{2}(2q-|S|)} \right\}. \qquad \text{(II.17)}$$

By (II.16,17), $\| V_q^{j-1} \|_{\substack{\emptyset \\ S},\Sigma_j,\leq n}$ is estimated in terms of $\| V_{q_v}^{j_v} \|_{\substack{\emptyset \\ S},\Sigma_{j_v+1},\leq n}$ of higher scale $j_v > j-1$. Therefore we can proceed by induction on the scale with the following inductive hypothesis

$$\| V_q^{j-1} \|_{\emptyset,\Sigma_j,\leq n} \leq |\lambda|^{\frac{q}{2}} M^{-\frac{j}{2}(3q-5)}, \qquad \text{(II.18)}$$

$$\| V_q^{j-1} \|_{S,\Sigma_j,\leq n} \leq \binom{2q}{|S|}^{-1} p(j)^{|S|} |\lambda|^{\frac{q}{2}} M^{-\frac{j}{2}(3q-\frac{3}{2}|S|-\frac{1}{4})} \prod_{k\in S} \left(\|f_k\|_{L^1} + \|f_k\|_{L^\infty} \right) \qquad \text{(II.19)}$$

where $p(j) = \prod_{i=j+1}^{0}(1+M^{\frac{1}{4}i}) \leq \prod_{i=-\infty}^{0}(1+M^{\frac{1}{4}i}) = c_3 < \infty$. Using (II.16,17), (II.18,19) is verified for λ sufficiently small[2]. In particular, we obtain

$$\int \prod_{i=1}^{2q} d\xi_i \, |f_i(\xi_i)| \, |W_q^r(\xi_1, \cdots, \xi_{2q})| = \|W_q^r\|_{\{1,\cdots,2q\}} \leq \sum_{j=r}^{0} \|V_q^j\|_{\{1,\cdots,2q\}}$$

$$= \sum_{j=r}^{0} \|V_q^j\|_{\{1,\cdots,2q\},\Sigma_{j+1}} \leq 2\, c_3^{2q} \, |\lambda|^{\frac{q}{2}} \prod_{k=1}^{2q} \Big(\|f_k\|_{L^1} + \|f_k\|_{L^\infty} \Big) . \tag{II.20}$$

This completes our outline of the proof of Theorem II.1 in the case without two and nontrivial four legged subgraphs. Now let us turn to the case where two legged subgraphs are allowed.

Renormalization of Two–Legged Subgraphs

So now, for $r+1 \leq j \leq 0$, let $\mathcal{W}^{j-1}$ be the effective potential without four legged subgraphs defined in Theorem II.1. Then

$$\mathcal{W}^{j-1}(\psi^{\leq j-1}) = \sum_{i=j-1}^{0} \mathcal{V}^i(\psi^{\leq j-1}) + \delta\mathcal{V}_r(\psi^{\leq j-1})$$

where

$$\mathcal{V}^0(\psi^{\leq j-1}) = -\mathcal{V}(\psi^{\leq j-1})$$

and, for $i < 0$,

$$\mathcal{V}^i(\psi^{\leq i}) = \mathcal{W}^i(\psi^{\leq i}) - \mathcal{W}^{i+1}(\psi^{\leq i})$$

$$= Q_4 \log \frac{1}{Y_{i+1}} \int e^{\mathcal{W}^{i+1}(\psi^{\leq i+1}) - \mathcal{W}^{i+1}(\psi^{\leq i})} d\mu_C^{i+1}(\psi^{i+1}) \tag{II.21}$$

and

$$Y_{i+1} = \int e^{\mathcal{W}^{i+1}(\psi^{i+1})} d\mu_C^{i+1}(\psi^{i+1}) .$$

The counterterm is given by

$$\delta\mathcal{V}_r(\psi) = \int \frac{d^3k}{(2\pi)^3} \delta e_r(\mathbf{k}, \lambda) \, \bar{\psi}(k) \psi(k)$$

where we write

$$\delta e_r(\mathbf{k}, \lambda) = \sum_{j=r}^{-1} \delta e_r^j(\mathbf{k}, \lambda) = \sum_{j=r}^{-1} \sum_{l=1}^{\infty} \delta e_{r,l}^j(\mathbf{k}) \, \lambda^l .$$

---

[2] First choose M big enough such that $c_2^{\sum_v q_v} M^{-\frac{1}{6}\sum_v q_v} \leq 1$ and then make λ small enough such that $c_1^{\sum_v q_v - q} \, |\lambda|^{\sum_v \frac{q_v}{2} - \frac{q}{2}} \leq 1$.

Define a localization operator $\mathbf{L}$ which annihilates all but quadratic monomials by

$$\mathbf{L}\int \frac{d^3k}{(2\pi)^3} G(k)\,\psi(k)\bar{\psi}(k) = \int \frac{d^3k}{(2\pi)^3} LG(k)\,\psi(k)\bar{\psi}(k)\,, \tag{II.22}$$

$$\mathbf{L}\mathcal{G}_q = 0 \quad \forall q \neq 1\,,$$

$$LG(k_0,\mathbf{k}) = G(0,\pi_F\mathbf{k})\,.$$

Here we assume that on a tubular neighbourhood $\mathcal{N}(F)$ of the Fermi curve $F = \{\mathbf{k}\in\mathbb{R}^2 \mid e(\mathbf{k}) = 0\}$ we can define a projection $\pi_F : \mathcal{N}(F) \to F$ such that

$$|e(\mathbf{k})| \leq M^j \;\Rightarrow\; |\pi_F\mathbf{k} - \mathbf{k}| \leq c\,M^j$$

where c is independent of j and $\mathbf{k}$. In the event that F is a circle of radius k_F, we simply let $\pi_F\mathbf{k} = k_F\frac{\mathbf{k}}{\|\mathbf{k}\|}$ and then $LG(k) = G(0, k_F\frac{\mathbf{k}}{\|\mathbf{k}\|}) = G(0,k_F)$ is independent of $\mathbf{k}$ by the rotation invariance of G.

The interaction for $\mathcal{V}^{j-1}$ is $\mathcal{W}^j$ which may be decomposed into a renormalized and a local part. Let $\mathbf{R} = 1 - \mathbf{L}$, then

$$\begin{aligned}\mathcal{W}^j(\psi^{\leq j}) &= \mathbf{R}\mathcal{W}^j(\psi^{\leq j}) + \mathbf{L}\mathcal{W}^j(\psi^{\leq j})\\ &= \sum_{i=j}^{0}\mathbf{R}\mathcal{V}^i(\psi^{\leq j}) + \sum_{i=j}^{-1}\mathbf{L}\mathcal{V}^i(\psi^{\leq j}) + \delta\mathcal{V}_r(\psi^{\leq j})\\ &= \sum_{i=j}^{0}\mathbf{R}\mathcal{V}^i(\psi^{\leq j}) + \int \frac{d^3k}{(2\pi)^3}\Big(\sum_{i=j}^{-1} LV_1^i(\mathbf{k},\lambda) + \delta e_r(\mathbf{k},\lambda)\Big)\,\psi^{\leq j}(k)\bar{\psi}^{\leq j}(k)\,.\end{aligned}$$

provided we define $\delta e_r(\mathbf{k},\lambda)$ to be invariant under $\mathbf{L}$.

If the quadratic part appears as a two legged vertex in the computation of $\mathcal{V}^{j-1}$, the Power Counting Lemma II.2 gives a factor (neglecting sectors)

$$\Big\|\sum_{i=j}^{-1} LV_1^i + \delta e_r\Big\|_{\emptyset} \times M^{-j}$$

which is big. On the other hand, the $\|\cdot\|_\emptyset$ norm of a two legged graph without two and four legged subgraphs is bounded by M^j which is a small number. Therefore one should choose δe_r such that $\sum_{i=j}^{-1} LV_1^i + \delta e_r$ can be treated as a kernel of a two legged graph of scale $\leq j-1$. That is, choose $\delta e_r = -\sum_{i=r}^{-1} LV_1^i$ or, if $LV_1^i(\mathbf{k},\lambda) = \sum_{l=1}^{\infty} LV_{1,l}^i(\mathbf{k})\lambda^l$,

$$\delta e_{r,l}^i(\mathbf{k}) = -LV_{1,l}^i(\mathbf{k}) \quad \forall r \leq i \leq -1,\; l \geq 1. \tag{II.23}$$

Note that $\mathcal{V}_1^i$ is defined through a functional integral whose integrand contains $\mathcal{W}^{i+1}$ and hence

$$\sum_{m=i+1}^{-1} LV_1^m + \delta e_r = -\sum_{m=r}^{i} LV_1^m = \sum_{m=r}^{i}\sum_{l=0}^{\infty} \delta e_{r,l}^m(\mathbf{k})$$

Thus (II.23) is of the form

$$\delta e^i_{r,l} = f^i_{r,l}\Big((\delta e^m_{r,l'})_{\substack{r\le m\le -1\\ 1\le l'\le l-1}} \Big) \tag{II.23a}$$

In particular, for $l = 1$, (II.23) is

$$\int \frac{d^3k}{(2\pi)^3} \delta e^i_{r,1}(\mathbf{k})\, \psi^{\le i}(k)\bar\psi^{\le i}(k) = -\mathbf{L}\int \Big(V(\psi^{\le i+1}) - V(\psi^{\le i}) \Big) d\mu_{C^{i+1}}$$

where V is the initial quartic interaction. The counterterm $\delta\mathcal{V}_r$ does not appear on the right hand side since there is at least one contraction. Thus $e^i_{r,l=1}$ is determined for all i. Iterating (II.23a) determines $e^i_{r,l}$ for $l > 1$, without the need for any estimates.

With this choice of δe_r, $\mathcal{W}^j$ becomes

$$\mathcal{W}^j(\psi^{\le j}) = \sum_{i=j}^{0} \mathbf{R}\mathcal{V}^i(\psi^{\le j}) - \sum_{i=r}^{j-1} \int \frac{d^3k}{(2\pi)^3} LV^i_1(\mathbf{k},\lambda)\, \psi^{\le j}(k)\bar\psi^{\le j}(k)\,. \tag{II.24}$$

Now, scales below j also appear on the right hand side of (II.24). That means that the estimates (II.16,17) for V^{j-1}_q contain scales below j, too. So we can no longer proceed by induction on scale. Nevertheless, there is another way to construct δe_r which allows one to prove the bounds on V^{j-1}_q by induction (on iteration steps, which play a rôle similar to the order of perturbation theory).

Lemma II.5 *Let $r > -\infty$ be the infrared cutoff and let $\underline{\mathcal{U}}$ be a vector of effective potentials $\underline{\mathcal{U}} = (\mathcal{U}^0, \mathcal{U}^{-1}, \cdots, \mathcal{U}^r)$. Define a sequence of vectors $\underline{\mathcal{U}}^k = (\mathcal{U}^{k,0}, \mathcal{U}^{k,-1}, \cdots, \mathcal{U}^{k,r})$ by*

$$\underline{\mathcal{U}}^0(\psi^{\le 0}) = (-\lambda\mathcal{V}(\psi^{\le 0}), 0, \cdots, 0)$$

$$\mathcal{U}^{k+1,0} = -\lambda\mathcal{V} \quad \forall k \ge 0$$

and for all $r + 1 \le j \le 0$, $k \ge 0$

$$\mathcal{U}^{k+1,j-1}(\psi^{\le j-1}) = Q_4 \log\frac{1}{I^k_j} \quad \times$$

$$\int e^{\sum_{i=j}^{0} \mathbf{R}\mathcal{U}^{k,i}(\psi^{\le j};\psi^{\le j-1}) - \sum_{i=r}^{j-1} \mathbf{L}\mathcal{U}^{k,i}(\psi^{\le j};\psi^{\le j-1})} d\mu_{C^j}\,, \tag{II.25}$$

$$I^k_j = \int e^{\sum_{i=j}^{0} \mathbf{R}\mathcal{U}^{k,i}(\psi^j) - \sum_{i=r}^{j-1} \mathbf{L}\mathcal{U}^{k,i}(\psi^j)} d\mu_{C^j}$$

where we have used the abbreviation $\mathcal{U}^{k,i}(\psi^{\le j};\psi^{\le j-1}) = \mathcal{U}^{k,i}(\psi^{\le j}) - \mathcal{U}^{k,i}(\psi^{\le j-1})$. Here $\mathbf{L}$ is the localization operator (II.22) and $\mathbf{R} = 1 - \mathbf{L}$.

Let n be an arbitrary natural number and P_n be the operator which projects out the nth order coefficient with respect to λ. Then one has

$$\forall s \ge 0 \qquad P_n\underline{\mathcal{U}}^{2n+s} = P_n\underline{\mathcal{U}}^{2n-1}$$

and, if $\mathcal{V}^j$ *are the potentials defined in (II.21) such that* $\mathcal{W}^r = \sum_{i=r}^{-1} \mathcal{V}^i + \delta\mathcal{V}_r$,

$$\forall r \le j \le -1 \qquad P_n\,\mathcal{U}^{2n,j} = P_n\,\mathcal{V}^j$$

where we used $\underline{\mathcal{U}}^{2n}$ *instead of* $\underline{\mathcal{U}}^{2n-1}$ *since it makes the formulas shorter. In particular, the coefficients up to* n^{th} *order of the the solution* δe_r *of the equations (II.23) are given by*

$$\forall 1 \le l \le n \qquad \delta e_{r,l} = \sum_{i=r}^{-1} \delta e_{r,l}^i = -\sum_{i=r}^{-1} LU_{1,l}^{2n,i}$$

if

$$\mathcal{U}^{2n,i}(\psi) = \sum_{\substack{q=1\\ q\neq 2}}^{\infty}\sum_{l=1}^{\infty} \lambda^l \int d\underline{\xi}\, U_{q,l}^{2n,i}(\xi_1,\cdots,\xi_{2q})\,\psi(\xi_1)\cdots\psi(\xi_{2q})\,.$$

Altogether, if W_q^r *are the kernels of the effective potential without four legged subgraphs defined in Theorem II.1, we have*

$$\sum_{l=1}^{n} \lambda^l W_{q,l}^r = \sum_{l=1}^{n}\sum_{i=r}^{0} \lambda^l U_{q,l}^{2n,i} - \delta_{q,1}\sum_{l=1}^{n}\sum_{i=r}^{-1} \lambda^l LU_{1,l}^{2n,i}\,. \tag{II.26}$$

Remark. *There is an analogous version of this lemma for the construction of the full effective potential, including four legged subgraphs. The only modifications to be made are to omit the operator* Q_4 *in the definition of* $\mathcal{U}^{k+1,j-1}$ *and to define* $\mathcal{U}^{k+1,0} = -\lambda\mathcal{V}$ *for all* $k \ge 0$.

By (II.26) we have to bound $U_{q,l}^{2n,i}$ and $LU_{1,l}^{2n,i}$. But (II.25) expresses $\underline{\mathcal{U}}^{k+1}$ in terms of $\underline{\mathcal{U}}^k$. That is, we do a single scale expansion in (II.25) and then proceed by induction on k.

The single scale expansion is done as above. It is generated by the integration by parts formula Lemma II.4. We Taylor expand the fields at the beginning of each step of partial integration and use sectors of length $M^{\frac{j}{2}}$. We apply the Combinatorial Tree Lemma II.3 to control the various sums produced in the expansion. Sector sums are estimated by the Sector Counting Lemma II.3' of [FMRT1]. The result is

$$|||U_q^{k+1,j-1}|||_{\emptyset,\Sigma_j,\le n} \le \sum_{\gamma\in\mathcal{P}_{n,q}} |||K_\gamma^{k+1,j-1}|||_{\emptyset,\Sigma_j,\le n} \le$$

$$\sup_{\gamma\in\mathcal{P}_{n,q}} \Big\{ c_2^{\sum_v q_v} c_1^{\sum_v q_v - q} \prod_{v\in V\setminus V_2} \Big(M^{\frac{j}{2}(q_v-2)} \sum_{j_v=j}^{-1} M^{\frac{j_v+1}{2}(2q_v-3)} |||U_{q_v}^{k,j_v}|||_{\emptyset,\Sigma_{j_v+1},\le n} \Big) \times$$

$$\prod_{v\in V_{2,R}} \Big(\sum_{j_v=j}^{-1} |||\,|x|\,U_1^{j_v}|||_{\emptyset,\le n} \Big) \prod_{v\in V_{2,L}} \Big(\sum_{j_v=r}^{j-1} |||U_1^{k,j_v}|||_{\emptyset,\le n} M^{-j} \Big)$$

$$M^{-\frac{j}{2}(q-2)} M^{-\frac{j}{2}(2q-3)} \Big\} . \tag{II.27}$$

We now briefly explain the power counting of the renormalized and counterterm two legged vertices. If a renormalized vertex of scale j_v contributes to $K_\gamma^{k+1,j-1}$, one obtains a factor[3] $\sup_{k \in \mathrm{supp} C^j(k)} |RU_1^{k,j_v}(k)|$ which is estimated as follows

$$\begin{aligned} |RU_1^{k,j_v}(k)| &= |U_1^{k,j_v}(k) - LU_1^{k,j_v}(k)| = |U_1^{k,j_v}(k_0, \mathbf{k}) - U_1^{k,j_v}(0, \pi_F \mathbf{k})| \\ &\le |\partial_{k_0} U_1^{k,j_v}(\tilde{k})| \cdot |k_0| + |\nabla_{\mathbf{k}} U_1^{k,j_v}(\tilde{\tilde{k}})| \cdot |\mathbf{k} - \pi_F \mathbf{k}| \\ &\le \| |x_0| U_1^{k,j_v} \|_\emptyset \cdot |k_0| + \| |\mathbf{x}| U_1^{k,j_v} \|_\emptyset \cdot |\mathbf{k} - \pi_F \mathbf{k}| \\ &\le 2 \| |x| U_1^{k,j_v} \|_\emptyset \cdot M^j \qquad \text{on the support of } C^j(k) . \end{aligned}$$

A counterterm vertex LU_1^{k,j_v} is simply estimated by

$$\sup_{k \in \mathrm{supp} C^j(k)} |LU_1^{k,j_v}(k)| \le \sup_k |U_1^{k,j_v}(0, \pi_F \mathbf{k})| \le \| U_1^{k,j_v} \|_\emptyset .$$

Assigning an M^{-j} to each counterterm vertex as in (II.27), one can say that there is an additional M^j for each two legged vertex. Furthermore, by conservation of momentum, there are no sector sums coming with a two legged vertex. Therefore we obtain (compare the discussion following (II.15))

$$M^{\frac{3}{2}j(\sum_v q_v - q - \sum_v 1 + 1)} M^{-j(\sum_v 1 - 1)} \prod_{v \in V \setminus V_2} M^{-\frac{j-(j_v+1)}{2}(2q_v - 3)} \prod_{v \in V_2} M^j =$$

$$\prod_{v \in V \setminus V_2} \left(M^{\frac{j}{2}(q_v - 2)} M^{\frac{j_v+1}{2}(2q_v - 3)} \right) M^{-\frac{j}{2}(q-2)} M^{-\frac{j}{2}(2q-3)} ,$$

which is the power counting of (II.27).

If some legs of $U_q^{k+1,j-1}$ are integrated against test functions, one obtains

$$\| U_q^{k+1,j-1} \|_{S, \Sigma_j, \le n} \le$$

$$\sum_{\gamma \in \mathcal{P}_{n,q}} \| K_\gamma^{k+1,j-1} \|_{S, \Sigma_j, \le n} \le \sup_{\gamma \in \mathcal{P}_{n,q}} \Big\{ c_2^{\sum_v q_v - |S|} c_1^{\sum_v q_v - q}$$

$$\prod_{v \in V_{int} \setminus V_2} \left(M^{\frac{j}{2}(q_v - 2)} \sum_{j_v = j}^{-1} M^{\frac{j_v+1}{2}(2q_v - 3)} \| U_{q_v}^{k,j_v} \|_{\emptyset, \Sigma_{j_v+1}, \le n} \right) \times$$

$$\prod_{v \in V_{ext} \setminus V_2} \left(M^{\frac{j}{2}(q_v - \frac{1}{2}|S_v|)} \sum_{j_v = j}^{-1} M^{\frac{j_v+1}{2}(2q_v - |S_v|)} \| U_{q_v}^{k,j_v} \|_{S_v, \Sigma_{j_v+1}, \le n} \right) \times$$

[3] One has to be a little careful in going to momentum space, since the Taylor operators $(x - c) \cdot \nabla_x$ break translation invariance, it still comes down to the above estimate.

$$\prod_{v\in V_{2,R}}\Big(\sum_{j_v=j}^{-1} |||\,|x|U_1^{k,j_v}|||_{\emptyset,\le n}\Big)\prod_{v\in V_{2,C}}\Big(\sum_{j_v=r}^{j-1} |||U_1^{k,j_v}|||_{\emptyset,\le n}M^{-j}\Big)\times$$

$$M^{-\frac{j}{2}(q-\frac{1}{2}|S|)}M^{-\frac{j}{2}(2q-|S|)}\prod_{f\ at\ V_{2,ext}}\|f\|_{L^1}\Big\}. \tag{II.28}$$

which may be compared to (II.17).

We are now in a position to state the induction hypothesises on $\underline{\mathcal{U}}^k$. They are

$$|||U_q^{k,j-1}|||_{\emptyset,\Sigma_j,\le n} \le |\lambda|^{\frac{q}{2}}M^{-\frac{j}{2}(3q-5)} \tag{II.29}$$

$$|||\,|x|^w\,U_1^{k,j-1}|||_{\emptyset,\le n} \le |\lambda|^{\frac{1}{2}}M^{-wj}M^{\frac{3}{2}j}, \qquad w\in\{0,1\} \tag{II.30}$$

$$|||U_q^{k,j-1}|||_{S,\Sigma_j,\le n} \le \binom{2q}{|S|}^{-1}p(j)^{|S|}\,|\lambda|^{\frac{q}{2}}\,M^{-\frac{j}{2}(3q-\frac{3}{2}|S|-\frac{1}{4})}\prod_{k\in S}\Big(\|f_k\|_{L^1}+\|f_k\|_{L^\infty}\Big) \tag{II.31}$$

Using (II.27,28), they are verified for $\underline{\mathcal{U}}^{k+1}$. In particular, (II.29-30) hold for $k=2n$. Then we can apply (II.26) of Lemma II.5 to conclude

$$\begin{aligned}
\int\prod_{i=1}^{2q}d\xi_i\,|f_i(\xi_i)|\,|W_q^r(\xi_1,\cdots,\xi_{2q})| &= |||W_q^r|||_{\{1,\cdots,2q\}}\\
&\le \sum_{i=r}^{-1}|||U_q^{2n,i}|||_{\{1,\cdots,2q\}}+\delta_{q,1}\sum_{i=r}^{-1}|||LU_q^{2n,i}|||_{\{1,2\}}\\
&= \sum_{i=r}^{-1}|||U_q^{2n,i}|||_{\{1,\cdots,2q\},\Sigma_{i+1}}+\delta_{q,1}\sum_{i=r}^{-1}|||LU_q^{2n,i}|||_{\{1,2\},\Sigma_{i+1}}\\
&\le 4\,c_3^{2q}\,|\lambda|^{\frac{q}{2}}\prod_{k=1}^{2q}\Big(\|f_k\|_{L^1}+\|f_k\|_{L^\infty}\Big).
\end{aligned}$$

This completes our discussion of Theorem II.1.

3 Four–Legged Diagrams

Spin plays no role in this section. So we suppress it. Feynman diagrams in this model have lines

k  $= \dfrac{1}{ik_0-e(\mathbf{k})}$

and vertices

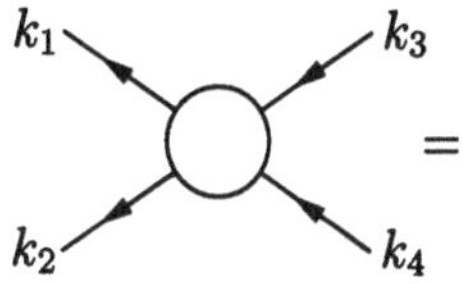

$$= (2\pi)^{d+1}\delta(k_1 + k_2 - k_3 - k_4)\lambda \langle k_1, k_2|\mathrm{V}|k_3, k_4\rangle$$

For example

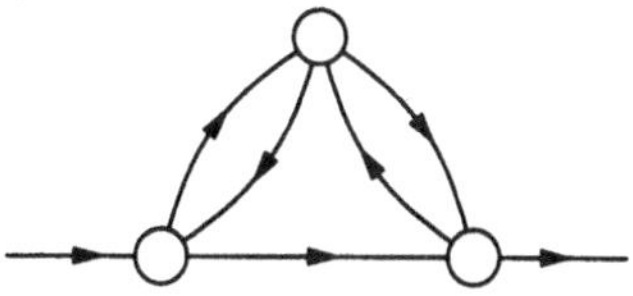

is one graph contributing to the proper self energy. This is a three loop graph. Choosing the loops as in

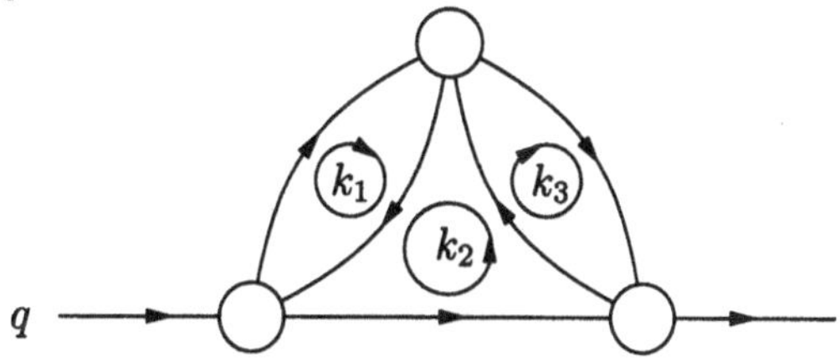

we see that the value of this graph is

$$\int đk_1\, đk_2\, đk_3 \frac{1}{i(k_1)_0 - e(\mathbf{k}_1)} \frac{1}{i(k_1+k_2)_0 - e(\mathbf{k}_1+\mathbf{k}_2)}$$
$$\frac{1}{i(k_2+k_3)_0 - e(\mathbf{k}_2+\mathbf{k}_3)} \frac{1}{i(k_3)_0 - e(\mathbf{k}_3)} \frac{1}{i(k_2+q)_0 - e(\mathbf{k}_2+\mathbf{q})}$$
$$\langle k_1+k_2, k_3|\mathrm{V}|k_1, k_2+k_3\rangle \langle k_1, k_2+q|\mathrm{V}|q, k_1+k_2\rangle \langle k_2+k_3, q|\mathrm{V}|k_2+q, k_3\rangle$$

It is not clear that this integral converges. The domain of integration is compact, because of the ultraviolet cutoff, but the integrand is singular.

To check for convergence one does "naive power counting" bounds. In field theory propagator singularities occur at points. Then power counting just comes down to some simple dimensional analysis. Here there are singularities on curves, like $(k_1)_0 = 0,\ \mathbf{k}_1 \in F$. We have to have a simple yet precise way of measuring whether the integrand is large a lot. To do so we decompose the propagator

$$C(k) = \frac{1}{ik_0 - e(\mathbf{k})}$$
$$= \sum_{j=-\infty}^{0} C^{(j)}$$

where

$$C^{(j)}(k) = \frac{1}{ik_0 - e(\mathbf{k})}\chi\left(2^j \le |ik_0 - e(\mathbf{k})| < 2^{j+1}\right)$$

Note, the perhaps bizarre, convention that j is negative. As j tends to *minus* infinity, 2^j approaches zero and, on the support of $C^{(j)}$, $|ik_0 - e(\mathbf{k})|$ approaches zero. Naive power counting just uses

Lemma III.1 *Let d be arbitrary and Hypothesis I be satisfied. Then*

a)

$$\|C^{(j)}\|_\infty = \sup_k |C^{(j)}(k)| \le 2^{-j}$$

b)

$$\|C^{(j)}\|_1 = \int \not{d}k \; |C^{(j)}(k)| \le \text{const}\, 2^j$$

Proof: Part a) is obvious because, by construction, $|ik_0 - e(\mathbf{k})| \ge 2^j$ on the support of $C^{(j)}(k)$.

For part b) observe that

$$\begin{aligned} &\mathrm{vol}\left\{ \, k = (k_0, \mathbf{k}) \;\middle|\; C^{(j)}(k) \ne 0 \, \right\} \\ &\le \mathrm{vol}\left\{ \, k_0 \;\middle|\; |k_0| \le 2^{j+1} \, \right\} \mathrm{vol}\left\{ \, \mathbf{k} \in B \;\middle|\; |e(\mathbf{k})| \le 2^{j+1} \, \right\} \\ &\le 2^{j+2} \mathrm{vol}\left\{ \, \mathbf{k} \in B \;\middle|\; |e(\mathbf{k})| \le 2^{j+1} \, \right\} \end{aligned}$$

The set $\left\{ \, \mathbf{k} \in B \;\middle|\; |e(\mathbf{k})| \le 2^{j+1} \, \right\}$ consists of a shell of thickness $O(2^j)$ around F and hence has volume bounded by const 2^j so that

$$\mathrm{vol}\left\{ \, k = (k_0, \mathbf{k}) \;\middle|\; C^{(j)}(k) \ne 0 \, \right\} \le \text{const}\, 2^{2j} \tag{III.1}$$

and

$$\begin{aligned} \|C^{(j)}\|_1 = \int \not{d}k \; |C^{(j)}(k)| &\le \sup_k |C^{(j)}(k)| \mathrm{vol}\left\{ \, k = (k_0, \mathbf{k}) \;\middle|\; C^{(j)}(k) \ne 0 \, \right\} \\ &\le \text{const}\, 2^j \end{aligned}$$

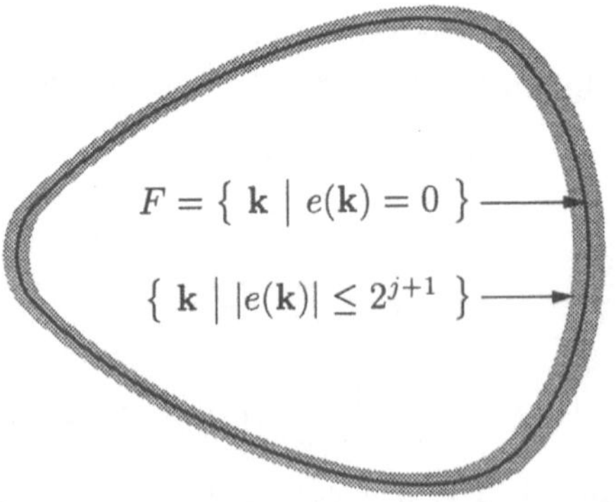

■

We remark that the smoothness condition $\nabla e(\mathbf{k}) \neq 0$ of Hypothesis I was used to get the volume bound (III.1). The corresponding volume for the Hubbard model at half filling is $|j|2^{2j}$ which leads to $\|C^{(j)}\|_1 \leq \text{const}\,|j|2^{j}$.

The analog of Lemma III.1 for the infrared Φ_4^4 model is $\|C^{(j)}\|_\infty \leq 2^{-2j}$, $\|C^{(j)}\|_1 \leq \text{const}\,2^{2j}$. The replacement $j \to 2j$ can be viewed simply as a change of units. So it is not too surprising that Lemma III.1 implies [FT2, FMRT1] that models satisfying Hypotheses I and IV obey bounds typical of strictly renormalizable models in the infrared regime. Two legged are linearly divergent and must be renormalized. Four legged subdiagrams are marginal and all other subdiagrams are convergent. As is normal for infrared models, the two legged counterterm is finite and the marginality of four legged subdiagrams does not require a counterterm. The four legged subdiagrams are divergent only for certain exceptional momenta and then only logarithmically divergent. These logarithmic singularities are integrable and hence do not prevent diagrams from being well defined. But they can cause the values of diagrams containing many four legged subdiagrams to be anomalously large through

$$\int \not{d}k \ln^n |k| \sim n!$$

Normally, under these circumstances one of two possibilities occur. The renormalization group flow of the four point function is either asymptotically free or is to a nontrivial fixed point and is accompanied by some interesting physics, like mass generation or symmetry breaking. We shall now see that under Hypotheses I-IV, the bounds which give marginality of four legged subdiagrams are not saturated. Four legged subdiagrams are in fact convergent. The models behave more like superrenormalizable models than strictly renormalizable ones.

To be concrete, we'll first do the naive power counting bound explicitly on one simple, but very important, graph – the particle-particle bubble

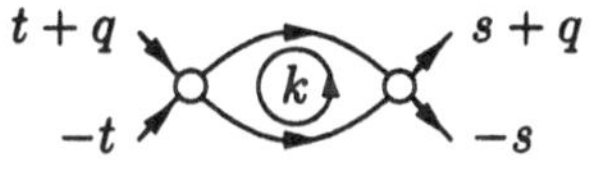

If the total momentum entering from the left is q, the value of this graph is

$$B(s,t,q) = \int \not{d}k\; C(-k+q)C(k)\,\langle -k+q, k|\mathrm{V}|t+q, -t\rangle\,\langle s+q, -s|\mathrm{V}| -k+q, k\rangle$$

Decomposing the two propagators into scales and then bounding the integral by the supremum of the integrand times the volume of the support of the integrand, we have

$$|B(s,t,q)| =$$

$$\left|\sum_{j_1,j_2\leq 0} \int_B \not{d}k\; C_{j_1}C_{j_2}\,\langle -k+q, k|\mathrm{V}|t+q, -t\rangle\,\langle s+q, -s|\mathrm{V}| -k+q, k\rangle\right|$$

$$\leq \sum_{j_1,j_2\leq 0} \|V\|_\infty^2 2^{-j_1-j_2}\,\times$$

$$\mathrm{vol}\{\ k \in \mathcal{B}\ |\ |ik_0 - e(\mathbf{k})| \le 2^{j_1+1}, |i(-k+q)_0 - e(-\mathbf{k}+\mathbf{q})| \le 2^{j_2+1}\ \} \le$$

$$\sum_{j_1,j_2\le 0} \|V\|_\infty^2 2^{-j_1-j_2} 2^{\min\{j_1,j_2\}} \mathrm{vol}\{\ \mathbf{k} \in \mathrm{B}\ |\ |e(\mathbf{k})| \le 2^{j_1+1} \quad |e(-\mathbf{k}+\mathbf{q})| \le 2^{j_2+1}\ \} \tag{III.2}$$

Even without using Hypotheses II and III we can bound the volume

$$\mathrm{vol}\{\ \mathbf{k} \in \mathrm{B}\ |\ |e(\mathbf{k})| \le 2^{j_1+1} \quad |e(-\mathbf{k}+\mathbf{q})| \le 2^{j_2+1}\ \}$$

$$\le \min\left\{\mathrm{vol}\{\ \mathbf{k} \in \mathrm{B}\ |\ |e(\mathbf{k})| \le 2^{j_1+1}\ \}, \mathrm{vol}\{\ \mathbf{k} \in \mathrm{B}\ |\ |e(-\mathbf{k}+\mathbf{q})| \le 2^{j_1+1}\ \}\right\}$$

$$\le \mathrm{const}\ \min\left\{2^{j_2}, 2^{j_2}\right\} = \mathrm{const}\ 2^{\min\{j_1,j_2\}} \tag{III.3}$$

This gives

$$\sup_{s,t,q} |B(s,t,q)| \le \sum_{j_1,j_2\le 0} \mathrm{const}\ \|V\|_\infty^2 2^{-j_1-j_2} 2^{2\min\{j_1,j_2\}}$$

$$= \sum_{j_1,j_2\le 0} \mathrm{const}\ \|V\|_\infty^2 2^{-|j_1-j_2|}$$

$$= \sum_{j_1\le 0} \mathrm{const}\ \|V\|_\infty^2$$

which diverges logarithmically. Recall that 2^j has the units of energy.

In the event that $e(\mathbf{k}) = e(-\mathbf{k})$, violating Hypothesis III, and $\mathbf{q} = 0$ we have

$$\mathrm{vol}\{\ \mathbf{k} \in \mathrm{B}\ |\ |e(\mathbf{k})| \le 2^{j_1+1} \quad |e(-\mathbf{k}+\mathbf{q})| \le 2^{j_2+1}\ \}$$

$$= \mathrm{vol}\{\ \mathbf{k} \in \mathrm{B}\ |\ |e(\mathbf{k})| \le 2^{\min\{j_1,j_2\}+1}\ \}$$

$$= O\left(2^{\min\{j_1,j_2\}+1}\right)$$

and (III.3) is saturated. In this case $q = 0$ really is an exceptional momentum for $B(q)$ which really does have a logarithmic singularity at $q = 0$.

We now turn on Hypotheses II,III and show that then the above bound is not saturated and that four legged subgraphs are really convergent so that the model really acts superrenormalizable. By Hypotheses III and analyticity (or even with just Hypothesis II if $\mathbf{q} \ne 0$) the Fermi curve F can only meet the reflected translated Fermi curve $-F+\mathbf{q}$ transversely or with a tangency of some finite order. Hence there is an $\epsilon > 0$ such that

$$\mathrm{vol}\{\ \mathbf{k} \in \mathrm{B}\ |\ |e(\mathbf{k})| \le 2^{j_1+1} \quad |e(-\mathbf{k}+\mathbf{q})| \le 2^{j_2+1}\ \} \le \mathrm{const}\ 2^{\min\{j_1,j_2\}} 2^{\epsilon\max\{j_1,j_2\}}$$

Here const $2^{\min\{j_1,j_2\}}$ is the thickness of each component of the intersection of the two shells and const $2^{\epsilon\max\{j_1,j_2\}}$ is a bound on the length of each component.

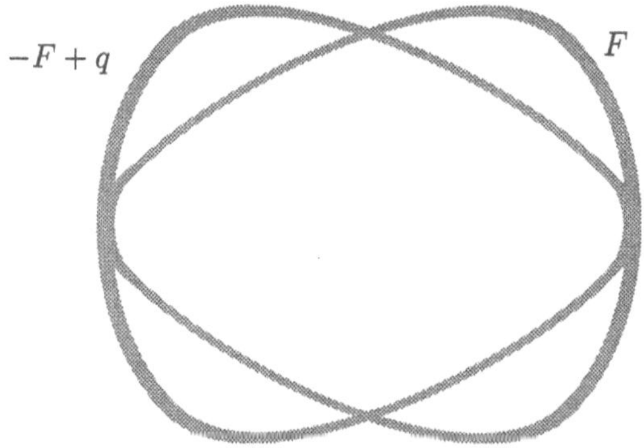

Substituting (III.3a) into (III.2) gives

$$\sup_{s,t,q}|B(s,t,q)| \le \sum_{j_1,j_2\le 0} \text{const}\,\|V\|_\infty^2 2^{-j_1-j_2} 2^{2\min\{j_1,j_2\}} 2^{\epsilon\max\{j_1,j_2\}}$$

$$= \sum_{j_1,j_2\le 0} \text{const}\,\|V\|_\infty^2 2^{-|j_1-j_2|} 2^{\epsilon\max\{j_1,j_2\}}$$

$$= \sum_{j\le 0} \text{const}\,\|V\|_\infty^2 2^{\epsilon j}$$

$$< \infty$$

When Hypothesis III is turned on the particle-particle bubble becomes uniformly bounded.

Of course the particle-particle bubble is just one graph. As a second example we consider the second most important graph in our class of models – the particle hole bubble

The value of this bubble is

$$B_2(s,t,q) = \int d\!\!\!{}^-k\; C(k+q)C(k)\,\langle t-q,k|\mathrm{V}|k,t\rangle\,\langle s,k|\mathrm{V}|k+q,s+q\rangle$$

When Hypothesis II is satisfied and when q is bounded away from zero, we can apply the same argument as in the particle-particle bubble, now using the fact that shells around F and $F+q$ have small intersections.

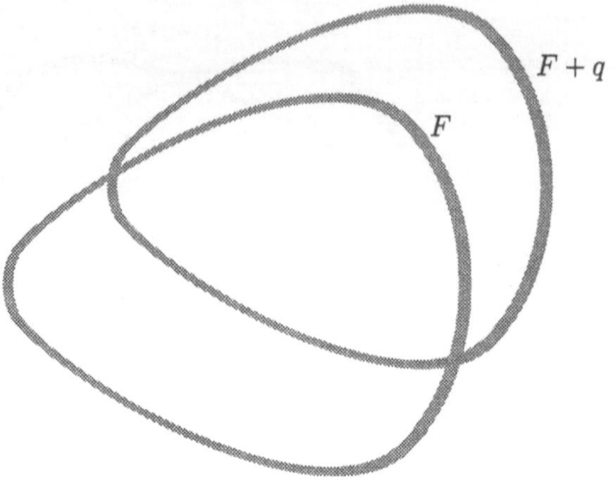

As an illustration of what happens when q is small, consider $q=0$. Then, changing variables to

$$x = k_0$$
$$y = e(\mathbf{k})$$

and some angular variable and performing the integral over the angular variable, we have

$$B_2(s,t,0) = \int d\!\!{}^-k \, \frac{1}{[ik_0 - e(\mathbf{k})]^2} \langle t,k|\mathrm{V}|k,t\rangle \langle s,k|\mathrm{V}|k,s\rangle$$
$$= \int dxdy \, \frac{1}{[ix-y]^2} I(x,y)$$

with $I(x,y)$ being some C^∞ function. Making the further change of variables to polar coordinates

$$B_2(s,t,0) = \int drd\theta \, \frac{r}{i[re^{i\theta}]^2} I(r\cos\theta, r\sin\theta)$$
$$= \int drd\theta \, \frac{r}{i[re^{i\theta}]^2} [I(0,0) + O(r)]$$

The potentially logarithmically divergent term

$$\int drd\theta \, \frac{1}{ire^{2i\theta}} I(0,0)$$

vanishes because

$$\int_0^{2\pi} d\theta \; e^{in\theta} = 0$$

for all nonzero integers n. Hence $B_2(s,t,0)$ is bounded.

Higher order graphs fall into two categories. There are strings of bubbles, like

and

that can be treated as above. And there are graphs like

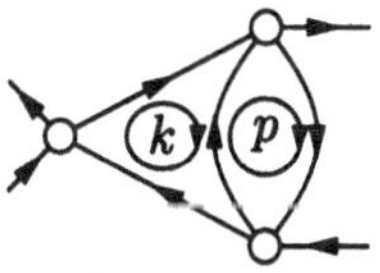

which have overlapping loops. Because the k-loop and the p-loop share a line, all three propagators $C^{(j_1)}(k)C^{(j_2)}(p)C^{(j_3)}(p-k)$ appear in the integrand. The supports properties of these propagators constrains the domain of integration to

$$\{\ (\mathbf{k},\mathbf{p}) \in \mathrm{B}\times\mathrm{B}\ \big|\ |e(\mathbf{k})| \le 2^{j_1+1},\ |e(\mathbf{p})| \le 2^{j_2+1},\ |e(\mathbf{p}-\mathbf{k})| \le 2^{j_3+1}\ \}$$

The third condition gives some "volume improvement" over naive power counting. See [FKLT1, FKST1]. So all four legged subdiagrams have convergent, rather than marginal, power counting.

References

[FKLT1] . Feldman, H. Knörrer, D. Lehmann and E. Trubowitz, in preparation.

[FKST1] . Feldman, H. Knörrer, M. Salmhofer and E. Trubowitz, Renormalization Theory for Many-Fermion Systems: One-Band Models with Non-Nested Fermi Surfaces, in preparation.

[FMRT1] . Feldman, J. Magnen, V. Rivasseau and E. Trubowitz, An Infinite Volume Expansion for Many Fermion Green's Functions, Helvetica Physica Acta, **65** (1992) 679-721.

[FMRT2] . Feldman, J. Magnen, V. Rivasseau and E. Trubowitz, Fermionic Many-Body Models, in *Mathematical Quantum Theory I: Field Theory and Many-Body Theory*, J. Feldman, R. Froese and L. Rosen eds, CRM Proceedings Lecture Notes.

[FMRT3] . Feldman, J. Magnen, V. Rivasseau and E. Trubowitz, Two Dimensional Many Fermion Systems as Vector Models, Europhysics Letters, **24** (1993) 521-526.

[FMRT4] . Feldman, J. Magnen, V. Rivasseau and E. Trubowitz, An Intrinsic 1/N Expansion for Many Fermion Systems, Europhysics Letters, **24** (1993) 437-442.

[FT1] . Feldman and E. Trubowitz, Perturbation Theory for Many Fermion Systems, Helvetica Physica Acta **63** (1990) 156-260.

[FT2] . Feldman and E. Trubowitz, The Flow of an Electron-Phonon System to the Superconducting State, Helvetica Physica Acta, **64** (1991) 214-357.

Higher order graphs fall into two categories. There are strings of bubbles, like

and

that can be summed as above. And there are graphs like

The Self-Avoiding Walk in Four Dimensions*

Steven E. Golowich

Department of Mathematics,
Princeton University, Princeton NJ 08544

Abstract
Recent progress on the self-avoiding walk on a four dimensional hypercubic lattice is presented. Methods for determining the long-time behavior of such walks by studying the Green's function are described.

1 Introduction

We study a nearest-neighbor weakly self-avoiding random walk, in continuous time, on the hypercubic lattice in four dimensions. This is defined by putting a measure on the space of simple random walks. For the simple process, the probability of traveling from 0 to x in time T satisfies

$$\begin{aligned} P_0(x,T) &= E(\mathbf{1}_{\omega(T)=x}|\omega(0)=0) \\ &= \left(e^{T\Delta}\right)_{0,x} \end{aligned}$$

if the jumping rate is chosen appropriately. Here $\omega : [0,T] \to \mathbb{Z}^d$ is a sample path. In order to define the self-avoiding walk, we first note that a measure of the self-intersections of a walk ω of length T is

$$\frac{1}{2}\int dx\, \tau_x^2 = \int_{0\le s\le t\le T} ds\, dt\, \mathbf{1}_{\omega(s)=\omega(t)},$$

where $\tau_x = \int_0^T ds \mathbf{1}_{\omega(s)=x}$ is the local time spent at site x. We use this to discourage walks from intersecting themselves via

$$P_\lambda(x,t) = E\left(e^{-\frac{1}{2}\lambda\int dx\,\tau_x^2}\mathbf{1}_{\omega(t)=x}|\omega(0)=0\right).$$

As it is defined here, P_λ needs further normalization to be a probability density, but this form is the most convenient to work with.

* Lecture given at the Workshop on Constructive Results in Field Theory, Statistical Mechanics and Condensed Matter Physics held at Ecole Polytechnique, Palaiseau, France, from July 25–27, 1994.

The SAW has long been studied via heuristic methods. Renormalization group calculations predict that, in $d = 4$ dimensions, the mean-square end-to-end distance behaves like [dG]

$$\begin{aligned}\langle \omega(T)^2 \rangle_\lambda &= \frac{1}{N} \sum_x x^2 P_\lambda(x,T) \\ &\sim T \log^{1/4} T, \end{aligned} \qquad (1)$$

as $T \to \infty$. Here $N = \sum_x P_\lambda(x,T)$ is the normalization. Our goal is to rigorously verify this prediction. In this talk we present results concerning the Green's function for this SAW which we expect to extend to yield (1). We will state the results precisely later; first, we describe what type of estimates on the Green's function we need in order to extract knowledge about walks of fixed length.

The interacting Green's function is defined as the Laplace transform of P_λ:

$$\begin{aligned}G_\lambda(x,a) &= \int_0^\infty dt\, e^{-at} P_\lambda(x,t) \\ &= \int_0^\infty dt\, E\left(e^{-\int dx\, (a\tau_x + \frac{1}{2}\lambda\tau_x^2)} \mathbf{1}_{\omega(t)=x} | \omega(0) = 0\right).\end{aligned}$$

The parameter a is called the killing rate. P_λ is recovered by inverting the transform:

$$P_\lambda(x,T) = \frac{1}{2\pi i} \int_\gamma da\, e^{aT} G_\lambda(x,a).$$

We will deform the contour γ to a convenient form below. First, one expects that as a function of a, the Green's function G_λ will have a singularity at a point $a_c = O(-\lambda)$. (Iagolnitzer and Magnen have, in fact, constructed G_λ at $a = a_c$ for a closely related model). Our strategy is to prove that G_λ is analytic in a region $\mathcal{D}$ consisting of the piece of $\mathbb{C}$ to the right of two rays, starting at a_c, and extending to ∞ at angles of $\pm(\frac{\pi}{2} + \delta)$, for some $\delta > 0$ (see Figure I).

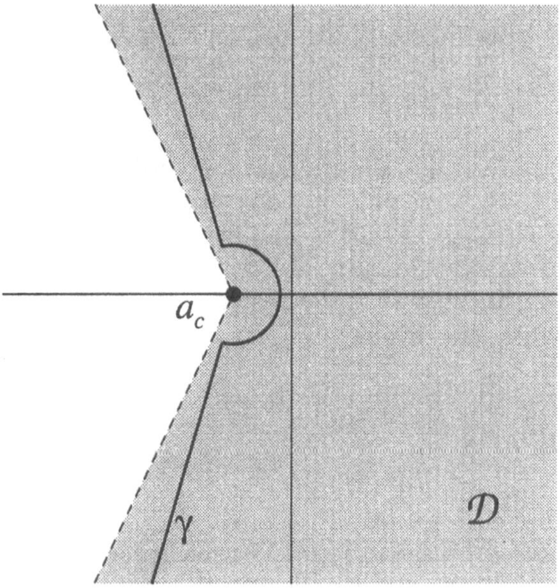

Figure I
The shaded region is $\mathcal{D}$, the set in which we need estimates and analyticity of G_λ. The contour γ is used in inverting the Laplace transform.

Inside this set, G_λ should obey the estimate

$$G_\lambda(x,\beta) = (1+O(\lambda))G_0(x,\beta_{\text{eff}}) + G_{\text{rem}}$$

where we have defined $\beta = a - a_c$, and $\beta_{\text{eff}} \sim \beta(-\log\beta)^{-1/4}$. G_{rem} is a small remainder. Now we proceed in a manner similar to Brydges and Imbrie in addressing the hierarchical version of this problem [BI], and deform the contour γ to be an arc of radius $|\beta| = T^{-1}$ and two rays extending off to ∞ that satisfy $\arg\beta = \pm(\frac{\pi}{2}+\epsilon)$. The purpose of this deformation is that, if we write G_λ as $(1+O(\lambda))G_0(x,\beta(\log T)^{-1/4}) + G'_{\text{rem}}$, then the G_0 term gives the leading behavior in $\langle\omega(T)^2\rangle_\lambda$, while we can show the remainder will only be a small correction by using the exponential decay from e^{aT} along the two rays of γ.

We now proceed to the heart of the problem, that of controlling the Green's function in the appropriate region of $\mathbb{C}$.

The first step is to use the identity

$$e^{-\frac{1}{2}\lambda\int dx\,\tau_x^2} = \int d\mu_I(\sigma) e^{i\sqrt{\lambda}\int dx\,\tau_x\sigma_x} \tag{2}$$

to rewrite G_λ as

$$G_\lambda(x,a) = \int d\mu_I(\sigma)\int_0^\infty dt\, E\left(e^{-\int dx\,\tau_x(a-i\sqrt{\lambda}\sigma_x)}\mathbf{1}_{\omega(t)=x}|\omega(0)=0\right)$$

$$= \int d\mu_I(\sigma) \int_0^\infty dt \left(e^{-t(-\Delta+a-i\sqrt{\lambda}\sigma)}\right)_{0x}$$

$$= \int d\mu_I(\sigma)(-\Delta + a - i\sqrt{\lambda}\sigma)^{-1}_{0x}. \tag{3}$$

In the first line, we see that the consequence of using (2) is that, instead of an interacting walk, we are seeing a noninteracting walk moving in an imaginary random potential $i\sqrt{\lambda}\sigma_x$. In the second line we have used the Feynman-Kac formula, to arrive at the form (3).

At this point, there are two choices for analyzing G_λ. One is to continue rewriting G_λ in yet another guise:

$$G_\lambda(x,a) = \int d\mu_I(\sigma) \int (d\Phi)\, \varphi_0 \bar{\varphi}_x\, e^{-(\Phi,(-\Delta+a-i\sqrt{\lambda}\sigma)\bar{\Phi})}$$

$$= \int d\mu_C(\Phi)\, \varphi_0 \bar{\varphi}_x\, e^{-\lambda \int dx\, \Phi_x^4},$$

where $C = (-\Delta + a)^{-1}$, and $d\mu_C(\Phi)$ is a combined bosonic-fermionic Gaussian measure; the field Φ is a superfield comprised of two bosonic and two Fermionic degrees of freedom per site x. This is the supersymmetric field theory representation of the SAW, which has been used successfully to treat various aspects of the hierarchical SAW [BEI, BI, GI].

The other option is to use essentially (3) as the starting point of the analysis. This was the tack taken by [IM] in constructing the Edwards model at the critical point. In perturbation theory the two approaches are the same (they have the same diagrammatic structure), but various aspects of the nonperturbative construction differ between the two pictures. Our approach is to adapt the phase space expansion of [IM] to the problem of controlling G_λ for appropriate values of a. New problems are introduced because we must consider values of a that are non-real, and it is the solution to these that we now describe.

First, it will be convenient to rewrite G_λ once more by defining

$$C = (-\Delta + a)^{-1}$$
$$A = B = \sqrt{C}$$
$$g = \sqrt{\lambda}$$

and finally

$$G_\lambda(x,a) = \int d\mu_I(\sigma) \left(B(1 - igA\sigma B)^{-1} A\right)_{0x}. \tag{4}$$

2 The Expansion

In this lecture we describe the construction of G_λ for all $a \in \mathcal{D}'$, where $\mathcal{D}' = \{z \in \mathbb{C} : |z| > \nu, |\arg z| < \frac{\pi}{2} + \delta\}$. Here ν can be small, but still $O(1)$ (say $1/100$). Later we will indicate how we expect to extend this result to all of $\mathcal{D}$.

We begin by briefly explaining the basic ideas of the [IM] expansion, and then indicate what changes have to be made to handle complex killing rates. So for now we will assume that $a = O(1) \in \mathbb{R}$. The expansion begins by paving the lattice with hypercubes Δ, with sides of length $O(1)$. We then introduce a set of variables $h_{\Delta,\Delta'} \in [0,1]$, one for each pair of cubes, and new operators $A(h), B(h)$ that interpolate between $\sqrt{C}$ when all the h's are one, and in which two cubes are completely decoupled if the corresponding $h_{\Delta,\Delta'}$ is zero. So we take

$$A(h;x,y) = A(x,y)h_{\Delta,\Delta'} \text{ if } \begin{matrix} x \in \Delta \\ y \in \Delta' \end{matrix} \text{ or } \begin{matrix} x \in \Delta' \\ y \in \Delta \end{matrix}$$

(also $h_{\Delta,\Delta} = 1$ for all Δ). The expansion is then generated by

$$G_\lambda(x,a) = \prod_{\substack{\text{pairs} \\ (\Delta,\Delta')}} (I_{\Delta,\Delta'} + R_{\Delta,\Delta'})G_\lambda(h;x,a)$$

where $G_\lambda(h)$ is defined by substituting $A(h), B(h)$ for the A, B inside the inverse in the definition of G_λ (4), and

$$I_{\Delta,\Delta'}(\cdot) = (\cdot)\big|_{h_{\Delta,\Delta'}=0}$$

$$R_{\Delta,\Delta'}(\cdot) = \int_0^1 dh_{\Delta,\Delta'} \frac{d}{dh_{\Delta,\Delta'}}(\cdot).$$

Expand the product to find

$$G_\lambda(x,a) = \sum_{\substack{\text{sets of pairs} \\ P}} \prod_{(\Delta,\Delta')\in P} R_{\Delta,\Delta'} \int d\mu(\sigma)\Big(B \underbrace{\left(1 - igA(h)\sigma B(h)\right)^{-1}}_{\equiv D^{-1}(h)} A\Big)_{0x}.$$

In order to organize the results of this expansion, we comment that the action of a derivative on $D^{-1}(h)$ is

$$\frac{d}{dh_{\Delta,\Delta'}} D^{-1}(h) = D^{-1}(h)\Delta A\Delta'\sigma B(h)D^{-1}(h) + (\text{three similar terms}).$$

After carrying out all the derivatives in the $R_{\Delta,\Delta'}$'s, we are left with many fields σ in between the $D^{-1}(h)$'s. These are turned into $\partial/\partial\sigma$'s by the identity covariance of the Gaussian measure. We can also insert the definition of $B(h)$ and $A(h)$ wherever they appear, to arrive at the following form:

$$G_\lambda(x,a) = \sum_P \sum \int d\mu(\sigma)\big(BD^{-1}(h)V_1D^{-1}(h)V_2 \cdots V_kD^{-1}(h)A\big)_{0x}$$

where

$$V_i = \begin{pmatrix} h_{\Delta,\Delta'} \\ \text{or} \\ 1 \end{pmatrix} \Delta' A \Delta B \Delta'' \begin{pmatrix} h_{\Delta,\Delta''} \\ \text{or} \\ 1 \end{pmatrix}.$$

The unlabeled sum is over all sets of V_i's the expansion can produce. At this point, for each term in the expansion with k V_i's, there are $k!$ different terms consisting of permutations of the positions of the V_i's. These can be collected by use of the Cauchy integral formula:

$$\begin{aligned} G_\lambda(x,a) &= \sum_P \sum \int d\mu(\sigma) \prod \frac{\partial}{\partial \gamma_i} \Big(D(h) + \sum_i \gamma_i V_i \Big)^{-1} \Big|_{\gamma_i = 0\ \forall i} \\ &= \sum_P \sum \int d\mu(\sigma) \frac{1}{2\pi i} \oint_{|\gamma_i| = r_i} \prod \frac{d\gamma_i}{\gamma_i^2} \Big(D(h) + \sum_i \gamma_i V_i \Big)^{-1}. \end{aligned} \tag{1}$$

The contours γ_i are taken to be circles of radius r_i; the r_i are chosen to be as large as they can be while also fulfilling the condition that

$$\Big\| \sum_i \gamma_i V_i \Big\| < \frac{1}{2}. \tag{2}$$

Due to the exponential decay of the kernels of A and B, it is clear that we can choose the r_i to have some exponential increase.

The reader is referred to [IM] for a detailed proof of convergence; for now we will just comment on two necessary features.

1) If we take two cubes Δ, Δ' to be connected if the corresponding variable $h_{\Delta,\Delta'}$ has been differentiated, then for a given choice of P the cubes comprising P break into potentially many connected components. The important point is that $D^{-1}(h)$ factorizes in the following sense. Suppose the cubes of P break into two connected components, Λ_1 and Λ_2, and we define every other individual Δ to be a connected component. Then
$$D^{-1}(h) = D^{-1}(h)|_{\Lambda_1} \oplus D^{-1}(h)|_{\Lambda_2} \oplus D^{-1}(h)|_{\Delta_1} \oplus \cdots.$$
Here D_Λ means the restriction of D to the subspace Λ. In other words, $D^{-1}(h)$ cannot connect two distinct connected components. This is also true of each operator V_i. Hence, every term in the expansion with more than one connected component vanishes identically. This should not surprise us, if we think about the $n \to 0$ or supersymmetric representations of the SAW. What happens in these representations is that all diagrams that have a closed loop (the vacuum diagrams) vanish. This is the same phenomenon as the above.
2) Intuitively, the proof of convergence of the expansion must rely on the fact that the V_i's that connect distant cubes will be very small due to the exponential decay of A and B. In order to take advantage of this, it is necessary to show that $\|D^{-1}(h)\|$ is not too large. Referring back to (1), we see that $\|D^{-1}(h)\| \le O(1)$ along with (2) implies that
$$\left\| \Big(D(h) + \sum_i \gamma_i V_i \Big)^{-1} \right\| \le O(1).$$
Then the factor of $\prod r_i^{-1}$, from $\prod d\gamma_i\, \gamma_i^{-2}$, gives exponentially decaying factors for each r_i, given the exponential increases that we put in the r_i's. These ultimately allow for the proof of convergence.

3 Complex a

The condition $\|D^{-1}(h)\| \leq 1$ is easy to see for the case of real a, due to the fact that the random potential in which the walk is moving is purely imaginary. For non-real a the situation is more complicated, and we must modify the expansion in order to prove the norm condition on $D^{-1}(h)$.

Let us first consider the case of all variables $h_{\Delta,\Delta'}$ set equal to one. We use the Cauchy-Schwarz inequality to show that, for any function $\varphi \neq 0$

$$\|D\varphi\| \geq \frac{\left|((A^{-1})^{\dagger}B\varphi,\varphi) - ig(B\varphi,\sigma B\varphi)\right|}{\|(A^{-1})^{\dagger}B\varphi\|}. \tag{1}$$

Because $A = (-\Delta + a)^{-1/2}$ is unitarily diagonalizable, we have that $(A^{-1})^{\dagger}B = U$, a unitary matrix. Furthermore, it is easy to see that the spectrum of U is contained in the set $\{z \in \mathbb{C} : |z| = 1 \text{ and } -\arg a \leq \arg z < 0\}$ (assuming $\Im a > 0$; reverse the inequalities otherwise). So $(U\varphi,\varphi) = w^*\|\varphi\|^2$, where w is a convex combination of eigenvalues of U. Examining the numerator in (1), we see that we are adding a purely imaginary number to $w^*\|\varphi\|^2$, and dividing by $\|U\varphi\| = \|\varphi\|$. We would like the result to be bounded below by $O(1)$, but it does not appear to be because $(B\varphi,\sigma B\varphi)$ can be any real number, due to the random field σ. The solution is to rotate the contours in the σ integrals by some angle ρ. Then a complex number with argument $\frac{\pi}{2} + \rho$ is added to $w^*\|\varphi\|^2$; the modulus of the result is bounded below by $O(1)$ if we choose $\rho = O(1)$.

So far we have not introduced the decoupled versions of A and B. If we did this in the same manner as before, the proof for the norm condition on $D^{-1}(h)$ would no longer work since we do not have any control over $A^{-1}(h)$. We rectify this by enlarging the size of the hypercubes Δ by a factor α, which will be taken large, and by not decoupling nearest neighbor blocks. The result of this is that the interpolating operator $A(h)$ can be written

$$A(h) = A(1+G) = (1+H)A$$

where $\|G\|$ and $\|H\|$ are both bounded by $O(\alpha^{-1})$. The result of these definitions is that w can have added to it a small piece of size $O(\alpha^{-1})$. Clearly this does not substantially change the above argument, and we can conclude that

$$\|D^{-1}(h)\| \leq O(1). \tag{2}$$

Equation (2) has not come without a price, however. As it stands, $D^{-1}(h)$ no longer factorizes across connected components, which is a crucial property. The solution to this is to "turn off" the interactions in parts of the lattice; interacting pieces separated by voids with no interactions will factorize properly. More precisely, before we begin the cluster expansion we introduce a set of variables $\{s_\Delta\} \in [0,1]$, one for each Δ, and write

$$D^{-1}(s) = \left(1 + igA\left(\sum_{\Delta} s_\Delta \Delta\right)\sigma B\right)^{-1}.$$

We then expand via

$$\prod_{\Delta}(I_\Delta + R_\Delta)G_\lambda(s),$$

where

$$I_\Delta(\cdot) = (\cdot)\big|_{s_\Delta=0}$$
$$R_\Delta(\cdot) = \int_0^1 ds_\Delta \frac{d}{ds_\Delta}(\cdot).$$

If a variable s_Δ has been differentiated, we say Δ has been "painted." The unpainted blocks have had the interaction turned off. We then proceed as above, introducing the interpolating $A(h)$ and $B(h)$, and do the cluster expansion. If we change our connection rules so that all painted blocks have a corridor of blocks around them to which they are defined to be connected, and re-form our connected components, then it is not hard to see that $D^{-1}(s,h)$ factorizes properly.

The upshot of the procedure we have described is the following

Theorem 1. *Fix $d > 2$. The Green's function for the weakly self-avoiding walk on $\mathbb{Z}^d$ is analytic in the region $a \in \mathbb{C}$ satisfying $|a| > \epsilon$, $|\arg a| < 3\pi/4 - b$, for $\epsilon > 0$ and $0 < b < \pi/4$, with $\lambda < \lambda_0$ sufficiently small, depending on ϵ. In this region it can be written*

$$G_\lambda(x,a) = G_0(x,a) + G_{\text{rem}}(x,a)$$

where $G_0(x,a) = (-\Delta + a)^{-1}_{0,x}$ and the remainder satisfies

$$|G_{\text{rem}}(x,a)| \le \frac{K_1\lambda}{1+|a|}e^{-\nu_a|x|}$$

where $K_1 = K_1(\epsilon)$, and $\nu_a = K_2 \ln(1 + |a|^{1/2})$ with $K_2 = O(1)$. K_1 and K_2 are constants.

So far we could have worked in any $d > 2$. To get good estimates in the limit $a \to a_c$, we will of course have to restrict to four or more dimensions and use a phase space expansion combined with a renormalization group analysis similar to the one developed in [IM]. Briefly, the renormalization group arguments are analogous to those of ϕ_4^4 theory [FMRS,R], with the infrared asymptotic freedom of the model being the crucial ingredient for success. The control of low-momentum fields works differently in the Edwards model than in ϕ_4^4 theory; a new method of resummation is introduced in [IM] that replaces the "domination" procedure of [FMRS,R].

The expansion we described, for the case of complex $O(1)$ killing rate, also applies to a single slice of this phase space expansion. Using these methods along with the ideas of [IM], we expect to obtain estimates proving non-diffusive behavior of the SAW in four dimensions.

References

[BEI] Brydges, D., Evans, S. N., and Imbrie, J. Z.: Self-Avoiding Walk on a Hierarchical Lattice in Four Dimensions. *Ann. Prob.* **20**, 82–124 (1992)

[BI] Brydges, D. and Imbrie, J. Z.: End-to-End Distance for a 4-Dimensional Hierarchical Self-Avoiding Walk. Preprint (1993)

[dG] de Gennes, P. G.: Exponents for the excluded volume problem as derived by the Wilson method. *Phys. Lett. A* **38**, 339–340 (1972)

[FMRS] Feldman, J., Magnen, J., Rivasseau, V., and Sénéor, R.: Construction and Borel Summability of Infrared Φ_4^4 by a Phase Space Expansion. *Commun. Math. Phys.* **109**, 437–480 (1987)

[GI] Golowich, S. E., and Imbrie, J. Z.: The Broken Supersymmetry Phase of a Self-Avoiding Random Walk. *Commun. Math. Phys.*, to appear (1994)

[IM] Iagolnitzer, D., and Magnen, J.: Polymers in a Weak Random Potential in Dimension Four: Rigorous Renormalization Group Analysis. *Commun. Math. Phys.*, to appear (1994)

[R] Rivasseau, V.: From Perturbative to Constructive Field Theory. Princeton: Princeton University Press (1991)

Charge Correlations for the Two–Dimensional Coulomb Gas

Tom R. Hurd *

Department of Mathematics and Statistics
McMaster University
Hamilton, Ontario L8S 4K1

Abstract

This paper is a summary of mathematical results contained in [14] concerning integer charge correlations for the Coulomb gas/sine-Gordon system in two dimensions. For $\beta = T^{-1} < 8\pi$ and small activity z, the UV problem is considered in a finite volume. A new proof is given of the fact that the pressure $p^{>m}(\beta, z)$, renormalized up to order m in perturbation theory, is analytic in z for $\beta < \beta_m = 8\pi(1 - 1/m)$. Higher correlations are treated and proven to be analytic in z for all $\beta < 8\pi$. The mth threshold value β_m appears as the value at which the exponent of the short distance power law of the mth subleading contribution to any correlation changes nonanalytically. In the Kosterlitz-Thouless phase $\beta > 8\pi$, the IR problem is treated with a fixed UV cutoff. The existing framework for the pressure is extended to all higher correlations. For the two point function, it is shown that at length scales larger than $\mathcal{O}(|z|^{-1/(\beta/4\pi-2)})$ the free field power law $|x-y|^{-\beta/2\pi}$ at long distances crosses over to a slower power law $|x-y|^{-4}$. This verifies a conjecture of Fröhlich and Spencer [12].

1 Introduction

The two dimensional Coulomb gas/sine-Gordon system is a classic model in mathematical physics. There exists a large body of discussion of its special properties, including landmark papers such as that of Kosterlitz and Thouless [15], Coleman [6], Zamolodchikov [19], and Sklyanin et al. [18]. On the level of constructive quantum field theory, there is a somewhat smaller body of results: these include the works of [11], [12], [13], [3],[1],[16].

My aim here is to extend the constructive renormalization group (RG) program developed in a series of papers [7],[9],[8],[10],[2] which began with a foundational paper [5] by Brydges and Yau. I shall explain how the BY method, with some adaptations based on ideas in [4], leads to a rather complete picture of perhaps the most important aspect of the model, namely the asymptotics of integer charge correlations, both in the ultraviolet (UV) and infrared (IR).

* Research supported by the Natural Sciences and Engineering Research Council of Canada.

We consider the Coulomb gas with activity z at temperature $T = \beta^{-1}$ on a finite torus $\Lambda(M) = \mathbf{R}^2/(L^M\mathbf{Z})^2$ where $L > 1, M$ are integers. Let the Coulomb potential be the inverse Laplacian on $\Lambda(M)$ with a short distance cutoff at scales $\sim L^{-N}$:

$$v_{M,N}(x) = L^{-2M} \sum_{\substack{p\in\Lambda(M)^* \\ p\neq 0}} p^{-2} e^{ipx}\, e^{-L^{-2N}p^2}, \tag{1}$$

where $\Lambda(M)^*$ denotes the dual lattice $(2\pi L^{-M}\mathbf{Z})^2$. Then the system is described by the grand canonical partition function

$$Z_{CG}(\beta, z, v_{M,N}) = \sum_{n,\mathbf{q}} \int_{\Lambda(M)^n} d\mathbf{x}\ \frac{z^n}{n!}\ \exp(-\beta E_{n,v_{M,N}}(\mathbf{x};\mathbf{q}))$$

where the Coulomb energy is given by

$$E_{n,v_{M,N}}(\mathbf{x};\mathbf{q}) = \sum_{a<b} q_a q_b v_{M,N}(x_a - x_b)$$

This can be expressed equivalently in terms of the sine-Gordon model:

$$Z_{CG}(\beta, z, v_{M,N}) = Z_{SG}(2z, \beta v_{M,N})$$

where

$$Z_{SG}(\zeta, \beta v_{M,N}) = \int d\mu_{\beta v_{M,N}}(\phi) e^{\zeta \int :\cos\phi:}. \tag{2}$$

Here $\mu_{\beta v_{M,N}}(\phi)$ is the Gaussian measure with covariance $\beta v_{M,N}(x,y)$ on a certain Sobolev space $\mathcal{H}(\Lambda(M))$, perturbed by the interaction $e^{\zeta\int :\cos\phi:}$. The Wick dots denote Wick ordering with respect to μ.

I am interested in two problems. The first is the UV problem of taking the $N \to \infty$ limit for fixed volume parameter M. The Kosterlitz-Thouless argument suggests that this limit should exist for any $\beta < 8\pi$ and small activity $\zeta = 2z$. We have indeed proved this: apart from a finite number of vacuum energy counterterms, the Wick ordering in (2) is the only renormalization necessary.

The second aspect of the model is the IR problem of taking $M \to \infty$ while holding N fixed. For this, (2) is not quite the right starting point because the Coulomb potential $v_{M,N}(x)$ diverges as $M \to \infty$. We should "zero" the Coulomb potential at the origin, i.e. take

$$w_{M,N}(x) = v_{M,N}(x) - v_{M,N}(0).$$

At the same time, it turns out to be necessary to suppress non-neutral charge configurations. Making use of the identities

$$\begin{aligned}\sum_{n,\mathbf{q}} \int_{\Lambda(M)^n} d\mathbf{x}\ \frac{z^n}{n!}\ \exp(-\beta v_{M,N}(0)(\sum_a q_a)^2)\ \exp(-\beta E_{n,w_{M,N}}(\mathbf{x};\mathbf{q}))& \\ = Z_{CG}(\beta, e^{\beta v_{M,N}(0)/2} z, v_{M,N})& \\ = \int d\mu_{\beta v_{M,N}}(\phi) e^{\zeta\int \cos\phi}.& \end{aligned} \tag{3}$$

and noticing that $v_{M,N}(0) = \mathcal{O}(N+M)\ \log L > 0$, we see that the SG model naturally leads to a suppression of non-neutral configurations. It appears, as indeed we shall show, that the *non-Wick-ordered* sine-Gordon model is an appropriate starting point for the IR problem. The difference between (2) and (3) is a divergent factor

$$\tau_{M,N} = e^{\beta v_{M,N}(0)/2} \sim L^{\beta(M+N)/4\pi} \tag{4}$$

multiplying the coupling constant ζ.

Now consider a configuration $\Sigma = \{\xi_1, \ldots, \xi_Q\}$ of Q external unit charges $\xi_a = (x_a, q_a)$, $q_a = \pm 1$, fixed at points x_a in $\Lambda(M)$. To this we associate a partition function Z^Σ where the Coulomb energies coming from the external charges are included. The *integer charge truncated correlations* are defined to be:

$$\begin{cases} S^{q_1}_{x_1} = Z^{-1} Z^{q_1}_{x_1} = (2z)^{-1} \times \text{density} & Q = 1 \\ S^{q_1 q_2}_{x_1 x_2} = Z^{-1} Z^{q_1 q_2}_{x_1 x_2} - Z^{-2} Z^{q_1}_{x_1} Z^{q_2}_{x_2} & Q = 2 \\ \text{etc.} & Q > 2 \end{cases}$$

Now, for the UV problem, we define a generating function

$$Z(\boldsymbol{\lambda}) = \int d\mu_{\beta v_{M,N}}(\phi) \prod_{a=1}^{Q} (1 + \lambda_a :e^{iq_a\phi(x_a)}:) e^{\zeta \int :\cos\phi:} \tag{5}$$

For the IR problem, we define

$$Z(\boldsymbol{\lambda}) = \int d\mu_{\beta v_{M,N}}(\phi) \prod_{a=1}^{Q} (1 + \lambda_a e^{iq_a\phi(x_a)}) e^{\zeta \int \cos\phi} \tag{6}$$

Then, in both cases, the integer charge truncated correlations are given by differentiating with respect to $\boldsymbol{\lambda}$:

$$\begin{cases} S^{q_1}_{x_1} = \left(\frac{\partial}{\partial\lambda_1}\right) \log Z(\boldsymbol{\lambda})\Big|_{\lambda=0} & Q = 1 \\ S^{q_1 q_2}_{x_1 x_2} = \left(\frac{\partial^2}{\partial\lambda_1\partial\lambda_2}\right) \log Z(\boldsymbol{\lambda})\Big|_{\lambda=0} & Q = 2 \cdot \\ \text{etc.} & Q > 2 \end{cases}$$

Fröhlich and Seiler ([11], p 899), have observed that for the sine-Gordon quantum field theory, "... these fields (i.e. $: e^{iq\phi} :$) are actually much more natural than the field ϕ." Field correlations such as $\langle \phi(x)\phi(y) \rangle$ have been treated in the sine-Gordon model in [10].

In this review, I will describe how the recent refinements of the renormalization group (RG) in constructive quantum field theory can be extended to give a rather complete picture of integer charge correlations.

Sections 2 and 3 will describe how the main theorems described in §5,6 (Theorems 5 and 9) lead to detailed results on the behavior of correlations, for the UV problem with $\beta < 8\pi$ and the IR problem for $\beta > 8\pi$, respectively. I will discuss how these new results relate to and improve on previously existing results.

In section 4, I will sketch the Brydges and Yau RG method and show how it extends to correlation functions. Then in sections 5 and 6, I will state Theorems 5 and 9. These are the main technical results dealing with the UV problem and IR problem respectively. In §6, I provide a thumbnail sketch of the proof of the vacuum IR result (Theorem 6) to give a flavour of how these things go. For the complete proofs of the results on correlations, the reader is directed to the original paper [14].

2 Pressure and Correlations for $\beta < 8\pi$

In this temperature range, we concentrate on the UV limit $N \to \infty$ in a fixed volume $\Lambda(0)$ (equally we could work in infinite volume by introducing a mass term in the Gaussian measure). We fix the RG rescaling parameter L to be a large integer, take a parameter $0 < \epsilon < 1/2$, and take a complex activity parameter $\zeta = 2z$ which is "small", i.e. $|\zeta| \le \bar{\zeta}(\beta, \epsilon, L)$.

The simplest consequence of the RG analysis of §5 is a formula (13) for the pressure as a sum over length scales L^{-i}, $i = 0, 1, \ldots, N$. In the $N \to \infty$ limit, this has the form

$$p(\zeta) = \sum_{i=0}^{\infty} \Omega_i(\zeta).$$

¿From Theorem 5, we find that each quantity Ω_i is analytic in $\zeta_i = L^{-2i}\tau_{i,0}\zeta \sim L^{(\beta/4\pi-2)i}\zeta$ with radius $\bar{\zeta}$, and is bounded there by

$$L^{2i}|\zeta_i|^{2-\epsilon}.$$

This result is somewhat stronger than the similar inequality (35) of [10].

Because of analyticity, we can immediately derive bounds on the power series about $\zeta = 0$. Write

$$p(\zeta) = \sum_{i=0}^{m-1} \zeta^i p^{(i)} + p^{\ge m}$$

where

$$\begin{aligned} \zeta^m p^{(m)} &= \frac{1}{2\pi i} \oint \frac{ds}{s^{m+1}} p(s\zeta) \\ p^{\ge m} &= \frac{1}{2\pi i} \oint \frac{ds}{s^m (s-1)} p(s\zeta) \end{aligned}$$

and similar expressions for $\Omega^{(m)}, \Omega^{\ge m}$. For each Ω_i, we can bound the integral over the contour $|s| = (L^{-2i}\tau_{i,0}|\zeta|)^{-1}\bar{\zeta} \gg 1$, and conclude

Theorem 1. *1. For any even integer $m \geq 2$, the mth order contribution to the pressure is bounded by*

$$|p^{(m)}| \leq \bar{\zeta}^{2-m-\epsilon} \sum_{i=0}^{\infty} L^{[2-m(2-\beta/4\pi)]i} \tag{7}$$

which converges for $\beta < \beta_m = 8\pi(1-1/m)$.
2. The mth renormalized pressure $p^{\geq m} = \sum_{a=m}^{\infty} p^{(m)}$ is bounded by

$$|p^{\geq m}| \leq 2\,|\zeta|^m\,\bar{\zeta}^{2-m-\epsilon} \sum_{i=0}^{\infty} L^{[2-m(2-\beta/4\pi)]i}$$

which converges for $\beta < \beta_m$.

Thus for $\beta \in [\beta_m, \beta_{m+2})$, the interaction needs only a vacuum renormalization up to counterterms of order m. This result is an independent and very much simplified proof of what is essentially Theorem 1.0 of [17].

A similarly comprehensive picture holds for the general truncated integer charge correlations. Let Σ be a configuration of $Q > 1$ external charges where the interparticle separations are all of order L^{-I} for some integer $I > 0$, i.e.

$$L^{-(I+1)} < dist(x_i, x_j) < L^{-I}$$

for all $i \neq j$. The RG analysis of §5 leads to an expansion

$$S^{\Sigma} = \sum_{i=0}^{I} D_{\Sigma}\Omega_i.$$

Theorem 5 implies that the quantities $D_{\Sigma}\Omega_i$ are analytic functions of ζ_i with radius $\bar{\zeta}$, bounded there by

$$Q!(c'L^{(\beta/4\pi)})^{(i-I)}\left(cL^{[\beta/4\pi+\epsilon(2-\beta/4\pi)]I}\right)^Q$$

for some quantities $c = c(\beta, \epsilon, L)$ and $c' = \mathcal{O}(1)$. By Cauchy inequalities, the mth order perturbative contribution is bounded by

$$|S_{\Sigma}^{(m)}| \leq Q!\left(\frac{|\zeta|}{\bar{\zeta}}\right)^m c^Q\, L^{[\beta/4\pi+\epsilon(2-\beta/4\pi)]IQ} \sum_{i=0}^{I}(c'L^{(\beta/4\pi)})^{(i-I)}L^{-(2-\beta/4\pi)im}$$

Thus

Theorem 2. *There is a number $C_2(\beta, \epsilon, L)$ such that for any integer $m \geq 0$*

$$|S_{\Sigma}^{(m)}|, |S_{\Sigma}^{\geq m}| \leq C_2^Q Q!\left(\frac{|\zeta|}{\bar{\zeta}}\right)^m d(\Sigma)^{-\Delta(\Sigma)}$$

where $d(\Sigma) = L^{-I}$ and the exponent of the short distance power law is

$$\begin{cases} \frac{(Q-1)\beta}{4\pi} + \epsilon Q(2-\frac{\beta}{4\pi}) & \text{for } \beta \leq \beta_{m+1} \\ \frac{Q\beta}{4\pi} - m(2-\frac{\beta}{4\pi}) + \epsilon Q(2-\frac{\beta}{4\pi}) & \text{for } \beta > \beta_{m+1} \end{cases} \tag{8}$$

This result is new. In fact, I have not been able to find any perturbative or constructive statement in the literature on the short distance asymptotics of general correlations. A direct comparison with perturbative bounds obtained by extending the tree expansion technique of [1] suggests that the above exponents for each $S^{(m)}$ are in fact sharp if we take $\epsilon = 0$.

Example: Two-point function, $\beta \in (\beta_8, \beta_9] = (7\pi, 72\pi/9]$:

$$\begin{cases} S^{(0)} \sim d(\Sigma)^{-[2\beta/4\pi]} \\ S^{(2)} \sim d(\Sigma)^{-[2\beta/4\pi - 2(2-\beta/4\pi)]} \\ S^{(4)} \sim d(\Sigma)^{-[2\beta/4\pi - 4(2-\beta/4\pi)]} \\ S^{(6)} \sim d(\Sigma)^{-[2\beta/4\pi - 6(2-\beta/4\pi)]} \\ S^{(\geq 8)} \sim d(\Sigma)^{-[\beta/4\pi + 2\epsilon(2-\beta/4\pi)]} \end{cases}$$

In general, for $\beta \in (\beta_m, \beta_{m+1}]$ we shall see a finite linear sequence of exponents, which at the mth level stabilize at the value $\frac{(Q-1)\beta}{4\pi}$. We can see that the thresholds β_m are values at which the exponent of the short distance power law of the mth order two-point function changes non-smoothly. This crossover phenomenon can be expected on the basis of a perturbative analysis. Note that correlations remain smooth even at the thresholds.

3 Results for $\beta > 8\pi$

In this low temperature range, we consider the IR problem, with a fixed UV cutoff at the unit distance scale. The volume is taken to be $\Lambda(M)$, where M is arbitrarily large. The main theorem for this problem is stated in §6: Here I describe the picture of the pressure and correlations which follows from it. Again, the parameter ζ is chosen "small", i.e. $|\zeta| \leq \bar{\zeta}(\beta, \epsilon, L)$.

Apology: In order to be consistent with the UV picture, scale labels i, j are negative in the IR regime and correspond to large length scales $L^{|i|}$.

Theorem 3. *The finite volume pressure is given by a sum over scales:*

$$p(\Lambda(M)) = \sum_{j=0}^{M} \Omega_{-j}.$$

Each term $\Omega_{-j}(\zeta)$ is analytic for $|\zeta| < \bar{\zeta}$, and bounded there by $L^{-2|j|} D\delta^{|j|}$, where $\delta < 1/2$ and D is a small constant. Therefore the pressure itself is analytic in ζ and since it vanishes to first order in ζ, is bounded by $O(|\zeta|^2)$, uniformly in M.

Now consider the general truncated integer charge correlations. As in §2, I suppose that the configuration Σ is such that the interparticle separations are all of the same order $L^{|I|}$ for some negative integer I. The theorem implies that there is a crossover size

$$L^{|I_0|} \sim |\zeta|^{\frac{-1}{\beta/4\pi-2}}$$

where the power law decay rate shifts. At distances shorter than this scale, the power is that of the free field $|x-y|^{-\beta/2\pi}$; a larger scales, the decay is the faster $|x-y|^{-4}$.

The RG leads to an expansion

$$S^{\Sigma} = \sum_{j=|I|}^{M} D_{\Sigma}\Omega_{-j}.$$

Theorem 6 implies that each $D_{\Sigma}\Omega_{-j}$ is analytic in ζ with radius $\bar{\zeta}$, and bounded by

$$\begin{cases} Q!(c'L^{-2})^{|i-I|}\left(c_1 L^{-\beta|I|/4\pi}\right)^Q & \text{if } |I| \le |I_0| \\ Q!(c'L^{-2})^{|i-I|}\left(c_2|\zeta|L^{-2|I|}\right)^Q & \text{if } |I| \ge |I_0| \end{cases}$$

for some quantities $c_i = c_i(\beta, \epsilon, L)$ and $c' = \mathcal{O}(1)$.

Thus in all cases, the sum over j converges uniformly in M.

Theorem 4. *The truncated correlation of a configuration Σ of $Q > 1$ points with separation $d(\Sigma) = L^{|I|}$ is bounded above by*

$$c^Q Q! d(\Sigma)^{-\Delta(\Sigma)}$$

where

$$\Delta(\Sigma) = \begin{cases} \frac{\beta}{4\pi}Q & \text{if } |I| \le |I_0| \\ 2Q & \text{if } |I| \ge |I_0| \end{cases}$$

The value of c is

$$\begin{cases} c_1 & \text{if } |I| \le |I_0| \\ c_2|\zeta| & \text{if } |I| \ge |I_0| \end{cases}$$

This result verifies a conjecture which goes back to [12]. They state that the two-point correlator should decay at large distances like the dipole-dipole correlation in a dipole gas, averaged over the dipole direction (i.e. like $|x|^{-4}$).

If we combine these upper bounds with an analysis of low order perturbation theory, we presumably obtain the same power laws as lower bounds on correlations. For example, an explicit calculation should show that the second order contribution to the two point correlator has the exact power law $|x|^{-4}$, and that for small ζ the higher order contributions are negligible.

4 RG Maps for Charge Correlations

The remainder of the paper will sketch the method by which the results of §2,3 are proved. We consider in more detail the UV problem $N \to \infty$ in a finite volume $\Lambda(0)$, for $\beta < 8\pi$. We take a configuration $\Sigma = \{\xi_1, \ldots, \xi_Q\}$ of Q external unit charges $\xi_a = (x_a, q_a)$, $q_a = \pm 1$, located at points x_a in $\Lambda(0)$. The truncated Q-point correlation function is defined to be:

$$S^{\Sigma} = \frac{\partial^Q}{\partial\lambda_1 \ldots \partial\lambda_Q}|_{\lambda=0} \log Z(\boldsymbol{\lambda}), \quad \boldsymbol{\lambda} = (\lambda_1, \ldots, \lambda_Q),$$

where

$$Z(\boldsymbol{\lambda}) = \int d\mu_{\beta v_{N,0}}(\phi) \prod_{a=1}^{Q} (1 + \lambda_a \tau_{N,0} e^{iq_a \phi(L^N x_a)}) \mathcal{Z}_N^N(\phi). \tag{9}$$

The functional integral has been rewritten on the rescaled volume $\Lambda(N) = L^N \Lambda(0)$. The "Boltzmannian" or "Gibb's Factor" $\mathcal{Z}_N^N$ is given by

$$\mathcal{Z}_N^N(\phi) = \exp[\zeta_N \int_{\Lambda(N)} \cos\phi],$$

where $\zeta_N = L^{-2N} \tau_{N,0} \zeta$ with

$$\tau_{N,0} = e^{\beta v_{N,0}(0)/2} \sim L^{\beta N/4\pi}$$

being just the Wick-ordering constant. The truncated 2-point function, for example, is

$$S_{x_1 x_2}^{q_1 q_2} = \tau_{N,0}^2 \left\langle e^{iq_1 \phi(L^N x_1)} e^{iq_2 \phi(L^N x_2)} \right\rangle_{\beta,N,\zeta}$$
$$-\tau_{N,0}^2 \left\langle e^{iq_1 \phi(L^N x_1)} \right\rangle_{\beta,N,\zeta} \left\langle e^{iq_2 \phi(L^N x_2)} \right\rangle_{\beta,N,\zeta}$$

We define the following operators on linear functions of $\boldsymbol{\lambda}$: for each $\sigma \subset \{1, \ldots, Q\}$, let

$$\lambda_\sigma = \prod_{a \in \sigma} \lambda_a \tag{10}$$

$$D_\sigma \cdot = \prod_{a \in \sigma} \frac{\partial \cdot}{\partial \lambda_a}\Big|_{\lambda=0} \tag{11}$$

$$P_\sigma = \lambda_\sigma \, D_\sigma \tag{12}$$

The identity operator on linear functions of $\boldsymbol{\lambda}$ can be written

$$P = \sum_\sigma P_\sigma = P_\emptyset + P_>.$$

For any Σ, the correlation function S^Σ is given by

$$D_\Sigma \log Z.$$

Our analysis is thus concerned with generating a convergent formula for $Z(\boldsymbol{\lambda})$.

Now I shall give a brief sketch of the RG maps developed for the partition function in the sequence of papers [5],[7],[10]. The reader should ideally refer to the last of these papers for a complete description.

The BY method generates a sequence of measures $d\nu_i$ on certain function spaces over the volumes $\Lambda(i)$, for $i = 0, 1, \ldots, N$. The measures of interest are always weak perturbations

$$d\nu_i(\phi) = d\mu_{\beta v_{i,0}}(\phi)\ \mathcal{Z}_i(\Lambda(i), \phi)$$

of a specific sequence of Gaussian measures $d\mu_{\beta v_{i,0}}$ whose covariances $\beta v_{i,0}(x, y)$ are given by (1). The Gibb's factor is taken in the form of a polymer expansion:

$$\begin{aligned}\mathcal{Z}_i^N(\Lambda(i), \phi) &= \sum_{X_1, \ldots, X_L} \prod_j K_i(X_j, \phi) \\ &\equiv \mathcal{E}xp[\Box + K_i](\Lambda(i), \phi)\end{aligned}$$

where the sum is over disjoint collections of "polymers", a polymer X being a union of closed unit squares with corners lying on the integer lattice in $\Lambda(i)$. The "$\mathcal{E}xp$" notation may be thought of as shorthand for the polymer expansion. The collection of polymer activities $K = \{K_i(X, \phi), X \text{ a polymer}\}$ is called an *analytic functional.*

The analytic functionals are to be regarded as lying in Banach spaces $\mathcal{B}_i$, whose norms $\|\cdot\|_i$ are given weights parametrized by certain quantities:

1. the large field parameter $\kappa_i > 0$;
2. the large set parameter $A > 0$;
3. the large derivative parameters $\mathbf{h}_i = (h_{0i}, h_{1i})$.

Now the RG map $\mathcal{R}_i$ is a specific nonlinear functional which takes a ball in $\mathcal{B}_i$ into $\mathcal{B}_{i-1}$. $\mathcal{R}$ is quite naturally the composition of three maps $\mathcal{S} \circ \mathcal{E} \circ \mathcal{F}$, called scaling, extraction and fluctuation. Taken all together, the map $\mathcal{R}$ has the following defining property. Let the *fluctuation covariance* be defined by the Fourier components $c_i(p) = v_{i,0}(p) - L^2 v_{i-1,0}(Lp)$ for $i > 1$ and $c_1(p) = v_{1,0}(p)$ for $i = 1$. Then for any $K_i \in \mathcal{B}_i$ small enough, $K_{i-1} = \mathcal{R}(K_i)$ is such that

$$\mathcal{E}xp[\Box + K_{i-1}](\Lambda(i-1), \phi) = e^{\Omega_i} \mu_{\beta c_i} * \mathcal{E}xp[\Box + K_i](\Lambda(i), \phi_L)$$

Here, $\mu *$ denotes Gaussian convolution, Ω_i is a certain carefully chosen constant, and ϕ_L denotes the rescaled field $\phi_L(x) = \phi(L^{-1}x)$. This defining property implies in particular that

$$\int d\mu_{\beta v_{i,0}} \mathcal{Z}_i = e^{\Omega_i} \int d\mu_{\beta v_{i-1,0}} \mathcal{Z}_{i-1}$$

The central idea of the RG is that by iteration starting from the measure with

$$\mathcal{Z}_N^N(\phi) = \exp \zeta_N \int_{\Lambda(N)} \cos\phi = \mathcal{E}xp[\Box + K_N^N](\phi),$$

the UV regularized partition function can be written

$$Z_{SG}(\zeta,\beta,v_{0,N}) = \left(\prod_{j>i}^{N} e^{\Omega_j^N}\right) \int d\mu_{\beta v_{i,0}} \mathcal{Z}_i^N$$

for any integer $i = 1, 2, \ldots, N$, or for $i = 0$

$$Z_{SG}(\zeta,\beta,v_{0,N}) = \left(\prod_{j>0}^{N} e^{\Omega_j^N}\right) \mathcal{Z}_0^N(\Lambda(0), \phi = 0).$$

The regularized pressure is

$$p(\zeta,\beta,v_{0,N}) = \sum_{j=1}^{N} \Omega_j^N + \log \mathcal{Z}_0^N. \tag{13}$$

Now the UV problem can be solved by controlling the limit

$$\mathcal{Z}_i = \lim_{N\to\infty} \mathcal{Z}_i^N$$

Now, following a similar idea introduced in [4], we can develop the analogous treatment for the λ-dependent partition function, by extending the activities K_N^N to λ-dependent activities so that

$$\mathcal{E}xp[\square + K_N^N(\boldsymbol{\lambda})](\Lambda(N)) = \prod_{a=1}^{Q} (1 + \lambda_a \tau_{N,0} e^{iq_a\phi(L^N x_a)}) \mathcal{Z}_N^N(\phi). \tag{14}$$

There is an essential distinction between vacuum and non-vacuum activities: $P_\emptyset K_N^N(X)$ is translation invariant, whereas $P_\Sigma K_N^N(X)$ is zero unless X contains all of the points in the set $L^N\Sigma$. This "pinning" is preserved under the RG and leads at each scale to a good power counting factor of L^{-2} for non-vacuum activities.

We define a new renormalization group map

$$\mathcal{R}_P = \mathcal{S} \circ \mathcal{E}_P \circ \mathcal{F} : K_i \to K_{i-1} = \mathcal{R}_P(K_i)$$

such that

$$\int d\mu_{v_{i,0}} \mathcal{E}xp(\square + K_i(\boldsymbol{\lambda}))(\Lambda(i)) = P\left(e^{\Omega_i} \int d\mu_{v_{i-1,0}} \mathcal{E}xp(\square + K_{i-1}(\boldsymbol{\lambda})\right)(\Lambda(i-1)).$$

It turns out that the only modifications needed to define $\mathcal{R}_P$ are to the extraction step $\mathcal{E}_P$, and lead to no new difficulties.

Applying this iteration N times to the formula (9) leads to an expansion for $Z(\boldsymbol{\lambda})$:

$$Z(\boldsymbol{\lambda}) = P\left(\left(\prod_{i=1}^{N} e^{\Omega_i^N}\right) \mathcal{E}xp(\square + K_0^N)(\Lambda(0))\right)$$

Since K_0^N is defined on a unit block,

$$\mathcal{E}xp(\square + K_0^N)(\Lambda(0)) = 1 + P_\emptyset K_0^N(\Delta) + P_> K_0^N(\Delta, \boldsymbol{\lambda}) = Pe^{\Omega_0^N(\lambda)}$$

where

$$\Omega_0^N(\boldsymbol{\lambda}) = P\log(1 + P_\emptyset K_0^N(\Delta) + P_> K_0^N(\Delta, \boldsymbol{\lambda})).$$

Then

$$Z(\boldsymbol{\lambda}) = P\left(\prod_{i=0}^{N} e^{\Omega_i^N}\right)$$

which gives the desired formula for any truncated correlation:

$$S^\Sigma = \sum_{i=0}^{N} \left(D_\Sigma \Omega_i^N\right). \tag{15}$$

5 The Main Result for $\beta < 8\pi$

We note a simplifying property of the expansion (15): since extractions are only made from "small" sets, $D_\Sigma \Omega_i^N = 0$ if $L^i \Sigma$ is not contained in some small set. This means that for two or more distinct points Σ, the expansion (15) has a finite number of terms which depends on the geometry of the points.

For simplicity, we make an assumption on the configuration Σ that the interparticle distances are all of the order L^{-I} for some integer I:

$$L^{-(I+1)} < dist(x_i, x_j) < L^{-I}$$

for all $i < j$. Then the upper limit of the sum (15) is $i = I$.

Norm parameters are chosen as in the vacuum case. The only change is to take a larger parameter $A = L^4$ in the large set regulator: this leads a smaller maximum allowed value $\bar{\zeta}$ of $|\zeta|$.

The main theorem now generalizes our earlier theorem ([10], Theorem 4.4) on the vacuum activities to the many point activities:

Theorem 5. *There is a number $\bar{\zeta}(\epsilon, L, \beta)$ and a geometric constant c_1 such that the following properties hold, uniformly in the UV cutoff N, and hence in the $N \to \infty$ limit for all integers $0 \le j \le i$:*

1. *(analyticity) The activities $D_\Sigma K_i^N$ and extracted parts $D_\Sigma \Omega_i^N$ are analytic functions of ζ_j for $|\zeta_j| \le \bar{\zeta}$.*
2. *(vacuum activities) The vacuum activities $D_\emptyset K_i^N$ and extracted parts $D_\emptyset \Omega_i^N$ satisfy the bounds:*

$$\|D_\emptyset K_i^N\|_{i-j} \le |\zeta_i|^{1-\epsilon} \tag{16}$$

$$|D_\emptyset \Omega_i^N| \le L^{2i} |\zeta_i|^{2-\epsilon} \tag{17}$$

3. (one point function) For any $\Sigma = \{(x,q)\}$,

$$\|D_\Sigma K_i\|_{i-j} \le 2AL^{\beta i/4\pi}\ e^{h_{0i-j}} \tag{18}$$

$$|D_\Sigma \Omega_i^N| \le 2L^{\beta i/4\pi}\ e^{h_{00}} \tag{19}$$

4. (multi-point functions) For any Σ *with* $|\Sigma| = Q > 1$

$$\|D_\Sigma K_i\|_{i-j} \le \begin{cases} Q!\left(2AL^{\beta i/4\pi}\ e^{h_{0i-j}}\right)^Q & \text{if } i \ge I \\ Q!(c_1 L^{(\beta/4\pi)})^{(i-I)}\left(2AL^{\beta I/4\pi}\ e^{h_{0I-j}}\right)^Q & \text{if } i \le I \end{cases}$$

and

$$|D_\Sigma \Omega_i^N| \le \begin{cases} 0 & \text{if } i > I \\ Q!(c_1 L^{(\beta/4\pi)})^{(i-I)}\left(2AL^{\beta I/4\pi}\ e^{h_{0I}}\right)^Q & \text{if } i \le I \end{cases} \tag{20}$$

Proof. The proof, given in [14], follows with moderate changes the proof of the vacuum result in [10]. □

This result leads rather quickly to the claimed bounds of §2.

6 Sketch of the Method and Results for $\beta > 8\pi$

For the Kosterlitz-Thouless phase $\beta > 8\pi$, one takes the initial measure $d\nu_0$ to be a weak perturbation of the Gaussian measure on a large volume $\Lambda(M)$ with covariance $v_{M,0}$ (with short distance cutoff $N = 0$). The RG map is defined in a similar way to the map defined for $\beta < 8\pi$. For each negative integer $i = 0, -1, \ldots, -(M-1)$, it takes activities K_i on volume $\Lambda(M+i)$ to activities on volume $\Lambda(M+i-1)$. There is one important difference in the definition of $\mathcal{R}$: the extraction step removes a factor $e^{\delta\sigma_i \int_{\Lambda(M+i)} (d\phi)^2}$ from $\mathcal{E}xp[\square + K_i]$, in addition to the vacuum extraction Ω_i. These wave function renormalizations are accumulated in the covariances

$$\hat{v}_i(p)^{-1} = p^2(e^{p^2} + \sigma_i)$$

where $\sigma_i = \sum_{j=1}^{|i|} \delta\sigma_{-j}$. Changing the normalization of the Gaussian measure leads to an additional contribution Ω' to the vacuum term Ω_i. To adequately control the extra renormalization cancellations, it is now necessary to consider functionals of $(\phi, \partial_i\phi, \partial_i\partial_j\phi)$, and the corresponding derivative regulators $\mathbf{h} = (h_0, h_1, h_2)$.

The following parameters lead to a contractive estimate for $\mathcal{R}$. Here, c_2, c_3, c_4, c_5 are certain $\mathcal{O}(1)$ geometric constants. We fix $\epsilon > 0$ and a large integer L so that

$$\delta \equiv c_2 \max(L^{2-(1-\epsilon)\beta/4\pi}, L^{-2}) < 1/2.$$

We take the regulators as follows:

1. G_i as before, with

$$\kappa_i = c_3(1 + \sum_{j=1}^{|i|} \delta^j);$$

2. Γ as before, with constant $A = \mathcal{O}(L^{2+\epsilon})$;
3. $\mathbf{h}_i = h_i(c_4 L, L, L^2)$ where $h_i = c_5\beta(2 - \sum_{j=1}^{|i|} \delta^j)$.

Note that $\frac{1}{2}h_0 < h_\infty < h_i \leq h_0$ and $\kappa_0 \leq \kappa_i < \kappa_\infty < 2\kappa_0$.

Theorem 6. *There is a constant $\bar{\zeta}$ depending on the above parameters such that for $|\zeta| < \bar{\zeta}$ the sequence of activities $K_0, K_{-1}, \ldots, K_{-M}$ is defined, as well as the associated constants $\Omega_i, \delta\sigma_i$. They satisfy the following bounds, uniformly in the volume parameter M, for $i = 0, -1, \ldots, -M$:*

$$\begin{aligned} \|K_i\|_i &\leq D\delta^{|i|}; \\ |\Omega_i| &\leq L^{-2|i|} D\delta^{|i|}; \\ |\delta\sigma_i| &\leq h_{1i}^{-2} D\delta^{|i|}, \end{aligned}$$

where $D = Ae^{2c_4c_5\beta L}|\zeta|$.

Remark: This theorem is a restatement of Theorem 5.1, the main result of [7], which incorporates the corrected regulators described in the erratum contained in [10].

The theorem is a consequence of the following two propositions on the IR RG map.

Proposition 7 *Suppose the wave function renormalization satisfies $|\sigma_i| < \epsilon$. Then the constants and regulators chosen above are such that the linearization of $\mathcal{R}$ at $K = 0$ satisfies the following bound:*

$$\|\mathcal{R}_1 K\|_{i-1} \leq \delta/2 \|K\|_i.$$

Proposition 8 *Suppose the wave function renormalization satisfies $|\sigma_i| < \epsilon$. Then there is a number $\bar{K}$ (depending on β, L, ϵ) and a geometric constant c_6 such that $\mathcal{R}$ is an analytic functional of K for $\|K\|_i \leq \bar{K}\delta^{|i|}$. For such K,*

$$\|\mathcal{R}(K)\|_{i-1} \leq c_6 L^2 \|K\|_i.$$

The extractions satisfy the bounds

$$\begin{aligned} |\Omega(K)| &\leq L^{-2|i|}\|K\|_i \\ |\delta\sigma(K)| &\leq h_{1i}^{-2}\|K\|_i. \end{aligned}$$

Proof of Theorem 6: We begin the induction from $i = 0$ to $i = -M$ by noting that

$$\|K_0\|_0 \leq 2Ae^{h_{00}}|\zeta| \equiv 2D.$$

Next, we write

$$\mathcal{R}(K) = \mathcal{R}_1 K + \oint \frac{ds}{s^2(s-1)} \mathcal{R}(sK)$$

and estimate the two terms. The first term is bounded by $\delta/2\|K\|_i$, by Proposition 7. The second term can be taken over the contour $|s| = \bar{K}/D \gg 1$, and by Proposition 8 has the bound

$$(c_6 L^2 \bar{K} \delta^{|i|})(D/\bar{K})^2$$

which is less than $D\delta^{|i|}/2$ if D (i.e. $|\zeta|$) is taken small enough. One can check that $|\sigma_i| < \epsilon$ remains satisfied. □

The decay of correlations at large distance is governed by the least irrelevant eigenvalue of the linearized RG map. Wave function terms decay like L^{-2} and induce a $|x-y|^{-4}$ power law in the two-point function. Charge q terms decay like $L^{-\beta|q|/4\pi}$: the charge one terms induce a power law $|x-y|^{-\beta/2\pi}$. Certainly for $\beta > 8\pi$ it is the wave function terms which dominate the asymptotics. However, the picture is complicated, because wave function terms enter only in second order perturbation theory. This means that for small coupling constant there is a crossover at a scale $I \sim I_0$ with

$$I_0 \sim \frac{1}{(\log L)(2-\beta/4\pi)} \log(L^2 e^{h_{00}}|\zeta|)$$

For $|I| \leq |I_0|$ one sees the power law $|x-y|^{-\beta/2\pi}$ induced by the charge 1 terms, while for $|I| > |I_0|$ one sees the $|x-y|^{-4}$ law of the wave function terms.

All of this is a consequence of the following theorem on general correlations, whose proof, given in [14], amounts to a moderate extension of the above vacuum result.

Theorem 9. *There is a number $\bar{\zeta}(\epsilon, L, \beta)$ and a geometric constant c_1 such that the following properties hold, uniformly in the volume $\Lambda(M)$, and hence in the thermodynamic limit for all $i = 0, -1, \ldots, -M$:*

1. *(analyticity) The activities $D_\Sigma K_i$ and extracted parts $D_\Sigma \Omega_i$ are analytic functions of ζ for $|\zeta| \leq \bar{\zeta}$.*
2. *(one point function) For any $\Sigma = \{(x,q)\}$,*

$$\|D_\Sigma K_i\|_i \leq \begin{cases} 2AL^{-\beta|i|/4\pi}\, e^{h_{0i}} & \text{if } |i| \leq |I_0| \\ 2AL^2 L^{-2|i|}\, e^{2h_{0i}}|\zeta| & \text{if } |i| \geq |I_0| \end{cases} \tag{21}$$

$$|D_\Sigma \Omega_i^N| \leq \begin{cases} 2AL^{-\beta|i|/4\pi}\, e^{h_{0i}} & \text{if } |i| \leq |I_0| \\ 2AL^2 L^{-2|i|}\, e^{2h_{0i}}|\zeta| & \text{if } |i| \geq |I_0| \end{cases} \tag{22}$$

3. *(multi-point functions) Let Σ be a configuration of $Q > 1$ points with linear separation of order $L^{|I|}, I < 0$.*

   (a) Suppose $|I| \le |I_0|$. Then

$$\|D_\Sigma K_i\|_i \le \begin{cases} Q!\left(2AL^{-\beta|i|/4\pi}\, e^{h_{0i}}\right)^Q & \text{if } |i| < |I| \\ Q!(c_1L^{-2})^{|i-I|}\left(2AL^{-\beta|I|/4\pi}\, e^{h_{0I}}\right)^Q & \text{if } |i| \ge |I| \end{cases}$$

   and

$$|D_\Sigma \Omega_i^N| \le \begin{cases} 0 & \text{if } |i| < |I| \\ Q!(c_1L^{-2})^{|i-I|}\left(2AL^{-\beta|I|/4\pi}\, e^{h_{0I}}\right)^Q & \text{if } |i| \ge |I| \end{cases} \tag{23}$$

   (b) Suppose $|I| \ge |I_0|$.

$$\|D_\Sigma K_i\|_i \le \begin{cases} Q!\left(2AL^{-\beta|i|/4\pi}\, e^{h_{0i}}\right)^Q & \text{if } |i| < |I_0| \\ Q!\left(2AL^2L^{-2|i|}\, e^{2h_{0i}}|\zeta|\right)^Q & \text{if } |I_0| \le |i| \le |I| \\ Q!(c_1L^{-2})^{|i-I|}\left(2AL^2L^{-2|I|}\, e^{2h_{0I}}|\zeta|\right)^Q & \text{if } |i| \ge |I| \end{cases}$$

   and

$$|D_\Sigma \Omega_i^N| \le \begin{cases} 0 & \text{if } |i| < |I| \\ Q!(c_1L^{-2})^{|i-I|}\left(2AL^2L^{-2|I|}\, e^{2h_{0I}}|\zeta|\right)^Q & \text{if } |i| \ge |I| \end{cases} \tag{24}$$

7 Acknowledgements

I am indebted to G. Benfatto, whose insight and key observations precipitated the present work, and to G. Keller and D. Brydges for telling me at an early stage about their approach to correlations. I would also like to thank V. Rivasseau and K. Gawedzki for their interest, comments and support, and to acknowledge the support of the following institutions where some of this work was done: Institut des Hautes Etudes Scientifiques, France; Ecole Polytechnique, France; and Università di Roma II.

References

1. G. Benfatto, G. Gallavotti, and F. Nicolò. On the massive sine-Gordon equation in the first few regions of collapse. *Commun. Math. Phys.*, 83:387–410, 1982.
2. D. Brydges, J. Dimock, and T.R. Hurd. The short distance behavior of ϕ_3^4. McMaster University preprint, 1994.
3. D. Brydges and P. Federbush. Debye screening. *Commun. Math. Phys.*, 73:197–246, 1980.
4. D. Brydges and G Keller. Correlation functions of general observables in dipole-type systems I: accurate upper bounds. University of Virginia preprint, 1994.
5. D. Brydges and H. T. Yau. Grad φ perturbations of massless Gaussian fields. *Commun. Math. Phys.*, 129:351–392, 1990.

6. S. Coleman. Quantum sine-Gordon equation as the massive Thirring model. *Phys. Rev.*, D11:2088–2097, 1975.
7. J. Dimock and T. R. Hurd. A renormalization group analysis of the Kosterlitz-Thouless phase. *Commun. Math. Phys.*, 137:263–287, 1991.
8. J. Dimock and T. R. Hurd. A renormalization group analysis of correlation functions for the dipole gas. *J. Stat. Phys.*, 66:1277–1318, 1992.
9. J. Dimock and T. R. Hurd. A renormalization group analysis of QED. *J. Math. Phys.*, 33:814–821, 1992.
10. J. Dimock and T. R. Hurd. Construction of the two-dimensional sine-Gordon model for $\beta < 8\pi$. *Commun. Math. Phys.*, 156:547–580, 1993.
11. J. Fröhlich and E. Seiler. The massive Thirring-Schwinger model QED_2. Convergence of perturbation theory and particle structure. *Helv. Phys. Acta.*, 49:889–924, 1976.
12. J. Fröhlich and T. Spencer. On the statistical mechanics of classical Coulomb and dipole gases. *J. Stat. Phys.*, 24:617–701, 1980.
13. J. Fröhlich and T. Spencer. The Kosterlitz-Thouless transition in two-dimensional Abelian spin systems and the Coulomb gas. *Commun. Math. Phys.*, 81:527–602, 1981.
14. T. R. Hurd. Charge correlations for the Coulomb gas. McMaster University preprint, 1994.
15. J. M. Kosterlitz and D. J. Thouless. Ordering, metastability and phase transitions in two-dimensional systems. *J. Phys.*, C6:1181–1203, 1973.
16. D. H. U. Marchetti, A. Klein, and J. F. Perez. Power law fall off in the Kosterlitz-Thouless phase of a two-dimensional lattice Coulomb gas. *J. Stat. Phys.*, 60:137, 1990.
17. F. Nicolò, J. Renn, and A. Steinmann. On the massive sine-Gordon equation in all regions of collapse. *Commun. Math. Phys.*, 105:291–326, 1986.
18. E.K. Sklyanin, L. A. Takhtadzhyan, and L. D. Faddeev. Quantum inverse problem method I. *Theor. Math. Phys.*, 40:688–706, 1980.
19. A. A. Zamolodchikov. Factorized S-Matrices in two-dimensions as the exact solutions of certain relativistic quantum field theory models. *Ann. Phys.*, 120, 1979.

Weakly Self–Avoiding Polymers in Four Dimensions

Daniel Iagolnitzer[1],
Jacques Magnen[2]

1 Service de Physique Théorique
CE Saclay, 91191 Gif sur Yvette
2 Centre de Physique Théorique, UPR 14 du CNRS,
Ecole Polytechnique, 91128 Palaiseau Cedex
e-mail: magnen@orphee.polytechnique.fr

Abstract
The critical properties of weakly self avoiding random walks in four dimension are established by applying the renormalization group method to polymer chains in a random imaginary potential.

1 Introduction

Weakly self avoiding polymers can be described by two models which are both due to Edwards, whose links are recalled below:
- the weakly self-avoiding walks,
- the polymers in an imaginary random potential, often called the ϕ^4 model with zero components.

This work deals with the infrared limit of the second model in the critical dimension four. For ultraviolet results in dimension two and three, or infrared results in dimension bigger than four see for instance the references given in [IM].

The self-avoiding polymer Green's functions are formally given by:

$$F_m(x-y) = \int \frac{1}{-\Delta + i\,g\,(\tilde{\eta} * V)(x)} \; e^{i\;a(\eta,g,m)\int V(z)dz} \; e^{\int dz a(\eta,g,m)^2/2} \, d\mu(V)$$

$$= \int d\mu(V) \int du\, C(x-u)\, \frac{1}{1+i\,g\,C*V}(u,y)\; e^{i\;a(\eta,g,m)\int V(z)dz} \; e^{\int dz a(\eta,g,m)^2/2} \tag{1}$$

where $d\mu(V)$ is the Gaussian measure (of mean zero) whose covariance is the δ function, η is a fixed ultraviolet cutoff and $\tilde{C}(p) = \eta(p)/p^2$. The constant $a(\eta, g, m)$ will be fixed by requiring that:

$$F_m(x-y) \sim cst\; \frac{e^{-m|x-y|}}{|x-y|^2} \quad \text{as } |x-y| \to \infty \tag{2}$$

We study here only the most difficult case, that is $m = 0$ [IM]. At zero mass, this theory is infrared just renormalizable, asymptotically free in the infrared limit, and its perturbative expansion is the same as that of the N-component ϕ^4

theory in which one puts $N = 0$. To show the existence of F_0 we proceed in a way similar to the construction of ϕ_4^4 with an ultraviolet cutoff (see for instance [R] and references therein).

The model studied is the Laplace transform of a weakly self avoiding walk with an ultraviolet cutoff:

$$F_m(x-y) = \int_0^{+\infty} e^{-m^2\,T}\, G_T(x-y)\, dT\ , \tag{3}$$

where $G_T(x-y)$ is the average density, with respect to the Wiener measure, of the paths $\sigma(t)$ which run from x to y in a time between T and $T + dT$.

Formally one has

$$G_T(x-y) \;=\; lim_{g\to\infty}\ (e^{a\ T}) \int \prod_{0\le t\le T} d\sigma(t)$$

$$\delta\big(\sigma(0)-x\big)\ \delta\big(\sigma(T)-y\big) e^{-\int_0^T (d\sigma/dt)^2 dt} e^{-\int_0^T dt\ dt'\, g\ \delta_{reg}(\sigma(t)-\sigma(t'))}\ , \tag{4}$$

where δ_{reg} is a regularized δ function.

Hence, it seems conversely possible to derive results on the first Edwards model through an inverse Laplace transformation; in his contribution to this volume, S. Golowich [G], explains what are the necessary steps (what is essentially needed is to extend our result to imaginary masses). Alternatively a direct study of the weakly self-avoiding walk seems also possible.

2 The Result

The function $F_0(x-y)$ will be defined as the limit of a theory with cutoffs:

$$F_0(x-y) \;=\; lim_{\Lambda,\rho\to\infty}\ F_{0,\Lambda,\rho}(x-y)\ , \tag{5}$$

where the momentum cutoff, is of size $M^{-\rho}$, ρ being a positive integer, and $M > 1$ being a fixed number; Λ, the volume cutoff, is a box of side size M^ρ. We introduce the infrared cutoff by replacing C by C_ρ:

$$\tilde{C}_\rho = \frac{\eta(p) - \eta_\rho(p)}{p^2}\ ,$$

with

$$\eta_\rho(p) = \eta(p\ M^\rho)\ . \tag{6}$$

The propagator C is not integrable. Therefore it is not possible to use it to prove the existence of F_0 (uniformly in ρ) through a single cluster expansion. We introduce intermediary scales $j,\ 1 \le j \le \rho$:

$$\tilde{C}^j = \frac{\eta^j(p)}{p^2}\ ,\quad \eta^j(p) = \eta_{j-1}(p) - \eta_j(p)\ ,$$

so that

$$\tilde{C}_\rho = \sum_{j=1}^{\rho} \tilde{C}^j \tag{7}$$

and

$$|C^j(x-y)| \;\leq\; O(1)\; M^{-2j}\;\; e^{-M^{-j}|x-y|} \tag{8}$$

for a convenient choice of η.

The key idea is to perform a cluster expansion for each index j relatively to a lattice L_j of mesh M^j. Indeed a cell (j, Δ) defined by a momentum scale j and a cube $\Delta \in L_j$ corresponds roughly to one degree of freedom in phase space. A multiscale expansion in couplings between the degrees of freedom is a way to implement non perturbatively renormalization group ideas.

We introduce the volume cutoff Λ by replacing $V(x)$ by $\Lambda(x)\; V(x)$, where $\Lambda(x)$ is the characteristic function of the box Λ; we replace also $C^j(x-y)$ by

$$\int dz (C^j)^{1/2}(x-z)\; \Lambda(z)\; (C^j)^{1/2}(z-y)\;. \tag{9}$$

Introducing the notation

$$A^j(x-y) = B^j(x-y) = \Lambda(x)\; (C^j)^{1/2}(x-y)\; \Lambda(y)\;, \tag{10}$$

the equivalent of the bound on C^j is:

$$|A^j(x-y)| = |B^j(x-y)| \leq O(1)\; M^{-3j}\; e^{-M^{-j}|x-y|}\;. \tag{11}$$

We define the operator AVB through its kernel:

$$(AVB)(n,x;m,y) = \int dz A^n(x-z)\; V(z)\; B^m(z-y) \tag{12}$$

and the definition of $F_{0,\Lambda,\rho}(x-y)$ is

$$F_{0,\Lambda,\rho}(x-y) \;=\; \sum_{k,l=1}^{\rho} \int d\mu(V) \int du\, dv B^k(x-u)\, \frac{1}{1+i\; g\; AVB}(k,u;l,v)\; A^l(v-y)$$
$$\times\;\; e^{i\; \int_\Lambda dz a(\eta,g,\Lambda,z)V(z)}\;\; e^{\int_\Lambda dz\; a(\eta,g,\Lambda,z)^2/2} \tag{13}$$

Theorem *One can compute* $a(\eta, g, \Lambda, z)$ *with*

$$lim_{\Lambda\to\infty} a(\eta,g,\Lambda,z) \;=\; a(\eta,g) \quad independent\ of\ z$$

such that :

$$F_0(x-y) \;=\; lim_{\Lambda,\rho\to\infty}\; F_{0,\Lambda,\rho}(x-y)$$

exists. Its asymptotic expansion at large $|x-y|$ *proves, at any order, the validity of the perturbative renormalization group computations. In particular the behavior of* F_0 *as* $|x-y| \to \infty$ *is:*

$$\frac{cst(\lambda)}{|x-y|^2}\;(1 + \frac{a}{ln|x-y|} + \frac{b\; ln(ln(|x-y|))}{(ln\;|x-y|)^2} + \frac{O(1)}{(ln\;|x-y|)^2}) \tag{14}$$

The asymptotic behavior of the two point function is that of a free model and the corrections are logarithmically smaller, which is the characteristic of an asymptotically free model.

In the following we give an outline of the proof of the above theorem through a (multiscale) expansion which is conveniently described in terms of "polymers" made of scaled cubes. We will use in the following the word "polymer" in that sense.

3 The "Polymer" Picture

The cluster expansions for scale j are performed with respect to links between cubes of L_j created by the half propagators of index j, i.e. A^j or B^j's. By definition any $A^j(.,z)$, or $B^j(z,.)$ operator is considered as localized in the cube of L_j containing the variable of the potential term V to which they hook hook, i.e. if $z \in \Delta \in L_j$ it is localized in Δ.

The coupling between two different scales j and k can arise through a vertex $A^j\ V\ B^k$; it can arise also through a V propagator:

$$\int d\mu(V)\ (A^j\ V\ B^j)\ \ (A^k\ V\ B^k)\ . \tag{15}$$

Thus the expansion on the links between scales, which will be labeled as links between cubes of different L's, will include perturbing the vertices as well as the measure on V.

The result of the expansion is a sum of terms, each term labeled by a set of links between scaled cubes:

- two cubes of the same scale j are connected if they are linked by at least one $A^j(x,y)$ or $B^j(x,y)$ (i.e. with x in one cube and y in the other).

- two cubes $\Delta, \Delta'$of different scales j,k are linked if there is at least one point $z \in \Delta,\ \Delta'$ at which one half leg, at least, of each scale is hooked, e.g. $A^j(.,z)$, $K^k(z,.)$.

A multiscale "polymer" is a set of connected cubes.

A "subpolymer" of scales of index smaller or equal to j is a set of connected cubes of index smaller or equal to j. The internal legs of such a "subpolymer" are the A or B's localized in the "polymer" cubes. The external legs of a "subpolymer" of scales smaller or equal to j are the A or B's of index strictly larger than j hooked to points to which also internal legs are hooked.

A "subpolymer" of index j is a set of connected cubes of index larger or equal to j containing at least one cube of index j.

4 Outline of the Proof

We rely on a parallel with the infrared ϕ_4^4 model. The proof of the existence and behavior of the infinite volume limit of the correlation function of ϕ_4^4 with an ultra violet cutoff has four main aspects that we examine successively and we describe each time the corresponding step in our model.

4.1 The Use of Local Stability

This step uses the fact that $e^{-\phi^4} \leq 1$. Its analogue here is the fact that the operator $iAVB$ has all its eigenvalues pure imaginary (for AVB is self-adjoint) so that:

$$||\frac{1}{1+iAVB}|| \leq 1 \,. \tag{16}$$

4.2 The Perturbative Bounds

These bounds in the ϕ^4 case are obtained by using the Gaussian free measure. Here the equivalent is the use of analyticity. For example the following quantity (where the z_Δ are arbitrary points):

$$G(\gamma) = \frac{1}{1+i\ g\ AVB\ +\ \sum_j \sum_{\Delta\in L_j} \gamma_{j,\Delta} A^j(.,z_\Delta)\chi_\Delta(z_\Delta)B^j(z_\Delta,.)} \tag{17}$$

is jointly analytic in the variables γ's provided they lie in the domain

$$|\gamma_{j,\Delta}| \leq cstM^{2j} \,,$$

where G is uniformly bounded. As a consequence we have the bound:

$$\left|\left\{\prod_{j,\Delta} \frac{d}{d\gamma_{j,\Delta}}\ G(\gamma)\right\}_{\gamma\equiv 0}\right| \leq \prod_{j,\Delta} \left(cst\ \ M^{-2j}\right) . \tag{18}$$

The main features to apply this bound are:
- the analyticity is considered for half vertices i.e. without the V fields
- in each half vertex the indices of the A, B's are the same
- there is at most one half vertex of index j per cube of L_j.

To use such bounds we "contract" systematically all the V fields of the vertices obtained by previous cluster expansion steps. In the next section we introduce interpolation parameters h, we define$A(h)$, $B(h)$, and we perform perturbative steps in those parameters h's through a formula such as:

$$\frac{d}{dh}\int d\mu(V)\frac{1}{1+i\ g\ A(h)VB(h)} =$$

$$-\int d\mu(V)\frac{1}{1+i\ g\ A(h)VB(h)}\left(\frac{d}{dh}i\ g\ A(h)VB(h)\right)\frac{1}{1+i\ g\ A(h)VB(h)}\ , \tag{19}$$

which gives by "contraction" of the V in the numerator (the derived vertex)

$$\int dz\int d\mu(V)\frac{1}{1+i\ g\ A(h)VB(h)}\frac{d}{dh}i\ g\ A(h;.,z)B(h;z,.))\frac{1}{1+i\ g\ A(h)VB(h)}$$

$$i\ g\ A(h;.,z)B(h;z,.)\frac{1}{1+i\ g\ A(h)VB(h)} + \text{symmetric term} . \tag{20}$$

Therefore (19) is equal to:

$$\int dz \int d\mu(V) \frac{d}{d\gamma_1}\frac{d}{d\gamma_2}$$

$$\left.\frac{1}{1+ig\ A(h)VB(h)+\gamma_1\frac{d}{dh}igA(h;z)B(h;z)+\gamma_2 igA(h;z)B(h;z)}\right|_{\gamma_1=\gamma_2=0} . \tag{21}$$

Then we can bound (19) using the analyticity bound (18).

4.3 The Large Field Problem

Because in a ϕ^4 or AVB vertex all scale indices are not necessarily equal, the perturbative vertices do not always have the form (15). For example a perturbative half vertex of scale j can be of the form:

$$A^j \sum_{k>j} B^k , \tag{22}$$

in which case it is impossible to obtain an analyticity domain of size M^{2j}, because of the spatial decrease of the $A^k, k > j$ operators, which is too weak. In the case of ϕ^4, this problem was solved by using the fact that for large ϕ the dominant term in the integration measure is the quartic potential and not the Gausiian measure. We describe now successively the Gaussian bound, the quartic bound, and finally the analog of this quartic bound in the actual case of our polymer model.

The Gaussian Bound The power counting of a ϕ^j field is also M^{-j}. Now if we generate by perturbation a finite number p of low momentum fields of type $\sum_{l>j} \phi^l$ in each cube of L_j, then this can give as many as $p\ M^{4(l-j)}$ fields ϕ^l of scale l per cube of L_l. This gives $p\ M^{4(l-j)}$ fields corresponding roughly to the same degree of freedom. On the other hand bounds are obtained by summing on each cube independently; since each configuration of cubes then occurs $\left(M^{4(l-j)}\right)!$ times, we have to take this into account by dividing out such a factor. If we used the Gaussian bound, we would thus obtain, apart from the scaling factor M^{-l} a factor per field

$$\frac{1}{M^{4(l-j)}!} \int |x|^{p\ M^{4(l-j)}}\ e^{-x^2}\ dx \sim \left(cst(p)\ \ M^{(2\ -\ 4/p)(l-j)}\right)^{p\ M^{4(l-j)}} . \tag{23}$$

This leads for each low momentum field to a divergent weight for $p \geq 2$:

$$\sum_{l>j} M^{-l}\ M^{(2\ -\ 4/p)(l-j)} . \tag{24}$$

The use of a Cauchy-type bound in our model would lead also to a similar divergent bound.

The Quartic Bound The same bound using the $e^{-\phi^4}$ yields instead of (23)

$$\frac{1}{M^{4(l-j)}!}\int |x|^{p\ M^{4(l-j)}}\ e^{-x^4}\ dx \sim \left(cst(p)\ M^{(1\ -\ 4/p)(l-j)}\right)^{p\ M^{4(l-j)}} . \quad (25)$$

This gives, for each low-momentum field, a finite weight

$$\sum_{l>j} M^{-l}\ M^{(1\ -\ 4/p)(l-j)} = M^{-j} \sum_{l>j} M^{-\ 4(l-j)/p} \leq cst(p)\ M^{-j} . \quad (26)$$

and this bound is the same as the weight of a ϕ^j field.

The Analog of the Quartic Bound We describe now how to obtain the same result for our polymer model. (The method that we are going to describe can be adapted also to the ϕ^4 theory, where it looks like a different presentation presentation of the argument sketched above.)

Let us consider two half vertices AB such as those in the numerators of (20), localized in the same cube Δ of L_j. We have:

$$\sum_{k,l<j} \int d\mu(V)\ \chi_\Delta(z)\ \chi_\Delta(w)\ R(.,.;j,.)$$

$$\left(A^j(.,z)B^k(z,.)\right)\ R(k,.;j,.)\ \left(A^j(.,w)B^l(w,.)\right)\ R(l,.;.,.)$$

$$= \frac{1}{2}\sum_{k,\ l\ <j} \int d\mu(V)\ \chi_\Delta(z)\ \chi_\Delta(w)\ R(.,.;j,.)$$

$$\left[\left(A^j(.,z)\ B^j(w,.)\right)\ R(j,.;k,.)\ \left(A^k(.,z)B^l(w,.)\right) + \text{symmetric term}\right] R(l,.;.,.) , \quad (27)$$

where $R = (1\ +\ i\ g\ AVB)^{-1}$. We used the invariance by transposition ${}^t(AVB) = AVB$ and ${}^tA = B$

In this simplified description of the multiscale expansion we replace $A^k(.,z)$, $z \in \Delta \in L_j$, $k < j$, by $A^k(.,w)$, $w \in \Delta$. In fact :

$$A^k(.,z) = A^k(.,w) + \int_0^1 dt\ \ (\mathbf{z}-\mathbf{w}).\boldsymbol{\nabla} A^k(.,tz+(1-t)w) . \quad (28)$$

Since $|z-w| \leq M^j$ and since the derivative of A^k yields a factor M^{-k}, the remainder term is indeed smaller by a factor M^{j-k}.

In the same way we can replace, up to regularized terms,

$$\sum_{k<j} A^k(.,z) \sum_{l<j} B^l(z,.) \text{ by } \frac{1}{|\Delta|}\int_\Delta du \sum_{k<j} A^k(.,u) \sum_{l<j} B^l(u,.) . \quad (29)$$

Then the expression (27) is (up to more regular terms) equal to:

$$\frac{1}{2}\int d\mu(V)\,\chi_{\Delta}(z)\,\chi_{\Delta}(w)\,\frac{1}{|\Delta|}\int_{\Delta} du V(u)\,R\,A^{j}(.,z)\,B^{j}(w,.)\,R \tag{30}$$

as can be checked from contraction of the averaged V. Indeed in the difference between (27) and (30) at least one A or B of index larger than j in the numerators has been replaced by either a ∇A or ∇B or a A or B operator of index smaller or equal to j.

We summarize the result by saying that we replace the large index "fields" by averages of the ultralocal field V.

The power counting of $\frac{1}{|\Delta|}\int_{\Delta} du V(u)$ is that of a jth scale half vertex and because V is ultralocal there is no accumulation of V's:

$$\int d\mu(V)\prod_{\Delta\in L_j}\left(\int\left(\frac{1}{|\Delta|}\int_{\Delta} du V(u)\right)^{2}\right)^{1/2} = \prod_{\Delta\in L_j} M^{-2j}\,. \tag{31}$$

Thus one finds that the weight of each low momentum $\sum_{k>j} B^k$ is indeed M^{-j}.

4.4 The Convergence of the Expansion

We bound the contribution associated to each "polymer" (where the low momentum A's have been replaced by corresponding averages of V's) using the analyticity bounds of type (18) and the bound (31) on averages of V's. We perform the relative sums on cubes always with the best available half propagator decay (see (11)). As a result the expansion converges if there is no "subpolymer" with four or less external legs.

More precisely, if we consider in the expansion of $F_0(x-y)$ the sub-sum where the scale indices of each vertex are held fixed, we obtain the bound

$$\frac{cst}{|x-y|^2}(cst\;\; g)^{(\text{number of cubes})/4}\prod_{j}\prod_{\alpha\in I_j} M^{4-e_\alpha}\,, \tag{32}$$

where, at fixed j, α indexes the "subpolymers" of index j, and where e_α is the number of external legs of the "subpolymer" of index (j,α).

Thus if all e_α were bigger than four, then at each scale we would obtain a factor smaller than $(M^{-1/5})$ for each external leg, which would make the sum over scales for each leg convergent. The expansion would then converge provided g is small enough.

4.5 The Renormalization and the Effective Model

We proceed as in a ϕ^4 model to deal with the divergences due to the presence of two and four-point functions. The coupling constant corresponding to the ϕ^4 vertex is g^2 in our model. Up to inessential differences, this part of the argument is quite similar in the two models, so we simply recall the outline the argument for the ϕ^4 vertex. Let us denote $G_{j;4}(x_1, ..., x_4)$ a four-point "subpolymer" of index j, where the x's are the points at which the external legs are hooked. Always by the same argument $G_{j;4}$ is almost local for the scales of lower momenta, so we decompose it in a local singular part and a regularized part:

$$G_{j;4}(x_1, ..., x_4) = \delta g_j^2 + G_{j;4}^{reg}(x_1, ..., x_4) \ ,$$

$$\delta g_j^2(z) = \int_\Lambda dx_2 \ dx_3 \ dx_4 G_{j;4}(z, x_2, x_3, x_4) \ , \tag{33}$$

$G_{j;n}^{reg}$ being regularized for the same reason as in (28).

The effective ϕ^4 coupling at scale j is then

$$g^2 \ + \ \sum_{l=0}^{j} \delta g_l^2(z) \ , \tag{34}$$

and one finds that:

$$\delta g_l^2(z) = \ -\beta \ g_{l-1}^4 + O(g_{l-1}^6) + O(g_{l-1}^4 \ e^{- \ M^{-l} \ dist(z, \partial\Lambda)}) \tag{35}$$

where β is a positive constant which depends on the model but not on the cutoffs. So if we start with a small enough g (such that the first scale expansion converges) then $g_j(z)$ decreases as j increases. We obtain also that the thermodynamic limit of the flow equations exist: $lim_{\Lambda\to\infty} \ g_j^2(z) = g_j^2$, and

$$g_j^2 = \frac{1}{g^{-2} + \beta \ ln \ M^j + O(ln \ j)} \ . \tag{36}$$

Thus g_j^2 goes logarithmically to zero as the energy scale goes to zero. This means that the model is indeed asymptotically free in the infrared regime.

The argument is quite similar for the two point functions. One has to choose the parameter $a(\eta, g)$ such that the two point coupling goes to zero:

$$a(\eta, g, \Lambda, z) + \sum_{l=0}^{j} \delta a_l(\Lambda, z) = 0 \ ,$$

$$\delta a_l(\Lambda, z) = \int_\Lambda dx_2 \ G_{j;2}(z, x_2) \ . \tag{37}$$

As above for the two point function, $\lim_{\Lambda\to\infty} a(\eta, g, \Lambda, z) \ = \ a(\eta, g)$.

5 The Multiscale Expansion

In this section we give some details on the perturbative steps of the expansion.

5.1 The Cluster Expansions Between Cubes of the Same Index

As already indicated a lattice L_j is introduced for each index j. For each pair of cubes $\Delta, \Delta' \in L_j$, one introduces a variable $h_{\Delta,\Delta'}$, (with $h_{\Delta,\Delta} \equiv 1$) and one defines

$$A^j(h,x,y) = \sum_{\Delta\,\Delta'} h_{\Delta\,\Delta'} \;\; \chi_\Delta(x)\; A^j(x,y)\;\; \chi_\Delta(y)\;, \tag{38}$$

where χ_Δ is the characteristic function of the cube Δ.

Then $F_{0,\Lambda,\rho}(x-y;h)$ is obtained by replacing A, B's by the $A(h), B(h)'$s.

The cluster expansion consists in writing for each index j and each pair $\Delta, \Delta' \in L_j \cap \Lambda$:

$$F_{0,\Lambda,\rho}(x-y) = F_{0,\Lambda,\rho}(x-y;1) = F_{0,\Lambda,\rho}(x-y;0) + \int_0^1 dh\;(\frac{d}{dh}F_{0,\Lambda,\rho}(x-y;h)) \tag{39}$$

In this way one obtains a sum of terms. Each term is characterized by the set of pairs (Δ, Δ') for which there is an integration over $h_{\Delta,\Delta'}$. By definition, two cubes Δ, Δ' of the same index are said to be connected, or to belong to the same "polymer", if the remainder term with integration over $h_{\Delta,\Delta'}$ has been selected.

5.2 The Vertical Expansions Between Cubes of Different Indices

For each index j and each cube $\Delta \in L_j$ we introduce an interpolating variable v_Δ such that in $F_{0,\Lambda,\rho}(x-y;h;v)$, the indices higher than j are decoupled from the lower indices in Δ at $v_\Delta = 0$. More precisely:

$$F_{0,\Lambda,\rho}(x-y;h;v) \;=\; \sum_{k,l=1}^{\rho} \int d\mu(V;v) \int dudv$$

$$B^k(x-u;h)\;\frac{1}{1+i\,g\,(AVB)}(k,u;l,w;h;v)\;\;A^l(w-y;h)\;e^{i\int_\Lambda a(\eta,g,\Lambda,z)V(z)dz}$$

$$e^{\int_\Lambda dz a(\eta,g,\Lambda,z)^2/2}\;, \tag{40}$$

where (AVB) is now a v- and h-dependent operator. We introduce first the notation

$$v_j(z) = \sum_{\Delta\in L_j} \chi_\Delta(z)\; v_\Delta\;. \tag{41}$$

We define now $(AVB)(h,v)$. $V(z)$ is in fact $V(v,l,z)$, namely a Gaussian variable of measure $d\mu(V,v)$ with covariance

$$\int d\mu(V)V(v,n,z)\,V(v,m,z') \;=\; \delta(z-z') \prod_{inf(n,m)\,\leq\, j\,<\, sup(n,m)} v_j(z)\,, \tag{42}$$

and

$$(AVB)(k,u;l,w;h;v) =$$

$$\int dz \left\{ \prod_{inf(l,k) \leq j < sup(l,k)} v_j(z) \right\} A(h,u,z) \; V\Big(v, \; sup(l,k), \; z\Big) \; B(h,z,w) \; . \tag{43}$$

We have $F_{0,\Lambda,\rho}(x-y;h) = F_{0,\Lambda,\rho}(x-y;h;1)$. The vertical expansion consists in expanding $F(v)$ around $v = 0$ for each v_Δ up to order 5, in order to ensure that the two and four-point "subpolymers" are automatically factorized. In this way the maximal number p of low-momentum legs per cube created by this vertical expansion remains bounded by $3 \cdot 5 = 15$. This justifies taking p bounded in section 4.3.

References

[IM] Iagolnitzer, D. and Magnen, J. *Polymers in a weak random in dimension four: rigorous renormalization group analysis*, Commun. Math. Phys. 162, 85-121 (1994).

[R] Rivasseau, R. *From perturbative to constructive renormalization*, chapters III. 2, 3, Princeton University Press (1991).

[G] Golowich, S. *The self-avoiding walk in four dimensions*, in this volume.

Springer-Verlag and the Environment

We at Springer-Verlag firmly believe that an international science publisher has a special obligation to the environment, and our corporate policies consistently reflect this conviction.

We also expect our business partners – paper mills, printers, packaging manufacturers, etc. – to commit themselves to using environmentally friendly materials and production processes.

The paper in this book is made from low- or no-chlorine pulp and is acid free, in conformance with international standards for paper permanency.

Lecture Notes in Physics

For information about Vols. 1–406
please contact your bookseller or Springer-Verlag

Vol. 407: P. L. Smith, W. L. Wiese (Eds.), Atomic and Molecular Data for Space Astronomy. VII, 158 pages. 1992.

Vol. 408: V. J. Martínez, M. Portilla, D. Sàez (Eds.), New Insights into the Universe. Proceedings, 1991. XI, 298 pages. 1992.

Vol. 409: H. Gausterer, C. B. Lang (Eds.), Computational Methods in Field Theory. Proceedings, 1992. XII, 274 pages. 1992.

Vol. 410: J. Ehlers, G. Schäfer (Eds.), Relativistic Gravity Research. Proceedings, VIII, 409 pages. 1992.

Vol. 411: W. Dieter Heiss (Ed.), Chaos and Quantum Chaos. Proceedings, XIV, 330 pages. 1992.

Vol. 412: A. W. Clegg, G. E. Nedoluha (Eds.), Astrophysical Masers. Proceedings, 1992. XX, 480 pages. 1993.

Vol. 413: Aa. Sandqvist, T. P. Ray (Eds.); Central Activity in Galaxies. From Observational Data to Astrophysical Diagnostics. XIII, 235 pages. 1993.

Vol. 414: M. Napolitano, F. Sabetta (Eds.), Thirteenth International Conference on Numerical Methods in Fluid Dynamics. Proceedings, 1992. XIV, 541 pages. 1993.

Vol. 415: L. Garrido (Ed.), Complex Fluids. Proceedings, 1992. XIII, 413 pages. 1993.

Vol. 416: B. Baschek, G. Klare, J. Lequeux (Eds.), New Aspects of Magellanic Cloud Research. Proceedings, 1992. XIII, 494 pages. 1993.

Vol. 417: K. Goeke P. Kroll, H.-R. Petry (Eds.), Quark Cluster Dynamics. Proceedings, 1992. XI, 297 pages. 1993.

Vol. 418: J. van Paradijs, H. M. Maitzen (Eds.), Galactic High-Energy Astrophysics. XIII, 293 pages. 1993.

Vol. 419: K. H. Ploog, L. Tapfer (Eds.), Physics and Technology of Semiconductor Quantum Devices. Proceedings, 1992. VIII, 212 pages. 1993.

Vol. 420: F. Ehlotzky (Ed.), Fundamentals of Quantum Optics III. Proceedings, 1993. XII, 346 pages. 1993.

Vol. 421: H.-J. Röser, K. Meisenheimer (Eds.), Jets in Extragalactic Radio Sources. XX, 301 pages. 1993.

Vol. 422: L. Päivärinta, E. Somersalo (Eds.), Inverse Problems in Mathematical Physics. Proceedings, 1992. XVIII, 256 pages. 1993.

Vol. 423: F. J. Chinea, L. M. González-Romero (Eds.), Rotating Objects and Relativistic Physics. Proceedings, 1992. XII, 304 pages. 1993.

Vol. 424: G. F. Helminck (Ed.), Geometric and Quantum Aspects of Integrable Systems. Proceedings, 1992. IX, 224 pages. 1993.

Vol. 425: M. Dienes, M. Month, B. Strasser, S. Turner (Eds.), Frontiers of Particle Beams: Factories with $e^+ e^-$ Rings. Proceedings, 1992. IX, 414 pages. 1994.

Vol. 426: L. Mathelitsch, W. Plessas (Eds.), Substructures of Matter as Revealed with Electroweak Probes. Proceedings, 1993. XIV, 441 pages. 1994

Vol. 427: H. V. von Geramb (Ed.), Quantum Inversion Theory and Applications. Proceedings, 1993. VIII, 481 pages. 1994.

Vol. 428: U. G. Jørgensen (Ed.), Molecules in the Stellar Environment. Proceedings, 1993. VIII, 440 pages. 1994.

Vol. 429: J. L. Sanz, E. Martínez-González, L. Cayón (Eds.), Present and Future of the Cosmic Microwave Background. Proceedings, 1993. VIII, 233 pages. 1994.

Vol. 430: V. G. Gurzadyan, D. Pfenniger (Eds.), Ergodic Concepts in Stellar Dynamics. Proceedings, 1993. XVI, 302 pages. 1994.

Vol. 431: T. P. Ray, S. Beckwith (Eds.), Star Formation and Techniques in Infrared and mm-Wave Astronomy. Proceedings, 1992. XIV, 314 pages. 1994.

Vol. 432: G. Belvedere, M. Rodonò, G. M. Simnett (Eds.), Advances in Solar Physics. Proceedings, 1993. XVII, 335 pages. 1994.

Vol. 433: G. Contopoulos, N. Spyrou, L. Vlahos (Eds.), Galactic Dynamics and N-Body Simulations. Proceedings, 1993. XIV, 417 pages. 1994.

Vol. 434: J. Ehlers, H. Friedrich (Eds.), Canonical Gravity: From Classical to Quantum. Proceedings, 1993. X, 267 pages. 1994.

Vol. 435: E. Maruyama, H. Watanabe (Eds.), Physics and Industry. Proceedings, 1993. VII, 108 pages. 1994.

Vol. 436: A. Alekseev, A. Hietamäki, K. Huitu, A. Morozov, A. Niemi (Eds.), Integrable Models and Strings. Proceedings, 1993. VII, 280 pages. 1994.

Vol. 437: K. K. Bardhan, B. K. Chakrabarti, A. Hansen (Eds.), Non-Linearity and Breakdown in Soft Condensed Matter. Proceedings, 1993. XI, 340 pages. 1994.

Vol. 438: A. Pękalski (Ed.), Diffusion Processes: Experiment, Theory, Simulations. Proceedings, 1994. VIII, 312 pages. 1994.

Vol. 439: T. L. Wilson, K. J. Johnston (Eds.), The Structure and Content of Molecular Clouds. 25 Years of Molecular Radioastronomy. Proceedings, 1993. XIII, 308 pages. 1994.

Vol. 440: H. Latal, W. Schweiger (Eds.), Matter Under Extreme Conditions. Proceedings, 1994. IX, 243 pages. 1994.

Vol. 441: J. M. Arias, M. I. Gallardo, M. Lozano (Eds.), Response of the Nuclear System to External Forces. Proceedings, 1994, VIII, 293 pages. 1995.

Vol. 446: V. Rivasseau (Ed.), Constructive Physics. Results in Field Theory, Statistical Mechanics and Condensed Matter Physics. Proceedings, 1994. X, 338 pages. 1995.

New Series m: Monographs

Vol. m 1: H. Hora, Plasmas at High Temperature and Density. VIII, 442 pages. 1991.

Vol. m 2: P. Busch, P. J. Lahti, P. Mittelstaedt, The Quantum Theory of Measurement. XIII, 165 pages. 1991.

Vol. m 3: A. Heck, J. M. Perdang (Eds.), Applying Fractals in Astronomy. IX, 210 pages. 1991.

Vol. m 4: R. K. Zeytounian, Mécanique des fluides fondamentale. XV, 615 pages, 1991.

Vol. m 5: R. K. Zeytounian, Meteorological Fluid Dynamics. XI, 346 pages. 1991.

Vol. m 6: N. M. J. Woodhouse, Special Relativity. VIII, 86 pages. 1992.

Vol. m 7: G. Morandi, The Role of Topology in Classical and Quantum Physics. XIII, 239 pages. 1992.

Vol. m 8: D. Funaro, Polynomial Approximation of Differential Equations. X, 305 pages. 1992.

Vol. m 9: M. Namiki, Stochastic Quantization. X, 217 pages. 1992.

Vol. m 10: J. Hoppe, Lectures on Integrable Systems. VII, 111 pages. 1992.

Vol. m 11: A. D. Yaghjian, Relativistic Dynamics of a Charged Sphere. XII, 115 pages. 1992.

Vol. m 12: G. Esposito, Quantum Gravity, Quantum Cosmology and Lorentzian Geometries. Second Corrected and Enlarged Edition. XVIII, 349 pages. 1994.

Vol. m 13: M. Klein, A. Knauf, Classical Planar Scattering by Coulombic Potentials. V, 142 pages. 1992.

Vol. m 14: A. Lerda, Anyons. XI, 138 pages. 1992.

Vol. m 15: N. Peters, B. Rogg (Eds.), Reduced Kinetic Mechanisms for Applications in Combustion Systems. X, 360 pages. 1993.

Vol. m 16: P. Christe, M. Henkel, Introduction to Conformal Invariance and Its Applications to Critical Phenomena. XV, 260 pages. 1993.

Vol. m 17: M. Schoen, Computer Simulation of Condensed Phases in Complex Geometries. X, 136 pages. 1993.

Vol. m 18: H. Carmichael, An Open Systems Approach to Quantum Optics. X, 179 pages. 1993.

Vol. m 19: S. D. Bogan, M. K. Hinders, Interface Effects in Elastic Wave Scattering. XII, 182 pages. 1994.

Vol. m 20: E. Abdalla, M. C. B. Abdalla, D. Dalmazi, A. Zadra, 2D-Gravity in Non-Critical Strings. IX, 319 pages. 1994.

Vol. m 21: G. P. Berman, E. N. Bulgakov, D. D. Holm, Crossover-Time in Quantum Boson and Spin Systems. XI, 268 pages. 1994.

Vol. m 22: M.-O. Hongler, Chaotic and Stochastic Behaviour in Automatic Production Lines. V, 85 pages. 1994.

Vol. m 23: V. S. Viswanath, G. Müller, The Recursion Method. X, 259 pages. 1994.

Vol. m 24: A. Ern, V. Giovangigli, Multicomponent Transport Algorithms. XIV, 427 pages. 1994.

Vol. m 25: A. V. Bogdanov, G. V. Dubrovskiy, M. P. Krutikov, D. V. Kulginov, V. M. Strelchenya, Interaction of Gases with Surfaces. XIV, 132 pages. 1995.

Vol. m 26: M. Dineykhan, G. V. Efimov, G. Ganbold, S. N. Nedelko, Oscillator Representation in Quantum Physics. IX, 279 pages. 1995.

Vol. m 27: J. T. Ottesen, Infinite Dimensional Groups and Algebras in Quantum Physics. IX, 218 pages. 1995.